Geometry, Symmetries, and Classical Physics

Geometry, Symmetries, and Classical Physics

A Mosaic

Manousos Markoutsakis

CRC Press
Taylor & Francis Group
Boca Raton London New York

CRC Press is an imprint of the
Taylor & Francis Group, an **informa** business

First edition published 2022
by CRC Press
6000 Broken Sound Parkway NW, Suite 300, Boca Raton, FL 33487-2742

and by CRC Press
2 Park Square, Milton Park, Abingdon, Oxon, OX14 4RN

CRC Press is an imprint of Taylor & Francis Group, LLC

ISBN: 978-0-367-53523-0 (hbk)
ISBN: 978-0-367-54141-5 (pbk)
ISBN: 978-1-003-08774-8 (ebk)

DOI: 10.1201/9781003087748

Publisher's note: This book has been prepared from camera-ready copy provided by the authors.

To my parents

Contents

Part I: Geometric Manifolds

Part II: Mechanics and Symmetry

Part III: Symmetry Groups and Algebras

Part IV: Classical Fields

Part V: Riemannian Geometry

Part VI: General Relativity and Symmetry

Part VII: Appendices

Preface

Classical theoretical physics is a remarkably coherent and beautiful subject. The particular viewpoint which we adopt in this book is that symmetry principles play a decisive role in the foundations of the theory. Indeed, the core of classical theoretical physics can be derived from the following three symmetry principles:

- *The underlying spacetime symmetry along with its representations defines the basic elements (i.e. the fields) of the theory.*
- *The dynamical evolution of the theory is encoded in the action principle.*
- *The fundamental interactions of the theory are determined by a gauge symmetry principle.*

This book provides a systematic discussion of classical spacetime and gauge symmetries and the associated physical invariances. The book covers geometric manifolds, the foundational continuous symmetry groups of classical mechanics and classical fields, as well as symmetries in geometry and general relativity. The beauty of theoretical physics derives in large part from the beauty of mathematics. Therefore, the book develops the central notions of differentiable manifolds, Lie groups and algebras, and Riemannian geometry from the outset and in the necessary conceptual depth. In addition, several nontrivial and exciting topics are covered, such as the discussion of conformal symmetry, Weyl symmetry, and the discussion of conserved quantities in general relativity, to name a few. The treatment of all topics is technically complete and this kind of presentation is necessary in order to develop a true understanding. It is indeed one of the primary goals of this text to provide a high degree of transparency.

The content of the book is divided into six parts and an appendix. In Part I, we begin with the mathematical foundations of differentiable manifolds, tangent spaces, and tensor fields. We introduce the metric and discuss geometric manifolds with a selection of relevant examples. We develop the machinery of differential forms and derive Stokes' theorem. We then introduce the notion of the Lie derivative and describe how it can be used to formulate symmetry on geometric manifolds.

Part II deals with classical dynamics and summarizes the Newtonian, Lagrangian and relativistic formulations of particle mechanics. We invoke the action principle and establish the Euler-Lagrange equations of motion. We then derive and apply Noether's theorem for nonrelativistic and for relativistic mechanics. The relation between symmetry and conservation represents the recurring theme.

Part III covers the algebraic aspects of symmetry. First, we provide the mathematical background on Lie groups, Lie algebras, and representations as needed for subsequent purposes. We develop in detail the rotation group, the Euclidean group, the Galilei group, the Lorentz group, the Poincaré group, and the conformal group. We disclose how relativistic symmetry leads to the existence of Weyl and Dirac spinors. For all spacetime symmetries, we systematically derive the generators of the Lie group in their field representation and the associated commutator relations.

Part IV is about classical field theory. The action principle is again the starting point, this time for field Lagrangians. We provide the examples of real and complex scalar fields, spinor fields and the Maxwell field. Noether's theorem for fields is derived and we deduce the conserved quantities. An important notion introduced here is that of the energy-momentum tensor in the canonical version and the symmetric version based on Belinfante's prescription. We derive the concrete conserved quantities for each of the spacetime transformations contained in the full conformal group. Then we discuss the notion of a conformally invariant field theory and give some examples. Finally, we examine the interaction of fields, with the gauge symmetry as the guiding principle.

Part V covers Riemannian geometry and related symmetry aspects. We first introduce the covariant derivative and the notion of connection on a differentiable manifold. After introducing Riemannian curvature, we discuss symmetry properties as well as the Ricci decomposition of the Riemann tensor. We then revisit symmetry on geometric manifolds and identify the properties of a manifold at maximal symmetry. Then Weyl rescalings and the associated Weyl-Schouten theorem are studied. Finally, we consider differentiable transformations from an algebraic point of view and introduce the corresponding infinite-dimensional group and algebra.

In Part VI, first the basic conceptions of general relativity as the most important classical theory of gravity are covered. In addition to Einstein's field equations and the Schwarzschild solution, the concept of an asymptotically flat spacetime is discussed. The subsequent complete treatment of the Lagrangian formalism in general relativity includes matter fields as well as the metric field and the case of manifolds with a boundary. Within the Lagrangian framework, the metric energy-momentum tensor is introduced in a natural way. We provide a thorough discussion of internal diffeomorphisms and the associated identically conserved Noether currents. In the last chapter, we discuss locally and globally conserved quantities in general relativity. In particular, the Komar integral quantities are introduced and discussed. The last section addresses the question of how Weyl rescaling symmetry can be achieved. Among other things, we discuss here the conformally coupled scalar field.

In the appendix we summarize the conventions used and some relevant mathematical results. This includes a detailed exposition of tensor algebra, matrix groups, Dirac delta distribution, Poisson and wave equations, calculus of variations, spheres in arbitrary dimensions, and hypersurfaces. We also provide a fairly complete collection of formulae for Weyl rescalings of the tensor fields that are relevant to us. An overview of all major spacetime symmetries concludes this part.

This book has a special focus on conformal symmetry and Weyl rescaling symmetry in $d \geq 3$ dimensions. Conformal symmetry in $d = 2$ is not treated in detail, since it is a topic of its own. The discussion of conformal symmetry begins with rigorous mathematical definitions, but is then translated into a more practical form. Topics discussed include conformality between manifolds, the Lie group and Lie algebra of conformal transformations, and the notion of conformal symmetry for Lagrangian field theories with concrete examples. Moreover, we cover Weyl rescalings of geometries, the Weyl-Schouten theorem describing the conditions for achieving conformal flatness, the implications of Weyl rescalings in general relativity, and finally the conformally coupled scalar field.

The book is designed to be self-contained, but it is assumed that the reader has a solid knowledge of linear algebra, analysis in several dimensions, classical mechanics, electromagnetism, and special relativity. In addition, a basic knowledge of general relativity is an advantage. Mathematical notation is kept consistent throughout the book. Central mathematical notions are first introduced rigorously, and then the emphasis is on their application. We make extensive use of the tensor indices notation, as this is the most economical way to formulate complex tensor equations. The Lie derivative and the covariant derivative are defined in an algebraic way, using symmetry principles as a guide. In writing this text, I have avoided using too involved mathematical constructions, such as the pullback of maps or the Hodge duality. Instead, we use the local transformations and the explicit expressions that employ the epsilon tensor. Thematically, we do not treat discrete symmetries such as space or time inversion, since these symmetries acquire their proper relevance within quantum theory. Furthermore, the Hamiltonian formalism unfortunately had to be omitted entirely in order to keep the size of the book manageable.

The bibliography at the end of each chapter and at the end of the book provides the curious reader with references to explore topics in greater depth. We list three types of references. First, there are original articles that address specific topics. Then there are textbooks that provide a wealth of further developments and applications that could not be covered in the limited space available. Finally, there is a selection of classic reference texts that have stood the test of time and continue to provide valuable insights today. Scattered throughout the text, the reader will encounter the sign (*exercise*). At these points, the reader is encouraged to tackle a straightforward exercise to solidify understanding.

I would like to take this opportunity to express my gratitude to the editorial team at CRC Press for making this book possible. My special thanks go to Editorial Assistant, Dr. Kirsten Barr and to Acquiring Editor, Rebecca Davies for supporting the book concept and for guiding me through the final stages of writing. My thanks also go to Shashi Kumar for helping me with the many intricacies within the TeX system. Finally, I would like to thank my family for their patience and steady support during the writing process.

I

Geometric Manifolds

1

Manifolds and Tensors

In this chapter we introduce the foundational notions of differentiable manifolds, vectors at a point, vector fields, and tensor fields. We begin with a summary of basic facts about differentiation in $\mathbb{R}^D$, since we aim to transfer the known calculational concepts to the case of differentiable manifolds. We provide the general definition of a differentiable manifold, introduce coordinates, and discuss diffeomorphisms. The directional derivative leads us to the algebraic definition of a vector as an element of the tangent space at a point. In the next step, we move from vectors at a point to vector fields defined over the entire manifold. Finally, we generalize to the multilinear structure and introduce general tensor fields on manifolds.

1.1 Differentiation in Several Dimensions

Euclidean Space $\mathbb{E}^D$

One of the cornerstones of classical physics is the use of the continuum of real numbers $\mathbb{R}$, or, for higher dimensions, the vector space $\mathbb{R}^D$ with integer dimensionality $D = 1, 2, 3, \ldots$. The elements of $\mathbb{R}^D$ are represented as column vectors

$$x = \begin{pmatrix} x^1 \\ \vdots \\ x^D \end{pmatrix}, \tag{1.1}$$

or, in abbreviated form, by the component notation x^k. The canonical basis vectors are given by $e_k = (0, \ldots, 1, \ldots, 0)^T$, where the entry 1 is at the kth row. The vector space $\mathbb{R}^D$ represents also a raw model for space in classical physics, provided we endow it with an additional structure. This structure is a *metric*, or equivalently, a *scalar product*, which for any two vectors x, y of $\mathbb{R}^D$ is defined by

$$\langle x, y \rangle \equiv \sum_{k=1}^{D} x^k y^k. \tag{1.2}$$

The scalar product, in turn, introduces a *norm* (or *length*) of a vector,

$$|x| \equiv \sqrt{\langle x, x \rangle} = \sqrt{(x^1)^2 + \cdots + (x^D)^2}. \tag{1.3}$$

DOI: 10.1201/9781003087748-1

If the origin of the vector space can be freely shifted, we have an *affine space* at hand and we can identify vectors with points. In this way, we can introduce the notion of *distance* $d(x, y)$ between any two points x, y of $\mathbb{R}^D$ as

$$d(x, y) \equiv |x - y|. \tag{1.4}$$

The above structure defines the D-dimensional *Euclidean space,* which we denote by $\mathbb{E}^D$. In general, we will not make a distinction between the linear $\mathbb{R}^D$ and the affine-linear $\mathbb{E}^D$.

Functions, Maps and Curves

Let us consider the linear spaces $\mathbb{R}^D$ and $\mathbb{R}^N$, with integer dimensions D, $N = 1, 2, 3, \ldots$, each space being equipped with its scalar product. Let us also consider open subsets U, V, etc. of these linear spaces on which we will define our maps. Open sets U are those for which every point $x \in U$ has a neighborhood that is completely contained in U. A *function f* is a map from an open subset $U \subset \mathbb{R}^D$ to the real numbers, i.e. $f : U \to \mathbb{R}$, $x \mapsto f(x)$. More generally, a *map F* assigns elements of $U \subset \mathbb{R}^D$ to elements of $\mathbb{R}^N$, i.e. $F : U \to \mathbb{R}^N$, $x \mapsto F(x)$, and the image is represented as a column vector,

$$F(x) = \begin{pmatrix} F^1(x) \\ \vdots \\ F^N(x) \end{pmatrix}, \tag{1.5}$$

or as $F^k(x)$ in the index notation. The special case $N = D$ corresponds to the case of a *vector field* on U. For $N = 1$, we recover the case of functions again. Another special case is for $D = 1$, where the elements of an open interval U of $\mathbb{R}$ are mapped to N-dimensional vectors. Then we speak about *curves*; i.e. a curve γ is a map $\gamma : U \to \mathbb{R}^N$, $t \mapsto \gamma(t)$.

Differentiability

Given a function $f : U \to \mathbb{R}$, $x \mapsto f(x)$, the *partial derivative* of $f(x)$ with respect to the variable x^k (for a certain index value k) at the point x is the limit

$$\frac{\partial f}{\partial x^k}(x) \equiv \lim_{t \to 0} \frac{f(x + te_k) - f(x)}{t}, \tag{1.6}$$

where e_k is the basis vector of $\mathbb{R}^D$ in the kth direction. The collection of all partial derivatives of a function constitutes the *gradient* of this function and is of great importance. The gradient $\mathrm{grad} f(x)$ of a function $f(x)$ is written as the row vector

$$\mathrm{grad} f \equiv \left(\frac{\partial f}{\partial x^1}, \ldots, \frac{\partial f}{\partial x^D} \right) \tag{1.7}$$

and belongs to the dual space of $\mathbb{R}^D$. If the gradient or, equivalently, all partial derivatives of a function exist, the function is said to be *differentiable*. Let us now approach the property of differentiability from a more conceptual point of view. We consider a map $F : U(\subset \mathbb{R}^D) \to \mathbb{R}^N$, $x \mapsto F(x)$. In principle, differentiability means that locally a linear approximation is possible. A map F is called *(totally) differentiable* at the point $x \in U$ if there is a linear map $\mathrm{D}F(x) : \mathbb{R}^D \to \mathbb{R}^N$, so that for $\xi \in \mathbb{R}^D$, $|\xi| \ll 1$, it is

$$F(x + \xi) = F(x) + \mathrm{D}F(x)\,\xi + o(|\xi|), \tag{1.8}$$

with

$$\lim_{\xi \to 0} \frac{o(|\xi|)}{|\xi|} = 0. \tag{1.9}$$

This means that the *remainder function* $o(|\xi|)$ as a power series must be higher than first order in $|\xi|$. The linear map $\mathrm{D}F(x)$ is called the *differential*, or *derivative*, of F at the point x. For functions $f(x)$, the symbol $\mathrm{d}f(x)$ is used, while for curves $\gamma(t)$, the symbol $\dot{\gamma}(t)$ is common. Obviously, the differential is represented by an $N \times D$-matrix. This matrix is easily seen to be the *Jacobian matrix* of F (*Carl Gustav Jacob Jacobi*) at the point x,

$$\mathrm{D}F = \left(\frac{\partial F^k}{\partial x^l} \right) \equiv \begin{pmatrix} \frac{\partial F^1}{\partial x^1} & \cdots & \frac{\partial F^1}{\partial x^D} \\ \vdots & \ddots & \vdots \\ \frac{\partial F^N}{\partial x^1} & \cdots & \frac{\partial F^N}{\partial x^D} \end{pmatrix}, \tag{1.10}$$

i.e. the differential is given by the ordered collection of all partial derivatives. The differential for functions $f(x)$ is simply the gradient $\mathrm{grad} f(x)$, while for curves $\gamma(t)$ it is the velocity vector $\dot{\gamma}(t)$. A map that is n-fold differentiable with continuous derivatives is called C^n. A map that can be differentiated arbitrarily often with continuous derivatives is called *smooth* and said to be C^∞. The set of all C^∞ functions on a open set $U \subset \mathbb{R}^D$ is denoted by $C^\infty(U)$ and similarly $C^\infty(\mathbb{R}^D)$ for the case the entire $\mathbb{R}^D$ is considered.

We recall here three central results regarding differentiation in $\mathbb{R}^D$ without proofs. The first one is the *chain rule*. If we have a nested map $F \circ G(x) = F(G(x))$ with $G : U(\subset \mathbb{R}^D) \to \mathbb{R}^N$ and $F : V(\subset \mathbb{R}^N) \to \mathbb{R}^M$, where G is differentiable in x and F is differentiable in $G(x)$, then the composed map $F \circ G$ is differentiable in x and its differential is given by

$$\mathrm{D}(F \circ G)(x) = \mathrm{D}F(G(x)) \cdot \mathrm{D}G(x). \tag{1.11}$$

In other words, the Jacobi matrix of the composed map is equal to the product of the Jacobi matrices of the single maps.

The second result deals with the question under which conditions a map can be locally inverted. The answer is given by the *inverse function theorem*, which in essence asserts that a map with an invertible Jacobi matrix can be inverted if we restrict the domain of definition in a suitable way. More precisely, let us consider a map $F : U(\subset \mathbb{R}^D) \to \mathbb{R}^D$ which is C^1 and the points x and $y = F(x)$ of $\mathbb{R}^D$. If the Jacobi matrix $\mathrm{D}F(x)$ is invertible, then there is an open subset $V \subset U$ containing x and an open subset $V' \subset \mathbb{R}^D$ containing y, so that F maps V one-to-one and onto V' and the inverse map $F^{-1} : V' \to V$ is also C^1. The Jacobi matrix of F^{-1} is given by

$$\mathrm{D}(F^{-1})(y) = (\mathrm{D}F(x))^{-1}, \tag{1.12}$$

i.e. by the inverse of the original Jacobi matrix. The Jacobi matrix $\mathrm{D}F$ of a map F from $\mathbb{R}^D$ to $\mathbb{R}^D$ is of square type and one can take the determinant of it, which we denote by $J(x)$,

$$J \equiv \det \left(\frac{\partial F^k}{\partial x^l} \right) \equiv \frac{\partial(F^1, \ldots, F^D)}{\partial(x^1, \ldots, x^D)} \equiv \left| \frac{\partial F}{\partial x} \right|, \tag{1.13}$$

and call the *Jacobian determinant*, or simply the *Jacobian* of F. As stated above, for an invertible map F it is, in coordinate notation,

$$\sum_{l=1}^{D} \frac{\partial F^k}{\partial x^l} \frac{\partial x^l}{\partial F^m} = \delta_m^k, \tag{1.14}$$

and thus we obtain the formula

$$\left| \frac{\partial F}{\partial x} \right| = \left| \frac{\partial x}{\partial F} \right|^{-1}. \tag{1.15}$$

The Jacobian determinant $J(x)$ is a crucial quantity in volume integrals, tensor densities, and in conformal transformations.

The third result is about how we can approximate a function by a *Taylor expansion* (*Brook Taylor*). Suppose we have a C^2 function $f : U(\subset \mathbb{R}^D) \to \mathbb{R}$ and a vector $\xi \in \mathbb{R}^D$, with $|\xi| \ll 1$. Then the function can be approximated by the Taylor expansion in the form

$$f(x + \xi) = f(x) + \sum_{k=1}^{D} \frac{\partial f}{\partial x^k}(x)\, \xi^k + \frac{1}{2} \sum_{k,l=1}^{D} \frac{\partial^2 f}{\partial x^k \partial x^l}(x)\, \xi^k \xi^l + o(|\xi|^2), \qquad (1.16)$$

with the remainder function $o(|\xi|^2)$ fulfilling

$$\lim_{\xi \to 0} \frac{o(|\xi|^2)}{|\xi|^2} = 0. \qquad (1.17)$$

The Taylor expansion for a C^n function is similar and employs the respective higher order partial derivatives.

Directional Derivative

We consider a function $f(x)$ defined on $\mathbb{R}^D$ and a vector v of $\mathbb{R}^D$ with unit length, $|v| = 1$. We also consider the straight line given by the expression $x + tv$, with a real parameter t. The *directional derivative*, denoted $v_x f(x)$, of the function $f(x)$ at the point x in the direction of v is the real number defined by

$$v_x f \equiv \left. \frac{\mathrm{d}}{\mathrm{d}t} f(x + tv) \right|_{t=0} = \lim_{t \to 0} \frac{f(x + tv) - f(x)}{t}. \qquad (1.18)$$

The definition is almost exactly the one for the partial derivative, with the only difference that the direction is not along one of the prime axes of $\mathbb{R}^D$, but along the vector v. By using the chain rule, we can see that the directional derivative can be expressed as

$$v_x f = v^k \frac{\partial f}{\partial x^k}(x) = \langle v, \mathrm{grad} f(x) \rangle. \qquad (1.19)$$

Note that we have used the summation convention above. In modern differential geometry, actually one takes a new point of view and considers the directional derivative along v to be a differential operator of the form

$$v^k \frac{\partial}{\partial x^k}, \qquad (1.20)$$

which acts on functions $f(x)$. This is a central idea when considering vectors on differentiable manifolds. A tangent vector at the point x is simply an expression of the form

$$v = v^k \frac{\partial}{\partial x^k}, \qquad (1.21)$$

with the local coordinate basis $\{\partial/\partial x^k\}$ and the specific vector components v^k. The directional derivative, defined as an operator acting on functions of the space $C^\infty(\mathbb{R}^D)$, obeys the algebraic properties of linearity

$$v_x(af + bg) = a(v_x f) + b(v_x g), \qquad (1.22)$$

and *Leibniz rule*

$$v_x(fg) = (v_x f)g(x) + f(x)(v_x g), \qquad (1.23)$$

for any real numbers a, b and any functions $f(x)$ and $g(x)$ (*Gottfried Wilhelm von Leibniz*). The two above algebraic properties will actually be our starting point for the definition of tangent vectors. As a technical remark, note that in the definition 1.18 of the directional derivative we employed a straight line $\gamma(t) = x + tv$, for which $\dot{\gamma}(0) = v$ holds. If we were to take any other curve $\gamma(t)$ with $\dot{\gamma}(0) = v$, the result of the directional derivative would be exactly the same. In other words, the directional derivative encompasses an entire equivalence class of curves within its definition. Two curves are considered equivalent, in this context, if their velocities $\dot{\gamma}(0)$ at the point $t = 0$ coincide.

1.2 Differentiable Manifolds

Notion of a Manifold

We think of a *differentiable manifold*, denoted by a calligraphic $\mathcal{M}$, as a point set which locally looks like $\mathbb{R}^D$ but globally may have a completely different form. In addition, all points of a differentiable manifold should be describable by D-tuples of real numbers, the coordinates of the points. The raw model of a differentiable manifold is a two-dimensional surface embedded in Euclidean space. The properties arising from this simple description are sufficient in many situations. Nonetheless, this description is conceptually too special, since it assumes that we have a means to measure distances between points, as it is possible in $\mathbb{R}^D$. In general, however, a differentiable manifold has no distance measure defined. The metric structure allowing the measurement of distances is an additional structure, which a manifold may or may not have. Another conceptual trap arises from the notion that the manifold is embedded in a higher-dimensional space. In fact, we want to have a definition of the manifold that is intrinsic and need not to refer to a higher dimensional space. Moreover, we require that a differentiable manifold includes closeness and differentiability as basic properties from the outset.

Before we proceed to the general definition, some elementary notions from topology are needed. Given an arbitrary point set M, a *topology* on this set is a collection of subsets of M, called *open sets*, with the property that unions of open sets are open and finite intersections of open sets are also open. We also require that M itself and the empty set $\varnothing$ are considered open sets. A *topological space* is a set M with a topology. Given a point p of M, a *neighborhood* of p is an open set containing p. A topological space is called a *Hausdorff space* (*Felix Hausdorff*) if for every pair of points p and q one finds neighborhoods of them that are disjoint. A topology essentially defines a notion of closeness between the points of the set M. A map $F : M \to N$ between two topological spaces M, N is called *continuous* if the inverse image of any open set of N is an open set of M. A map $F : M \to N$ between two topological spaces is a *homeomorphism* if it is continuous and has a continuous inverse. Two topological spaces with a homeomorphism between them are considered topologically equivalent.

A *differentiable* (or *smooth*) *manifold* $\mathcal{M}$ is a topological space with a *differentiable* (or *smooth*) *structure*, which means:

- $\mathcal{M}$ has a family of pairs $\{(U_i, \psi_i)\}$, with open subsets U_i of $\mathcal{M}$ and maps $\psi_i : U_i \to \mathbb{R}^D$.
- The union of the open subsets covers the entire set $\mathcal{M}$, i.e. $\bigcup_i U_i = \mathcal{M}$. The maps $\psi_i : U_i \to \mathbb{R}^D$ are homeomorphisms from U_i to open subsets of $\mathbb{R}^D$.
- For any pair of subsets U_i, U_j with non-vanishing intersection, $U_i \cap U_j \neq \varnothing$, the map $\psi_j \circ \psi_i^{-1}$ from the subset $\psi_i(U_i \cap U_j)$ of $\mathbb{R}^D$ to the subset $\psi_j(U_i \cap U_j)$ of $\mathbb{R}^D$ is C^∞, in the usual sense of analysis on $\mathbb{R}^D$.

Each pair (U_i, ψ_i) is called a *coordinate chart*. The open subset U_i is called a *coordinate patch* and the map ψ_i is called a *coordinate map*. Within a coordinate chart (U_i, ψ_i), every point p of the manifold is mapped to a D-tuple of numbers $\psi_i(p)$ of $\mathbb{R}^D$. This D-tuple $\left(\psi_i^1(p), \dots, \psi_i^D(p)\right)$ comprises the *coordinates* of the point p and can be considered as a representation of the point p of $\mathcal{M}$ in the linear space $\mathbb{R}^D$. In this sense, a manifold is locally, within a patch, homeomorphic to a subset of Euclidean space. Globally, however, the manifold has in general a different structure than Euclidean space. The *dimension* of the manifold $\mathcal{M}$ is the integer number D of coordinates needed to describe a point. The definition above also specifies how two different sets of coordinates relate to each other. For a subset $U_i \cap U_j \neq \varnothing$ of $\mathcal{M}$, we can use the coordinate representation of $\psi_i(U_i \cap U_j)$ or the one of $\psi_j(U_i \cap U_j)$. If we change the coordinates with the map $\psi_j \circ \psi_i^{-1}$, then this must be done in a smooth way. This ensures that when we move within the manifold, the assignment of coordinates happens smoothly throughout. By taking a holistic view, we can consider the collection of all chosen coordinate charts $\{(U_i, \psi_i)\}$, which is called an *atlas* covering the manifold. An atlas is not unique to a differentiable manifold. Given two different atlases, they are consistent if their union is also an atlas according to the definition of a manifold. The atlas of a manifold which is the union of all possible ones, is called the *maximal atlas*. It should be noted that in physics it is more common to use the term *coordinate system* or *reference system* instead of coordinate chart.

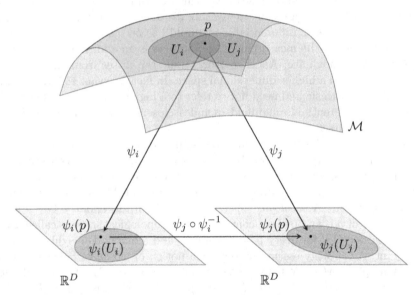

Figure 1.1: Manifold and coordinates

Let us provide a few examples of differentiable manifolds. The vector space $\mathbb{R}^D$, the Euclidean space $\mathbb{E}^D$, and any open subspace of these spaces is a differentiable manifold. One specialty of $\mathbb{R}^D$ and $\mathbb{E}^D$ is that it is possible to cover these by a single, global coordinate system, which of course is the Cartesian coordinate system. The group $GL(n, \mathbb{R})$ of all invertible real $n \times n$-matrices M, as a subset of $\mathbb{R}^{n^2}$,

$$GL(n, \mathbb{R}) \equiv \{M \in Mat(n, \mathbb{R}) \mid \det M \neq 0\}, \tag{1.24}$$

is a differentiable manifold. Further, the 1-dimensional unit circle S^1, the 2-dimensional unit sphere S^2

$$S^2 \equiv \{x \in \mathbb{R}^3 \mid |x| = 1\}, \tag{1.25}$$

and, more generally, the n-dimensional unit sphere S^n are all differentiable manifolds. In order to cover these manifolds by coordinates, however, we need more than one coordinate system. The 2-dimensional torus $S^1 \times S^1$, as a Cartesian set product, is also a manifold. Generally, given any two differentiable manifolds $\mathcal{M}$ and $\mathcal{N}$ with respective dimensions D and N, their Cartesian product $\mathcal{M} \times \mathcal{N}$ is also a differentiable manifold with dimension $D + N$ and a coordinate representation by $(D + N)$-tuples. Pictorially, we can grasp such a product manifold by imagining that to every point of $\mathcal{M}$ a copy of the entire manifold $\mathcal{N}$ is attached.

A special type of manifolds $\mathcal{M}$ are those with a boundary. These manifolds consist of their boundary $\partial \mathcal{M}$ and their interior $\mathrm{Int}(\mathcal{M}) \equiv \mathcal{M} \setminus \partial \mathcal{M}$, which are both manifolds in their own right, as defined above. In order to define correctly the notion of a manifold with boundary, we consider the *closed upper half-space* in D dimensions

$$\bar{H}^D \equiv \left\{ (x^1, \ldots, x^D) \in \mathbb{R}^D \mid x^D \geq 0 \right\}. \tag{1.26}$$

The *boundary* $\partial \bar{H}^D$ of the closed upper half-space $\bar{H}^D$ is the point set

$$\partial \bar{H}^D \equiv \left\{ (x^1, \ldots, x^{D-1}, 0) \in \mathbb{R}^D \right\}. \tag{1.27}$$

A D-dimensional *differentiable manifold with boundary* $\mathcal{M}$ is defined as before, with the exception that it contains an atlas with two types of coordinate charts. A coordinate chart either maps to $\mathbb{R}^D$ or a coordinate chart maps to $\bar{H}^D$. All other requirements for manifolds remain as stated before. The *boundary* $\partial \mathcal{M}$ of the manifold consists of all *boundary points* p of $\mathcal{M}$ which are mapped to the tuples of the form $(x^1, \ldots, x^{D-1}, 0)$. The *interior* $\mathrm{Int}(\mathcal{M})$ of the manifold consists of the so-called *regular points* which are mapped to general tuples $(x^1, \ldots, x^D)$. We can see that the boundary is a $(D-1)$-dimensional manifold, while the interior is a D-dimensional manifold. The closed real interval $[a, b]$, for example, is a manifold with a boundary consisting of the two points a, b (indeed a discrete set has manifold dimension zero), while the open interval (a, b) represents the interior. An arbitrary manifold $\mathcal{M}$ can be considered to be a manifold with boundary $\partial \mathcal{M} = \varnothing$ and $\mathrm{Int}(\mathcal{M}) \equiv \mathcal{M}$. Manifolds with a boundary will be relevant when we discuss the Stokes integral formula.

Now that we know what a differentiable manifold is, let us emphasize again that a manifold has no measure of distance by itself. A differentiable manifold encodes only the notions of closeness (through its topology) and differentiability (through its coverage by an atlas). The points of the manifold can be described by D-tuples, which represent the assigned coordinates and are to be thought of as suitable labels of the points. Coordinates are not unique and can be changed. Coordinates provide a description of the points, so that we can use standard methods of analysis of $\mathbb{R}^D$, such as differentiation and integration.

Coordinates

Let us look more closely to the notion of coordinates. For the space $\mathbb{R}^D$, we simply identify each of its points with its coordinates and consequently we write out a D-tuple for both. For a general manifold $\mathcal{M}$, we need to make the distinction between a point and its coordinates clearer. To this end, coordinates are introduced to be functions. In the case $\mathbb{R}^D$, a *coordinate function* x^μ is the map that assigns to a vector $a \in \mathbb{R}^D$ its μth coordinate a^μ, i.e. the coordinate function $x^\mu : \mathbb{R}^D \to \mathbb{R}$ is the projection defined as

$$x^\mu(a) = a^\mu. \tag{1.28}$$

In the case of a general manifold $\mathcal{M}$, we first make a choice of a coordinate chart (U, ψ). Suppose the point p of $\mathcal{M}$ has the coordinates $\psi(p) = a$. Then the coordinate function is considered to act locally $x^\mu \circ \psi : U \to \mathbb{R}$ as

$$x^\mu(\psi(p)) = a^\mu. \tag{1.29}$$

In abuse of notation, the coordinate functions are thought to be functions of the points of the manifold itself and we write $x^\mu(p) = a^\mu$ then. This type of equation makes sense if we remember that it is valid within the choice of a coordinate chart. We speak about *local coordinates* x^μ then. We must point out that *the local coordinates x^μ, $\mu = 1, \ldots, D$, are not vectors*. The local coordinates x^μ are real D-tuples describing points of a manifold. There is no vector space structure defined for general coordinates.

When we consider a point p of a manifold and we change from one coordinate chart (U, ψ) to another one (U', ψ'), the coordinate values change from $\psi(p) = x = (x^1, \ldots, x^D)$ to $\psi'(p) = x' = (x'^1, \ldots, x'^D)$. These coordinate values are related to each other by

$$(x'^1, \ldots, x'^D) = (x'^1(x), \ldots, x'^D(x)), \tag{1.30}$$

and inversely by

$$(x^1, \ldots, x^D) = (x^1(x'), \ldots, x^D(x')). \tag{1.31}$$

According to the third condition in the definition of a manifold, these functional relationships are smooth. Hence, we can apply standard methods of analysis in $\mathbb{R}^D$ to study this local *coordinate transformation*. The Jacobian $J = \det(\partial x'/\partial x)$, for instance, is non-vanishing within the overlap of the two chosen coordinate patches. Coordinate transformations will be a central and recurring topic in this book.

The choice of local coordinates allows us to define an *orientation* on a manifold. A manifold $\mathcal{M}$ is called *orientable* if there exists an atlas $\{(U_i, \psi_i)\}$ so that for every point $p \in \mathcal{M}$ the coordinate transformations functions $\psi_j \circ \psi_i^{-1}$ for all possible charts for p have a positive Jacobian determinant, $\det(\partial x'/\partial x) > 0$. Here we use the symbols $x = \psi_i(p)$ and $x' = \psi_j(p)$. An orientable manifold has two opposite global orientations, the "positive" and the "negative" one, where the name assignment is based on convention. This is the generalization of the notion of positive (right hand) orientation and negative (left hand) orientation in $\mathbb{R}^3$.

Submanifolds

The idea of a submanifold should be intuitively clear, although the formal definition is somewhat technical if done in a coordinate-free fashion. Here we will restrict ourselves to a practical coordinate-based definition. The idea is the following: a manifold is a set which essentially has D degrees of freedom. If we want to define a subset with $n \leq D$ degrees of freedom, we need to have $D - n$ conditions that implement the restriction in a smooth way. Let us formulate this. Consider a D-dimensional manifold $\mathcal{M}$ and a subset $\mathcal{S}$ of $\mathcal{M}$. The set $\mathcal{S}$ is called an *n-dimensional submanifold of $\mathcal{M}$* if it can be described in local coordinates by $D - n$ equations of the form

$$\left. \begin{aligned} f^1(x^1, \ldots, x^D) &= 0 \\ &\vdots \\ f^{D-n}(x^1, \ldots, x^D) &= 0 \end{aligned} \right\}. \tag{1.32}$$

We require for the $D - n$ differentiable real functions $f^1, \ldots, f^{D-n}$ that the corresponding Jacobian matrix $(\partial f^\alpha/\partial x^\mu)$ has in all points the maximal rank, equal $D - n$. In fact, the submanifold $\mathcal{S}$ itself is an n-dimensional differentiable manifold. We call a one-dimensional submanifold a *curve* and a two-dimensional submanifold a *surface*. A submanifold with dimensionality $D - 1$, defined by one scalar equation $f(x^1, \ldots, x^D) = 0$, is called a *hypersurface*.

Maps between Manifolds

We have seen that we can deal with the points of a manifold by using a coordinate representation of them. The idea of using coordinates applies also when we study maps between manifolds. Let us start with a function $f : \mathcal{M} \to \mathbb{R}$ defined on a manifold. Instead of looking at the points of the manifold, we use a coordinate representation ψ and view the function $f \circ \psi^{-1}$ that maps points of $\mathbb{R}^D$ to $\mathbb{R}$. It is standard practice to treat the function f as if it were a function of the coordinates and write $f(x)$ or $f(x^1, \ldots, x^D)$ for its values.

If we have a map $F : \mathcal{M} \to \mathcal{N}$ between two differentiable manifolds $\mathcal{M}$ and $\mathcal{N}$, we can define a local representative map $\varphi \circ F \circ \psi^{-1}$ by choosing a coordinate representation ψ on $\mathcal{M}$ and a representation φ on $\mathcal{N}$. Once again, the original map is identified with its local representation and we use the simplified notation $F(x) = F(x^1, \ldots, x^D)$ for its values. Within coordinates, we say that a map F is smooth if its local coordinate representation is smooth in the sense of $\mathbb{R}^D$. A smooth map $F : \mathcal{M} \to \mathcal{N}$ which is such that its inverse F^{-1} exists and is also smooth is called a *diffeomorphism*. Two manifolds which can be related to each other by a diffeomorphism are considered equivalent from the differentiability point of view. Every diffeomorphism can be seen from two different points of view. Either the diffeomorphism defines a coordinate transformation and leaves the manifold untouched, which corresponds to the *passive view*. Or the diffeomorphism is viewed as a deformation of the manifold itself, which corresponds to the *active view*. It is important that we keep a clear understanding on how we interpret and use diffeomorphisms in applications.

The smooth functions $f : \mathcal{M} \to \mathbb{R}$ on a given manifold $\mathcal{M}$ constitute a real vector space denoted by $C^\infty(\mathcal{M})$. For any two functions f and g of $C^\infty(\mathcal{M})$ and any real number a, the vector space operations are defined pointwise by $(f + g)(p) = f(p) + g(p)$ and $(af)(p) = a\, f(p)$. Of course, we can restrict functions on any open subset U of a given manifold and consider only the vector space of smooth functions on U, which is then denoted by $C^\infty(U)$.

1.3 Tangent Structure, Vectors and Covectors

Tangent Space

We have seen that within $\mathbb{R}^D$ each vector defines a directional derivative, which in turn has certain algebraic properties. We now take these algebraic properties as the starting point to define vectors on differentiable manifolds. Given a manifold $\mathcal{M}$ and a point p of the manifold, we consider the space of functions $C^\infty(\mathcal{M})$.* A *tangent vector at the point p*, denoted X_p, is a map $X_p : C^\infty(\mathcal{M}) \to \mathbb{R}$, $f \mapsto X_p(f)$, which satisfies *linearity* and the *Leibniz rule*. I.e. for any two functions $f(x)$, $g(x)$ of $C^\infty(\mathcal{M})$ and any real numbers a, b, it is

$$X_p(af + bg) = a\, X_p(f) + b\, X_p(g), \tag{1.33}$$

and

$$X_p(fg) = X_p(f)\, g(p) + f(p)\, X_p(g). \tag{1.34}$$

Now the set of all tangent vectors at the point p constitutes a real vector space, denoted as $T_p\mathcal{M}$, and called *the tangent space of $\mathcal{M}$ at the point p*. The vector space operations are naturally defined by $(X_p + Y_p)(f) = X_p f + Y_p f$ for any two vectors X_p, Y_p, and $(aX_p)(f) = a\, X_p f$ for any real number a. It is common to use the simplified notation $X_p f$ without brackets.

---

*We may consider alternatively the function space $C^\infty(U)$ for an open set U containing p.

We would like to emphasize that tangent vectors are not elements of the manifold, as in the case of $\mathbb{R}^D$. In the figure 1.2 we give an illustration of a manifold $\mathcal{M}$ and its tangent space $T_p\mathcal{M}$. The vectors X_p and Y_p are elements of the tangent space. We like to imagine

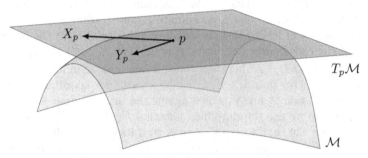

Figure 1.2: Manifold and tangent space

that the point p of the manifold defines the origin of the linear space $T_p\mathcal{M}$. This pictorial representation stems from planes tangent to surfaces, but our tangent space definition is completely independent of any notion of an embedding in a higher-dimensional space.

Vectors in Local Coordinates

We can write any vector X_p of the tangent space $T_p\mathcal{M}$ as a linear combination of basis vectors. A natural basis is defined if we choose local coordinates x^μ around the point p. We can assume that the coordinates of p have the particular value $x^\mu(p) = 0$ at the point p, i.e. p is mapped to the origin of the tangent space $T_p\mathcal{M}$. We consider an arbitrary function $f \in C^\infty(\mathcal{M})$ that depends on the coordinates x and takes the values $f(x) = f(x^1, \ldots, x^D)$. We Taylor-expand this function around the origin $x^\mu(p) = 0$ and obtain, to first order

$$f(x) = f(0) + x^\mu \left. \frac{\partial f}{\partial x^\mu} \right|_{x(p)}. \tag{1.35}$$

Note that we use the summation convention here and henceforth. Now we let X_p operate on the function f itself. By using the Leibniz rule, we have

$$X_p f = X_p(f(0)) + X_p(x^\mu) \left. \frac{\partial f}{\partial x^\mu} \right|_p + x^\mu(p)\, X_p \left(\left. \frac{\partial f}{\partial x^\mu} \right|_p \right). \tag{1.36}$$

It is $X_p(c) = 0$ for any constant c. We note further that $x^\mu(p) = 0$. Thus, it is

$$X_p = X_p(x^\mu) \left. \frac{\partial}{\partial x^\mu} \right|_p. \tag{1.37}$$

The real numbers $X^\mu \equiv X_p(x^\mu)$ are the *vector components* of the vector X_p in the local coordinates x^μ. Hence, we can write

$$X_p = X^\mu \left. \frac{\partial}{\partial x^\mu} \right|_p. \tag{1.38}$$

This is the basis expansion we are looking for. The D vectors

$$\left. \frac{\partial}{\partial x^\mu} \right|_p \tag{1.39}$$

constitute a basis of the tangent space $T_p\mathcal{M}$, the so-called *local coordinate basis* (or *natural basis*). It is reassuring that we have derived the same natural basis of differential operators as in the case of $\mathbb{R}^D$. Moreover, again in analogy to $\mathbb{R}^D$, the real number $X_p f$ given by

$$X_p f = X^\mu \left.\frac{\partial f}{\partial x^\mu}\right|_p \tag{1.40}$$

is the *directional derivative* of f at the point p along X_p. We can always choose another set of local coordinates around the point p, let us call them x'^μ. Then the tangent vector X_p can be written as

$$X_p = X'^\mu \left.\frac{\partial}{\partial x'^\mu}\right|_p, \tag{1.41}$$

where the new components are given by $X'^\mu = X_p(x'^\mu)$. Because of the chain rule, it is

$$\frac{\partial}{\partial x'^\mu} = \left.\frac{\partial x^\nu}{\partial x'^\mu}\right|_p \frac{\partial}{\partial x^\nu}, \tag{1.42}$$

which here represents the *transformation law for basis vectors*. Thus, the components of the vector X_p transform as

$$X'^\mu = \left.\frac{\partial x'^\mu}{\partial x^\nu}\right|_p X^\nu. \tag{1.43}$$

This is the well-known *contravariant transformation law for vector components*, now expressed for the general case of manifolds.

Differential of a Map

The notions of tangent space and vectors at a point give us the means to linearize maps between manifolds. Consider a D-dimensional manifold $\mathcal{M}$, an N-dimensional manifold $\mathcal{N}$, and a smooth map $F : \mathcal{M} \to \mathcal{N}$ between them. Then we can define at each point p of $\mathcal{M}$ a linear map $DF_p : T_p\mathcal{M} \to T_{F(p)}\mathcal{N}$, $X_p \mapsto DF_p(X_p)$ between the corresponding tangent spaces in the following way. To each vector X_p of $T_p\mathcal{M}$, the vector

$$DF_p(X_p) \equiv \left(\left.\frac{\partial F^\alpha}{\partial x^\nu}\right|_p X^\nu \right) \left.\frac{\partial}{\partial y^\alpha}\right|_{F(p)} \tag{1.44}$$

of $T_{F(p)}\mathcal{N}$ is assigned. Here F^α is the αth component of the representative function of F between $\mathbb{R}^D$ and $\mathbb{R}^N$, $\alpha = 1, \ldots, N$. The local coordinates $\{x^\mu\}$ and $\{y^\alpha\}$ in the respective spaces define the basis vectors. The linear map DF_p in coordinates is defined simply by the Jacobian matrix of the map F. The map DF generalizes the concept of the differential, known from calculus in $\mathbb{R}^D$ to the case of differentiable manifolds. The naming for DF varies and we will call it simply the *differential of the map F*. Mathematicians prefer the notation F^* and speak about the *pushforward*, but we will not use this terminology. Let us remark that one can define the differential in a coordinate-free manner, see for example [12].

Vectors Tangent to Curves

We have introduced the tangent space by using the algebraic properties of a differential operator. There is an alternative way to introduce the tangent space, which is based on the idea of vectors being tangent to curves. On our manifold $\mathcal{M}$ we consider smooth *1-parameter curves*. These are maps $\gamma : I \to \mathcal{M}$, $t \mapsto \gamma(t)$ assigning to a real parameter t of the interval

$I \subset \mathbb{R}$ a point p of the manifold $\mathcal{M}$. Two curves γ_1 and γ_2 crossing at a point p of $\mathcal{M}$ for the parameter value $t = 0$ as $\gamma_1(0) = \gamma_2(0) = p$ are said to be *tangent* at the point p iff

$$\frac{\mathrm{d}x^{\mu}}{\mathrm{d}t}(\gamma_1(0)) = \frac{\mathrm{d}x^{\mu}}{\mathrm{d}t}(\gamma_2(0)), \tag{1.45}$$

for all $\mu = 1, \ldots, D$ in local coordinates. In fact, this definition is independent of the coordinate choice. Now two curves crossing the point p are defined to be equivalent if they are tangent to each other. This leads to the notion of a vector as the *equivalence class* of all curves being tangent at the point p. This definition, despite being geometric and visual in its character, can be used also in the case of infinite-dimensional manifolds.

In figure 1.3 we illustrate the concept of the equivalence class, denoted $[\gamma]$, of tangent curves γ at a point $\gamma(0)$. The tangent vector $[\gamma]$ belongs to the tangent space $T_p\mathcal{M}$. The

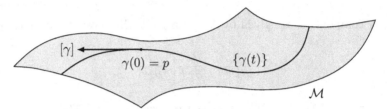

Figure 1.3: Tangent vector on curve

action of a tangent vector $X_p = [\gamma]$ on functions f defined on the manifold $\mathcal{M}$ is once again given by the directional derivative, which takes now the form

$$X_p f = \frac{\mathrm{d}}{\mathrm{d}t} f(\gamma(t)) \bigg|_{t=0}. \tag{1.46}$$

We have encountered this expression already in the case of $\mathbb{R}^D$. Let us view the function f and the curve γ in local coordinates. By using the chain rule, we obtain for the directional derivative the expression

$$X_p f = \frac{\mathrm{d}x^{\mu}}{\mathrm{d}t}(\gamma(t)) \bigg|_{t=0} \frac{\partial f}{\partial x^{\mu}} \bigg|_p. \tag{1.47}$$

In other words, the vector $X_p = [\gamma]$ is given as a differential operator with its components X^{μ} being

$$X^{\mu} = \frac{\mathrm{d}x^{\mu}}{\mathrm{d}t}(\gamma(0)). \tag{1.48}$$

We call $\mathrm{d}x^{\mu}(\gamma(0))/\mathrm{d}t$ the components of the *velocity* and they uniquely define the tangent vector $X_p = [\gamma]$, given here as an equivalence class of curves.

It is useful to look at an example. Let us consider the *unit helix curve* $\gamma(t)$ embedded in $\mathbb{R}^3$ and given in components by

$$\gamma(t) = \begin{pmatrix} \cos t \\ \sin t \\ t \end{pmatrix}, \tag{1.49}$$

with t being a positive real parameter. By using the Cartesian coordinates (x, y, z), the tangent vector $\dot{\gamma}(t)$ at the curve point $\gamma(t)$ is given as

$$\dot{\gamma}(t) = -\sin t \frac{\partial}{\partial x} + \cos t \frac{\partial}{\partial y} + \frac{\partial}{\partial z}. \tag{1.50}$$

The basis $\{\partial/\partial x, \partial/\partial y, \partial/\partial z\}$ is, in fact, the Cartesian basis.

Cotangent Space

Now that we have introduced tangent vectors, the definition of covectors is straightforward. The *cotangent space at the point $p \in \mathcal{M}$* is simply the algebraic dual* of the tangent space, $\tilde{T}_p\mathcal{M} \equiv (T_p\mathcal{M})^{\sim}$. It is comprised of the linear maps $\alpha_p : T_p\mathcal{M} \to \mathbb{R}$, $X_p \mapsto \alpha_p(X_p)$, called *covectors*, which assign to each vector X_p a real number $\alpha_p(X_p)$. The cotangent space is a linear space. It is common to use the notation $\langle \alpha_p, X_p \rangle \equiv \alpha_p(X_p)$ for the *pairing* of α_p and X_p, even if no scalar product is present. Given any smooth function f defined around a point p, we can define uniquely a new covector $\mathrm{d}f_p$ through the relation

$$\langle \mathrm{d}f_p, X_p \rangle = X_p f. \tag{1.51}$$

When we choose the function f to be the coordinate function x^μ, the above formula says

$$\langle \mathrm{d}x_p^\mu, X_p \rangle = X^\mu, \tag{1.52}$$

i.e. the covector $\mathrm{d}x_p^\mu$ picks the μth component of the vector X_p. By specializing to the case where the vector is a basis vector, we obtain

$$\left\langle \mathrm{d}x_p^\mu, \left.\frac{\partial}{\partial x^\nu}\right|_p \right\rangle = \delta_\nu^\mu. \tag{1.53}$$

This means that the set $\{\mathrm{d}x_p^\mu\}$ of D covectors at the point p constitutes a basis of the cotangent space $\tilde{T}_p\mathcal{M}$ that is dual to the coordinate basis of $T_p\mathcal{M}$. Every element α_p of the cotangent space can be written as linear combination as

$$\alpha_p = \alpha_\mu \, \mathrm{d}x_p^\mu, \tag{1.54}$$

with the *covector components* α_μ in local coordinates given by

$$\alpha_\mu = \left\langle \alpha_p, \left.\frac{\partial}{\partial x^\mu}\right|_p \right\rangle. \tag{1.55}$$

The pairing $\langle \alpha_p, X_p \rangle$ in local coordinates is calculated as the contraction

$$\langle \alpha_p, X_p \rangle = \alpha_\mu X^\mu. \tag{1.56}$$

Considering a covector given as $\mathrm{d}f_p$, its coordinate basis expansion is easily seen to be (*exercise 1.1*)

$$\mathrm{d}f_p = \left.\frac{\partial f}{\partial x^\mu}\right|_p \mathrm{d}x_p^\mu. \tag{1.57}$$

This expression is consistent with the notion of the differential of a function, as known from analysis in $\mathbb{R}^D$. In particular, the expression $\mathrm{d}x_p^\mu$ has two meanings. On one hand, it represents a basis vector of the cotangent space $\tilde{T}_p\mathcal{M}$. On the other hand, it is truly the differential of the coordinate function x^μ evaluated at the point p, in the sense of analysis in $\mathbb{R}^D$. If we have an equation of the form $f(x^1, \dots, x^D) = 0$ involving the coordinate functions x^μ and differentiate, we obtain an equation of the form $h(x^1, \dots, x^D, \mathrm{d}x^1, \dots, \mathrm{d}x^D) = 0$. The expressions $\mathrm{d}x^\mu$ can then be interpreted in both of their meanings. We will make use

---

*Some authors use an asterisk for denoting the algebraic dual. We use the tilde so that we can reserve the asterisk for the complex conjugation.

of this fact when we study metrics on manifolds. We can view a covector α_p in two different local coordinate systems,

$$\alpha_p = \alpha_\mu \, dx_p^\mu = \alpha'_\mu \, dx_p'^\mu. \tag{1.58}$$

For relating the two component representations α_μ and α'_μ to each other, we note that

$$dx_p'^\mu = \left.\frac{\partial x'^\mu}{\partial x^\nu}\right|_p dx_p^\nu, \tag{1.59}$$

which is the expression of the *transformation law for basis covectors*. This relation is equivalent to the transformation law of covector components

$$\alpha'_\mu = \left.\frac{\partial x^\nu}{\partial x'^\mu}\right|_p \alpha_\nu. \tag{1.60}$$

This formula is the known *covariant transformation law for covector components*, expressed here for differentiable manifolds.

1.4   Vector Fields and the Commutator

Vector Fields

In the previous section, we have introduced vectors and covectors defined at each point of the manifold. The next natural step is to view fields of vectors (or covectors) as maps assigning to each point of the manifold a vector (or covector) in a smooth way. Now we make this notion more concrete and start with the vector case. Given a manifold $\mathcal{M}$, let us consider the union of all tangent vector spaces $T_p\mathcal{M}$ for all points p of the manifold as

$$T\mathcal{M} \equiv \bigcup_{p \in \mathcal{M}} T_p\mathcal{M} \tag{1.61}$$

and call this set the *tangent bundle of $\mathcal{M}$*. Now a *vector field* is a map $X : \mathcal{M} \to T\mathcal{M}$ assigning to each point p of the manifold a vector X_p in a smooth way. There are different ways to express the smoothness and one of these can be to demand that the vector field components $X^\mu(x)$ as functions of local coordinate variables x^μ depend smoothly upon them. From now on, we will use the simplified notation

$$\partial_\mu = \frac{\partial}{\partial x^\mu} \tag{1.62}$$

for the local basis vectors, varying from point to point in the manifold. There is no reference to the point p anymore. In terms of such a local basis $\{\partial_\mu\}$, a vector field X is written as

$$\boxed{X = X^\mu(x) \, \partial_\mu} \tag{1.63}$$

The transformation rule for the locally varying components X^μ of the vector field is written as

$$\boxed{X'^\mu(x') = \frac{\partial x'^\mu}{\partial x^\nu} X^\nu(x)} \tag{1.64}$$

Pictorially, we can imagine a vector field as in the sketch 1.4 below. We attach each local vector at its corresponding point of the manifold. However, we should remember that each one of these vectors actually belongs to a different tangent space.

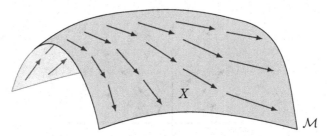

Figure 1.4: Vector field

To give a concrete example, consider the 2-dimensional manifold $\mathbb{R}^2 \setminus \{0\}$ and the vector field

$$X = \frac{x}{(x^2 + y^2)^{3/2}} \frac{\partial}{\partial x} + \frac{y}{(x^2 + y^2)^{3/2}} \frac{\partial}{\partial y} \tag{1.65}$$

in Cartesian coordinates (x, y). This is the vector field generated by a positive point charge at the coordinate origin. We can transform to the coordinate system employing polar coordinates (r, θ), which are defined by

$$\left. \begin{array}{rcl} x &=& r \cos \theta \\ y &=& r \sin \theta \end{array} \right\}, \tag{1.66}$$

or, inversely, by

$$\left. \begin{array}{rcl} r &=& \sqrt{x^2 + y^2} \\ \theta &=& \arctan (y/x) \end{array} \right\}. \tag{1.67}$$

The new basis vectors are found to be

$$\frac{\partial}{\partial r} = \frac{\partial x}{\partial r} \frac{\partial}{\partial x} + \frac{\partial y}{\partial r} \frac{\partial}{\partial y} = \cos \theta \frac{\partial}{\partial x} + \sin \theta \frac{\partial}{\partial y}, \tag{1.68}$$

and

$$\frac{\partial}{\partial \theta} = \frac{\partial x}{\partial \theta} \frac{\partial}{\partial x} + \frac{\partial y}{\partial \theta} \frac{\partial}{\partial y} = -r \sin \theta \frac{\partial}{\partial x} + r \cos \theta \frac{\partial}{\partial y}. \tag{1.69}$$

The vector field X in the new basis reads

$$X = \frac{1}{r^2} \frac{\partial}{\partial r}, \tag{1.70}$$

as expected. (*exercise 1.2*) In figure 1.5 we provide a pictorial illustration of this r^{-2}-law vector field. The locally varying basis vector ∂_r is pointing away from the origin.

Covector Fields

Conceptually, the case of covector fields is very similar to vector fields. First, we consider the *cotangent bundle* $\tilde{T}\mathcal{M}$ of the manifold $\mathcal{M}$ as the union of all cotangent spaces,

$$\tilde{T}\mathcal{M} \equiv \bigcup_{p \in \mathcal{M}} \tilde{T}_p \mathcal{M}. \tag{1.71}$$

A *covector field* is a map $\alpha : \mathcal{M} \to \tilde{T}\mathcal{M}$ which assigns to each point p of the manifold a covector α_p in a smooth way. Smoothness is attained if the covector field components $\alpha_\mu(x)$ depend smoothly on the local coordinates x. In terms of a local dual basis $\{dx^\mu\}$, a covector field α is written as

$$\boxed{\alpha = \alpha_\mu(x)\, dx^\mu} \tag{1.72}$$

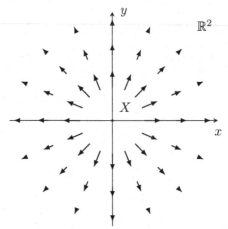

Figure 1.5: Two-dimensional vector field $X = r^{-2}\partial_r$

The transformation law for the components $\alpha_\mu(x)$ reads

$$\alpha'_\mu(x') = \frac{\partial x^\nu}{\partial x'^\mu}\,\alpha_\nu(x) \tag{1.73}$$

For the directional derivative of a function $f(x)$ on the manifold $\mathcal{M}$ along the vector field X, the basic relation

$$Xf = \mathrm{d}f(X) = \langle \mathrm{d}f, X \rangle \tag{1.74}$$

holds. Considering a smooth curve in the manifold $\gamma : I \to \mathcal{M}$, $t \mapsto \gamma(t)$, for which all its tangent vectors coincide with a vector field X on the points along the curve, the directional derivative of a function $f(x)$ along the curve is given as

$$\mathrm{d}f(X) = \frac{\mathrm{d}}{\mathrm{d}t} f(\gamma(t)). \tag{1.75}$$

Finally, the duality between vector and covector basis elements is expressed as

$$\langle \mathrm{d}x^\mu, \partial_\nu \rangle = \delta^\mu_\nu \tag{1.76}$$

Active and Passive Transformations

We define the notions of active and passive transformations by considering vector fields here, although the ideas apply equally well to covector fields. Consider a manifold $\mathcal{M}$ and its points p, which are mapped to coordinates $x = \psi(p)$ under a certain coordinate system ψ. If we change the coordinate system to another one ψ', then we obtain new coordinates $x' = \psi'(p)$ for each point of the manifold. This corresponds to a *passive transformation*. Any vector field X defined over the manifold remains unchanged. The vector field can be written in components in either of the two coordinate systems as

$$X = X^\mu(x)\partial_\mu = X'^\mu(x')\partial'_\mu. \tag{1.77}$$

This means that the basis changes, $\partial_\mu \mapsto \partial'_\mu$, and the components change, $X^\mu(x') \mapsto X'^\mu(x)$, but the vector field remains as it is. The new coordinates are given by

$$X'^\mu(x') = \frac{\partial x'^\mu}{\partial x^\nu} X^\nu(x). \tag{1.78}$$

In the above constellation, one talks about the *passive view* of transformations. In contrast, an *active transformation* changes the points of the manifold itself, while the coordinate system used remains unchanged. An active transformation is a diffeomorphism of the manifold $\mathcal{M}$ that assigns to each point p a new point p'. For both manifold points we use the same coordinate system ψ, so that we have the coordinates $x = \psi(p)$ for the original point and $x' = \psi(p')$ for the new point. The change in the points induces a change in any vector field X defined over the manifold. The active transformation maps the vector field $X = X^\mu(x)\partial_\mu$ to the new vector field X', which is given as

$$X' = X'^\mu(x')\partial_\mu, \tag{1.79}$$

with the components

$$X'^\mu(x') = \frac{\partial x'^\mu}{\partial x^\nu} X^\nu(x). \tag{1.80}$$

This is the very same formula as before, but now within a different context. Here we talk about the *active view* of transformations. In the following, we will always describe how a considered transformation is meant to act.

Algebra of Vector Fields

Technically, a vector field X defines a map that assigns to each smooth function f on the manifold a new smooth function Xf. Thus, we can view the vector field as a map $X : C^\infty(\mathcal{M}) \to C^\infty(\mathcal{M})$, $f \mapsto Xf$ defined by $(Xf)(p) \equiv X_p f$ for all points p on the manifold. In this way, we can let multiple vector fields $X, Y, \ldots$ act on smooth functions f. The composition of two vector fields X and Y denoted XY is naturally given as $(XY)f \equiv X(Yf)$. Interestingly, the composition XY is linear in f, but it does not fulfill the Leibniz rule and consequently it is not a vector field. However, the combination $XY - YX$ is linear and satisfies the Leibniz rule. (*exercise 1.3*) The notation

$$[X,Y] \equiv XY - YX \tag{1.81}$$

is used here and is called the *commutator* of the vector fields X and Y, or, the *Lie bracket* (*Marius Sophus Lie*). The Leibniz rule reads then

$$[X,Y](fg) = ([X,Y]f)g + f([X,Y]g). \tag{1.82}$$

By using a coordinate representation, the commutator $[X,Y]$ of two vector fields X, Y can be expressed as (*exercise 1.4*)

$$[X,Y]^\mu = X^\nu \partial_\nu Y^\mu - Y^\nu \partial_\nu X^\mu. \tag{1.83}$$

From an algebraic point of view, the commutator itself $[\cdot, \cdot]$, besides being bilinear in its arguments X, Y, is also *antisymmetric*,

$$[X,Y] = -[Y,X], \tag{1.84}$$

and obeys the *Jacobi identity*,

$$[[X,Y],Z] + [[Y,Z],X] + [[Z,X],Y] = 0, \tag{1.85}$$

for any three vector fields X, Y, Z. A real vector space equipped with a product that is bilinear, antisymmetric and satisfies the Jacobi identity is called a *real Lie algebra*, see also the definitions in appendix B.1. This means that the space of all vector fields X on a manifold $\mathcal{M}$, in combination with the commutator, constitutes a real Lie algebra, denoted $\mathcal{X}\mathcal{M}$. The dual space $\widetilde{\mathcal{X}}\mathcal{M}$ contains all covector fields α on the manifold.

1.5   Tensor Fields on Manifolds

Tensors at a Point

We proceed now to the definition of tensors on manifolds. The reader should recall the algebraic concepts described in appendix B.1. For any single point p of the manifold $\mathcal{M}$, we can build (m, n)-*type tensors* T_p at the point p, which are elements of the D^{m+n}-dimensional tensor product space $\mathcal{T}_p^{(m,n)}\mathcal{M}$, where the latter is given as

$$\mathcal{T}_p^{(m,n)}\mathcal{M} \equiv \left(\overset{m}{\bigotimes} T_p\mathcal{M} \right) \otimes \left(\overset{n}{\bigotimes} \widetilde{T}_p\mathcal{M} \right). \tag{1.86}$$

The tensor product space at the point p is constructed from multiple copies of the tangent space $T_p\mathcal{M}$ and the cotangent space $\widetilde{T}_p\mathcal{M}$. Every (m, n)-type tensor T_p of the space $\mathcal{T}_p^{(m,n)}\mathcal{M}$ is a multilinear map

$$T_p : \underbrace{\widetilde{T}_p\mathcal{M} \times \cdots \times \widetilde{T}_p\mathcal{M}}_{m\text{-fold}} \times \underbrace{T_p\mathcal{M} \times \cdots \times T_p\mathcal{M}}_{n\text{-fold}} \to \mathbb{R} \tag{1.87}$$

that assigns to each $(m + n)$-tuple $(\alpha_1, \ldots, \alpha_m, X_1, \ldots, X_n)$ of m covectors and n vectors at the point p the real number $T_p(\alpha_1, \ldots, \alpha_m, X_1, \ldots, X_n)$. A tensor of the space $\mathcal{T}_p^{(m,n)}\mathcal{M}$ is called m-*fold contravariant and n-fold covariant*. Whenever we choose a local coordinate system x^μ around the point p, we induce a local basis in the tensor space $\mathcal{T}_p^{(m,n)}\mathcal{M}$ consisting of the D^{m+n} elements

$$\partial_{\mu_1}\big|_p \otimes \cdots \otimes \partial_{\mu_m}\big|_p \otimes \mathrm{d}x_p^{\nu_1} \otimes \cdots \otimes \mathrm{d}x_p^{\nu_n}, \tag{1.88}$$

which are constructed as tensor products of m basis vectors $\partial_\mu\big|_p$ and n basis covectors $\mathrm{d}x_p^\nu$. Hence, any tensor T_p at the point p is written as a linear combination

$$T_p = T^{\mu_1 \cdots \mu_m}{}_{\nu_1 \ldots \nu_n} \cdot \partial_{\mu_1}\big|_p \otimes \cdots \otimes \partial_{\mu_m}\big|_p \otimes \mathrm{d}x_p^{\nu_1} \otimes \cdots \otimes \mathrm{d}x_p^{\nu_n}. \tag{1.89}$$

The local components of the tensor T_p are given as

$$T^{\mu_1 \cdots \mu_m}{}_{\nu_1 \ldots \nu_n} = T_p \left(\mathrm{d}x_p^{\mu_1}, \ldots, \mathrm{d}x_p^{\mu_m}, \partial_{\nu_1}\big|_p, \ldots, \partial_{\nu_n}\big|_p \right). \tag{1.90}$$

When we say "local" we mean that the tensor T_p is defined as an element of a vector space at the point p. The components of T_p are defined by a choice of basis at p. However, we can move out of the point p within the manifold and ask if we can define a tensorial object, which has T_p or its components $T^{\mu_1 \cdots \mu_m}{}_{\nu_1 \ldots \nu_n}$ as local values at the point p. This chain of thoughts leads to the notion of a tensor field, to which we now turn.

Tensor Fields

An (m, n)-*type tensor field* ϕ on the manifold $\mathcal{M}$ is a smooth assignment of an (m, n)-type tensor ϕ_p at each point p of the manifold. Once again smoothness can be imposed by requiring that the components of the tensor field depend smoothly on the local coordinates x^μ. Formally, we write

$$\phi : \mathcal{M} \to \mathcal{T}^{(m,n)}\mathcal{M} \equiv \bigcup_{p \in \mathcal{M}} \mathcal{T}_p^{(m,n)}\mathcal{M}, \tag{1.91}$$

using the (m, n)-*tensor bundle* $\mathcal{T}^{(m,n)}\mathcal{M}$ as the range. Tensor fields ϕ provide a most important mathematical notion in order to formulate classical physical laws. A tensor field equation $F = 0$, with a certain tensor field F, is the same in all coordinate systems, exactly as we would require from a universally valid (physical) law. In terms of notation, in our applications we will occasionally write $\phi(x)$ to display the local dependency of the tensor field ϕ. This is, strictly speaking, an abuse of notation, since the tensor field ϕ is independent on any coordinate choice. Still, this notation is useful to make the distinction between rigid tensors (at a point) and tensor fields (defined for each point of a manifold) clear. The dependence on local coordinates x will also be displayed for the components of tensor fields. As soon as local coordinates x are chosen, the corresponding locally varying (m, n)-tensor basis

$$\partial_{\mu_1} \otimes \cdots \otimes \partial_{\mu_m} \otimes dx^{\nu_1} \otimes \cdots \otimes dx^{\nu_n} \tag{1.92}$$

is introduced. In terms of this basis, the (m, n)-tensor field ϕ is expanded as

$$\boxed{\phi = \phi^{\mu_1 \cdots \mu_m}{}_{\nu_1 \ldots \nu_n}(x)\, \partial_{\mu_1} \otimes \cdots \otimes \partial_{\mu_m} \otimes dx^{\nu_1} \otimes \cdots \otimes dx^{\nu_n}} \tag{1.93}$$

where the tensor field components $\phi^{\mu_1 \cdots \mu_m}{}_{\nu_1 \ldots \nu_n}(x)$ are given as

$$\phi^{\mu_1 \cdots \mu_m}{}_{\nu_1 \ldots \nu_n}(x) = \phi\,(dx^{\mu_1}, \ldots, dx^{\mu_m}, \partial_{\nu_1}, \ldots, \partial_{\nu_n}). \tag{1.94}$$

Now let us examine what happens when we change the coordinates of a particular point in the manifold. An active (or passive) transformation from x^μ to x'^μ, given by

$$x'^\mu = \frac{\partial x'^\mu}{\partial x^\nu} x^\nu, \tag{1.95}$$

leads to the following general transformation of the tensor field components:

$$\boxed{\phi'^{\mu_1 \cdots \mu_m}{}_{\nu_1 \ldots \nu_n}(x') = \frac{\partial x'^{\mu_1}}{\partial x^{\rho_1}} \cdots \frac{\partial x'^{\mu_m}}{\partial x^{\rho_m}} \frac{\partial x^{\sigma_1}}{\partial x'^{\nu_1}} \cdots \frac{\partial x^{\sigma_n}}{\partial x'^{\nu_n}} \phi^{\rho_1 \cdots \rho_m}{}_{\sigma_1 \ldots \sigma_n}(x)} \tag{1.96}$$

This is the *tensor field transformation law*, as employed in classical tensor analysis for the basic definition of a tensor field.

Let us remark that, similar to tensors at a point, tensor fields over a manifold can be also regarded as multilinear maps. As explained before, a tensor at a point assigns to a tuple of vectors and covectors at that point a real number. In analogy, a tensor field over a manifold assigns to a tuple of vector fields and covector fields a smooth function. The straightest way to see this, is to use the component representation of a tensor field ϕ. The components $\phi^{\mu_1 \cdots \mu_m}{}_{\nu_1 \ldots \nu_n}(x)$, defined over the entire manifold, are used to assign to a tuple $(\alpha_1, \ldots, \alpha_m, X_1, \ldots, X_n)$ of m covector fields and n vector fields the smooth function

$$\phi^{\mu_1 \cdots \mu_m}{}_{\nu_1 \ldots \nu_n}(x) \cdot \alpha_{1 \mu_1}(x) \cdots \alpha_{m \mu_m}(x) \cdot X_1^{\nu_1}(x) \cdots X_n^{\nu_n}(x). \tag{1.97}$$

This means that the tensor field ϕ is a multilinear map of the form

$$\phi: \underbrace{\widetilde{\mathcal{X}}\mathcal{M} \times \cdots \times \widetilde{\mathcal{X}}\mathcal{M}}_{m\text{-fold}} \times \underbrace{\mathcal{X}\mathcal{M} \times \cdots \times \mathcal{X}\mathcal{M}}_{n\text{-fold}} \to C^\infty(\mathcal{M}). \tag{1.98}$$

Multilinearity here means especially that ϕ is homogeneous in any scalar smooth function $f(x)$ appearing in the arguments, so that it is $\phi(\ldots, fX, \ldots) = f\,\phi(\ldots, X, \ldots)$.

Tensor Densities

Extending the purely algebraic constructions of *tensor densities*, as explained in appendix B.1, to the case of tensor fields on manifolds is straightforward. Per definition, a *tensor density field* T with components $T^{\mu_1\cdots\mu_m}{}_{\nu_1\ldots\nu_n}(x)$ transforms as

$$T'^{\mu_1\cdots\mu_m}{}_{\nu_1\ldots\nu_n}(x') = J(x)^w \frac{\partial x'^{\mu_1}}{\partial x^{\rho_1}} \cdots \frac{\partial x'^{\mu_m}}{\partial x^{\rho_m}} \frac{\partial x^{\sigma_1}}{\partial x'^{\nu_1}} \cdots \frac{\partial x^{\sigma_n}}{\partial x'^{\nu_n}} T^{\rho_1\cdots\rho_m}{}_{\sigma_1\ldots\sigma_n}(x), \qquad (1.99)$$

with the Jacobian determinant

$$J(x) \equiv \det\left(\frac{\partial x'^{\mu}(x)}{\partial x^{\nu}}\right). \qquad (1.100)$$

and the *weight* w. In addition to tensor density fields, the so-called *numerical tensors* are important for conceptual aspects and for practical applications. The numerical tensors are the *generalized Kronecker delta* $\delta^{\mu_1\cdots\mu_n}_{\nu_1\ldots\nu_n}$ and the *Levi-Civita epsilon symbol* $\varepsilon_{\mu_1\ldots\mu_D}$. They extend the algebraic definitions from appendix B.1 to the case of manifolds. The numerical tensors not only retain their components invariant under coordinate transformations, but they keep their component values constant throughout all the points of the manifold. This means that under arbitrary diffeomorphisms of the manifold, it is always for the Kronecker delta

$$\delta'^{\mu_1\cdots\mu_n}_{\nu_1\ldots\nu_n} = \delta^{\mu_1\cdots\mu_n}_{\nu_1\ldots\nu_n}, \qquad (1.101)$$

and, similarly, for the epsilon symbol

$$\varepsilon'_{\mu_1\ldots\mu_D} = \varepsilon_{\mu_1\ldots\mu_D}. \qquad (1.102)$$

We should notice that the epsilon symbol has as many components as the dimensionality D of the manifold. In addition to the epsilon symbol $\varepsilon_{\mu_1\ldots\mu_D}$, which is a $(0, D)$-tensor density field of weight $w = 1$, one defines the *Levi-Civita epsilon tensor* $\epsilon_{\mu_1\ldots\mu_D}$ by

$$\epsilon_{\mu_1\ldots\mu_D} \equiv \sqrt{|g|}\, \varepsilon_{\mu_1\ldots\mu_D}, \qquad (1.103)$$

where $g \equiv \det(g_{\mu\nu})$ is the determinant of a symmetric $(0, 2)$-tensor field $g_{\mu\nu}(x)$, called the *metric*, whose primary task is to define a geometry on the manifold, as we will see in the next chapter. With the above definition, the epsilon tensor $\epsilon_{\mu_1\ldots\mu_D}$ represents an absolute $(0, D)$-tensor field. In the following, we will use both, the epsilon symbol and the epsilon tensor, depending on which of the two is more appropriate for the specific application.

Further Reading

Our discussion of differentiable manifolds focused primarily on the needs in classical physics. The reader interested to encounter a mathematically rigorous treatment of manifolds and tangent spaces can turn, for example, to Boothby [12]. A modern text on differential manifolds that is closer to the approach needed in physics is provided by Renteln [73]. The various equivalent ways for introducing the notion of the tangent vector are carefully discussed by Isham [44]. For a timeless presentation of classical tensor analysis one can consult the book of Eisenhart [26].

2

Geometry and Integration on Manifolds

In this chapter we introduce the metric as an additional structure, whose central role is to define the geometry of a manifold. After a survey of the basic properties of the metric, we discuss the induced geometry of submanifolds and of hypersurfaces in particular. Subsequently, we consider the conditions of isometry and conformality between different geometric manifolds. We provide several examples of classical geometries that are relevant for later purposes. In the second part of this chapter, we develop the rudiments of integration on manifolds. We first introduce differential forms and the exterior derivative so that we can define integration measures for curves, hypersurfaces, and volumes in a unified way. The invariance of volume integrals is shown. In the last section, we discuss the integral theorem of Stokes, the Gauss divergence theorem, and a special version of Stokes' theorem for antisymmetric tensor fields.

2.1 Geometry and Metric

Pythagorean Theorem

So far we have introduced manifolds with their notions of closeness and differentiability. However, we have not explained how to define a geometry on a manifold. Geometry literally means "measurement of land". To do so, one needs measures such as distance, area, and angle. These measures of geometry are not contained per default in the basic manifold structure. One needs to introduce them as additional structures. The origins of geometry are probably as old as mankind, but geometry as a branch of pure mathematics took its first shape within *Euclidean geometry* (*Euclid of Alexandria*) during the 4th century BC. From the body of knowledge in Euclidean geometry, we will focus here only on the *Pythagorean theorem* (*Pythagoras of Samos*), since it is sufficient to guide us to a general notion of a metric on manifolds. The Pythagorean theorem is a statement about geometric squares in the plane being attached to right-angled triangles. It states:

> *For a right-angled triangle, the sum of the squares on the two sides enclosing the right angle is equal the square on the hypotenuse.*

DOI: 10.1201/9781003087748-2

For the ancient Greek geometers, this was understood as a statement about geometric objects, in this case about the squares. At that time, the Pythagorean theorem was not formulated algebraically, nor was it primarily a statement about numbers. Today, the interpretation of the Pythagorean theorem can be within geometric notions or within algebraic notions. Correspondingly, there are proofs of this theorem leaning either to geometric methods, or algebraic methods, or a mixture of both. The shortest way to see the validity of the Pythagorean theorem is by virtue of the construction below, see for instance [41]. Here

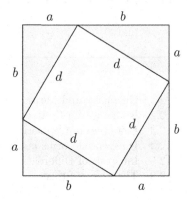

Figure 2.1: A graphical proof of the Pythagorean theorem

we note that the area of the outer square (with the side length $a + b$) minus the area of the inner square (with the side length d) is four times the area of the right-angled triangle with the short sides a and b. Thus, we arrive at the relation

$$d^2 = a^2 + b^2. \tag{2.1}$$

Interpreted geometrically, this represents a relation between the three *geometric squares* attached to a right-angled triangle. The above relation has equally well an algebraic meaning, namely that the *arithmetic square* d^2 of the distance d is the sum of the *arithmetic squares* a^2 and b^2 of the right-angled sides a and b. Consequently, one can express the distance d by

$$d = \sqrt{a^2 + b^2}. \tag{2.2}$$

This formula prescribes how to measure distances in a system of right-angled coordinates.

Cartesian Geometry

We move ahead to the 17th century AD and the development of analytic geometry by *René Descartes*. Descartes introduced coordinates as numeric labels representing geometric points and thus opened the door for the analytical treatment of geometric questions. The starting point is the Euclidean space $\mathbb{E}^D$, where each geometric point is represented by a unique D-tuple of numbers, the right-angled *Cartesian coordinates* of this geometric point. If two points p and q of Euclidean space are represented by the D-tuples $v = (v^1, \ldots, v^D)$ and $w = (w^1, \ldots, w^D)$ respectively, then the *distance* $d(p, q)$ between these points is given by

$$d(p, q) = \sqrt{(v^1 - w^1)^2 + \cdots + (v^D - w^D)^2}. \tag{2.3}$$

Equivalently, one can introduce the *norm* (or *length*) $|v|$ of a vector v, being the distance between the coordinate origin O and the end point p of the vector v. The value of the norm is

$$|v| = \sqrt{(v^1)^2 + \cdots + (v^D)^2}. \tag{2.4}$$

The *geometric angle* φ between any two non-zero vectors v and w is given by

$$\cos\varphi = \frac{v^1 w^1 + \cdots + v^D w^D}{|v|\,|w|}. \tag{2.5}$$

The value range for φ is $0 \leq \varphi \leq \pi$. All the above coordinate representations of distances and angles can be condensed in one central object defining these formulae. It is the *Euclidean metric*, which analytically is given by the D-dimensional unit matrix, or equivalently, by the Kronecker delta δ_{kl}. The Euclidean metric is equivalent to a *real scalar product* (a *real symmetric bilinear form*) $\langle \cdot, \cdot \rangle$ in $\mathbb{E}^D$. For any two vectors v and w, their scalar product $\langle v, w \rangle$ is given by

$$\langle v, w \rangle = \delta_{kl} v^k w^l = v^1 w^1 + \cdots + v^D w^D. \tag{2.6}$$

By using the scalar product, we can write the norm of a vector as

$$|v| = \sqrt{\langle v, v \rangle}, \tag{2.7}$$

while the angle φ between any two non-zero vectors v and w becomes

$$\cos\varphi = \frac{\langle v, w \rangle}{|v|\,|w|}. \tag{2.8}$$

The basic geometric measures above are fully encoded in the metric, so that one can identify the metric as the crucial structure needed in order to define geometry.

Metric Tensor Field

Now the question arises how we can introduce properly a metric (or, equivalently, a scalar product) on a differentiable manifold $\mathcal{M}$. Since a manifold is not a linear space, we will introduce the metric initially for each single point p and the corresponding tangent space $T_p\mathcal{M}$. Then we will require that the metric changes smoothly from point to point. In other words, the metric is supposed to be a smooth tensor field on the manifold. Based on these considerations, we define a *Riemannian metric* (*Georg Friedrich Bernhard Riemann*) on the manifold $\mathcal{M}$ to be a positive definite and symmetric $(0,2)$-tensor field

$$g : \mathcal{M} \;\rightarrow\; \bigcup_{p \in \mathcal{M}} \tilde{T}_p\mathcal{M} \otimes \tilde{T}_p\mathcal{M}, \tag{2.9}$$

that assigns to each point p of the manifold a $(0,2)$-type tensor g_p acting on the corresponding tangent space $T_p\mathcal{M}$,

$$g_p : T_p\mathcal{M} \times T_p\mathcal{M} \;\rightarrow\; \mathbb{R}, \quad (X_p, Y_p) \mapsto g_p(X_p, Y_p). \tag{2.10}$$

We require the metric tensor to have the following two crucial properties: first for any pair (X_p, Y_p) of vectors the *symmetry* property

$$g_p(X_p, Y_p) = g_p(Y_p, X_p) \tag{2.11}$$

holds. Secondly *positive definiteness* holds, which means that

$$g_p(X_p, X_p) > 0, \tag{2.12}$$

for all vectors $X_p \neq 0$. For the inclusion of the metric of spacetimes, we need to replace the condition of positive definiteness by the slightly weaker condition of *non-degeneracy*. Non-degeneracy means that if

$$g_p(X_p, Y_p) = 0, \tag{2.13}$$

for all vectors X_p (of the specific tangent space $T_p\mathcal{M}$), then it must be necessarily $Y_p = 0$. The corresponding non-degenerate metric is called a *pseudo-Riemannian metric*. A manifold with a metric, denoted $(\mathcal{M}, g)$, is called a *geometric manifold*. More specifically, it is called a *Riemannian manifold* if the metric is positive definite. It is called a *pseudo-Riemannian manifold* if the metric is only non-degenerate. In local coordinates $\{x^\mu\}$ of the manifold, the *metric tensor field* g is given as

$$\boxed{g = g_{\mu\nu}(x)\, \mathrm{d}x^\mu \otimes \mathrm{d}x^\nu} \tag{2.14}$$

where the *metric tensor components* $g_{\mu\nu}(x)$ are defined as

$$g_{\mu\nu}(x) \equiv g(\partial_\mu, \partial_\nu) \tag{2.15}$$

and satisfy the symmetry condition

$$g_{\mu\nu}(x) = g_{\nu\mu}(x). \tag{2.16}$$

In many situations $g_{\mu\nu}(x)$ can be handled as a $D \times D$-matrix that depends on local coordinates x^μ. Due to its symmetry, $g_{\mu\nu}(x)$ has $D(D+1)/2$ independent components. The (active or passive) transformation from coordinate variables $\{x^\mu\}$ to new ones $\{x'^\mu\}$ within a fixed tangent space $T_p\mathcal{M}$ induces a transformation of the metric tensor components exactly as known from the previous chapter:

$$\boxed{g'_{\mu\nu}(x') = \frac{\partial x^\rho}{\partial x'^\mu} \frac{\partial x^\sigma}{\partial x'^\nu}\, g_{\rho\sigma}(x)} \tag{2.17}$$

The matrix $g_{\mu\nu}(x)$ has a maximal rank, due to its non-degeneracy (or positive definiteness) and as such it has an inverse, which we denote as $g^{\mu\nu}(x)$, i.e. with upper indices. It is

$$g^{\mu\nu}g_{\nu\rho} = g_{\rho\nu}g^{\nu\mu} = \delta^\mu_\rho, \tag{2.18}$$

for all points in the manifold. The matrix $g^{\mu\nu}(x)$ provides the tensor components of the *inverse metric* g^{-1}, which has the local basis expansion

$$g^{-1} = g^{\mu\nu}(x)\, \partial_\mu \otimes \partial_\nu \tag{2.19}$$

where the components $g^{\mu\nu}(x)$ are defined as

$$g^{\mu\nu}(x) = g^{-1}(\mathrm{d}x^\mu, \mathrm{d}x^\nu). \tag{2.20}$$

Another quantity we will use is the *determinant of the metric*, $\det(g)$, which actually always exists, due to the non-degeneracy, and is usually also denoted by the symbol $g \equiv \det(g_{\mu\nu})$. We note that for the inverse metric g^{-1} it is $\det(g^{-1}) = (\det(g))^{-1}$.

The symmetric matrix $g_{\mu\nu}(x)$ can be diagonalized with real diagonal entries. If the metric is positive definite, all diagonal entries are positive reals. If the metric is only non-degenerate, then there are positive and negative diagonal entries. By a suitable rescaling of the basis vectors, one can achieve that the diagonal entries are all $+1$ in the Riemannian case, and ± 1 in the pseudo-Riemannian case. So a strictly Riemannian metric can be brought into the Euclidean form $g_{\mu\nu} = \delta_{\mu\nu}$ in each point. However, the transformation of the basis vectors in order to bring the metric in the diagonal form will be different from point to point in the manifold. Generally, there exists no global transformation of the basis vectors in the manifold being able to diagonalize the metric everywhere. In the pseudo-Riemannian case there will be a number p of entries equal $+1$ and a number n of entries equal -1. The pair (p, n) is constant throughout the pseudo-Riemannian manifold and is called the *signature*, or, the *index* of the metric. A metric for which only one entry has a different sign than the others is called a *Lorentzian metric*. The prime example in four dimensions is provided by Minkowski space and is introduced in Section 2.3.

Metric defines Geometry

With a metric tensor field at hand, we obtain for each tangent space a metric structure. For each point p of the manifold $\mathcal{M}$ and each two vectors X, Y of the associated tangent space $T_p\mathcal{M}$, their scalar product is defined as

$$\langle X, Y \rangle \equiv g(X, Y). \tag{2.21}$$

Here we have left out the explicit mentioning of the point p. By employing coordinates, the scalar product can be written as

$$\langle X, Y \rangle = g_{\mu\nu} X^\mu Y^\nu. \tag{2.22}$$

The norm of a vector X is consequently

$$|X| = \sqrt{\langle X, X \rangle}. \tag{2.23}$$

In pseudo-Riemannian manifolds, there are non-zero vectors for which the norm vanishes. These vectors are called *null vectors* (or *isotropic vectors*). The geometric angle φ between any two non-zero vectors X and Y is defined again as

$$\cos\varphi = \frac{\langle X, Y \rangle}{|X|\,|Y|}. \tag{2.24}$$

By using the metric on a manifold, we can measure and thus compare vectors of the same tangent space, but we cannot compare vectors belonging to different tangent spaces. In order to achieve this in a proper way, one needs a prescription how to carry out this comparison. This is provided by the notion of connection, which we will encounter in Chapter 18.

In classical notation, the metric g is written as an "infinitesimal distance squared" as

$$\boxed{\mathrm{d}s^2 = g_{\mu\nu}(x)\,\mathrm{d}x^\mu\mathrm{d}x^\nu} \tag{2.25}$$

where we have used the *symmetrized product* $\mathrm{d}x^\mu\mathrm{d}x^\nu$, which is defined as

$$\mathrm{d}x^\mu\mathrm{d}x^\nu \equiv \frac{1}{2}\left(\mathrm{d}x^\mu \otimes \mathrm{d}x^\nu + \mathrm{d}x^\nu \otimes \mathrm{d}x^\mu\right) = \mathrm{d}x^\nu\mathrm{d}x^\mu. \tag{2.26}$$

Within this classical notation, the basis elements $\mathrm{d}x^\mu$ are interpreted as "infinitesimal displacements". The geometric formula 2.25 represents the generalization of the distance formula of Cartesian analytic geometry.

Now we would like to define a measure of "distance" between points in a Riemannian manifold. Before we do so, we first define the length of a curve in a Riemannian manifold. Suppose we have a C^1 curve $\gamma : I \to \mathcal{M}$, $t \mapsto \gamma(t)$, mapping the parameter t from the interval $I = [t_1, t_2] \subset \mathbb{R}$ to points $\gamma(t)$ in the manifold $\mathcal{M}$, as indicated in figure 2.2.

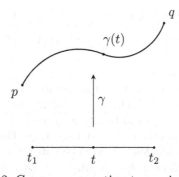

Figure 2.2: Curve γ connecting two points p and q

The start and end points are $\gamma(t_1) = p$ and $\gamma(t_2) = q$, respectively. The tangent vector to the curve is denoted $\dot\gamma(t)$ and is given in components by

$$\dot\gamma(t) = \frac{\mathrm{d}x^\mu}{\mathrm{d}t}(t)\left.\frac{\partial}{\partial x^\mu}\right|_{\gamma(t)}. \tag{2.27}$$

Now we define the *length l of the curve* to be the integral

$$l \equiv \int_\gamma \mathrm{d}s, \tag{2.28}$$

with the *line element* $\mathrm{d}s = |g_{\mu\nu}\,\mathrm{d}x^\mu\mathrm{d}x^\nu|^{1/2}$. In order to calculate the length of the curve l, we can use the explicit parametrization and evaluate the integral as

$$l = \int_{t_1}^{t_2}\left|g_{\mu\nu}(x(t))\frac{\mathrm{d}x^\mu}{\mathrm{d}t}(t)\frac{\mathrm{d}x^\nu}{\mathrm{d}t}(t)\right|^{\frac12}\mathrm{d}t = \int_{t_1}^{t_2}|\dot\gamma(t)|\,\mathrm{d}t. \tag{2.29}$$

The value of the line integral is independent of the parametrization, exactly as we would expect from a geometric object like a curve length. Indeed, if $t \mapsto s = s(t)$ is a new parametrization with $\mathrm{d}s/\mathrm{d}t \neq 0$ throughout all values of t, then one can see that the line integral with the parameter s has the same value. (*exercise 2.1*) One speaks about a *reparametrization invariance*. We should remark here that the definitions for the line integral above depend crucially on the scalar product of the Riemannian manifold. Later in this chapter, we will define line integrals without using a metric.

The *distance $d(p,q)$ between two points p and q* in the manifold is defined as the minimum curve length over all possible curves $\gamma(p,q)$ having p and q as end-points, i.e.

$$d(p,q) \equiv \min_{\gamma(p,q)}\int_\gamma \mathrm{d}s. \tag{2.30}$$

The actual calculation of the extremal curve providing the minimal length between two given points is, in general, a nontrivial problem and part of the calculus of variations. We will tackle this question later in the discussion of Riemannian geometry in Chapter 18. In any case, the above definition of distance is explicit enough to allow us to derive the characteristic properties of a distance function. For any points p, q, and r in the manifold $\mathcal{M}$ it is

$$
\begin{aligned}
&d(p,p) = 0, && \text{trivial distance,} \\
&d(p,q) \geq 0, && \text{positive definiteness,} \\
&d(p,q) = d(q,p), && \text{symmetry,} \\
&d(p,r) + d(r,q) \geq d(p,q), && \text{triangle inequality.}
\end{aligned}
\tag{2.31}
$$

These are exactly the distance formulae as known from Euclidean geometry.

Raising and Lowering of Indices
The primary role of the metric tensor g on a manifold is to define the geometry. However, it allows additionally an important algebraic equivalence between vectors and covectors at each point of the manifold. The idea is to assign to each vector X, for each point of the manifold, the corresponding covector $g(X, \cdot)$ at that point. Obviously, this can be extended to vector fields and covector fields defined over the entire manifold. In this way, the metric tensor defines a vector space isomorphism between $T_p\mathcal{M}$ and the algebraic dual $\widetilde{T}_p\mathcal{M}$ at each point p. In the same manner, the metric defines an isomorphism between the space

of vector fields $\mathcal{X}\mathcal{M}$ and the space of covector fields $\widetilde{\mathcal{X}}\mathcal{M}$. Practically, when we choose coordinates, we assign to the vector

$$X = X^\mu \partial_\mu \tag{2.32}$$

the covector

$$g(X, \cdot) = g_{\mu\nu} \, dx^\mu(X) \otimes dx^\nu = g_{\mu\nu} X^\mu dx^\nu. \tag{2.33}$$

The components of this new covector are given by

$$X_\mu \equiv g_{\mu\nu} X^\nu, \tag{2.34}$$

which we call the *covariant components* of X. We say that the index μ has been *lowered* by the metric tensor. The inverse is also possible. Given a covector

$$\omega = \omega_\mu \, dx^\mu, \tag{2.35}$$

one can define an associated vector by *raising* the index by

$$\omega^\mu \equiv g^{\mu\nu} \omega_\nu, \tag{2.36}$$

and obtain the *contravariant components* of ω. In a coordinate-independent fashion, the associated vector is written as $g^{-1}(\omega, \cdot)$, since it is

$$g^{-1}(\omega, \cdot) = g^{\mu\nu} \, \partial_\mu(\omega) \otimes \partial_\nu = g^{\mu\nu} \omega_\mu \partial_\nu. \tag{2.37}$$

More generally, the metric tensor allows to *raise and lower indices* of arbitrary (m, n)-type tensors in the obvious way. Finally, let us mention that the metric induces also a scalar product between arbitrary (m, n)-type tensors. Given two tensors of the same type $T^{\mu_1 \cdots \mu_m}{}_{\nu_1 \ldots \nu_n}$ and $S^{\rho_1 \cdots \rho_m}{}_{\sigma_1 \ldots \sigma_n}$ their scalar product is naturally defined as

$$\langle T, S \rangle \equiv g_{\mu_1 \rho_1} \cdots g_{\mu_m \rho_m} \, g^{\nu_1 \sigma_1} \cdots g^{\nu_n \sigma_n} \, T^{\mu_1 \cdots \mu_m}{}_{\nu_1 \ldots \nu_n} \, S^{\rho_1 \cdots \rho_m}{}_{\sigma_1 \ldots \sigma_n}. \tag{2.38}$$

Here all indices have been contracted. Of course, one can define metric contractions with arbitrary tensors, not necessarily of the same type, where a number of indices remain uncontracted. These uncontracted indices are called *free*.

Geometry of a Submanifold

Suppose we have a D-dimensional geometric manifold $\mathcal{M}$ and an n-dimensional submanifold $\mathcal{S}$ of $\mathcal{M}$, see definition in Section 1.2. The metric $g_{\mu\nu}(x)$ on $\mathcal{M}$ induces in a natural way a metric $h_{\alpha\beta}(y)$ on the submanifold $\mathcal{S}$. Here we use the coordinates x^μ in $\mathcal{M}$ and the coordinates y^α in $\mathcal{S}$, with the index range $\alpha = 1, \ldots, n$. The coordinates y^α in $\mathcal{S}$ are expressible by the manifold coordinates x^μ, i.e. $y^\alpha = y^\alpha(x)$. Inversely, we can use the coordinates x^μ in the form $x^\mu = x^\mu(y)$ for describing the points in $\mathcal{S}$. In order to determine the natural metric on the submanifold $\mathcal{S}$, we consider displacements

$$dx^\mu|_\mathcal{S} = \frac{\partial x^\mu}{\partial y^\alpha} dy^\alpha \tag{2.39}$$

within $\mathcal{S}$. The *induced metric* h on the submanifold $\mathcal{S}$ is obtained by restricting the original metric $g = g_{\mu\nu}(x) \, dx^\mu dx^\nu$ of the manifold $\mathcal{M}$ to the submanifold $\mathcal{S}$. This definition gives

$$h \equiv g|_\mathcal{S} = \frac{\partial x^\mu}{\partial y^\alpha} \frac{\partial x^\nu}{\partial y^\beta} \, g_{\mu\nu}(x) \, dy^\alpha dy^\beta. \tag{2.40}$$

Written out in components, the induced metric h is

$$h = h_{\alpha\beta}(y)\, \mathrm{d}y^{\alpha} \mathrm{d}y^{\beta} \tag{2.41}$$

and therefore we can immediately express the metric components $h_{\alpha\beta}(y)$ through the metric components $g_{\mu\nu}(x)$ as

$$h_{\alpha\beta}(y) = \frac{\partial x^{\mu}}{\partial y^{\alpha}} \frac{\partial x^{\nu}}{\partial y^{\beta}}\, g_{\mu\nu}(x). \tag{2.42}$$

The induced metric h transfers the geometry of the parent manifold $\mathcal{M}$ to the submanifold $\mathcal{S}$. We will discuss examples of this concept in Section 2.3.

Geometry of a Hypersurface

In the case the submanifold $\mathcal{S}$ is of dimension $n = D - 1$, it is called a hypersurface and is commonly denoted Σ. The hypersurface Σ can be defined by one scalar equation

$$f(x^{1}, \ldots, x^{D}) = 0. \tag{2.43}$$

Let us consider coordinates $(y^{1}, \ldots, y^{D-1})$ for the points of the hypersurface Σ and the displacements $\mathrm{d}x^{\mu}|_{\Sigma}$ within Σ, as defined before. The differential of the above equation yields

$$\mathrm{d}f = \partial_{\mu}f\, \mathrm{d}x^{\mu}|_{\Sigma} = 0. \tag{2.44}$$

From the vantage point of the manifold $\mathcal{M}$, the displacements $\mathrm{d}x^{\mu}|_{\Sigma}$ are all tangential to Σ, so that the vector field $n^{\mu}(x)$ defined by

$$n^{\mu} \equiv g^{\mu\nu}\partial_{\nu}f \tag{2.45}$$

is normal to the hypersurface Σ. In other words, for any vector field $t^{\mu}(x)$, being tangential to the hypersurface Σ, we have $g_{\mu\nu}n^{\mu}t^{\nu} = 0$. In the case the norm $g_{\mu\nu}n^{\mu}n^{\nu}$ is non-zero throughout the manifold $\mathcal{M}$, we can normalize the vector field $n^{\mu}(x)$ and introduce the *unit normal vector field* $N^{\mu}(x)$ as

$$N^{\mu} \equiv \frac{n^{\mu}}{\sqrt{|g_{\mu\nu}n^{\mu}n^{\nu}|}}, \tag{2.46}$$

so that we have

$$N^{2} = g_{\mu\nu}N^{\mu}N^{\nu} = \operatorname{sgn}(n^{2}) = \pm 1. \tag{2.47}$$

The sign value in the last equation depends in part on the signature of the metric $g_{\mu\nu}(x)$, see our previous definition, but this is not crucial for the moment. With the unit normal vector field $N^{\mu}(x)$ at hand, we can define a symmetric $(0, 2)$-tensor field $h_{\mu\nu}(x)$ as

$$h_{\mu\nu} \equiv g_{\mu\nu} - N^{2}N_{\mu}N_{\nu}, \tag{2.48}$$

which is called the *projection operator* with respect to the hypersurface Σ. Note that $h_{\mu\nu}(x)$ is defined in D dimensions. The projection operator $h_{\mu\nu}(x)$ on the manifold $\mathcal{M}$ is closely related to the induced metric $h_{\alpha\beta}(y)$ on the hypersurface Σ. It can easily be shown that the contraction of the projection operator with vectors normal to the hypersurface is zero, while the contraction with tangential vectors maps them to their duals. In addition, the formula for the induced metric 2.42 can be written as (*exercise 2.2*)

$$h_{\alpha\beta}(y) = \frac{\partial x^{\mu}}{\partial y^{\alpha}} \frac{\partial x^{\nu}}{\partial y^{\beta}}\, h_{\mu\nu}(x). \tag{2.49}$$

On the other hand, the projection operator is not a metric tensor, since the matrix $h_{\mu\nu}(x)$ is, in general, not invertible. The task of the projection operator is to project any given vector of the manifold to the hypersurface Σ and this will be used later in Section 22.3.

2.2   Isometry and Conformality

Isometry of Manifolds

Two topological spaces are considered to be the same, if they are related by a homeomorphism. Two differential manifolds are considered to be the same, if they are related by a diffeomorphism. Can we define a similar equivalence relation between two Riemannian manifolds, where each one possesses a different geometry? We denote a Riemannian manifold by a pair $(\mathcal{M}, g)$ showing explicitly the metric. If we have two distinct geometric manifolds, $(\mathcal{M}, g)$ and $(\mathcal{N}, a)$, with their respective metric tensors g and a, they are considered geometrically equivalent, if there exists a diffeomorphism $F : \mathcal{M} \to \mathcal{N}$, such that the norms of vectors and the angles between vectors are the same under the respective metrics. In other words, the map F must be such that

$$\boxed{g_p(X, Y) = a_{F(p)}(\mathrm{D}F_p(X), \mathrm{D}F_p(Y))} \tag{2.50}$$

for all points p of $\mathcal{M}$ and for all tangent vectors X, Y of $T_p\mathcal{M}$. The point $F(p)$ belongs to the manifold $\mathcal{N}$ and the map $\mathrm{D}F_p$ is the differential map induced by F at the point p. By using the coordinate representations of the vectors and their images, we obtain readily the geometry-preservation condition, now expressed for the metric components,

$$g_{\mu\nu}(x) = \frac{\partial F^\rho}{\partial x^\mu}(x)\frac{\partial F^\sigma}{\partial x^\nu}(x)\, a_{\rho\sigma}(F(x)). \tag{2.51}$$

Again x represents the coordinates of the point p, while $F(x)$ represents the coordinates of the point $F(p)$. A map F with the above geometry-preserving property is called an *isometry*.

> *Two geometric manifolds represent the same geometry if they are related by an isometry.*

The relations 2.50 and 2.51 express the *isometry condition*. Let us point out that the two manifolds and the two respective metrics are not at all transformed here, they stay as they are. What happens is that we identify a diffeomorphism, which establishes an equivalence of the two geometries. The isometry property defines an equivalence relation between Riemannian manifolds: if $(\mathcal{M}, g)$ is isometric to $(\mathcal{N}, a)$, and $(\mathcal{N}, a)$ is isometric to $(\mathcal{P}, h)$, then $(\mathcal{M}, g)$ is isometric to $(\mathcal{P}, h)$. A special situation is to consider the set of all isometries of a given Riemannian manifold $(\mathcal{M}, g)$ to itself. These isometries constitute a group: the composition of any two isometries is again an isometry. The associativity property and the existence of neutral and inverse elements are all obvious. In this constellation, these isometries act as so-called *symmetry transformations* on the manifold. We will pick up this line of thoughts again later when we talk about symmetries of manifolds.

Conformality of Manifolds

We have just introduced the notion of an isometry, which expresses how two given geometries are essentially the same. Lengths and angles are invariant measures under isometry maps. Now we turn to maps between geometric manifolds, which require only the invariance of angles. Lengths are not required to be invariant anymore. Such a differentiable map between geometric manifolds is called a *conformal map*, or *conformism*. More precisely, given two geometric manifolds $(\mathcal{M}, g)$ and $(\mathcal{N}, a)$, they are called *conformally related*, if there is a diffeomorphism $F : \mathcal{M} \to \mathcal{N}$, such that the condition

$$\boxed{\Psi^2(p)\, g_p(X, Y) = a_{F(p)}(\mathrm{D}F_p(X), \mathrm{D}F_p(Y))} \tag{2.52}$$

holds. $p \in \mathcal{M}$ and $F(p) \in \mathcal{N}$ are manifold points. $X, Y \in T_p\mathcal{M}$ and $\mathrm{D}F_p(X), \mathrm{D}F_p(Y) \in T_{F(p)}\mathcal{N}$ are corresponding tangent vectors. The real function Ψ, called the *conformal factor*, is defined over the entire manifold $\mathcal{M}$ and fulfills $\Psi > 0$. The strict positivity is needed to ensure that the inverse $1/\Psi^2$ always exists. In coordinates the condition 2.52 becomes

$$\Psi^2(x)\, g_{\mu\nu}(x) = \frac{\partial F^\rho}{\partial x^\mu}(x)\frac{\partial F^\sigma}{\partial x^\nu}(x)\, a_{\rho\sigma}(F(x)). \tag{2.53}$$

The conformal factor $\Psi(x)$ changes the lengths of tangent vectors at each point. However, the angles between vectors remain the same. The equations 2.52 and 2.53 express the *conformality condition*. Isometries are special cases of conformal maps, for which $\Psi = 1$ holds throughout. The conformality between geometric manifolds defines an equivalence relation for these manifolds. In the special setup of a conformal map $F : \mathcal{M} \to \mathcal{M}$ of a manifold $\mathcal{M}$ to itself, the map F will keep infinitesimally the "shape" of the geometry. In Chapter 14 we provide examples of conformal maps.

2.3   Examples of Geometries

Exploring Geometries

Now after having introduced many general concepts, it is instructive to get an impression how a concrete metric can look like and what type of geometry it represents. In the following, we will review examples of metrics defining *flat geometries*, *spheric geometries*, and *hyperbolic geometries*, which represent the three *classical geometries*. We will look at the expression for the line element and discuss how distances are determined in each case. Technically, we can derive new geometries from existing ones. We have developed how this is achieved for a submanifold of a given geometric manifold, by using the induced metric. It presupposes that we can embed the manifold of interest into a larger geometric manifold, typically Euclidean space. This is the default method how to proceed and in practical terms it requires a differentiation. A second, simpler method is to embed the geometry of interest into Euclidean space again and use suitable coordinates for the submanifold. By imposing a defining condition for the geometry of interest, we can reduce the number of used coordinates and reach to the sought metric. Finally, there is a third method, which uses once again the embedding and the defining equation of the submanifold. The defining equation in the form $f(x^1, \ldots, x^D) = 0$ is differentiated and thus yields a condition containing the basis elements $\{\mathrm{d}x^1, \ldots, \mathrm{d}x^D\}$, which in turn is used to derive the desired metric. We will illustrate these methods in the following examples.

Euclidean Metric

A D-dimensional geometric manifold has the *Euclidean metric* if this metric is given by

$$\mathrm{d}s^2 = \left(\mathrm{d}x^1\right)^2 + \cdots + \left(\mathrm{d}x^D\right)^2. \tag{2.54}$$

The prime example is the 3-dimensional *Euclidean space* $\mathbb{E}^3$ modeling classical physical space. The metric is then written

$$\mathrm{d}s^2 = \mathrm{d}x^2 + \mathrm{d}y^2 + \mathrm{d}z^2. \tag{2.55}$$

with the Cartesian coordinates (x, y, z). The distance between two points is given by

$$\Delta s^2 = \Delta x^2 + \Delta y^2 + \Delta z^2, \tag{2.56}$$

where the deltas denote the differences of the respective coordinates of the points. Of course, one is free to use non-Cartesian coordinates.

Lorentzian Metric

A manifold of dimension $d = 1 + D$ is said to have a flat *Lorentzian metric* (*Hendrik Antoon Lorentz*) if the metric is given by[*]

$$ds^2 = \left(dx^0\right)^2 - \left(dx^1\right)^2 - \cdots - \left(dx^D\right)^2. \tag{2.57}$$

The prime example in $d = 1 + 3$ dimensions is the metric of flat *Minkowski space* $\mathbb{M}_4$ (*Hermann Minkowski*), modeling the special relativistic spacetime $(\mathbb{E}^4, \eta_{\mu\nu})$, with $\eta_{\mu\nu} = \mathrm{diag}(1, -1, -1, -1)$. We write

$$ds^2 = dt^2 - dx^2 - dy^2 - dz^2, \tag{2.58}$$

with t being the time variable (in suitable physical units where the speed of light is set $c = 1$) and x, y, z being the Cartesian space coordinates. The same expression for the metric appears also in the curved spacetime of general relativity, but only as a special case. In a general spacetime, the metric has off-diagonal entries $g_{\mu\nu} \neq 0$ for $\mu \neq \nu$. The general relativistic metric can be put into diagonal form in each point of the spacetime manifold by a suitable and unique locally varying coordinate transformation. This will be discussed within Riemannian geometry in Chapter 19. Considering the case of Minkowski space again, the distance between two points, called *events*, is given by

$$\Delta s^2 = \Delta t^2 - \Delta x^2 - \Delta y^2 - \Delta z^2. \tag{2.59}$$

It is apparent that the distance between two spacetime points can be zero even if they do not coincide or are equal zero. We will discuss the physical interpretation of Minkowski spacetime later in Chapter 6 on relativistic mechanics.

Metric of 2-Dimensional Cylinder

We want to derive the metric of a 2-dimensional *cylinder* of radius a, denoted $C^2(a)$. To this end, we consider the cylinder to be embedded in Euclidean space $\mathbb{E}^3$. The cylinder point set is

$$C^2(a) \equiv \left\{ (x, y, z) \in \mathbb{E}^3 \mid x^2 + y^2 = a^2 \right\}. \tag{2.60}$$

In 3-dimensional Euclidean space we can use the *cylindrical coordinates* r, φ, z for the description of any geometric point. Generally, the relation between Cartesian and cylindrical coordinates is

$$\left. \begin{array}{rcl} x &=& r \cos\varphi \\ y &=& r \sin\varphi \\ z &=& z \end{array} \right\}. \tag{2.61}$$

The new coordinates are the *radial coordinate* r, the *angular coordinate* φ, and the *axial* (or *height*) *coordinate* z. The range of φ is $0 \leq \varphi < 2\pi$. From the above coordinate relations, we can derive the 3-dimensional Euclidean metric written in cylindrical coordinates as (*exercise 2.3*)

$$ds^2 = dr^2 + r^2 d\varphi^2 + dz^2. \tag{2.62}$$

Now we realize that for the cylinder manifold with radius $r = a$ the radial coordinate can be held fixed and it is then $dr = 0$. So we deduce that the metric of a cylinder with radius a is

$$ds^2 = a^2 d\varphi^2 + dz^2. \tag{2.63}$$

---

[*]We use the signature $(+ - \cdots -)$, see also appendix A.2.

This is the desired result for the cylinder metric, expressed by the two coordinates φ and z. We can derive the same result, if we simply follow the recipe of the induced metric. (*exercise 2.4*) The metric can be written in matrix form as

$$(g_{\mu\nu}) = \begin{pmatrix} g_{\varphi\varphi} & g_{\varphi z} \\ g_{z\varphi} & g_{zz} \end{pmatrix} = \begin{pmatrix} a^2 & 0 \\ 0 & 1 \end{pmatrix}. \tag{2.64}$$

The distance between any two points on the cylinder is given by the length of the *helix curve* connecting the two points. The helix appears "curved" when we take the 3-dimensional Euclidean point of view. If we roll out the cylinder, the helix becomes a straight line. In fact, the cylinder itself is a flat manifold with zero curvature.

Metric of 2-Dimensional Sphere

We proceed to the first example of a non-flat geometry, the 2-dimensional *sphere* $S^2(a)$, with radius a, considered as a geometric manifold.[*] We will derive the metric of the 2-sphere by the different methods mentioned before. We consider $S^2(a)$ as a submanifold of 3-dimensional Euclidean space $\mathbb{E}^3$ and use the known metric in order to derive the metric of the 2-sphere. As a point set, the 2-sphere is given by

$$S^2(a) \equiv \left\{ (x, y, z) \in \mathbb{E}^3 \mid x^2 + y^2 + z^2 = a^2 \right\}. \tag{2.65}$$

Let us first introduce suitable coordinates. In 3-dimensional Euclidean space, we can use the *spherical* (or *polar*) *coordinates* r, θ, φ for describing arbitrary geometric points. The general relation between Cartesian and spherical coordinates is, per convention,

$$\left. \begin{aligned} x &= r \sin\theta \cos\varphi \\ y &= r \sin\theta \sin\varphi \\ z &= r \cos\theta \end{aligned} \right\}. \tag{2.66}$$

The *radial coordinate* r measures the distance from the coordinate origin. The angle θ, starting from the z-axis, is called the *polar angle coordinate*. The angle φ, starting from the x-axis and increasing toward the y-axis, is called the *azimuthal angle coordinate*. The range of these angle coordinates is $0 \leq \theta \leq \pi$ and $0 \leq \varphi < 2\pi$. From the above coordinate relations we can deduce the 3-dimensional Euclidean metric, written in spherical coordinates as (*exercise 2.5*)

$$ds^2 = dr^2 + r^2(d\theta^2 + \sin^2\theta \, d\varphi^2). \tag{2.67}$$

This metric describes arbitrary points in Euclidean space. However, the sphere of radius a, being embedded with its center at the coordinate origin, obeys $r = a$ and so we can set $dr = 0$, reducing the number of coordinates for the description. The resulting expression for the metric of $S^2(a)$ is

$$ds^2 = a^2(d\theta^2 + \sin^2\theta \, d\varphi^2), \tag{2.68}$$

with θ and φ being the used coordinates. The same result is obtained by employing the formula for the induced metric. (*exercise 2.6*) Written out as matrix, the metric of the 2-sphere is

$$(g_{\mu\nu}) = \begin{pmatrix} g_{\theta\theta} & g_{\theta\varphi} \\ g_{\varphi\theta} & g_{\varphi\varphi} \end{pmatrix} = \begin{pmatrix} a^2 & 0 \\ 0 & a^2 \sin^2\theta \end{pmatrix}. \tag{2.69}$$

---

[*]We reserve the symbol S^2 for the *unit sphere* with radius one, i.e. $S^2 = S^2(1)$.

We can actually use another suited pair of coordinates to describe the 2-sphere. These coordinates can be the radial coordinate r and the azimuthal angle φ, for example. We start with the defining equation of the sphere in Cartesian coordinates,

$$x^2 + y^2 + z^2 = a^2. \tag{2.70}$$

Differentiation of this equation allows us to eliminate $\mathrm{d}z$ in the Euclidean metric and to write the metric as

$$\mathrm{d}s^2 = \mathrm{d}x^2 + \mathrm{d}y^2 + \frac{(x\,\mathrm{d}x + y\,\mathrm{d}y)^2}{a^2 - x^2 - y^2}. \tag{2.71}$$

Now we transform to planar spherical coordinates r and φ in the (x, y)-plane with $x = r\cos\varphi$ and $y = r\sin\varphi$. By expressing the Cartesian x, y, $\mathrm{d}x$, $\mathrm{d}y$ by the spherical r, φ, $\mathrm{d}r$, $\mathrm{d}\varphi$, we obtain the desired expression for the metric, (*exercise 2.7*)

$$\mathrm{d}s^2 = \frac{a^2}{a^2 - r^2}\,\mathrm{d}r^2 + r^2\,\mathrm{d}\varphi^2. \tag{2.72}$$

In matrix notation, this metric is written as

$$(g_{\mu\nu}) = \begin{pmatrix} g_{rr} & g_{r\varphi} \\ g_{\varphi r} & g_{\varphi\varphi} \end{pmatrix} = \begin{pmatrix} \frac{a^2}{a^2 - r^2} & 0 \\ 0 & r^2 \end{pmatrix}. \tag{2.73}$$

The distance between any two points on the 2-sphere is given by the length of the arc segment of a *great circle* connecting the two points.

Metric of 3-Dimensional Sphere

The metric of the 3-dimensional *hypersphere* (or simply 3-*sphere*) of radius a, denoted $S^3(a)$, is determined very similarly to the 2-dimensional case. Again, an embedding in the Euclidean space, in this case the 4-dimensional $\mathbb{E}^4$, is used:

$$S^3(a) \equiv \left\{ (x, y, z, w) \in \mathbb{E}^4 \mid x^2 + y^2 + z^2 + w^2 = a^2 \right\}. \tag{2.74}$$

A possible choice of *hyperspherical coordinates* in four dimensions is

$$\left. \begin{aligned} x &= r\sin\theta_2 \sin\theta_1 \cos\varphi \\ y &= r\sin\theta_2 \sin\theta_1 \sin\varphi \\ z &= r\sin\theta_2 \cos\theta_1 \\ w &= r\cos\theta_2 \end{aligned} \right\}. \tag{2.75}$$

The *radial coordinate* r determines the distance from the origin. There are three angle coordinates now, two *polar angles* θ_2 and θ_1, taking the values $0 \leq \theta_2, \theta_1 \leq \pi$ and one *azimuthal angle* φ, taking the values $0 \leq \varphi < 2\pi$. The angle θ_2 starts from the positive direction of the w-axis, whereas the angle θ_1 starts from the positive direction of the z-axis. The azimuthal angle φ again starts from the positive x-axis and increases toward the positive y-axis. As before, we deduce the 4-dimensional Euclidean metric in hyperspherical coordinates as (*exercise 2.8*)

$$\mathrm{d}s^2 = \mathrm{d}r^2 + r^2\left\{ (\mathrm{d}\theta_2)^2 + \sin^2\theta_2\left[(\mathrm{d}\theta_1)^2 + \sin^2\theta_1 (\mathrm{d}\varphi)^2 \right] \right\}. \tag{2.76}$$

As a result, the metric of the 3-sphere $S^3(a)$ is derived as

$$\mathrm{d}s^2 = a^2\left\{ (\mathrm{d}\theta_2)^2 + \sin^2\theta_2\left[(\mathrm{d}\theta_1)^2 + \sin^2\theta_1 (\mathrm{d}\varphi)^2 \right] \right\}. \tag{2.77}$$

The above expression for the metric uses only angle coordinates. As in the lower-dimensional case, we can express the metric by the radial coordinate and two angle coordinates instead. Again we use the defining equation of the sphere, in the present case

$$x^2 + y^2 + z^2 + w^2 = a^2, \tag{2.78}$$

in order to eliminate the auxiliary coordinate w. After a little algebra, we arrive at the result (*exercise 2.9*)

$$ds^2 = \frac{a^2}{a^2 - r^2}\, dr^2 + r^2\, d\Phi^2. \tag{2.79}$$

The squared angle differential $d\Phi^2$ is given by the expression

$$d\Phi^2 = (d\theta_1)^2 + \sin^2\theta_1 (d\varphi)^2, \tag{2.80}$$

containing only the angle coordinates θ_1 and φ. The metric for the 3-dimensional sphere, as written in 2.79 is very useful in the discussion of cosmological models. The discussed 2-sphere and 3-sphere are examples of so-called *spheric geometries* with positive constant curvature. The geometry of a sphere can be extended to D dimensions and this is displayed in appendix B.7.

Weyl Rescaling

A *conformal metric* is one which is related to the Euclidean metric by a locally varying positive factor. Starting from the 2-dimensional Euclidean space, we can define a new *conformal geometry* by introducing a metric of the form

$$ds^2 = \Omega^2(x, y)\left(dx^2 + dy^2\right), \tag{2.81}$$

with $\Omega(x, y) > 0$. In full analogy is the D-dimensional case, with a conformal metric defined as

$$ds^2 = \Omega^2(x)\left[\left(dx^1\right)^2 + \cdots + \left(dx^D\right)^2\right]. \tag{2.82}$$

More generally, starting from an arbitrary given metric g of the geometric manifold $(\mathcal{M}, g)$, one can define a new metric $\widehat{g}$ by applying the pointwise multiplication with $\Omega^2(x)$ in the form

$$\boxed{\widehat{g}_{\mu\nu}(x) \equiv \Omega^2(x)\, g_{\mu\nu}(x)} \tag{2.83}$$

The scalar function $\Omega(x) > 0$ is the so-called *conformal factor*. This transformation of the metric is called a *Weyl rescaling (Hermann Klaus Hugo Weyl)*. The result of the Weyl rescaling is the geometric manifold $(\mathcal{M}, \widehat{g})$. Given a geometric manifold $(\mathcal{M}, g)$, the Weyl rescaling transformations of the metric constitute an abelian group denoted $\mathrm{Weyl}(\mathcal{M}, g)$. We must point out that the Weyl rescaling procedure is not a coordinate transformation of the points of the manifold. The Weyl rescaling procedure defines an entirely new metric, describing a new geometry. As a concrete example of a 2-dimensional conformal geometry, let us view the one given by

$$ds^2 = \frac{dx^2 + dy^2}{\left(1 + x^2 + y^2\right)^2}. \tag{2.84}$$

This metric is well-defined for all real values of the coordinates x and y and it is not isometric to Euclidean geometry. Other important cases of conformal geometries are the so-called hyperbolic geometries and in the following we provide two prime examples.

Two Models of Hyperbolic Geometry

In the 19th century the mathematicians *Carl Friedrich Gauss, Nikolai Ivanovich Lobachevsky,* and *Janos Bolyai* discovered *hyperbolic geometry*. Hyperbolic geometry is footed on the same axioms as Euclidean geometry, except of the 5th axiom about parallel lines. The Euclidean axiom about parallels states:

> *Given a straight line and a point outside that line, there is exactly one straight line that passes through that point and remains parallel to the given straight line.*

In contrast to the above, hyperbolic geometry allows an *infinite* number of parallel lines to pass through the given point. A concrete model of a 2-dimensional hyperbolic geometry according to *Eugenio Beltrami* and *Jules Henri Poincaré* is the *upper half-plane model*, abbreviated *UHP*, defined as the plane

$$H^2 \equiv \left\{ (x,y) \in \mathbb{R}^2 \mid y > 0 \right\}, \tag{2.85}$$

equipped with the metric

$$ds^2 = \frac{1}{y^2} \left(dx^2 + dy^2 \right). \tag{2.86}$$

For constantly decreasing values of y, the distance ds becomes larger and larger. The line $y = 0$ does not belong to the upper half-plane model and lies at infinity. We will discuss geodesics and curvature of the upper half-plane model in detail later. Here we will just illustrate how its geodesic curves, i.e. the curves of shortest distance, look like from Euclidean point of view. This means we consider the position and shape of the geodesic curves as they appear on our 2-dimensional sheet of paper. There are two types of geodesic curves here: segments of *half circles*, whose origin lies on the line $y = 0$, and segments of *vertical lines*, with $x = $ const. In figure 2.3 we provide an illustration of a selected set of geodesic curves.

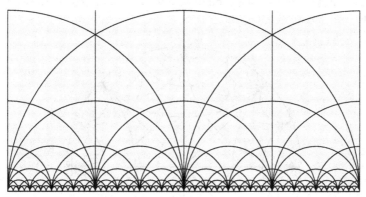

Figure 2.3: A selection of geodesic curves in the UHP model H^2

The upper half-plane model can be extended to D dimensions by considering the space

$$H^D \equiv \left\{ (x^1, \ldots, x^D) \in \mathbb{R}^D \mid x^D > 0 \right\}, \tag{2.87}$$

and by implementing the metric

$$ds^2 = \frac{1}{(x^D)^2} \left[(dx^1)^2 + \cdots + (dx^D)^2 \right]. \tag{2.88}$$

One speaks of the *upper half-space model* then.

There is an alternative model describing the same 2-dimensional hyperbolic geometry as the UHP model. This goes under the name *Poincaré disk model*. The manifold consists of the 2-dimensional open disk of unit radius

$$P^2 \equiv \left\{ (u,v) \in \mathbb{R}^2 \mid u^2 + v^2 < 1 \right\}, \tag{2.89}$$

equipped with the metric

$$ds^2 = 4 \frac{du^2 + dv^2}{\left(1 - u^2 - v^2\right)^2}. \tag{2.90}$$

The Poincaré disk model represents a geometry isometric to the UHP model. To make the connection between the two representations, one considers the smooth and invertible map $F : H^2 \to P^2$, $(x,y) \mapsto F(x,y) = (u,v)$ between the two manifolds, given by

$$u = \frac{2x}{x^2 + (1+y)^2}, \qquad v = \frac{x^2 + y^2 - 1}{x^2 + (1+y)^2}, \tag{2.91}$$

with the inverse

$$x = \frac{2u}{u^2 + (1-v)^2}, \qquad y = \frac{1 - u^2 - v^2}{u^2 + (1-v)^2}. \tag{2.92}$$

It is a straightforward calculation to show that the isometry condition 2.51 is fulfilled. (*exercise 2.10*) Thus, the UHP metric and the Poincaré disk metric describe the same geometry. The geodesic curves in the Poincaré disk model are *arc segments of circles which meet the disk boundary at a right angle*. In figure 2.4 we display some of these geodesic curves to give a glimpse of how this geometry behaves. The Poincaré disk boundary lies at infinite distance for any of the interior points.

Figure 2.4: A selection of geodesic curves in the Poincaré disk model P^2

With the introduction of the UHP and Poincaré model, we have provided the most basic definitions for hyperbolic manifolds. The reader interested in gaining a deeper understanding of this beautiful subject of geometry may consult, for example, the book of Ratcliffe [72].

2.4 Differential Forms and the Exterior Derivative

Measuring Volumes

We have discussed the notions of angle, length and distance, which crucially depend on the existence of a metric. Similarly, we would like to measure also volumes within manifolds. Having a D-dimensional manifold $\mathcal{M}$, how can we measure the volume of a region U of the manifold? Surely we want our definition of the volume to be independent of any coordinate choice. We will see soon, that the fundamental entities one needs to employ to measure volumes are antisymmetric tensor fields. Remarkably, volume integrals on manifolds can be defined without any reference to a metric. However, when a metric is present, as it is for a geometric manifold, the formulae are enriched by additional properties. We would like to measure the surface content of $(D-1)$-dimensional hypersurfaces as well. To this end, we view these hypersurfaces as manifolds in their own right and measure their $(D-1)$-dimensional "volume" respectively.

It is helpful to look how integration is carried out in the well-known case of calculus on $\mathbb{R}$. We consider a real function $f : [a,b] \to \mathbb{R}$, $x \mapsto f(x)$. The differential $\mathrm{d}f(x)$ of the function at the point x is given as

$$\mathrm{d}f(x) = \frac{\mathrm{d}f}{\mathrm{d}x}(x)\,\mathrm{d}x. \tag{2.93}$$

For real functions, usually the Riemann or the Lebesgue integral is introduced (*Henri L. Lebesgue*). With this integral notion, the differential can be integrated readily over the interval (i.e. the "volume") $[a,b]$ to yield

$$\int_a^b \mathrm{d}f(x) = \int_a^b \frac{\mathrm{d}f}{\mathrm{d}x}(x)\,\mathrm{d}x = f(b) - f(a). \tag{2.94}$$

This is the *fundamental theorem of calculus*. It states that the integral of the differential of a function over the interval $[a,b]$ is equal to the difference of the function values at the boundary points b and a. We will see soon that this theorem is generalized to D-dimensional manifolds through the theorem of Stokes.

The other important theorem from calculus on $\mathbb{R}$ is the *integral transformation formula*. Consider a real function $h : [a,b] \to \mathbb{R}$, $x' \mapsto h(x')$ and suppose we have a differentiable and invertible one-to-one relation $x' = \varphi(x)$ and $x = \varphi^{-1}(x')$, with a new variable x. The variable x runs through the real interval $\left[\varphi^{-1}(a), \varphi^{-1}(b)\right]$. Then the integral of the function $h(x')$ over the interval $[a,b]$ can be calculated by using the variable x as

$$\int_a^b h(x')\,\mathrm{d}x' = \int_{\varphi^{-1}(a)}^{\varphi^{-1}(b)} h\big(\varphi(x)\big) \frac{\mathrm{d}\varphi}{\mathrm{d}x}(x)\,\mathrm{d}x. \tag{2.95}$$

This integral transformation rule is a direct consequence of the chain rule in differentiation. The above basic facts of calculus will lead us to the generalizations needed for the case of manifolds.

Differential Forms

Let us ask now what the basic elements are which we can integrate on a manifold. Technically, we need to find suitable integration measures and the first one we are looking for is the "volume". We know from the algebraic theory of tensors that alternating tensors and determinants define volumes, see appendix B.1. An antisymmetric $(0, D)$-tensor can be seen as a map assigning to a set of D vectors a well-defined volume measure, while the determinant provides the scalar value of the volume spanned by these D vectors. This is a clear indication of how we can proceed in the case of manifolds. The algebraic notions and elements are replaced by locally varying ones, for each point in the manifold. We start with a general D-dimensional differential manifold $\mathcal{M}$, which may have a metric or not. The cotangent space at each point p is denoted $\widetilde{T}_p\mathcal{M}$. For any integer n with $0 \leq n \leq D$ and for each single point p we consider the linear space of antisymmetric $(0, n)$-tensors. According to our notation, we write $\bigwedge^n \left(\widetilde{T}_p\mathcal{M} \right)$ for this space. We call the antisymmetric $(0, n)$-tensors from now on *n-forms* (or *differential forms of degree n*) *at the point p*. An n-form at each point p is an antisymmetric multilinear map assigning to each n-tuple of local vectors a real number. A step further, and in full analogy to Section 1.5, we introduce fields and consider all points of the manifold. This means that we define *n-forms* α over the whole manifold as maps

$$\alpha : \mathcal{M} \;\rightarrow\; \bigcup_{p \in \mathcal{M}} \left(\overset{n}{\bigwedge} \widetilde{T}_p\mathcal{M} \right), \tag{2.96}$$

assigning in a smooth way to each point of the manifold an n-form at that point. A differential form of order 0 is simply a scalar function on the manifold. The *exterior product* of n-forms is defined exactly as in the algebraic case, for each point in the manifold. If we choose a coordinate system $\{x^\mu\}$ on the manifold, an n-form α can be written as

$$\alpha = \sum_{\mu_1, \dots, \mu_n} \alpha_{\mu_1 \dots \mu_n}(x) \, \mathrm{d}x^{\mu_1} \otimes \cdots \otimes \mathrm{d}x^{\mu_n} \tag{2.97}$$

with respect to the tensor basis $\{\mathrm{d}x^{\mu_1} \otimes \cdots \otimes \mathrm{d}x^{\mu_n}\}$. The components $\alpha_{\mu_1 \dots \mu_n}(x)$ of the n-form are totally antisymmetric. Note that, for the sake of clarity, we write out all the sums explicitly in this section. By employing the antisymmetrized basis elements

$$\mathrm{d}x^{\mu_1} \wedge \cdots \wedge \mathrm{d}x^{\mu_n} = n! \, \mathrm{Alt}(\mathrm{d}x^{\mu_1} \otimes \cdots \otimes \mathrm{d}x^{\mu_n}), \tag{2.98}$$

instead of the tensor basis, the n-form can be written as

$$\alpha = \frac{1}{n!} \sum_{\mu_1, \dots, \mu_n} \alpha_{\mu_1 \dots \mu_n}(x) \, \mathrm{d}x^{\mu_1} \wedge \cdots \wedge \mathrm{d}x^{\mu_n}, \tag{2.99}$$

in which the combinatorial factor $n!$ appears explicitly, compare with the definitions in appendix B.1. A third alternative component representation is obviously

$$\alpha = \sum_{\mu_1 < \dots < \mu_n} \alpha_{\mu_1 \dots \mu_n}(x) \, \mathrm{d}x^{\mu_1} \wedge \cdots \wedge \mathrm{d}x^{\mu_n}. \tag{2.100}$$

As soon as the order n of the form in determined, the indices structure is fixed to be $1 \leq \mu_1 < \dots < \mu_n \leq D$. Here it makes sense to simplify the notation. We can use the symbol M for the ordered index tuple $(\mu_1, \dots, \mu_n)$ and the symbol $<$ within the sum to imply the strictly monotonic increase of the index values. Thus, we can write the basis expansion 2.100 more compactly as

$$\alpha = \sum_< \alpha_M(x) \, \mathrm{d}x^M. \tag{2.101}$$

This condensed notation is useful when calculations become longer. For the *degree* of the n-form α we write $\deg \alpha = n$.

Let us look at a few simple examples in $\mathbb{R}^3$ with the Cartesian coordinates* (x, y, z). A 1-form $\omega^{(1)}$ is given by an expression

$$\omega^{(1)} = A\,\mathrm{d}x + B\,\mathrm{d}y + C\,\mathrm{d}z. \tag{2.102}$$

A 2-form $\omega^{(2)}$ is given by

$$\omega^{(2)} = P\,\mathrm{d}y \wedge \mathrm{d}z + Q\,\mathrm{d}z \wedge \mathrm{d}x + R\,\mathrm{d}x \wedge \mathrm{d}y. \tag{2.103}$$

Lastly, a 3-form $\omega^{(3)}$ on $\mathbb{R}^3$ is written as

$$\omega^{(3)} = \Omega\,\mathrm{d}x \wedge \mathrm{d}y \wedge \mathrm{d}z, \tag{2.104}$$

consisting of only one term. A 4-form, and all forms of higher degree on $\mathbb{R}^3$, vanish identically. We should point out that even these simple examples do not use any metric properties of $\mathbb{R}^3$.

The Exterior Derivative

Let us consider an n-form α defined on a general manifold. We can introduce a derivation notion for the n-form, the so-called *exterior derivative*, which yields the $(n+1)$-form $\mathrm{d}\alpha$, which is written formally in coordinates as

$$\mathrm{d}\alpha = \sum_{\rho < \mu_1 < \ldots < \mu_n} (\mathrm{d}\alpha)_{\rho\mu_1\ldots\mu_n}(x)\,\mathrm{d}x^\rho \wedge \mathrm{d}x^{\mu_1} \wedge \cdots \wedge \mathrm{d}x^{\mu_n}, \tag{2.105}$$

and is defined by the formula

$$\boxed{\mathrm{d}\alpha \equiv \sum_{\mu_1 < \ldots < \mu_n} \mathrm{d}\alpha_{\mu_1\ldots\mu_n} \wedge \mathrm{d}x^{\mu_1} \wedge \cdots \wedge \mathrm{d}x^{\mu_n}} \tag{2.106}$$

where the quantity $\mathrm{d}\alpha_{\mu_1\ldots\mu_n}(x)$ is the usual differential of the component $\alpha_{\mu_1\ldots\mu_n}(x)$,

$$\mathrm{d}\alpha_{\mu_1\ldots\mu_n} = \sum_\mu \frac{\partial \alpha_{\mu_1\ldots\mu_n}}{\partial x^\mu}\,\mathrm{d}x^\mu. \tag{2.107}$$

Obviously, the exterior derivative coincides with the usual differential in the case of functions, which are simply 0-forms. It can be shown that the components of the exterior derivative can be computed as $(\mathrm{d}\alpha)_{\rho\mu_1\ldots\mu_n} = (n+1)\,\partial_{[\rho}\alpha_{\mu_1\ldots\mu_n]}$. (*exercise 2.11*) We have defined the exterior derivative within coordinates, but the definition is actually independent of them. Phrased differently, the exterior derivative is, like an n-form, an invariant quantity under differentiable changes of the coordinates. Suppose we change from the coordinates x^μ to the coordinates x'^μ. Then for the components $\alpha_{\mu_1\ldots\mu_n}(x)$ it is

$$\alpha_{\mu_1\ldots\mu_n} = \frac{\partial x'^{\nu_1}}{\partial x^{\mu_1}} \cdots \frac{\partial x'^{\nu_n}}{\partial x^{\mu_n}}\,\alpha'_{\nu_1\ldots\nu_n}. \tag{2.108}$$

Thus, we get

$$\frac{\partial \alpha_{\mu_1\ldots\mu_n}}{\partial x^\mu} = \frac{\partial}{\partial x^\mu}\left(\frac{\partial x'^{\nu_1}}{\partial x^{\mu_1}} \cdots \frac{\partial x'^{\nu_n}}{\partial x^{\mu_n}}\right)\alpha'_{\nu_1\ldots\nu_n} + \frac{\partial x'^{\nu_1}}{\partial x^{\mu_1}} \cdots \frac{\partial x'^{\nu_n}}{\partial x^{\mu_n}}\frac{\partial \alpha'_{\nu_1\ldots\nu_n}}{\partial x^\mu}. \tag{2.109}$$

---

*Depending on convenience we denote Cartesian coordinates either as x, y, z or as x^1, x^2, x^3.

The first term in the rhs does not contribute anything in the sum 2.106, since it leads to terms containing indices pairs, which are symmetric in the partial derivatives but antisymmetric in the exterior products. Consequently, we obtain

$$d\alpha = \frac{1}{n!} \sum_{\substack{\nu,\nu_1,\ldots,\nu_n \\ \mu,\mu_1,\ldots,\mu_n}} \frac{\partial \alpha'_{\nu_1\ldots\nu_n}}{\partial x'^\nu} \frac{\partial x'^\nu}{\partial x^\mu} \frac{\partial x'^{\nu_1}}{\partial x^{\mu_1}} \cdots \frac{\partial x'^{\nu_n}}{\partial x^{\mu_n}} \, dx^\mu \wedge dx^{\mu_1} \wedge \cdots \wedge dx^{\mu_n}, \qquad (2.110)$$

and, since the basis element $dx^{\mu_1} \wedge \cdots \wedge dx^{\mu_n}$ transforms as a $(n,0)$-tensor field, it is

$$d\alpha = \sum_{\nu_1 < \ldots < \nu_n} d\alpha'_{\nu_1\ldots\nu_n} \wedge dx'^{\nu_1} \wedge \cdots \wedge dx'^{\nu_n}. \qquad (2.111)$$

Thus, we have shown that the exterior differentiation is a coordinate-independent operation.

As a simple example let us calculate the exterior derivative of the 1-form

$$\omega = \sum_\nu \omega_\nu dx^\nu. \qquad (2.112)$$

The exterior derivative is here

$$d\omega = \sum_{\mu,\nu} \partial_\mu \omega_\nu \, dx^\mu \wedge dx^\nu = \sum_{\mu<\nu} \left(\partial_\mu \omega_\nu - \partial_\nu \omega_\mu \right) dx^\mu \wedge dx^\nu. \qquad (2.113)$$

Let us now derive the characteristic properties of the exterior derivative. Obviously the exterior derivative is linear. Moreover, it obeys a specific product rule. Given an m-form α and an n-form β, we can view their exterior product $\alpha \wedge \beta$. The exterior derivative of this product is then

$$d\left(\alpha \wedge \beta\right) = d\alpha \wedge \beta + (-1)^{\deg \alpha} \alpha \wedge d\beta, \qquad (2.114)$$

where $\deg \alpha = m$ is the degree of the m-form α. For the proof we first consider the case of two functions (i.e. 0-forms), in which the usual product rule of differentiation applies and thus provides the evidence. Now if we have two forms

$$\alpha = \sum_< \alpha_M dx^M \quad \text{and} \quad \beta = \sum_< \beta_N dx^N, \qquad (2.115)$$

their exterior product is

$$\alpha \wedge \beta = \sum_< \alpha_M \beta_N \, dx^M \wedge dx^N. \qquad (2.116)$$

According to the definition of the exterior product and the usual product rule for functions, it is

$$d\left(\alpha \wedge \beta\right) = \sum_< \left[\left(d\alpha_M\right)\beta_N + \alpha_M \left(d\beta_N\right) \right] \wedge dx^M \wedge dx^N, \qquad (2.117)$$

or, after reshuffling,

$$d\left(\alpha \wedge \beta\right) = \sum_< \left[d\alpha_M \wedge dx^M \wedge \beta_N dx^N + (-1)^{\deg\alpha} \alpha_M dx^M \wedge d\beta_N \wedge dx^N \right], \qquad (2.118)$$

which is exactly what we wanted to prove. Another important property of any n-form α is

$$d(d\alpha) = 0. \qquad (2.119)$$

The proof is left to the reader. (*exercise 2.12*) Any form ω for which $d\omega = 0$ holds is called *closed*. A form ω is called *exact*, if there exists a form α for which $\omega = d\alpha$. Every exact form is necessarily closed. The inverse is true within so-called star-shaped regions, and is known as *Poincaré's lemma*.

Let us see how exterior differentiation works in the case of $\mathbb{R}^3$ with Cartesian coordinates. For a scalar function $f(x, y, z)$, the exterior derivative is the 1-form

$$\mathrm{d}f = \frac{\partial f}{\partial x}\,\mathrm{d}x + \frac{\partial f}{\partial y}\,\mathrm{d}y + \frac{\partial f}{\partial z}\,\mathrm{d}z. \tag{2.120}$$

For the 1-form $\omega^{(1)} = A\,\mathrm{d}x + B\,\mathrm{d}y + C\,\mathrm{d}z$ the exterior derivative is the 2-form

$$\mathrm{d}\omega^{(1)} = \left(\frac{\partial C}{\partial y} - \frac{\partial B}{\partial z}\right)\mathrm{d}y \wedge \mathrm{d}z + \left(\frac{\partial A}{\partial z} - \frac{\partial C}{\partial x}\right)\mathrm{d}z \wedge \mathrm{d}x + \left(\frac{\partial B}{\partial x} - \frac{\partial A}{\partial y}\right)\mathrm{d}x \wedge \mathrm{d}y. \tag{2.121}$$

For the 2-form $\omega^{(2)} = P\,\mathrm{d}y \wedge \mathrm{d}z + Q\,\mathrm{d}z \wedge \mathrm{d}x + R\,\mathrm{d}x \wedge \mathrm{d}y$ exterior differentiation gives the 3-form

$$\mathrm{d}\omega^{(2)} = \left(\frac{\partial P}{\partial x} + \frac{\partial Q}{\partial y} + \frac{\partial R}{\partial z}\right)\mathrm{d}x \wedge \mathrm{d}y \wedge \mathrm{d}z. \tag{2.122}$$

Finally, for the 3-form $\omega^{(3)} = \Omega\,\mathrm{d}x \wedge \mathrm{d}y \wedge \mathrm{d}z$ the exterior derivative vanishes identically. In the three above formulae we recognize the *gradient*, the *curl*, and the *divergence* known from classical calculus. Exterior differentiation generalizes these operators to arbitrary dimensions and coordinates. The identity $\mathrm{d}(\mathrm{d}\alpha) = 0$ contains the known formulae $\mathrm{curl}\,\mathrm{grad} = 0$ and $\mathrm{div}\,\mathrm{curl} = 0$ of vector calculus.

The Various Roles of $\mathrm{d}x^\mu$

At this point let us pause for a moment and look at the quantity $\mathrm{d}x^\mu$, which has appeared in multiple occasions. The quantity $\mathrm{d}x^\mu$, with x^μ being the coordinate function, takes on different roles depending on the context:

- It is the *differential* of the coordinate function x^μ, in the sense of $\mathbb{R}^D$.
- It is the *dual basis* of the cotangent space, when the coordinate system x^μ is chosen.
- It is the *exterior derivative* of the coordinate function x^μ.
- It is the *integration measure* for the coordinate variable x^μ, as we will see soon.

Hence, $\mathrm{d}x^\mu$ can be used in different ways, depending on the problem at hand. The effectiveness of differential geometry indeed relies to some extend on a suited set of definitions and notations.

2.5   Integrals of Differential Forms

Integral Transformation Formula in $\mathbb{R}^D$

After the introduction of differential forms and their basic properties, we are now ready to proceed to the definition of integrals. Let us first recall a key property of integrals in $\mathbb{R}^D$. Given a function $f : V(\subset \mathbb{R}^D) \to \mathbb{R}$, $x' \mapsto f(x')$, the volume integral of the function over the domain V is provided by the expression

$$\int_V f(x'^1, \ldots, x'^D)\,\mathrm{d}x'^1 \cdots \mathrm{d}x'^D. \tag{2.123}$$

Now consider a differentiable and invertible one-to-one map $\varphi : U(\subset \mathbb{R}^D) \to V(\subset \mathbb{R}^D)$, $x \mapsto \varphi(x) = x'$. Instead of the variables $(x'^1, \ldots, x'^D)$, one can use another collection of variables

$(x^1, \ldots, x^D)$ for the calculation of the integral. The variables $(x^1, \ldots, x^D)$ take their values in the set $U = \varphi^{-1}(V)$. Then the integral is calculated as

$$\int_V f(x') \, \mathrm{d}x'^1 \cdots \mathrm{d}x'^D = \int_{\varphi^{-1}(V)} f(\varphi(x)) \left| \det\left(\frac{\partial \varphi(x)}{\partial x} \right) \right| \mathrm{d}x^1 \cdots \mathrm{d}x^D, \qquad (2.124)$$

with the Jacobian determinant $J(x)$ of the map φ given as

$$J(x) = \det\left(\frac{\partial \varphi(x)}{\partial x} \right). \qquad (2.125)$$

The equation 2.124 is the well-known *integral transformation formula*. By introducing the *volume integration measure*

$$\mathrm{d}x^1 \cdots \mathrm{d}x^D \qquad (2.126)$$

in $\mathbb{R}^D$, we can memorize equation 2.124 as the symbolic transformation rule

$$\mathrm{d}x'^1 \cdots \mathrm{d}x'^D = |J(x)| \, \mathrm{d}x^1 \cdots \mathrm{d}x^D. \qquad (2.127)$$

When we consider the case of manifolds in the next paragraph, this transformation rule will be obtained in a similar form and the integration variables x and x' will simply represent two sets of coordinates.

Integrals of D-Forms

For defining the integral of D-forms on manifolds, we will let ourselves be guided by the requirement of invariance. Only an integral which is invariant under coordinate transformations represents a satisfactory notion for integration. We also demand the previously discussed properties of integration in $\mathbb{R}^D$ to be included in the generalization. Let us consider a D-dimensional differentiable manifold $\mathcal{M}$ and a D-form α on the manifold,

$$\alpha = \frac{1}{D!} \alpha_{\mu_1 \ldots \mu_D}(x) \, \mathrm{d}x^{\mu_1} \wedge \cdots \wedge \mathrm{d}x^{\mu_D}. \qquad (2.128)$$

Note that we employ the summation convention again. The D-form α is a coordinate-independent quantity. The components $\alpha_{\mu_1 \ldots \mu_D}(x)$ transform as a $(0, D)$-tensor, while the product $\mathrm{d}x^{\mu_1} \wedge \cdots \wedge \mathrm{d}x^{\mu_D}$ exhibits the transformation behavior of a $(D, 0)$-tensor. The D-form can be written as

$$\alpha = \alpha(x) \, \mathrm{d}x^1 \wedge \cdots \wedge \mathrm{d}x^D, \qquad (2.129)$$

with the single component*

$$\alpha(x) = \frac{1}{D!} \varepsilon^{\mu_1 \ldots \mu_D} \alpha_{\mu_1 \ldots \mu_D}(x) \qquad (2.130)$$

and a single basis element $\mathrm{d}x^1 \wedge \cdots \wedge \mathrm{d}x^D$. The space of D-forms is indeed one-dimensional, see appendix B.1. In the representation 2.129 the component $\alpha(x)$ and the basis element $\mathrm{d}x^1 \wedge \cdots \wedge \mathrm{d}x^D$ transform as a scalar densities of opposite weights. Indeed, under a general coordinate transformation,

$$x^\mu \mapsto x'^\mu = \frac{\partial x'^\mu}{\partial x^\nu} x^\nu, \qquad (2.131)$$

---

*Note that we use the epsilon symbol $\varepsilon^{\mu_1 \ldots \mu_D}$ here, not the epsilon tensor. The epsilon symbol is a tensor density, see appendix B.1.

the basis element $\mathrm{d}x^1 \wedge \cdots \wedge \mathrm{d}x^D$ transforms as a scalar density with weight $w = 1$, i.e.

$$\mathrm{d}x'^1 \wedge \cdots \wedge \mathrm{d}x'^D = \det\left(\frac{\partial x'}{\partial x}\right) \mathrm{d}x^1 \wedge \cdots \wedge \mathrm{d}x^D, \tag{2.132}$$

while the component $\alpha(x)$ transforms as a scalar density with weight $w = -1$, i.e.

$$\alpha'(x') = \det\left(\frac{\partial x}{\partial x'}\right) \alpha(x). \tag{2.133}$$

After these remarks, we can now proceed to the definition. We consider a domain U in the manifold $\mathcal{M}$, over which we want to integrate. Furthermore we choose a positively oriented coordinate chart $\psi : U \to \mathbb{R}^D$, $p \mapsto \psi(p) = x$ for the points p in U. The *integral of a D-form* $\alpha = \alpha(x)\,\mathrm{d}x^1 \wedge \cdots \wedge \mathrm{d}x^D$ *over the domain U* is defined as

$$\boxed{\int_U \alpha \equiv \int_{\psi(U)} \alpha(x)\,\mathrm{d}x^1 \cdots \mathrm{d}x^D} \tag{2.134}$$

where the integral on the rhs is the usual Lebesgue / Riemann integral in $\mathbb{R}^D$. This definition is indeed coordinate-independent. Let us choose another coordinate chart $\psi' : U \to \mathbb{R}^D$, $p \mapsto \psi'(p) = x'$. For the coordinates x and x', the coordinate transformation is given by the diffeomorphism $\varphi = \psi' \circ \psi^{-1} : \psi(U) \to \psi'(U)$, $x \mapsto \varphi(x) = x'$ in $\mathbb{R}^D$, see Section 1.2. The integral of the D-form is now given as

$$\int_U \alpha \equiv \int_{\psi'(U)} \alpha'(x')\,\mathrm{d}x'^1 \cdots \mathrm{d}x'^D. \tag{2.135}$$

According to the integral transformation formula 2.124, the above rhs is equal

$$\int_{\psi'(U)} \alpha'(x')\,\mathrm{d}x'^1 \cdots \mathrm{d}x'^D = \int_{\psi(U)} \alpha'\big(x'(x)\big) \left| \det\left(\frac{\partial x'}{\partial x}\right) \right| \mathrm{d}x^1 \cdots \mathrm{d}x^D. \tag{2.136}$$

The integrand $\alpha'\big(x'(x)\big)$ transforms as shown in 2.133 and the Jacobian factors cancel each other in the case of orientation-preserving transformations. Thus, we obtain

$$\int_{\psi'(U)} \alpha'(x')\,\mathrm{d}x'^1 \cdots \mathrm{d}x'^D = \int_{\psi(U)} \alpha(x)\,\mathrm{d}x^1 \cdots \mathrm{d}x^D. \tag{2.137}$$

For transformations flipping the orientation, a minus sign appears. This means that the combined transformation rules for D-forms and integral variables ensure that the integral definition for D-forms delivers an invariant quantity, as desired.

Volume Integrals

Let us now define volumes on general manifolds and on Riemannian manifolds in particular. A possible but not invariant *volume form on a general differentiable manifold*, denoted dvol is

$$\mathrm{dvol} \equiv \mathrm{d}x^1 \wedge \cdots \wedge \mathrm{d}x^D. \tag{2.138}$$

However, the *invariant volume form on a geometric manifold* dvol_g is defined by the expression

$$\mathrm{dvol}_g \equiv \sqrt{|g|}\,\mathrm{d}x^1 \wedge \cdots \wedge \mathrm{d}x^D, \tag{2.139}$$

where $|g(x)|$ is the absolute value of the determinant of the metric $g_{\mu\nu}(x)$,

$$|g| \equiv |\det(g_{\mu\nu})|. \tag{2.140}$$

The volume form dvol_g is invariant under diffeomorphisms due to the factor $\sqrt{|g|}$. Indeed, by changing coordinates from x to x', the invariant volume form dvol_g transforms as

$$\mathrm{dvol}'_g = \sqrt{\left|\det(g'_{\mu\nu}(x'))\right|}\, \mathrm{d}x'^1 \wedge \cdots \wedge \mathrm{d}x'^D. \tag{2.141}$$

By inserting the single transformed quantities in the above rhs, we obtain the invariance

$$\mathrm{dvol}'_g = \mathrm{dvol}_g, \tag{2.142}$$

being valid for all orientation-preserving transformations. Given a region U as a subset of a Riemannian manifold, the geometric volume of U is measured by the integral

$$\mathrm{Vol}(U) \equiv \int_U \mathrm{dvol}_g. \tag{2.143}$$

When a scalar function $\phi(x)$ is defined on U, its *volume integral* over U is the expression

$$\int_U \phi\, \mathrm{dvol}_g = \int_{\psi(U)} \phi(x)\sqrt{|g|}\, \mathrm{d}x^1 \cdots \mathrm{d}x^D. \tag{2.144}$$

The range of integration $\psi(U) \subseteq \mathbb{R}^D$ is determined by choosing a suitable coordinate system ψ. Now that we have defined volume integrals on manifolds, we can proceed to the definition of integrals over submanifolds, focussing on curves and hypersurfaces.

Curve Integrals

A 1-dimensional curve C within a D-dimensional manifold represents a submanifold over which we can integrate. Consider a 1-form $\alpha = \alpha_\mu(x)\, \mathrm{d}x^\mu$ on a D-dimensional manifold. Then the *curve* (or *line*) *integral* of the 1-form over the integration region C is given by

$$\int_C \alpha \equiv \int_{t_1}^{t_2} \alpha_\mu\big(x(t)\big)\frac{\mathrm{d}x^\mu}{\mathrm{d}t}(t)\, \mathrm{d}t, \tag{2.145}$$

with a parametrization $t \mapsto x^\mu(t)$ of the curve. Although a specific parametrization is used, the actual integral value is independent of it. (*exercise 2.13*)

Hypersurface Elements

We first intend to generalize the Euclidean surface elements known from calculus in $\mathbb{R}^3$ to the case $\mathbb{R}^D$. By achieving this, we are actually already done for the general case of manifolds, because the same formulae apply, with the difference that in the general case the coordinates are local functions. Given the space $\mathbb{R}^3$ with the Cartesian coordinates (x^1, x^2, x^3), the surface element pointing to the 1-direction is $\mathrm{d}x^2 \wedge \mathrm{d}x^3$. Similarly, the surface element in 2-direction is $\mathrm{d}x^3 \wedge \mathrm{d}x^1$, and the surface element in 3-direction is $\mathrm{d}x^1 \wedge \mathrm{d}x^2$. Let us denote the surface element in 1-direction by $\mathrm{d}\sigma_1$ and write

$$\mathrm{d}\sigma_1 \equiv \mathrm{d}x^2 \wedge \mathrm{d}x^3. \tag{2.146}$$

This 2-form $\mathrm{d}\sigma_1$ can be obtained by exterior differentiation of the 1-form σ_1 given by

$$\sigma_1 \equiv \frac{1}{2}\,\varepsilon_{1kl}\, x^k \mathrm{d}x^l. \tag{2.147}$$

Geometrically, the 1-form σ_1 represents the triangle-shaped surface element which is formed between the coordinate origin O, the point (x^2, x^3), and the point $(x^2 + \mathrm{d}x^2, x^3 + \mathrm{d}x^3)$, see figure 2.5. Expressed as a determinant of the corresponding column vectors it is

$$\sigma_1 = \frac{1}{2}\det\begin{pmatrix} x^2 & x^2 + \mathrm{d}x^2 \\ x^3 & x^3 + \mathrm{d}x^3 \end{pmatrix}. \tag{2.148}$$

Summing up all these triangle surfaces yields an approximation of the surface content of the region under consideration. In this way, the exterior derivative $d\sigma_1$ provides the differential surface element in the $(2,3)$-plane. The analogous formulae apply for the other two Cartesian directions.

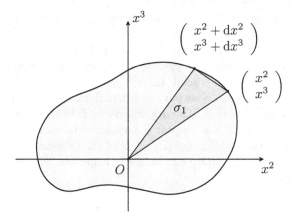

Figure 2.5: Plane surface element

Now the generalization to D dimensions is an algebraic task. The *hypersurface element* in μ-direction is the $(D-1)$-form $d\sigma_\mu$ given as (*exercise 2.14*)

$$d\sigma_\mu = \frac{1}{(D-1)!}\,\varepsilon_{\mu\mu_2\ldots\mu_D}\,dx^{\mu_2}\wedge\cdots\wedge dx^{\mu_D}. \tag{2.149}$$

The above rhs can be reshuffled to become the simpler looking expression

$$d\sigma_\mu = (-1)^{\mu-1}\,dx^1\wedge\cdots\wedge dx^{\mu-1}\wedge dx^{\mu+1}\wedge\cdots\wedge dx^D, \tag{2.150}$$

containing only one term. The set of all elements $d\sigma_\mu$, $\mu = 1,\ldots,D$, constitutes a basis for the space of $(D-1)$-forms. With respect to the index μ, the hypersurface element $d\sigma_\mu$ behaves like a covector density with weight $w = 1$, due to the epsilon symbol, which is a tensor density of the same weight. The formulae 2.149 and 2.150 for hypersurface elements are also valid for general manifolds, in which case the x^μ represent general coordinates.

2.6 Theorem of Stokes

Manifolds with Boundary and Orientation

Central to Stokes' theorem are manifolds $\mathcal{M}$ with a boundary $\partial\mathcal{M}$. Let us discuss here how a tangent space on the boundary can be constructed and how the normal vector field on the boundary looks like. The boundary points are labeled by coordinates of the form $(x^1,\ldots,x^{D-1},0)$. Consequently, the tangent space $T_p(\partial\mathcal{M})$ on each boundary point $p \in \partial\mathcal{M}$ is spanned by the possible, but not unique, basis $\{\partial_1,\ldots,\partial_{D-1}\}$ of $D-1$ vectors. In the notation here, we leave out the explicit reference to the point p. Stated differently, a tangent vector of $\partial\mathcal{M}$ is also a tangent vector of $\mathcal{M}$, given by a linear combination $X^\mu\partial_\mu$, but with the predefined value $X^D = 0$. For identifying a normal vector on each boundary point, we just need to ask which vector is not expressible by the basis vectors $\partial_1,\ldots,\partial_{D-1}$. This is clearly a vector proportional to ∂_D. The vector ∂_D itself points to values of an increasing x^D-coordinate, i.e. it points to the interior of the manifold $\mathcal{M}$. If we want to have an *outward pointing normal vector* on $\partial\mathcal{M}$, then this is given by $-a\,\partial_D$, with a being a positive real number.

Another structure to consider for Stokes' theorem is the orientation of a manifold. In Chapter 1, we have characterized the orientation by looking at the Jacobian determinant of any two coordinate charts of a point. The orientability condition can be expressed in a different way if we consider the basis vectors involved. Suppose we have a point $p \in \mathcal{M}$ and a pair of charts leading to the coordinates x and x'. Then, the condition that the Jacobian determinant $\det(\partial x'/\partial x)$ is strictly positive is exactly the condition that the determinant of the transformation matrix between the two bases $\{\partial_1, \ldots, \partial_D\}$ and $\{\partial'_1, \ldots, \partial'_D\}$ is strictly positive. So the condition of orientability of the manifold is translated to the orientability of all its tangent spaces. We need to be careful that whenever a basis such as $\{\partial_1, \ldots, \partial_D\}$ is used, the order of the vectors is crucial. Interchanging the order of any two vectors, will reverse the orientation of the basis. If we have an orientable manifold $\mathcal{M}$ with boundary $\partial\mathcal{M}$, then the boundary is also orientable. The orientation on $\partial\mathcal{M}$ is induced by the orientation of $\mathcal{M}$. A basis $\{\partial_1, \ldots, \partial_{D-1}\}$ of the tangent space on the boundary $\partial\mathcal{M}$ is, per definition, positively oriented if the basis $\{-\partial_D, \partial_1, \ldots, \partial_{D-1}\}$ of the tangent space on $\mathcal{M}$ is positively oriented. The validity of this definition can be easily inspected, e.g. by considering a closed surface in 3-dimensional space.

Stokes' Theorem

We are now ready to discuss the highlight of integration theory on manifolds, the *theorem of Stokes (Sir George Gabriel Stokes)*. We start with a D-dimensional oriented manifold $\mathcal{M}$. Within this manifold we consider a domain of integration $U \subseteq \mathcal{M}$ which is assumed to be a D-dimensional oriented manifold with a $(D-1)$-dimensional boundary ∂U. The orientation of U induces an orientation on the boundary ∂U. Moreover, the boundary is required to be closed, which means that it must be topologically compact. For any $(D-1)$-form α defined on U, the exterior derivative $\mathrm{d}\alpha$ is a D-form and the *Stokes integral formula* holds:

$$\int_U \mathrm{d}\alpha = \int_{\partial U} \alpha \tag{2.151}$$

The general proof of Stokes' theorem is rather technical and explained in many texts. Instead of repeating this here, we will restrict ourselves to a sketch of the proof for the case of a region U within the Euclidean plane $\mathbb{R}^2$. First we demonstrate the theorem for an oriented square surface S with the boundary ∂S, as depicted in the figure 2.6 below. We consider

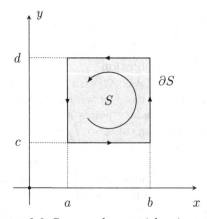

Figure 2.6: Square plaque with orientation

the 1-form

$$\alpha = A\,\mathrm{d}x + B\,\mathrm{d}y, \tag{2.152}$$

with the components $A(x, y)$ and $B(x, y)$ being dependent on the Cartesian coordinates (x, y). The exterior derivative of α is given by

$$\mathrm{d}\alpha = \left(\frac{\partial B}{\partial x} - \frac{\partial A}{\partial y}\right) \mathrm{d}x \wedge \mathrm{d}y. \qquad (2.153)$$

Consequently the surface integral of $\mathrm{d}\alpha$ over the square plaque S is

$$\int_S \mathrm{d}\alpha = \int_S \frac{\partial B}{\partial x} \,\mathrm{d}x \,\mathrm{d}y - \int_S \frac{\partial A}{\partial y} \,\mathrm{d}x \,\mathrm{d}y. \qquad (2.154)$$

Carrying out the integrals and reshuffling leads immediately to

$$\int_S \mathrm{d}\alpha = \int_a^b A(x, c) \,\mathrm{d}x + \int_c^d B(b, y) \,\mathrm{d}y + \int_a^b A(x, d) \,\mathrm{d}x + \int_c^d B(b, y) \,\mathrm{d}y. \qquad (2.155)$$

The last sum of integrals is exactly the line integral of α along the boundary ∂S. So we obtain

$$\int_S \mathrm{d}\alpha = \int_{\partial S} \alpha, \qquad (2.156)$$

which is the Stokes formula for an oriented square plaque S. In the next step, we view an arbitrary shaped 2-dimensional domain of integration U lying within the Euclidean plane $\mathbb{R}^2$ and possessing a closed boundary ∂U. The strategy is to approximate this domain U by a collection of square plaques. Overall, the domain U is approximated by an edged domain $\overline{U}$, as indicated in figure 2.7 below. The orientation of the domain U is imposed on the

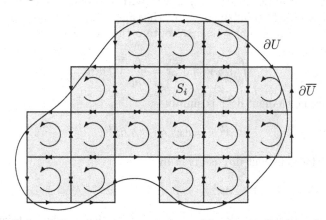

Figure 2.7: Integration domain approximated by a set of oriented squares S_i

edged domain $\overline{U}$. The integral of a D-form $\mathrm{d}\alpha$ over the domain U is approximated by an integral over the edged domain $\overline{U}$. Now the edged domain $\overline{U}$ can be considered to be the union of a set of square plaques S_i. Consequently the surface integral over $\overline{U}$ is the sum of the surface integrals over all oriented square elements S_i. Here we apply the Stokes formula to each one of these squares S_i. We realize that the line integral contributions along the sides of each two neighboring squares cancel each other exactly. So only those line integral contributions that represent the outside border line $\partial\overline{U}$ remain. We can refine the size, the number and the positions of the squares S_i, as needed, to approximate the given domain U. In the limit of infinitesimal square elements, we follow exactly the shape of the domain U. This leads finally to the Stokes integral formula 2.151 for the 2-dimensional Euclidean case. The general modern version of Stokes' theorem 2.151 contains all the classical vector analysis integral theorems. The classical Gauss divergence formula and the Stokes integral formula for antisymmetric tensors are particularly interesting for our later applications, so we discuss them in the following.

Gauss Divergence Formula for Vector Densities

We consider again a general manifold and recall the hypersurface element $d\sigma_\mu$, as given in 2.150, with its transformation property as a covector density with weight $w = 1$. If we now consider a vector density field $a^\mu(x)$ on the manifold with weight $w = -1$, the quantity

$$\alpha \equiv a^\mu \, d\sigma_\mu \tag{2.157}$$

is a scalar $(D-1)$-form. Its exterior derivative is given, after a simple calculation, as

$$d\alpha = (\partial_\mu a^\mu) \, dx^1 \wedge \cdots \wedge dx^D. \tag{2.158}$$

Hence, the Stokes integral formula 2.151 attains the form

$$\boxed{\int_U (\partial_\mu a^\mu) \, dx^1 \wedge \cdots \wedge dx^D = \oint_{\partial U} a^\mu \, d\sigma_\mu} \tag{2.159}$$

This is the *Gauss divergence formula for a vector density* $a^\mu(x)$ with weight $w = -1$. It is valid without any reference to a metric. This divergence formula is called also the *Gauss-Ostrogradsky theorem* in honor of its formulation by *Mikhail Vasilyevich Ostrogradsky*.

The divergence formula can be developed further whenever a metric $g_{\mu\nu}(x)$ is present. In that case, we can use the fact that $\sqrt{|g|}$ is a scalar density of weight $w = -1$ and write the vector density field $a^\mu(x)$ as $a^\mu = \sqrt{|g|}\, V^\mu$, with a proper vector field $V^\mu(x)$. This is explained in Section 18.3. Specializing even further and taking our manifold to be Euclidean space $\mathbb{E}^D$ with the metric $g_{kl}(x) = \delta_{kl}$, we obtain the divergence formula as it is usually known from calculus,

$$\int_U (\partial_k V^k) \, d^D x = \oint_{\partial U} V^k \, d\sigma_k. \tag{2.160}$$

The Euclidean volume element $d^D x$ is simply

$$d^D x = dx^1 \wedge \cdots \wedge dx^D, \tag{2.161}$$

while the hypersurface element $d\sigma_k$ is given by

$$d\sigma_k = (-1)^{k-1} \, dx^1 \wedge \cdots \wedge dx^{k-1} \wedge dx^{k+1} \wedge \cdots \wedge dx^D, \tag{2.162}$$

with the x^k being Cartesian coordinates here. As an application, let us take the vector field $V^k(x) = x^k$, since Cartesian coordinates can be identified with vector components in $\mathbb{E}^D$. Then it is $\partial_k x^k = D$ and we can determine the volume $\mathrm{Vol}(U)$ of a region $U \subset \mathbb{E}^D$ as

$$\mathrm{Vol}(U) = D^{-1} \oint_{\partial U} \sum_{k=1}^{D} (-1)^{k-1} \, dx^1 \wedge \cdots \wedge x^k \wedge \cdots \wedge dx^D. \tag{2.163}$$

This means that the measurement of the volume is reduced to a hypersurface integration.

Stokes Formula for Antisymmetric Tensor Densities

We proceed to an application of the Stokes integral formula 2.151, which will be important in Chapter 23. Let us consider a D-dimensional oriented manifold $\mathcal{M}$ and within it a $(D-1)$-dimensional hypersurface Σ having a closed boundary. The closed boundary, denoted $\partial\Sigma$, is itself a $(D-2)$-dimensional oriented manifold. We are going to apply Stokes' theorem and express the integral of a specific $(D-1)$-form over the hypersurface Σ through the equivalent integral of the associated $(D-2)$-form over the boundary $\partial\Sigma$. We have identified the hypersurface element $d\sigma_\mu$ of the hypersurface Σ already in 2.149. For the boundary manifold $\partial\Sigma$, the corresponding hypersurface element is the $(D-2)$-form $d\sigma_{\mu\nu}$ given as

$$d\sigma_{\mu\nu} = \frac{1}{(D-2)!}\, \varepsilon_{\mu\nu\mu_3\ldots\mu_D}\, dx^{\mu_3} \wedge \cdots \wedge dx^{\mu_D}. \tag{2.164}$$

With respect to the indices $\mu\nu$, the hypersurface element $d\sigma_{\mu\nu}$ behaves like a $(0,2)$-tensor density with weight $w=1$. Now if we take an *antisymmetric $(2,0)$-tensor density* $a^{\mu\nu}(x)$ with weight $w=-1$, we can build the contraction

$$\alpha \equiv a^{\mu\nu}\, d\sigma_{\mu\nu} \tag{2.165}$$

and this represents a scalar $(D-2)$-form. This $(D-2)$-form α can be integrated over the boundary $\partial\Sigma$ and we can equate the result with the integral of the $(D-1)$-form $d\alpha$ over the hypersurface Σ. In order to compute $d\alpha$, let us be definite and consider the relevant case of a manifold $\mathcal{M}$ with $D=4$. So we have

$$d\sigma_\mu = \frac{1}{3!}\, \varepsilon_{\mu\alpha\beta\gamma}\, dx^\alpha \wedge dx^\beta \wedge dx^\gamma \tag{2.166}$$

as the surface element of the 3-dimensional hypersurface Σ and

$$d\sigma_{\mu\nu} = \frac{1}{2}\, \varepsilon_{\mu\nu\beta\gamma}\, dx^\beta \wedge dx^\gamma \tag{2.167}$$

as the surface element of the 2-dimensional boundary $\partial\Sigma$. All indices run from 1 to 4. The quantity $\alpha \equiv a^{\mu\nu}\, d\sigma_{\mu\nu}$ is a 2-form and its exterior derivative $d\alpha$ is calculated as

$$d\alpha = \frac{1}{2}(\partial_\rho a^{\mu\nu})\, \varepsilon_{\mu\nu\beta\gamma}\, dx^\rho \wedge dx^\beta \wedge dx^\gamma. \tag{2.168}$$

Next, we write $\partial_\rho a^{\mu\nu} = \delta_\rho^\sigma \partial_\sigma a^{\mu\nu}$ and take the *Schouten identity* (*Jan Arnoldus Schouten*)

$$\delta_\rho^\sigma \varepsilon_{\mu\nu\beta\gamma} = \delta_\mu^\sigma \varepsilon_{\rho\nu\beta\gamma} + \delta_\nu^\sigma \varepsilon_{\mu\rho\beta\gamma} + \delta_\beta^\sigma \varepsilon_{\mu\nu\rho\gamma} + \delta_\gamma^\sigma \varepsilon_{\mu\nu\beta\rho} \tag{2.169}$$

for the Kronecker delta and the epsilon symbol into account, see appendix B.1. This yields

$$d\alpha = \frac{1}{2}\Big[\partial_\mu a^{\mu\nu}\varepsilon_{\rho\nu\beta\gamma} + \partial_\nu a^{\mu\nu}\varepsilon_{\mu\rho\beta\gamma} + \partial_\beta a^{\mu\nu}\varepsilon_{\mu\nu\rho\gamma} + \partial_\gamma a^{\mu\nu}\varepsilon_{\mu\nu\beta\rho}\Big] dx^\rho \wedge dx^\beta \wedge dx^\gamma. \tag{2.170}$$

By using the anticommutation properties of the terms in the brackets, we obtain

$$d\alpha = (\partial_\mu a^{\nu\mu}\varepsilon_{\nu\rho\beta\gamma} - \partial_\rho a^{\mu\nu}\varepsilon_{\mu\nu\beta\gamma})\, dx^\rho \wedge dx^\beta \wedge dx^\gamma, \tag{2.171}$$

which in turn is equivalent to

$$3\, d\alpha = (\partial_\mu a^{\nu\mu})\, \varepsilon_{\nu\rho\beta\gamma}\, dx^\rho \wedge dx^\beta \wedge dx^\gamma. \tag{2.172}$$

By taking into account the definition of $d\sigma_\mu$, we finally obtain

$$d\alpha = 2\, (\partial_\mu a^{\nu\mu})\, d\sigma_\nu. \tag{2.173}$$

If we apply this result for the 3-form $d\alpha$ to the Stokes integral formula 2.151, we obtain

$$\boxed{\int_\Sigma (\partial_\nu a^{\mu\nu})\, d\sigma_\mu = \frac{1}{2}\oint_{\partial\Sigma} a^{\mu\nu}\, d\sigma_{\mu\nu}} \tag{2.174}$$

This represents the *Stokes formula for an antisymmetric tensor density* $a^{\mu\nu}(x)$ with weight $w=-1$. This formula will be discussed further in Section 18.3 for the case of a geometric manifold and later be applied in the definition of conserved quantities in general relativity.

Further Reading

There are numerous differential geometry books available with high applicability to the questions of theoretical physics. For an accessible discussion of metrics, coordinates, and geometries as needed in physics, one may turn to the books of Dubrovin, Fomenko, Novikov [25], Frankel [30], and Nakahara [58]. For a review of hyperbolic and spherical geometry, the interested reader should consult Ratcliffe [72]. The reader who wants to work through an explicit proof of Stokes' theorem can turn to the classic text of Lovelock and Rund [55].

3

Symmetries of Manifolds

In this chapter, we address the question of how we can meaningfully define a notion of symmetry for geometric manifolds. First we consider general diffeomorphisms defined by vector fields and study how arbitrary tensor fields transform. This leads and to the central notion of the Lie derivative. To introduce a notion of symmetry for geometric manifolds, we consider the transformation behavior and the invariance of the metric under active diffeomorphisms. The symmetry transformations of a geometric manifold are identified with the isometries of the manifold. Analytically, the existence of a geometric symmetry is equivalent to the existence of so-called Killing vector fields. Besides isometries, conformal transformations are also considered and some basic examples are given.

3.1 Transformations and the Lie Derivative

Diffeomorphisms of Manifolds and Variations of Tensor Fields

We consider a geometric manifold $(\mathcal{M}, g)$ and diffeomorphisms $F : \mathcal{M} \to \mathcal{M}$ of the manifold. When we talked about manifolds, we have emphasized the distinction between a point p of the manifold, on one hand, and the coordinates $x = \psi(p)$ of the point, on the other hand. However, in order to simplify notation, henceforth we will make use of the coordinates notation x^μ to denote both, the coordinates and the geometric point. Similarly, when we study a vector field $V = V^\mu(x)\partial_\mu$ we will use the component notation $V^\mu(x)$ or V^μ to represent both, the components and the vector itself. In doing so, we must take care to distinct between *active transformations* of points and *passive transformations* of points, the later corresponding only to a transformation of the coordinates. We will be diligent in describing how transformations are meant to act. In the first two chapters we considered primarily the passive standpoint. In this and the following chapters we will concentrate on active transformations, i.e. actual changes of the geometric points. For arbitrary active transformations of tensor fields, we will employ the general transformation formula 1.96.

Let us consider general tensor fields $\phi^{\mu_1\cdots\mu_m}{}_{\nu_1\ldots\nu_n}(x)$ on the manifold, and use the simplified notation $\phi_i(x)$ for them. The *multi-index i* represents the entire set of contravariant and covariant tensor indices. Furthermore, let us consider a diffeomorphism $F : x \mapsto F(x) \equiv x'$ representing an active transformation of the points of the manifold. Written infinitesimally, the transformation is

$$x \mapsto x'(x) = x + \delta x(x). \tag{3.1}$$

DOI: 10.1201/9781003087748-3

We call $\delta x(x)$ the *variation* of the point x. The active transformation of the manifold points x induces a corresponding active transformation of the components of the tensor field $\phi_i(x)$, symbolically

$$\phi_i(x) \mapsto \phi_i'(x'), \tag{3.2}$$

where the new tensor components $\phi_i'(x')$ are given by the formula 1.96. Actually, there are two different notions of a variation of the field $\phi_i(x)$. The one is the *form* (or *function*) *variation*, $\delta\phi_i(x)$, defined as

$$\delta\phi_i(x) \equiv \phi_i'(x) - \phi_i(x). \tag{3.3}$$

The quantity $\delta\phi_i \equiv \phi_i' - \phi_i$ measures the changes of the function itself. The other type of variation is the so-called *total variation*, $\overline{\delta}\phi_i(x)$, defined as

$$\overline{\delta}\phi_i(x) \equiv \phi_i'(x') - \phi_i(x), \tag{3.4}$$

in which the combined change in the function *and* the points is taken into account. Using the infinitesimal form $x'(x) = x + \delta x(x)$, we find the following fundamental relation between the two types of variation

$$\boxed{\overline{\delta}\phi_i(x) = \delta\phi_i(x) + \partial_\mu \phi_i(x)\,\delta x^\mu(x)} \tag{3.5}$$

Here we have kept only the terms up to first order in $\delta x(x)$. The relation 3.5 represents a central result, which is used to define the various types of symmetry transformations. From this we can infer that the variation $\delta x^\mu(x)$ is a proper vector field. In the special case of a scalar field, $f(x)$, the total variation $\overline{\delta}f(x)$ vanishes,

$$\overline{\delta}f(x) = f'(x') - f(x) = 0, \tag{3.6}$$

and we conclude that its form variation $\delta f(x)$ is given as

$$\delta f(x) = -\partial_\mu f(x)\,\delta x^\mu(x). \tag{3.7}$$

The Lie Derivative

Let us now ask which measure would be a useful one for the change of a tensor field $\phi_i(x)$ under transformations of the manifold. Consider again the active transformation $F : x \mapsto x'$ in its infinitesimal form,

$$x'^\mu(x) = x^\mu + \epsilon X^\mu(x). \tag{3.8}$$

where we have written the variation $\delta x^\mu(x)$ as

$$\delta x^\mu(x) \equiv \epsilon X^\mu(x). \tag{3.9}$$

Here we use the real parameter ϵ, with $|\epsilon| \ll 1$, and the vector field $X^\mu(x)$ to define the transformation.* If we want to measure the rate of change, we are led to the Lie derivative. The *Lie derivative* of the tensor field $\phi_i(x)$ *with respect to the vector field* $X^\mu(x)$ is the tensor field $\mathcal{L}_X\phi_i(x)$, which has the same valence and is defined as

$$\boxed{\mathcal{L}_X\phi_i(x) \equiv \lim_{\epsilon \to 0} \frac{\phi_i(x) - \phi_i'(x)}{\epsilon}} \tag{3.10}$$

Note that the prime is on the term with the minus sign, in contrast to usual folklore in the definition of derivatives.

---

*Later we will write the variation more generally as $\delta x^\mu(x) = \epsilon^a X_a^\mu(x)$ using a set of r real parameters ϵ^a. The index a can be a tensorial index.

Formulae for the Lie Derivative

Let us now derive some useful formulae for the Lie derivative of scalars, vectors, covectors, tensors and tensor densities. We note that the transformation $x'^\mu(x) = x^\mu + \epsilon X^\mu(x)$ implies that

$$\frac{\partial x'^\mu}{\partial x^\rho}(x) = \delta^\mu_\rho + \epsilon \frac{\partial X^\mu}{\partial x^\rho}(x). \tag{3.11}$$

The inverse transformation, $x^\mu(x') = x'^\mu - \epsilon X^\mu(x(x'))$, leads to

$$\frac{\partial x^\sigma}{\partial x'^\nu}(x') = \delta^\sigma_\nu - \epsilon \frac{\partial X^\sigma}{\partial x'^\nu}(x(x')). \tag{3.12}$$

These basic relations will be used in the following. We start with the simplest case of a scalar field function $f(x)$. For a scalar field $f(x) = f'(x + \epsilon X)$ we apply the Taylor expansion for linear terms in ϵX and obtain the result

$$\boxed{\mathcal{L}_X f = X^\mu \partial_\mu f = X f} \tag{3.13}$$

This is exactly the directional derivative Xf of the function $f(x)$ along the vector field X. Comparing with the formula 3.5, we infer that for a scalar field $f(x)$ we can write

$$\mathcal{L}_{\delta x} f = -\delta f. \tag{3.14}$$

As expected, the Lie derivative is closely related to the form variation. We will investigate this relation at the end of this section more generally. But let us turn now to a $(1,1)$-tensor field and derive the formula for calculating the Lie derivative. A $(1,1)$-tensor field $\phi^\mu_\nu(x)$ will transform to $\phi'^\mu_\nu(x')$ like

$$\phi'^\mu_\nu(x') = \frac{\partial x'^\mu}{\partial x^\rho}(x) \frac{\partial x^\sigma}{\partial x'^\nu}(x') \phi^\rho_\sigma(x). \tag{3.15}$$

The lhs is Taylor-expanded around the initial point x,

$$\phi'^\mu_\nu(x + \epsilon X) = \phi'^\mu_\nu(x) + \epsilon X^\lambda(x) \left.\frac{\partial \phi'^\mu_\nu}{\partial x'^\lambda}(x')\right|_{\epsilon=0} + \mathcal{O}(\epsilon^2). \tag{3.16}$$

We note that for $\epsilon = 0$ the primes disappear,

$$\left.\frac{\partial \phi'^\mu_\nu}{\partial x'^\lambda}(x')\right|_{\epsilon=0} = \frac{\partial \phi^\mu_\nu}{\partial x^\lambda}(x). \tag{3.17}$$

The rhs yields, up to first order in ϵ,

$$\left(\delta^\mu_\rho + \epsilon \frac{\partial X^\mu}{\partial x^\rho}\right)\left(\delta^\sigma_\nu - \epsilon \frac{\partial X^\sigma}{\partial x'^\nu}\right) \phi^\rho_\sigma(x) =$$

$$\phi^\mu_\nu(x) + \epsilon \frac{\partial X^\mu}{\partial x^\rho} \phi^\rho_\nu(x) - \epsilon \frac{\partial X^\sigma}{\partial x'^\nu} \phi^\mu_\sigma(x) + \mathcal{O}(\epsilon^2). \tag{3.18}$$

By combining the results for the lhs and rhs, we obtain for the rate of change

$$\frac{\phi^\mu_\nu(x) - \phi'^\mu_\nu(x)}{\epsilon} = X^\lambda \frac{\partial \phi^\mu_\nu}{\partial x^\lambda}(x) - \frac{\partial X^\mu}{\partial x^\rho} \phi^\rho_\nu(x) + \frac{\partial X^\sigma}{\partial x'^\nu} \phi^\mu_\sigma(x) + \mathcal{O}(\epsilon). \tag{3.19}$$

Now we take the limit $\epsilon \to 0$ and note that in the third term on the rhs the variable x'^ν becomes x^ν. Thus, we can conclude to the desired formula for the Lie derivative of the mixed tensor field $\phi^\mu_\nu(x)$:

$$\boxed{\mathcal{L}_X \phi^\mu_\nu = X^\rho \partial_\rho \phi^\mu_\nu - \phi^\rho_\nu \partial_\rho X^\mu + \phi^\mu_\rho \partial_\nu X^\rho} \tag{3.20}$$

The formula for a general tensor field is completely analogous to the case of a $(1,1)$-tensor field, with the exception that one has to consider multiple co- and contravariant indices. We write the result for a tensor field $\phi^{\mu_1\cdots\mu_m}{}_{\nu_1\ldots\nu_n}(x)$ here:

$$
\begin{aligned}
\mathcal{L}_X \phi^{\mu_1\cdots\mu_m}{}_{\nu_1\ldots\nu_n} =\ & X^\rho \partial_\rho \phi^{\mu_1\cdots\mu_m}{}_{\nu_1\ldots\nu_n} \\
& - \phi^{\rho\cdots\mu_m}{}_{\nu_1\ldots\nu_n} \partial_\rho X^{\mu_1} - \cdots - \phi^{\mu_1\cdots\rho}{}_{\nu_1\ldots\nu_n} \partial_\rho X^{\mu_m} \\
& + \phi^{\mu_1\cdots\mu_m}{}_{\rho\ldots\nu_n} \partial_{\nu_1} X^\rho + \cdots + \phi^{\mu_1\cdots\mu_m}{}_{\nu_1\ldots\rho} \partial_{\nu_n} X^\rho .
\end{aligned}
\tag{3.21}
$$

In Section 3.3 we will employ the Lie derivative in order to express symmetry properties of manifolds and tensor fields. But for now, for the sake of completeness, let us write down two important special cases of the Lie derivative formula. For a vector field $V^\mu(x)$ the Lie derivative along another vector field $X^\nu(x)$ is given by

$$
\boxed{\mathcal{L}_X V^\mu = X^\nu \partial_\nu V^\mu - V^\nu \partial_\nu X^\mu}
\tag{3.22}
$$

This is exactly the μth component of the commutator $[X, V]$ of X and V, see Section 1.4. For a covector field $\omega_\mu(x)$ the Lie derivative along a vector field $X^\nu(x)$ is given by

$$
\boxed{\mathcal{L}_X \omega_\mu = X^\nu \partial_\nu \omega_\mu + \omega_\nu \partial_\mu X^\nu}
\tag{3.23}
$$

Let us now examine the relation between the Lie derivative $\mathcal{L}_X \phi_i(x)$ of an arbitrary tensor field $\phi_i(x)$ and its form variation $\delta\phi_i(x)$. According to the definitions 3.3 and 3.10, the following general relation holds:

$$
\mathcal{L}_X \phi_i = -\left.\frac{\partial(\delta\phi_i)}{\partial\epsilon}\right|_{\epsilon=0}.
\tag{3.24}
$$

The above formula is valid for form variations 3.3, being linear in the parameter ϵ, as well as for more general form variations, being nonlinear in ϵ. For infinitesimal form variations, being linear in ϵ, we can write equivalently

$$
\boxed{\mathcal{L}_{\delta x} \phi_i = -\delta\phi_i}
\tag{3.25}
$$

where we have $\delta x = \epsilon X$ again. The reader is encouraged to derive these fundamental relations. (*exercise 3.1*)

Lie Derivative of Tensor Densities

Here we will generalize the Lie derivative formula to the case of tensor density fields. We view the transformation $x' = x + \epsilon X$ and its Jacobian determinant $J(x)$ with the definition

$$
J = \det\left(\frac{\partial x'}{\partial x}\right).
\tag{3.26}
$$

For example, a $(1,1)$-tensor density field $T^\mu{}_\nu(x)$ will transform according to

$$
T'^\mu{}_\nu(x') = J(x)^w \frac{\partial x'^\mu}{\partial x^\rho}(x) \frac{\partial x^\sigma}{\partial x'^\nu}(x') T^\rho{}_\sigma(x),
\tag{3.27}
$$

where the Jacobian is raised to the wth power, the weight of the tensor density. For small ϵ, we can approximate up to first order and use the general determinant formula

$$
\det(1 + \epsilon A) = 1 + \epsilon \operatorname{Tr}(A) + \mathcal{O}(\epsilon^2).
\tag{3.28}
$$

By applying this formula, we obtain

$$J^w = 1 + \epsilon\, w\, \partial_\rho X^\rho + \mathcal{O}(\epsilon^2). \tag{3.29}$$

Henceforth the derivation of the Lie derivative formula is similar to the proper tensor field case. (*exercise 3.2*) The formula for the $(1,1)$-tensor density $T^\mu{}_\nu(x)$ is

$$\boxed{\mathcal{L}_X T^\mu{}_\nu = X^\rho \partial_\rho T^\mu{}_\nu - T^\rho{}_\nu \partial_\rho X^\mu + T^\mu{}_\rho \partial_\nu X^\rho - w\,(\partial_\rho X^\rho)\, T^\mu{}_\nu} \tag{3.30}$$

The last term represents the generalization for $w \neq 0$ and has the same form $-w\,(\partial \cdot X)\,T_i$ for a tensor density $T_i(x)$ of any valence. With $w = 0$ we recover proper tensor fields.

Algebraic Properties of the Lie Derivative

For vector fields X, Y, etc. on a manifold, the Lie derivatives $\mathcal{L}_X$, $\mathcal{L}_Y$, etc. have some noteworthy algebraic properties. We recall the commutator $[X,Y] = XY - YX$ of vector fields and the formula $\mathcal{L}_X Y = [X,Y]$ for the Lie derivative of a vector field. The Lie derivative acts linearly so that for any real numbers a, b and any vector fields V, W it is

$$\mathcal{L}_X(aV + bW) = a\,\mathcal{L}_X V + b\,\mathcal{L}_X W. \tag{3.31}$$

In addition, again for any two numbers a and b, the linearity rule

$$\mathcal{L}_{aX+bY} V = a\,\mathcal{L}_X V + b\,\mathcal{L}_Y V \tag{3.32}$$

holds. We introduce the *commutator of Lie derivatives* as

$$[\mathcal{L}_X, \mathcal{L}_Y] \equiv \mathcal{L}_X \mathcal{L}_Y - \mathcal{L}_Y \mathcal{L}_X. \tag{3.33}$$

The nested action of the Lie derivative poses no issue, since it does not change the valence of a tensor. The commutator $[\mathcal{L}_X, \mathcal{L}_Y]$ is bilinear in its arguments and antisymmetric. By using the antisymmetry of the commutator $[X,Y]$, we see immediately that it is

$$\mathcal{L}_X Y = -\mathcal{L}_Y X. \tag{3.34}$$

For an arbitrary tensor density field $T_i(x)$, the following relation holds (*exercise 3.3*)

$$\mathcal{L}_{[X,Y]} T_i = [\mathcal{L}_X, \mathcal{L}_Y]\, T_i. \tag{3.35}$$

The last identity combined with the Jacobi identity yields

$$\mathcal{L}_X [Y, Z] = [\mathcal{L}_X Y, Z] + [Y, \mathcal{L}_X Z]. \tag{3.36}$$

This is the *Leibniz rule* for the Lie derivative applied to the commutator product $[X,Y]$. The Lie derivative satisfies the Leibniz property with other types of products as well. For a 1-form ω, a vector field V and their natural pairing $\omega(V) = \langle \omega, V \rangle$ it is

$$\mathcal{L}_X \langle \omega, V \rangle = \langle \mathcal{L}_X \omega, V \rangle + \langle \omega, \mathcal{L}_X V \rangle. \tag{3.37}$$

For any two tensor fields ϕ, ψ and their tensor product $\phi \otimes \psi$ it is

$$\mathcal{L}_X (\phi \otimes \psi) = (\mathcal{L}_X \phi) \otimes \psi + \phi \otimes (\mathcal{L}_X \psi). \tag{3.38}$$

The same Leibniz rule applies also to the exterior product of forms. Finally, it can be readily shown that the commutator of Lie derivatives satisfies the *Jacobi identity*

$$[[\mathcal{L}_X, \mathcal{L}_Y], \mathcal{L}_Z] + [[\mathcal{L}_Y, \mathcal{L}_Z], \mathcal{L}_X] + [[\mathcal{L}_Z, \mathcal{L}_X], \mathcal{L}_Y] = 0. \tag{3.39}$$

The proof is left to the reader. (*exercise 3.4*) The Jacobi identity will appear again when we study another type of derivative, the covariant derivative.

3.2 Symmetry Transformations of Manifolds

General Considerations

How can we define the notion of "symmetry"? In general terms, symmetry is the property of staying "invariant" under a certain transformation. Correspondingly, a symmetry transformation is a transformation which leaves an object invariant. In our case the object is a mathematical or a physical structure. This is typically a space with some defining properties, or a specific quantity of interest, or a physical law expressed as an equation. If we want to study the symmetries of a space, we need a set of transformations acting on it. If there is a transformation that leaves the space with its defining properties unchanged, then we have a *symmetry transformation* at hand. In many cases the symmetry transformations are as characteristic as the defining properties of the space. One can reverse the procedure then and actually define the space based on its invariance under certain transformations. The spaces we consider in this section are geometric manifolds $(\mathcal{M}, g)$. Their defining structure is, foremost, the metric g. So what we are looking for are transformations of coordinates and tensor fields that leave the metric invariant. Ultimately, we are led to the central notions of isometry and conformality. We should note, however, that we can define different types of symmetries, where quantities other than the metric remain invariant. For instance, when we discuss the symmetry of physical theories, we will be concerned with the invariance of the action integral or the invariance of the equations of motion.

Isometry Transformations

As we saw in the previous chapter, two given geometric manifolds $(\mathcal{M}, g)$ and $(\mathcal{N}, a)$ are considered geometrically equivalent, if there exists an isometry map $F : \mathcal{M} \to \mathcal{N}$ between them. In that case the lengths and angles of vectors are preserved, when compared between the two manifolds. Now we have the situation of a single geometric manifold $(\mathcal{M}, g)$, where both structures, the manifold $\mathcal{M}$ and the metric g are held fixed. The question is now, how can we characterize the symmetry transformations that leave this geometric manifold $(\mathcal{M}, g)$ invariant? The answer is readily found: the symmetry transformations are exactly the isometry maps of $(\mathcal{M}, g)$ to itself. Let us formulate this in detail. In technical terms, we view an active transformation $F : \mathcal{M} \to \mathcal{M}$, $x \mapsto F(x) \equiv x'$ of the points of the manifold. The transformation F indices also the transformation of tensor fields and, in particular, the transformation of the metric. We denote the metric gained through the map F by g'. The new metric components $g'_{\mu\nu}$ play the role of the metric components $a_{\mu\nu}$ in the isometry condition 2.51. Now the crucial point is that this metric $g'_{\mu\nu}$ is identical to the original one, i.e.

$$g'_{\mu\nu}(x) = g_{\mu\nu}(x). \tag{3.40}$$

Thus, the defining condition of the *symmetry transformations* of the coordinates on the geometric manifold $(\mathcal{M}, g)$ reads

$$\boxed{\; g_{\mu\nu}(x) = \frac{\partial x'^{\rho}}{\partial x^{\mu}}(x) \, \frac{\partial x'^{\sigma}}{\partial x^{\nu}}(x) \, g_{\rho\sigma}(x') \;} \tag{3.41}$$

This should be understood as an implicit equation defining the symmetry transformation $x \mapsto x'$. Let us observe also that this is exactly the general formula 2.17 for the transformation of the metric components when the equality $g'_{\mu\nu}(x) = g_{\mu\nu}(x)$ holds. So according to the considerations above, the isometry map represents the most general symmetry transformation of the geometric manifold $(\mathcal{M}, g)$. We call these the *isometry transformations* of the geometric manifold. In summary we can state:

The symmetry transformations of a geometric manifold $(\mathcal{M}, g)$ are exactly the isometry transformations.

We will apply the isometry condition 3.41 later in various concrete cases, like rotations and Lorentz transformations. Let us move on to a more general situation in the next step and consider conformal structures.

Conformal Transformations

Starting again with a single manifold $(\mathcal{M}, g)$, we can view the diffeomorphic maps $F : \mathcal{M} \to \mathcal{M}$ of the manifold to itself that preserve the angles between vectors but do not preserve the lengths. The specific maps we need to consider here are obviously the conformal maps of the manifold $(\mathcal{M}, g)$ to itself. Assuming that the metric stays invariant as a function, i.e.

$$g'_{\mu\nu}(x) = g_{\mu\nu}(x), \tag{3.42}$$

the generic conformality relation 2.53 translates this to the condition

$$\Psi^2(x)\, g_{\mu\nu}(x) = \frac{\partial x'^\rho}{\partial x^\mu}(x)\frac{\partial x'^\sigma}{\partial x^\nu}(x)\, g_{\rho\sigma}(x') \tag{3.43}$$

This is the defining condition of the so-called *conformal transformations* of the coordinates of the geometric manifold $(\mathcal{M}, g)$. This should be understood as an implicit equation that defines the conformal transformation $x \mapsto x'$.

Apparently, we are not finding ourselves in the same situation as with isometry transformations. Can the conformality condition 3.43 still be understood in some way as an isometry-like condition and what would be the preserved structure? If we compare 3.43 with the general transformation rule 2.17, we obtain for the total change $g'_{\mu\nu}(x')$ of the metric the relation

$$g'_{\mu\nu}(x') = \Psi^{-2}(x(x'))\, g_{\mu\nu}(x'). \tag{3.44}$$

If $\Psi = 1$ holds, the last relation reduces to the basic invariance of the metric. On the other hand, the rhs of the last equation is of the form $\Omega^2(x')\, g_{\mu\nu}(x')$, if we make the identification $\Omega^2(x') \equiv \Psi^{-2}(x(x'))$. This motivates us to consider the notion of a *metric defined only up to a local scale factor*. In other words, instead of considering the manifold with one definite metric, we view the manifold with an equivalence class of metrics. Starting with the initial metric g of $(\mathcal{M}, g)$, we define the equivalence class $[g]$ of metrics as

$$[g] \equiv \left\{\Omega^2 g, \text{ for all real scalar functions } \Omega, \text{with } \Omega > 0\right\}. \tag{3.45}$$

The pair consisting of the manifold $\mathcal{M}$ and the equivalence class of metrics $[g]$ constitutes a so-called *conformal manifold* $(\mathcal{M}, [g])$. In this context we speak about a *conformal geometry*. The conformal manifold $(\mathcal{M}, [g])$ is the structure that stays invariant under conformal transformations, as specified in 3.43. We can summarize this in a statement:

The symmetry transformations of a conformal manifold $(\mathcal{M}, [g])$ are exactly the conformal transformations.

Let us remark that every given geometric manifold $(\mathcal{M}, g)$ can be used to construct the associated conformal manifold $(\mathcal{M}, [g])$ by allowing the metric to be scaled by smooth non-vanishing functions.

It is important to compare the just studied conformal transformations with the Weyl rescalings, which we have encountered in Section 2.3. Weyl rescalings do not depend on any coordinate changes. Weyl rescalings re-define from the outset the basic geometry, by introducing a scaling factor into the initial metric, as expressed in 2.83. Conformal transformations on a manifold with a fixed metric, in contrast, give rise to coordinate transformations. These conformal coordinate transformations could in turn be used to define a scaling between an initial and a final metric, as expressed in 3.44. However, conformal coordinate transformations do not serve the primary purpose of defining a new geometry. Hence, despite their interrelation, we will maintain the distinction between these two types of transformations.

3.3   Isometric and Conformal Killing Vectors

Killing Vectors

In the following, we move on to characterize isometry transformations and conformal transformations analytically by using the Lie derivative. We consider a geometric manifold. A transformation acting on the manifold is an isometry exactly if the condition 3.41 is met. Plugging in the infinitesimal transformation equation

$$x'^{\mu}(x) = x^{\mu} + \epsilon K^{\mu}(x), \tag{3.46}$$

defined by the vector field $K^{\mu}(x)$ and the real parameter ϵ, with $|\epsilon| \ll 1$, one obtains (*exercise 3.5*)

$$K^{\rho}\partial_{\rho}g_{\mu\nu} + g_{\rho\nu}\partial_{\mu}K^{\rho} + g_{\mu\rho}\partial_{\nu}K^{\rho} = 0. \tag{3.47}$$

By using the Lie derivative $\mathcal{L}_K$, this condition can be compactly rewritten as

$$\boxed{\mathcal{L}_K g_{\mu\nu} = 0} \tag{3.48}$$

and is called the *Killing equation* (*Wilhelm Karl Joseph Killing*). A vector field $K^{\mu}(x)$ fulfilling this condition is called a *Killing vector field* (abbreviated *KVF*). So when we start with a Killing vector field $K^{\mu}(x)$ and construct the transformation $x'(x) = x + \epsilon K(x)$, it will be automatically an isometry transformation of the geometric manifold. We will develop a concrete example of Killing vectors in the next section.

One can ask how many linearly independent Killing vectors exist for a manifold, in each point. We will tackle this problem later using tools of Riemannian geometry. We will see that for a D-dimensional geometric manifold there are at most $D(D+1)/2$ distinct Killing vectors. The Killing vector fields of a given geometric manifold $\mathcal{M}$ constitute a real Lie algebra, with respect to the commutator of vector fields, as we will show in Section 9.6. This is a subalgebra of the Lie algebra of all vector fields $\mathcal{X}\mathcal{M}$, which we have introduced in the first chapter.

Conformal Killing Vectors

Let us now consider conformal transformations on a manifold. A transformation acting on the manifold is a conformal transformation if the condition 3.43 is valid. Because we want to study small transformations, we write the conformal factor $\Psi(x)$ as

$$\Psi(x) = \exp \sigma(x), \tag{3.49}$$

using a real scalar function $\sigma(x) = \epsilon\tau(x)$, with $|\epsilon| \ll 1$. Expanding in powers of ϵ it is

$$\Psi(x) = 1 + \epsilon\tau(x) + \mathcal{O}(\epsilon^2). \tag{3.50}$$

By inserting the transformation $x' = x + \epsilon K$ into the condition 3.43, we obtain

$$\mathcal{L}_K g_{\mu\nu} = 2\tau g_{\mu\nu}. \tag{3.51}$$

Taking the trace of both sides, the scalar function $\tau(x)$ is derived to be

$$\tau = \frac{1}{2D}\left(K^\alpha g^{\beta\gamma}\partial_\alpha g_{\beta\gamma} + 2\partial_\alpha K^\alpha\right). \tag{3.52}$$

This yields the defining formula for a *conformal Killing vector field* (abbreviated *CKVF*),

$$\boxed{\mathcal{L}_K g_{\mu\nu} = \frac{1}{D}\left(K^\alpha g^{\beta\gamma}\partial_\alpha g_{\beta\gamma} + 2\partial_\alpha K^\alpha\right) g_{\mu\nu}} \tag{3.53}$$

The above equation is called the *conformal Killing equation* and expresses the condition that the vector field $K^\mu(x)$ must satisfy for the associated transformation to be a conformal symmetry of the manifold. We will discuss the conformal Killing equation again when we will approach conformal symmetry from a group-theoretical standpoint. In Section 9.6, we will demonstrate that the conformal Killing vectors constitute a real Lie algebra with respect to the commutator of vector fields. In the next section we discuss a special case of conformal transformations, the so-called scale transformations.

Lie Symmetry Condition for Arbitrary Tensors

The symmetry of a geometric manifold $(\mathcal{M}, g)$ is expressed by the condition $\mathcal{L}_K g_{\mu\nu} = 0$, in which $K^\mu(x)$ is a Killing vector field. For such a vector field $K^\mu(x)$, the metric $g_{\mu\nu}(x)$ remains invariant. We can generalize this notion of symmetry to an arbitrary tensor field $\phi_i(x)$ by demanding that the condition

$$\mathcal{L}_K \phi_i = 0 \tag{3.54}$$

holds. As usually, the vector field $K^\mu(x)$ defines the transformation $x' = x + \epsilon K$ on the coordinates. Let us see why the condition $\mathcal{L}_K \phi_i = 0$ is a suitable requirement to express invariance of the tensor field $\phi_i(x)$. Suppose we choose our coordinate system $\{\partial_\mu\}$ to be such that locally $\partial_1 = K$, which is something can always achieve. Then, the components of K are $K^1(x) = 1$ and $K^2(x) = \ldots = K^D(x) = 0$. The Lie derivative $\mathcal{L}_K \phi_i$ of the tensor field $\phi_i(x)$ reduces essentially to the partial derivative:

$$\mathcal{L}_K \phi_i = K^1 \frac{\partial \phi_i}{\partial x^1}. \tag{3.55}$$

If the tensor components $\phi_i(x)$ are independent of the coordinate x^1, the Lie derivative $\mathcal{L}_K \phi_i$ vanishes. This makes it clear that a vector field $K^\mu(x)$ with the property $\mathcal{L}_K \phi_i = 0$ in fact defines locally the directions for which the tensor field $\phi_i(x)$ remains invariant.

3.4   Euclidean and Scale Transformations

Euclidean Transformations

A first example of a geometric symmetry is provided within the 3-dimensional Euclidean space $\mathbb{E}^3$ equipped with the standard constant metric $\delta_{jk} = \text{diag}(1,1,1)$. The Killing equation for the sought after Killing vectors $K_j(x)$ is

$$\partial_j K_k + \partial_k K_j = 0. \tag{3.56}$$

One possible solution for this equation is $K^j(x) = c^j$, with a constant vector c^j corresponding to *translations*. We have the freedom of 3 parameters to define a translation. The Killing equation states that $K_j(x)$ is at most linear in the coordinates x^j, as it is

$$\partial_j \partial_k K_l = 0. \tag{3.57}$$

This last equation is derived from the Killing equation by differentiation and cyclic permutation of the indices. Hence, we can make the ansatz $K^j(x) = A^j{}_k x^k$. By inserting this into the Killing equation we obtain the condition $A_{jk} = -A_{kj}$. This type of transformation corresponds to *rotations*, with 3 free parameters. The translations and rotations combined correspond to 6 linearly independent Killing vectors. Written explicitly, the *Killing vector fields for translations* are

$$K_j = \delta_j^k \partial_k, \tag{3.58}$$

with the index $j = 1, 2, 3$ counting the translational degrees of freedom. The *Killing vector fields for rotations* have the form

$$K_j = \epsilon_{jkl} x^k \partial^l, \tag{3.59}$$

where the index $j = 1, 2, 3$ counts the rotational degrees of freedom. Actually, we have reached the maximum number of distinct symmetry transformations. The *Euclidean transformations*

$$x'^j = c^j + A^j{}_k x^k \tag{3.60}$$

are the most general symmetry transformations of Euclidean space $(\mathbb{E}^3, \delta)$. The Euclidean transformations, also called *motions*, constitute a group whose group-theoretical aspects will be discussed in Chapter 10.

Parametrizing Rotations

As a technical remark, for later use, we note here how rotations in the plane look like and how they can be parametrized. The graphics 3.1 below displays a counterclockwise rotation by an angle θ. This an active transformation of the point p. Obviously, the change of the coordinates of the point p can be written as a matrix multiplication as

$$\begin{pmatrix} x' \\ y' \end{pmatrix} = \begin{pmatrix} \cos\theta & -\sin\theta \\ \sin\theta & \cos\theta \end{pmatrix} \begin{pmatrix} x \\ y \end{pmatrix}. \tag{3.61}$$

We will use this convention of the angle parameter θ throughout this book.

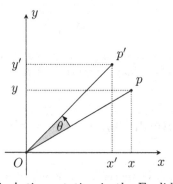

Figure 3.1: Active rotation in the Euclidean plane

Scale Transformations

Let us consider the case of the conformal manifold $(\mathbb{R}^D, [g])$, where each member metric $g_{\mu\nu}$ of the equivalence class $[g]$ is obtained from the Euclidean metric $\delta_{\mu\nu}$ by multiplication with a *constant* factor ω^2, with ω being a strictly positive real number, i.e.

$$g_{\mu\nu} = \omega^2 \delta_{\mu\nu}. \tag{3.62}$$

The corresponding conformal Killing equation is

$$\partial_\mu K_\nu + \partial_\nu K_\mu = 2\delta_{\mu\nu}. \tag{3.63}$$

The constant vector $K^\mu = c^\mu$ is, in general, not a solution of this equation. By applying the ansatz $K^\mu(x) = M^\mu{}_\nu x^\nu$, we obtain for the tensor $M^\mu{}_\nu$ the form

$$M_{\mu\nu} = \delta_{\mu\nu} + A_{\mu\nu}, \tag{3.64}$$

with the part $A_{\mu\nu}$ being antisymmetric, $A_{\mu\nu} = -A_{\nu\mu}$. The part $A_{\mu\nu}$ defines D-dimensional rotations, which are possible symmetry transformations of this manifold, corresponding to $D(D-1)/2$ free parameters. The part $\delta_{\mu\nu}$ allows only one multiplicative parameter. The associated Killing vector field is

$$K = x^\mu \partial_\mu. \tag{3.65}$$

This Killing vector is called the *dilatation vector* and it generates the *scale transformations*. A pure scale transformation, without a rotation, has the form

$$x'^\mu = sx^\mu, \tag{3.66}$$

with the real parameter s fixing the scale applied. Scale transformations are symmetry transformations of the conformal manifold $(\mathbb{R}^D, [g])$.

Further Reading

We have motivated and introduced the Lie derivative from an analytic point of view. A more geometric definition, using the notion of one-parameter groups of transformations, can be found in the books of Frankel [30] and Nakahara [58]. An accessible discussion of differential geometry with an emphasis on symmetry principles provides the text of Dubrovin, Fomenko, and Novikov [25].

II

Mechanics and Symmetry

4

Newtonian Mechanics

In this first chapter on classical physics, we consider mechanics as introduced by Newton. An essential component of classical Newtonian mechanics is time, which is postulated as independent of space and independent of the dynamics of the physical system. The resulting spacetime is the Galileian spacetime. We introduce inertial reference frames as the natural reference systems in which can we measure mechanical quantities and express mechanical laws. The formulation of the foundational Newtonian laws is followed by a discussion of systems of many particles and conserved quantities. In the last section, we discuss Newton's law of universal gravitation in its discrete and continuous formulation and demonstrate its equivalence to the shell theorem.

4.1 Galileian Spacetime

Galileian Spacetime

We begin our journey in physics with a definition of classical mechanical space and time. Physical *space* in Newtonian mechanics (*Sir Isaac Newton*) is independent of the physical processes taking place in it. This space is absolute in the sense that it is the same for every observer. We model this space by $\mathbb{E}^3$, the 3-dimensional Euclidean affine-linear space. This space has an infinite extension in all directions. We assume physical space to be *homogeneous* and *isotropic*. Homogeneous means that no particular point in space is singled out. Isotropic means that space possesses no preferred direction. *Time* in Newtonian mechanics is a quantity *independent of space*. Time flows independently of any physical process and its steady flow is the same for every observer. All time points are a priori equivalent, in other words, time is *homogeneous*. We model time as the continuum $\mathbb{R}$. The physical combination of space and time is called *spacetime*. In the present case of Newtonian mechanics, spacetime is given by the Cartesian product $\mathbb{R} \times \mathbb{E}^3$, with the metric

$$g_{kl} = \delta_{kl} = \text{diag}(1,1,1,1). \tag{4.1}$$

This classical spacetime is called *Galileian spacetime*, denoted $\mathbb{G}_4$ (*Galileo Galilei*). Galilei spacetime represents the "stage" on which all classical mechanical processes take place.

Observers and Inertial Reference Systems

In classical mechanics, an *observer* can carry out observations and measurements with an arbitrary precision. An observer measuring a physical system, and in particular space and

DOI: 10.1201/9781003087748-4

time, must choose a *reference system* (or *reference frame* or *coordinate system*) to carry out the measurements. Which are the appropriate, or natural, reference systems in classical mechanics? The natural reference systems are the ones in which the mechanical laws take their most general and simplest form. First of all, the sought after reference frames need to respect the structure of spacetime. This means that we seek for reference systems in which space is homogeneous and isotropic and in which time is homogeneous. These are the so-called *inertial reference systems*. Indeed, the mechanical laws can be formulated generally and in their simplest form by using inertial reference systems. For instance, inertial frames are the ones in which a particle experiencing no forces stays at rest or moves with constant velocity. Let us point out here that the existence of inertial frames is an empirical fact. An inertial frame can be constructed, to a very good approximation, by orienting it relative to the fixed stars visible to the eye. Per default, whenever we measure physical processes in classical mechanics, we will assume that we have chosen an inertial reference system. If a reference system S_2 moves with constant velocity with respect to an inertial reference system S_1, then the reference system S_2 is also an inertial system. Thus, an equivalence class of reference systems is defined, the class of the inertial reference systems.

Galilei Principle of Relativity

When we formulate physical laws, we want them to be universally valid and independent of our choices of reference systems. This is made precise in the *Galileian principle of relativity*, which states:

> *All mechanical laws have the same form in all inertial reference systems.*

This principle is a condensed form of the statement that mechanical laws are independent of the inertial observer, and that they are valid and unaltered at different places and at different times. The principle of relativity is very general and will be used again as a postulate in relativistic physics. In special relativity, we will require the invariance of *all* fundamental physical laws, not only of mechanical laws, for all inertial systems. A dramatic step further, in general relativity, the principle of relativity will be extended to include *all* reference systems, not only the inertial ones, thus allowing accelerated frames. There, we will demand that all physical laws are independent of any selection of reference system. We speak then of *general covariance*. Classical physics can be rendered to be generally covariant and later in this book we will show how this is achieved.

Kinematics: Position, Velocity and Acceleration

A technical question arises when measuring space, time and other physical quantities: we must clearly draw a distinction between a tensorial quantity, which is an invariant geometric object, on one hand, and the components or coordinates of this tensor on the other hand, which transform depending on the reference system chosen. We will use the boldface notation $\boldsymbol{x}$, $\boldsymbol{v}$, etc. for Euclidean vectors of $\mathbb{E}^3$ and x^k, v^k, etc. with $k = 1, 2, 3$ for their Cartesian coordinates in a reference frame. Classical mechanics uses the idealization of a *point particle*. The point particle has no extent in space and its position at any instant of time t is given by the *position vector* $\boldsymbol{x}(t)$. The rate of change of the position vector, given as the derivative with respect to the time coordinate t, is called the *velocity* $\boldsymbol{v}(t)$ of the point particle,

$$\boldsymbol{v}(t) \equiv \dot{\boldsymbol{x}}(t) \equiv \frac{\mathrm{d}\boldsymbol{x}}{\mathrm{d}t}(t). \tag{4.2}$$

The rate of change of the velocity is called the *acceleration* $\boldsymbol{b}(t)$ of the point particle,

$$\boldsymbol{b}(t) \equiv \dot{\boldsymbol{v}}(t) \equiv \ddot{\boldsymbol{x}}(t) \equiv \frac{\mathrm{d}^2\boldsymbol{x}}{\mathrm{d}t^2}(t). \tag{4.3}$$

For the description of motion, second-order differential equations are fully sufficient. In fact, most of the dynamical equations in physics are second-order differential equations.

Galilei Transformations

Since we now have introduced Galilean spacetime and inertial frames, we can ask how we can identify the transformations from one inertial frame to another inertial frame. They represent a well-defined freedom in our classical mechanical theory. We can describe the same mechanical system in one or in another inertial reference frame. This corresponds to a coordinate transformation, i.e. the passive view of transformations of spacetime. We can take, equally well, the active point of view for such spacetime transformations. Then we can ask which classical mechanical laws are invariant, and as such are acceptable, under corresponding active transformations.

In order to derive the transformation formula between inertial reference frames, let us start with the following simple situation. See the figure 4.1 below. We consider two inertial

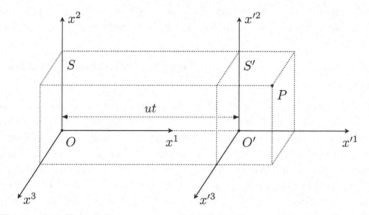

Figure 4.1: Event P as measured in the reference systems S and S'

reference frames S and S' having their three axes in parallel and aligned with the same orientation. We imagine that the reference frame S', as seen within the reference frame S, moves along the x^1-axis with the constant velocity $\boldsymbol{u} = u\,\boldsymbol{e}_1$. Further, we assume that the two reference frames completely coincide at the time point $t = t' = 0$. A particular spacetime point P, a so-called *event*, with spacetime coordinates (x^1, x^2, x^3, t) in the frame S has the spacetime coordinates (x'^1, x'^2, x'^3, t') in the frame S' given classically by

$$\left.\begin{aligned}
x'^1 &= x^1 - u\,t \\
x'^2 &= x^2 \\
x'^3 &= x^3 \\
t' &= t
\end{aligned}\right\} . \tag{4.4}$$

Written in vectorial form, this passive transformation reads

$$\left.\begin{aligned}
\boldsymbol{x}' &= \boldsymbol{x} - \boldsymbol{u}\,t \\
t' &= t
\end{aligned}\right\} . \tag{4.5}$$

This type of transformation is called a *Galilei boost*. Changing from the passive point of view to the active point of view, the Galilei boost of the position vector reads $\boldsymbol{x}' = \boldsymbol{x} + \boldsymbol{u}\,t$. Since there is *isotropy of space*, we can allow also *rotations* of the positions, $\boldsymbol{x}' = R\,\boldsymbol{x}$, described by rotation operators R, with the property $R^T R = 1$. In addition, because of *homogeneity of space*, we can allow *translations in space*, $x' = x + a$, described by constant translation vectors a. Finally, the time coordinate can be transformed as $t' = t + \tau$, with a constant *translation in time* τ, due to *homogeneity of time*. Thus, we reach to the sought after transformations between inertial frames, the *Galilei transformations* of spacetime:

$$\boxed{\begin{aligned} x &\mapsto & x' = R\,x + u\,t + a \\ t &\mapsto & t' = t + \tau \end{aligned}} \tag{4.6}$$

Let us note that the space part and the time part of a Galilei transformation are independent of each other and do not mix. We should also emphasize that all parameters of a Galilei transformation, R, u, a, and τ, are constants in time. In this book, we will not consider space reflections $x \mapsto -x$ (the so-called *parity transformations*) and *time reversal*, $t \mapsto -t$. Later in this chapter, however, we will study the behavior of mechanical systems under scale transformations.

Per construction, the Galilei transformations conserve the homogeneity of space and time and the isotropy of space. As such they leave the structure of Galilei spacetime intact. Formulated differently:

> *The Galilei transformations are the symmetry transformations of Galileian spacetime* $\mathbb{G}_4$.

The Galilei symmetry is not only a useful property to understand the structure of Galilei spacetime, but it also imposes constraints on how to construct a dynamical theory obeying this symmetry. Let us see how this works. Arguably, the primary goal in classical mechanics is to determine the position of every particle at each time point, i.e. to know the curve $x = x(t)$ for every constituent. We can ask if the use of the notions of position and velocity are enough to achieve this. Under a Galilei transformation, it is $dt' = dt$, so that $dx'/dt' = R\,dx/dt + u$ and $d^2x'/dt'^2 = R\,d^2x/dt^2$. We see that only the acceleration remains invariant under Galilei boosts and under translations in space and time. Positions and velocities are not invariant. Thus, if we want our mechanical laws to be invariant under general Galilei transformations, we must ensure that the dynamical equations will be comprised of second-order time derivatives and vectorial entities.

4.2　Newton's Laws of Mechanics

Newton's 1st Law: Inertia and Mass

Before we formulate the dynamics of classical mechanics, we should state what happens when no interaction takes place. This is described in *Newton's 1st law*, the *law of inertia*:

> *A particle, on which no net external forces act, remains at rest or in uniform motion.*

This means that such a point particle remains at rest or moves with constant velocity in the chosen inertial reference frame. This law is invariant under Galilei transformations. When a point particle, on which no forces act upon, is described by the position vector x with $\dot{x} = \text{const}$ in one inertial frame, then in another inertial frame it is $\dot{x}' = \text{const}$ too. The law of inertia may seem trivial at a first sight but it has deep connections to the notions of space and time and to the definition of what the inertial mass really is. For our purposes here, we define the *inertial mass* of a point particle to be simply an intrinsic property measuring the resistance of the particle against changes of its momentary state of motion. Mathematically, the inertial mass is a positive scalar quantity m and for simplicity we will assume it to be a constant in time.

Newton's 2nd Law: Dynamics

We come to the formulation of *Newton's 2nd law*, the *law of dynamics*, which is central to the theory of classical mechanics. It reads

$$m\,\ddot{\boldsymbol{x}}(t) = \boldsymbol{F}(\boldsymbol{x}(t), \dot{\boldsymbol{x}}(t), t) \qquad (4.7)$$

On the lhs of the equation, there is the product of the inertial mass m of the point particle and its momentary acceleration $\ddot{\boldsymbol{x}}(t)$. This equals the *force* $\boldsymbol{F}(\boldsymbol{x}(t), \dot{\boldsymbol{x}}(t), t)$ on the rhs, which is acting on the point particle. The force generally depends on the position and velocity of the point particle and it can have also an explicit time dependence. Newton's dynamical equation for a particle represents a system of three second-order ordinary differential equations (ODEs). Given the initial position and velocity of the particle, its motion is determined for all times. This is the *classical determinism*, which is a common attribute to the entire realm of classical physics. Newton's 2nd law can be viewed as a defining equation for the notion of force acting on the point particle. The 2nd law does not give the detailed form, nature, or origin of the force. Some examples of forces are the gravitational force, the electrostatic Coulomb force, the electromagnetic Lorentz force, central forces as gradients of potentials, frictional forces, and so on. When the force is known, then the practical task in mechanics is to determine the curve $\boldsymbol{x}(t)$ of the particle.

We define the *momentum* $\boldsymbol{p}(t)$ of a particle as the product of its inertial mass times its velocity, $\boldsymbol{p}(t) \equiv m\,\dot{\boldsymbol{x}}(t)$. With this definition, Newton's force law reads

$$\dot{\boldsymbol{p}}(t) = \boldsymbol{F}(\boldsymbol{x}(t), \dot{\boldsymbol{x}}(t), t), \qquad (4.8)$$

i.e. the rate of change of momentum equals the net force acting. The form 4.8 of Newton's law is actually a bit more general than in the form 4.7, since it includes the case the mass is variable in time.

Newton's 3rd Law: Actio = Reactio

Suppose we have two point particles, labeled a and b, interacting with each other. *Newton's 3rd law* states that, at any instant of time, if a particle b exerts a force $\boldsymbol{F}_{ab}$ on a particle a, then the particle a exerts a force $\boldsymbol{F}_{ba}$ on the particle b, which is the exact opposite,

$$\boldsymbol{F}_{ab} = -\boldsymbol{F}_{ba} \qquad (4.9)$$

If no other external forces exist for these two particles, then this bipartite system is closed and there is no net force remaining within it. We want to point out that Newton's 3rd law establishes a one-to-one correspondence between pairs of particles. Also, we emphasize that the interaction is meant to be instantaneous, there is no "signal delay" between the changes in one particle or the other.

Galilei Invariance of Newtonian Dynamics

Now we demonstrate that Newton's 2nd law is invariant under Galilei transformations, and as such it is an admissible physical law within Galilei spacetime. We consider a point particle with constant mass m and position vector $\boldsymbol{x}(t)$, the later being a solution of the equation of motion 4.7 with the force $\boldsymbol{F}(\boldsymbol{x}(t), \dot{\boldsymbol{x}}(t), t)$. Now the solution $\boldsymbol{x}(t)$ is actively Galilei-transformed to obtain

$$\boldsymbol{x}(t) \;\mapsto\; [\boldsymbol{x}(t)]' = R\,\boldsymbol{x}(t) + \boldsymbol{u}\cdot t + \boldsymbol{a}. \qquad (4.10)$$

If we use the time variable $t' = t + \tau$, instead of t, we obtain the fully transformed solution $\boldsymbol{x}'(t')$, given as

$$\boldsymbol{x}'(t') \equiv [\boldsymbol{x}(t(t'))]' = R\,\boldsymbol{x}(t' - \tau) + \boldsymbol{u}\cdot(t' - \tau) + \boldsymbol{a}. \qquad (4.11)$$

The question is now if the new curve $\boldsymbol{x}'(t')$ is a solution of Newton's 2nd law, with all quantities being Galilei-transformed. First we note that

$$\frac{\mathrm{d}^2}{\mathrm{d}t'^2} = \frac{\mathrm{d}^2}{\mathrm{d}t^2} \tag{4.12}$$

and thus

$$\frac{\mathrm{d}^2}{\mathrm{d}t'^2}\,\boldsymbol{x}'(t') = R\,\frac{\mathrm{d}^2}{\mathrm{d}t^2}\,\boldsymbol{x}(t). \tag{4.13}$$

Now we require that the Newtonian force transforms as

$$\boldsymbol{F}'(\boldsymbol{x}'(t'),\dot{\boldsymbol{x}}'(t'),t') = R\,\boldsymbol{F}(\boldsymbol{x}(t),\dot{\boldsymbol{x}}(t),t), \tag{4.14}$$

which is an expression of the vector property of the force. With this transformation behavior, we obtain the law

$$m\frac{\mathrm{d}^2}{\mathrm{d}t'^2}\,\boldsymbol{x}'(t') = \boldsymbol{F}'(\boldsymbol{x}'(t'),\dot{\boldsymbol{x}}'(t'),t') \tag{4.15}$$

for the transformed quantities. This is exactly what we wanted to prove. So we can state:

Newton's 2nd law is Galilei invariant.

This is our first example of a physical law being invariant under specific (spacetime) transformations. We will encounter many more examples later.

Mechanical Quantities

In mechanics, and in the rest of physics, one needs a few more quantities, so let us introduce them. In order to describe rotational motion around a given point, typically the origin of the reference frame, one introduces *moments*. For a single point particle with position $\boldsymbol{x}$ and momentum $\boldsymbol{p}$, the *angular momentum* $\boldsymbol{L}$ is defined as

$$\boldsymbol{L} \equiv \boldsymbol{x} \times \boldsymbol{p}, \tag{4.16}$$

using the cross product in three dimensions. If the particle experiences a force $\boldsymbol{F}$, then around the origin it experiences a *torque* (or *moment of force*) $\boldsymbol{N}$, defined as

$$\boldsymbol{N} \equiv \boldsymbol{x} \times \boldsymbol{F}. \tag{4.17}$$

Both, angular momentum and torque, depend on the origin of the reference frame chosen. The equation of motion for the angular momentum is

$$\frac{\mathrm{d}\boldsymbol{L}}{\mathrm{d}t} = \boldsymbol{N}, \tag{4.18}$$

in full analogy to Newton's 2nd law. A point particle in motion can be described by its velocity, its momentum and by its *kinetic energy* T, which is defined as

$$T \equiv \frac{1}{2}m\boldsymbol{v}^2. \tag{4.19}$$

The temporal change of the kinetic energy is given by

$$\frac{\mathrm{d}T}{\mathrm{d}t} = \boldsymbol{F} \cdot \boldsymbol{v}, \tag{4.20}$$

where $\boldsymbol{F}$ is the force applied to the particle. The quantity $\boldsymbol{F} \cdot \boldsymbol{v}$ on the rhs is called the *power* generated by the force. Besides kinetic energy, a particle can possess also a *potential energy* $V(x)$, which here is meant to depend solely on the particle's position, but not on the velocity or explicitly on time. The potential energy can be used to derive the applied force on the particle in the form

$$F = -\nabla V(x), \tag{4.21}$$

i.e. the force is minus the spatial gradient of the potential energy. Forces which can be derived in this way are called *conservative*. Sometimes, instead of potential energy one speaks more loosely about the potential, which however is a slightly different quantity. We will discuss this later within Newtonian gravity. The potential energy is required to stay unchanged under Galilei transformations. Furthermore, under scale transformations of the argument, the potential energy is required to be a homogeneous function. Let us consider such a *scale transformation* of the position vector,

$$x \mapsto x' = a\,x, \tag{4.22}$$

where a is a positive real parameter. The potential $V(x)$ is supposed to transform as

$$V(x) \mapsto V'(x') = V(a\,x) = a^{-d_V}\,V(x), \tag{4.23}$$

with the *scaling dimension* d_V of the function $V(x)$. We will explain this type of transformation behavior in more detail when we discuss conformal transformations in Chapter 14. For now, we are sufficiently prepared to study the scale dependency of Newtonian dynamics.

Scale Invariance

We consider *scale transformations* of space and time. We can ask what the equations of motion tell us when we are free to rescale the space and time quantities. First let us rescale space and time as

$$\left.\begin{array}{rcl} x' &=& a\,x \\ t' &=& b\,t \end{array}\right\}, \tag{4.24}$$

with the positive real parameters $a > 0$ and $b > 0$. Under such a transformation it is

$$\frac{\mathrm{d}^2 x'}{\mathrm{d}t'^2} = \frac{a}{b^2}\frac{\mathrm{d}^2 x}{\mathrm{d}t^2}. \tag{4.25}$$

We assume that the force is given as gradient of a potential,

$$F = -\frac{\partial}{\partial x}V(x), \tag{4.26}$$

with $V(x)$ being a homogeneous function with scaling dimension d_V, i.e.

$$V'(a\,x) = a^{-d_V}\,V(x). \tag{4.27}$$

Then we obtain for the transformed force

$$F' = -\frac{\partial}{\partial x'}V'(x') = a^{-1-d_V}\,F. \tag{4.28}$$

So, assuming the constancy of masses, the Newtonian dynamics is invariant if and only if

$$b = a^{1+\frac{1}{2}d_V}. \tag{4.29}$$

If we scale the lengths as $x' = a\,x$, we need to scale the time as $t' = a^{1+\frac{1}{2}d_V}t$, in order to keep the dynamics valid. For the ratios of the time variable and the space variables, we obtain the relation

$$\left(\frac{t'}{t}\right) = \left(\frac{x'}{x}\right)^{1+\frac{1}{2}d_V}. \tag{4.30}$$

This somewhat abstract result will be put into practice soon.

4.3   Systems of Particles and Conserved Quantities

System of Point Particles

We consider a *system of particles* consisting of N distinguishable mass points m_α, $\alpha = 1, \ldots, N$, interacting with each other and with the external world. The *state* of such a classical mechanical system is given by the set of all positions $\boldsymbol{x}_\alpha(t)$ and all velocities $\dot{\boldsymbol{x}}_\alpha(t)$ of all particles at any instant t of time. This includes the principal ability of an observer to get arbitrary precise information about the positions and velocities. Knowing the state of a mechanical system at one instant of time, one is principally able to calculate the state of the system for each other time point, in the future or in the past. Within the framework of classical mechanics, the system is completely deterministic. At a time when the physical world was thought to be a mechanical system, this type of determinism caused a clash with the notion of human free will.

Forces and Potential Energy

In order to determine the time evolution of our system of particles, we start with the Newtonian forces. Each particle m_α, $\alpha = 1, \ldots, N$, moves according to its equation of motion

$$m_\alpha \ddot{\boldsymbol{x}}_\alpha(t) = \boldsymbol{F}_\alpha(\boldsymbol{x}_1(t), \ldots, \boldsymbol{x}_N(t), \dot{\boldsymbol{x}}_1(t), \ldots, \dot{\boldsymbol{x}}_N(t), t). \tag{4.31}$$

The force $\boldsymbol{F}_\alpha$ applied on each particle can be split into two parts, an *external force* $\boldsymbol{F}_\alpha^{\text{ext}}$ and *internal force* $\boldsymbol{F}_\alpha^{\text{int}}$,

$$\boldsymbol{F}_\alpha = \boldsymbol{F}_\alpha^{\text{ext}} + \boldsymbol{F}_\alpha^{\text{int}}. \tag{4.32}$$

The external force $\boldsymbol{F}_\alpha^{\text{ext}}$ is required to depend only on the position $\boldsymbol{x}_\alpha$, but not on the positions and velocities of the other particles. When all external forces vanish, the system is called *closed*. The internal force $\boldsymbol{F}_\alpha^{\text{int}}$ is the sum of all forces $\boldsymbol{F}_{\alpha\beta}$ exerted from all particles $\beta \neq \alpha$ of the system on the particle of interest α, i.e.

$$\boldsymbol{F}_\alpha^{\text{int}} = \sum_{\beta \neq \alpha} \boldsymbol{F}_{\alpha\beta}. \tag{4.33}$$

The sum is over all values $\beta = 1, \ldots, N$, with $\beta \neq \alpha$. All single forces $\boldsymbol{F}_{\alpha\beta}$ exerted from particle β on particle α are supposed to be so-called *central forces*. This means that the two-particle forces $\boldsymbol{F}_{\alpha\beta}$ are of the form

$$\boldsymbol{F}_{\alpha\beta} = f_{\alpha\beta}(|\boldsymbol{x}_\alpha - \boldsymbol{x}_\beta|) \frac{\boldsymbol{x}_\alpha - \boldsymbol{x}_\beta}{|\boldsymbol{x}_\alpha - \boldsymbol{x}_\beta|}, \tag{4.34}$$

with scalar functions $f_{\alpha\beta} = f_{\beta\alpha}$, for all α and β. Evidently, for these central forces it is $\boldsymbol{F}_{\alpha\beta} = -\boldsymbol{F}_{\beta\alpha}$, as we would expect. Central forces can always be derived from a potential energy as gradients, which means that they are conservative forces. The precise relation is

$$\boldsymbol{F}_{\alpha\beta} = -\boldsymbol{\nabla}_\alpha V_{\alpha\beta}(|\boldsymbol{x}_\alpha - \boldsymbol{x}_\beta|), \tag{4.35}$$

where $\boldsymbol{\nabla}_\alpha = \partial/\partial\boldsymbol{x}_\alpha$ is the gradient operator with respect to the position $\boldsymbol{x}_\alpha$. We can see easily that $f_{\alpha\beta}(x) = -V'_{\alpha\beta}(x)$ and that $V_{\alpha\beta}(x) = V_{\beta\alpha}(x)$, for all α and β. The potential energy $V_{\alpha\beta}(x)$ is a scalar function under Galilei transformations and a homogeneous function under scale transformations. Each function $V_{\alpha\beta}$ provides a description of the interaction between the two particles α and β. Note that in the Newtonian framework, all interactions are instantaneous, despite the fact that we use the mathematical tool of a field function $V_{\alpha\beta}(x)$. For the internal force $\boldsymbol{F}_\alpha^{\text{int}}$ on the particle a, we can write

$$\boldsymbol{F}_\alpha^{\text{int}} = -\boldsymbol{\nabla}_\alpha \sum_{\beta \neq \alpha} V_{\alpha\beta}(|\boldsymbol{x}_\alpha - \boldsymbol{x}_\beta|). \tag{4.36}$$

Typically we encounter the situation where also the external force F_α^{ext} is conservative and given as a gradient,

$$F_\alpha^{\text{ext}} = -\boldsymbol{\nabla}_\alpha V_\alpha(\boldsymbol{x}_\alpha). \tag{4.37}$$

Finally, let us take the sum of all forces on all particles $\alpha = 1, \ldots, N$ of the system under consideration. The resulting *total force* F is

$$\boldsymbol{F} \equiv \sum_\alpha \boldsymbol{F}_\alpha = \sum_\alpha \boldsymbol{F}_\alpha^{\text{ext}} \equiv \boldsymbol{F}^{\text{ext}}. \tag{4.38}$$

This means that the total force F on the system is equal to the *total external force* F^{ext}, since the internal forces cancel each other exactly. In the following we will identify the fundamental quantities of the particle system which are conserved in time.

Total Momentum

We define the *total momentum* P of the system to be the sum over all single-particle momenta,

$$\boldsymbol{P} \equiv \sum_\alpha \boldsymbol{p}_\alpha. \tag{4.39}$$

By using the basic equation 4.31, the dynamical equation for the total momentum is readily obtained to be

$$\frac{d\boldsymbol{P}}{dt} = \boldsymbol{F}^{\text{ext}}. \tag{4.40}$$

I.e. the rate of change of the total momentum of the system is equal to the total external force exerted on the system. When the system is closed, the total momentum is conserved,

$$\boxed{\boldsymbol{F}^{\text{ext}} = \boldsymbol{0} \quad \Rightarrow \quad \boldsymbol{P} = \text{const}} \tag{4.41}$$

Total Angular Momentum

The *total angular momentum* L of the system is the sum of all single-particle angular momenta,

$$\boldsymbol{L} \equiv \sum_\alpha \boldsymbol{x}_\alpha \times \boldsymbol{p}_\alpha. \tag{4.42}$$

The dynamical equation for the total angular momentum is easily derived to be

$$\frac{d\boldsymbol{L}}{dt} = \boldsymbol{N}^{\text{ext}}, \tag{4.43}$$

provided that all internal forces of the system are central forces. The quantity N^{ext} is the *total external torque* applied on the system, and is defined as the sum of all single external momenta of force,

$$\boldsymbol{N}^{\text{ext}} \equiv \sum_\alpha \boldsymbol{x}_\alpha \times \boldsymbol{F}_\alpha^{\text{ext}}. \tag{4.44}$$

When the total external torque vanishes, e.g. when the system is closed, the total angular momentum is conserved in time,

$$\boxed{\boldsymbol{N}^{\text{ext}} = \boldsymbol{0} \quad \Rightarrow \quad \boldsymbol{L} = \text{const}} \tag{4.45}$$

Total Energy

For examining the energy of the total system, we use once again the equation of motion 4.31, which we contract with the velocity $\boldsymbol{v}_\alpha$ and where we sum over all particles, to obtain

$$\sum_\alpha m_\alpha \dot{\boldsymbol{v}}_\alpha \cdot \boldsymbol{v}_\alpha = \sum_\alpha \boldsymbol{F}_\alpha^{\text{ext}} \cdot \boldsymbol{v}_\alpha + \frac{1}{2} \sum_{\substack{\alpha,\beta \\ \alpha \neq \beta}} \boldsymbol{F}_{\alpha\beta} \cdot (\boldsymbol{v}_\alpha - \boldsymbol{v}_\beta). \tag{4.46}$$

A straightforward calculation shows that it is

$$\boldsymbol{F}_{\alpha\beta} \cdot (\boldsymbol{v}_\alpha - \boldsymbol{v}_\beta) = -\frac{\mathrm{d}}{\mathrm{d}t} V_{\alpha\beta}(|\boldsymbol{x}_\alpha - \boldsymbol{x}_\beta|). \tag{4.47}$$

By using this formula, we obtain the rate of change

$$\frac{\mathrm{d}E}{\mathrm{d}t} = \sum_\alpha \boldsymbol{F}_\alpha^{\text{ext}} \cdot \boldsymbol{v}_\alpha. \tag{4.48}$$

Here E denotes the *total energy* of the entire system of particles,

$$E \equiv T + V, \tag{4.49}$$

consisting of the *total kinetic energy* T of the system,

$$T \equiv \frac{1}{2} \sum_\alpha m_\alpha \boldsymbol{v}_\alpha^2, \tag{4.50}$$

plus the *total potential energy* V of the system,

$$V \equiv \sum_{\alpha < \beta} V_{\alpha\beta}(|\boldsymbol{x}_\alpha - \boldsymbol{x}_\beta|). \tag{4.51}$$

The equation 4.48 says that the rate of change of the total energy of the system of particles is equal to the *total power* generated by the external forces on the system. For a closed system, the energy is a constant in time,

$$\boxed{\boldsymbol{F}_\alpha^{\text{ext}} = 0, \ \forall \alpha \ \Rightarrow \ E = \text{const}} \tag{4.52}$$

In the case where the external forces are all conservative, we can take another view, enlarge our system, and incorporate the potential energies of the external forces into the total energy, which is then still conserved. In detail, we define the total energy H of the enlarged system as

$$H \equiv T + U, \tag{4.53}$$

with the total potential energy U now being

$$U \equiv \sum_\alpha V_\alpha(\boldsymbol{x}_\alpha) + \sum_{\alpha < \beta} V_{\alpha\beta}(|\boldsymbol{x}_\alpha - \boldsymbol{x}_\beta|). \tag{4.54}$$

Then, the conservation of energy reads

$$\frac{\mathrm{d}H}{\mathrm{d}t} = 0. \tag{4.55}$$

Note however that the total energy H is only defined up to an additive constant.

Galilei Momentum

The *center of mass* $\boldsymbol{X}$ of the N-particle system is defined as the weighted average position vector

$$\boldsymbol{X} \equiv \frac{1}{M} \sum_\alpha m_\alpha \boldsymbol{x}_\alpha, \tag{4.56}$$

with M being the *total mass* of the system,

$$M \equiv \sum_\alpha m_\alpha. \tag{4.57}$$

We define the *Galilei momentum* (also called the *center of mass momentum*) as

$$\boldsymbol{G} \equiv t\boldsymbol{P} - M\boldsymbol{X}, \tag{4.58}$$

where the time t in the definition is the one measured in the reference frame of choice. At $t = 0$ the Galilei momentum $\boldsymbol{G}$ is essentially the center of mass. In addition, the Galilei momentum can be written as the sum of all single-particle Galilei momenta. For the Galilei momentum $\boldsymbol{G}$, the dynamical equation is

$$\frac{\mathrm{d}\boldsymbol{G}}{\mathrm{d}t} = \boldsymbol{F}^{\text{ext}}, \tag{4.59}$$

being very similar to the rate of change of the total linear momentum. For a closed system, the Galilei momentum is conserved in time,

$$\boxed{\boldsymbol{F}^{\text{ext}} = \boldsymbol{0} \quad \Rightarrow \quad \boldsymbol{G} = \text{const}} \tag{4.60}$$

Further, in the case of a closed system, it is

$$\boldsymbol{P} = M\dot{\boldsymbol{X}}, \tag{4.61}$$

with a constant total momentum $\boldsymbol{P}$. Thus, the center of mass moves with constant velocity $\boldsymbol{P}/M$ within the frame of reference,

$$\boldsymbol{X}(t) = \boldsymbol{X}(0) + t\frac{\boldsymbol{P}}{M}. \tag{4.62}$$

Finally let us note that the Galilei momentum is invariant under Galilei boosts, $\boldsymbol{G}' = \boldsymbol{G}$. In summary, we have found ten quantities, $\boldsymbol{P}$, $\boldsymbol{L}$, E and $\boldsymbol{G}$, which remain constant during the course of time within a closed mechanical system. In the next chapter we relate these ten quantities to the ten independent Galilei transformations.

Mass Density and Current Density

For certain systems of point particles, it is useful to describe the mass distribution and its motion in space by density functions. These are field functions defined over all space and time. For a system of N point particles with masses m_α, $\alpha = 1, \ldots, N$, the *mass density* $\rho(\boldsymbol{x}, t)$ is defined as

$$\rho(\boldsymbol{x}, t) \equiv \sum_\alpha m_\alpha \delta(\boldsymbol{x} - \boldsymbol{z}_\alpha(t)), \tag{4.63}$$

where $\boldsymbol{z}_\alpha(t)$ is the trajectory of the αth particle. $\rho(\boldsymbol{x}, t)$ is a measure of mass per volume unit. The integral of the mass density over all space yields the total mass of the system. Similarly, one defines the *mass current density* $\boldsymbol{j}(\boldsymbol{x}, t)$ as the sum

$$\boldsymbol{j}(\boldsymbol{x}, t) \equiv \sum_\alpha m_\alpha \frac{\mathrm{d}\boldsymbol{z}_\alpha(t)}{\mathrm{d}t} \delta(\boldsymbol{x} - \boldsymbol{z}_\alpha(t)). \tag{4.64}$$

The 3-vector $j(x,t)$ is a measure of mass per time unit and per (perpendicular) area unit. The integral of the mass current density over all space is the total momentum of the particle system. Provided that the particles stay unchanged over the course of time, i.e. no particles are created or destroyed, the *continuity equation*

$$\boxed{\frac{\partial \rho}{\partial t} + \boldsymbol{\nabla} \cdot \boldsymbol{j} = 0}$$

(4.65)

holds. This equation is called also the *current conservation equation*. By integrating over a finite volume within the particle system, we see that the change of the mass within this volume equals the momentum flow through the surface. The detailed steps are left to the reader. (*exercise 4.1*) We will encounter the current conservation again within the relativistic context.

4.4 Gravitation and the Shell Theorem

Newton's Universal Gravitation

The *gravitational force* acts as an attractive force between any two material bodies possessing *gravitational masses* m_α and m_β and is proportional to the product of these masses and inversely proportional the square of the distance between them. We will use the experimental observation that the *inertial mass* of a body, which we used so far in the equations of motion, is equal to the *gravitational mass* of this body. If there is a point mass m_α at the position x_α and a point mass m_β at the position x_β, then the particle β will exert a gravitational force $\boldsymbol{F}_{\alpha\beta}$ on the particle α given by

$$\boxed{\boldsymbol{F}_{\alpha\beta} = -G_{\mathrm{N}} \frac{m_\alpha m_\beta}{|x_\alpha - x_\beta|^2} \boldsymbol{e}_{\alpha\beta}}$$

(4.66)

with the unit vector

$$\boldsymbol{e}_{\alpha\beta} \equiv \frac{x_\alpha - x_\beta}{|x_\alpha - x_\beta|}$$

(4.67)

pointing from β to α. This is *Newton's law of universal gravitation*, published in 1687 by Isaac Newton in his landmark work Philosophiae Naturalis Principia Mathematica. The constant of proportionality G_{N} appearing in the law is the *gravitational constant* and has the value $G_{\mathrm{N}} = 6,67408(31) \times 10^{-11} \, \mathrm{kg}^{-1} \, \mathrm{m}^3 \, \mathrm{s}^{-2}$ in SI units. The first measurement of the gravitational constant was carried out by *Henry Cavendish* in 1798. The value noted here is the latest one provided by the CODATA group in 2015. Obviously it is $\boldsymbol{F}_{\alpha\beta} = -\boldsymbol{F}_{\beta\alpha}$. The gravitational force is a central force and as such it is also conservative. For a purely gravitational closed system, the total angular momentum and the total energy are conserved quantities. The gravitational force of multiple bodies is linearly added to a total force vector. Due to this additivity and the $|x|^{-2}$ long-range character, the gravitational attraction, despite its smallness in strength, becomes the dominating force on macroscopic and on cosmic length scales. Let us remark also that the gravitational force as given in 4.66 transforms as we would expect under Galilei transformations.

One can take a different point of view and ask what is the acceleration of a mass probe m_α that experiences gravitational forces. If there is only one other mass m_β at the position x_β, then this acceleration, denoted $\boldsymbol{g}(x_\alpha)$, is

$$\boldsymbol{g}(x_\alpha) = -G_{\mathrm{N}} \frac{m_\beta}{|x_\alpha - x_\beta|^2} \boldsymbol{e}_{\alpha\beta}.$$

(4.68)

If the mass m_β and its position $\boldsymbol{x}_\beta$ are given, the above equation can be interpreted as the definition of a *field* $\boldsymbol{g}(\boldsymbol{x})$, for a certain region in space. We call $\boldsymbol{g}(\boldsymbol{x})$ the *Newtonian gravitational field strength* generated by the point mass m_β. Thus, the gravitational force acting on the particle m_α is

$$\boldsymbol{F}_{\alpha\beta} = m_\alpha\, \boldsymbol{g}(\boldsymbol{x}_\alpha). \tag{4.69}$$

Both, the gravitational force and the gravitational field, are obtained as gradients of specific quantities, to which we now turn.

Gravitational Potential

Let us first relate gravitational force and potential energy to each other. Looking at a particle m_α within a closed system of N gravitating particles, the *gravitational potential energy* $V_{\alpha\beta}$ of the single particle m_α interacting with all other particles m_β, $\beta \neq \alpha$, is

$$V_{\alpha\beta} = -G_{\mathrm{N}} \sum_{\substack{\beta=1 \\ \beta \neq \alpha}}^{N} \frac{m_\alpha m_\beta}{|\boldsymbol{x}_\alpha - \boldsymbol{x}_\beta|}, \tag{4.70}$$

i.e. the potential energy of the particle m_α is the sum of all potential energies between m_α and all other members m_β, $\beta \neq \alpha$, of the particle set. The potential energy is obviously negative, which corresponds to the property of gravity to be an attractive force. By using the gradient formula

$$\boldsymbol{\nabla}_\alpha \frac{1}{|\boldsymbol{x}_\alpha - \boldsymbol{x}_\beta|} = -\frac{\boldsymbol{e}_{\alpha\beta}}{|\boldsymbol{x}_\alpha - \boldsymbol{x}_\beta|^2}, \tag{4.71}$$

we obtain the desired relation between the gravitational force and gravitational potential energy for the particle m_α to be

$$\boldsymbol{F}_{\alpha\beta} = -\boldsymbol{\nabla}_\alpha V_{\alpha\beta}. \tag{4.72}$$

Now let us move on to the concept of the gravitational field. We consider a *mass probe*, call it m, to experience the gravitational attraction of a system of discrete particles m_β. Note that here we do not count the particle m to be part of the set of particles m_β. The acceleration $\boldsymbol{g}(\boldsymbol{x})$ that the mass probe m experiences under the gravitational influence, can be written as a gradient as

$$\boldsymbol{g}(\boldsymbol{x}) = -\boldsymbol{\nabla}\Phi(\boldsymbol{x}). \tag{4.73}$$

The scalar function $\Phi(\boldsymbol{x})$ is the *Newtonian gravitational potential* generated by the mass points m_β and is given by

$$\Phi(\boldsymbol{x}) = -G_{\mathrm{N}} \sum_\beta \frac{m_\beta}{|\boldsymbol{x} - \boldsymbol{x}_\beta|}. \tag{4.74}$$

The scalar field $\Phi(\boldsymbol{x})$ is interpreted as the gravitational potential energy per unit mass. In the case of a continuous mass distribution, described by a *mass density* $\rho(\boldsymbol{x})$, the gravitational potential becomes the integral expression

$$\Phi(\boldsymbol{x}) = -G_{\mathrm{N}} \int_{\mathbb{R}^3} \frac{\rho(\boldsymbol{y})}{|\boldsymbol{x} - \boldsymbol{y}|}\, \mathrm{d}^3 y\,. \tag{4.75}$$

One can recover the discrete formula, if the mass distribution is actually one for a discrete set of particles, like

$$\rho(\boldsymbol{x}) = \sum_\beta m_\beta\, \delta(\boldsymbol{x} - \boldsymbol{x}_\beta). \tag{4.76}$$

One can see easily that the gravitational potential satisfies the partial differential equation of Poisson, see appendix B.5,

$$\boxed{\nabla^2 \Phi(\boldsymbol{x}) = 4\pi G_{\mathrm{N}} \rho(\boldsymbol{x})} \tag{4.77}$$

This has to be understood as the *field equation of Newtonian gravitation*. The gravitational potential $\Phi(\boldsymbol{x})$ represents the *Newtonian gravitational field*. In the framework of Newtonian gravitation, it is common to consider the gravitational potential as not explicitly time-dependent.

In Section 4.2, we discussed how the potential energy $V(\boldsymbol{x})$ scales as a function of its argument. The gravitational potential $\Phi(\boldsymbol{x})$ is a homogeneous function too, with the scaling equation

$$\Phi(a\,\boldsymbol{x}) = a^{-1}\Phi(\boldsymbol{x}), \tag{4.78}$$

i.e. the scaling dimension is equal to one. Thus, if we have a solution of the equations of motion under the influence of Newtonian gravity, the solution scales like

$$\left(\frac{t'}{t}\right)^2 = \left(\frac{x'}{x}\right)^3. \tag{4.79}$$

But this is exactly *Kepler's third law of planetary motion* (*Johannes Kepler*), which states that the square of the orbital period of a planet around the sun is proportional to the cube of the major axis of its orbit. The major axis of an elliptical orbit is defined as the longest diameter of the ellipse. Kepler's third law makes a comparison of the ratios of the periods and distances for each pair of planets, i.e. for each pair of solutions of the equations of motion.

Newton's Shell Theorem

In the following, we will examine the gravitational field for spherical objects and obtain some surprising results. Specifically, we will calculate the gravitational potential and the force generated by a thin spherical shell, a homogeneously filled ball and, finally, an arbitrary spherically symmetric mass distribution. Throughout we will assume these objects to be rigid and fixed in space. The system of reference will be chosen with its origin to reside at the center of the spherical object.

First, we consider a *thin spherical shell* $S(R, \rho)$ with radius R and surface mass density (i.e. mass per unit area) ρ. The total mass M of this spherical shell is $M = 4\pi R^2 \rho$. We view a mass probe m at the distance r from the sphere's center, where r can be greater or smaller than the radius R. In the sketch 4.2 below, we depict the situation where the mass probe is at a point P outside the sphere. We will calculate the gravitational potential $\Phi(r)$ of the mass shell at the point P at a distance r from the center. To this end, we parametrize the points on the mass shell by the angles θ and φ with the ranges $0 \le \theta \le \pi$ and $0 \le \varphi < 2\pi$. What we will do is to add up all contributions of differential mass shell elements to the gravitational potential. It is convenient to consider the surface element $\mathrm{d}S = (R\sin\theta\,\mathrm{d}\varphi)(R\,\mathrm{d}\theta)$ and then integrate out the two angles. The integration over the angle φ is trivial, due to the axial symmetry of the setup. The differential mass $\mathrm{d}M$ of a surface element $\mathrm{d}S$ is

$$\mathrm{d}M = \rho R^2 \,\mathrm{d}\varphi \,\sin\theta\,\mathrm{d}\theta. \tag{4.80}$$

The differential potential $\mathrm{d}\Phi(r)$ generated by a surface element $\mathrm{d}^2 S$ at the point P is correspondingly

$$\mathrm{d}\Phi(r) = -G_{\mathrm{N}}\frac{\mathrm{d}M}{s}, \tag{4.81}$$

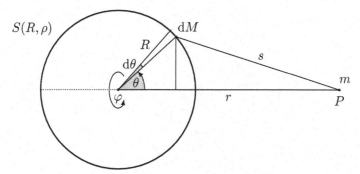

Figure 4.2: Mass probe m at the point P outside the spherical mass shell

with the distance s calculated as $s^2 = r^2 + R^2 - 2rR\cos\theta$. Now we integrate over the two angles, to cover the whole spherical shell and obtain for the potential at the point P

$$\Phi(r) = -2\pi G_{\mathrm{N}}\rho R^2 \int_0^\pi \frac{\sin\theta \, d\theta}{\sqrt{r^2 + R^2 - 2rR\cos\theta}}. \tag{4.82}$$

The integration over the angle φ yields the factor 2π above. The integral can be simplified by changing the integration variable to $u = \cos\theta$, thus obtaining

$$\Phi(r) = -\frac{G_{\mathrm{N}}M}{2} \int_{-1}^1 \frac{du}{\sqrt{r^2 + R^2 - 2rRu}}. \tag{4.83}$$

Now we need to distinct between two cases when evaluating the integral. For points P being outside or on the shell, $r \geq R$, the potential $\Phi(r)$ is

$$\Phi(r \geq R) = -\frac{G_{\mathrm{N}}M}{r}. \tag{4.84}$$

Thus, the gravitational force $F(r)$ on the mass probe m for points outside or on the shell is

$$F(r \geq R) = -\frac{G_{\mathrm{N}}Mm}{r^2}. \tag{4.85}$$

In other words, *for all points outside or on the spherical shell, this shell behaves as if its total mass were concentrated in its center.* For points P being inside the shell, $0 \leq r < R$, the result is

$$\Phi(r < R) = -\frac{G_{\mathrm{N}}M}{R} = \text{const}, \tag{4.86}$$

i.e. the gravitational potential keeps a constant value inside the spherical shell. Consequently, *for all points inside the shell, the gravitational force vanishes,*

$$F(r < R) = 0. \tag{4.87}$$

The above results for the gravitational attraction of a spherical shell constitute *Newton's shell theorem* derived for the first time in the Principia.

The next spherical object of our analysis is a *homogeneously filled ball* $B(R, \rho)$ with radius R and volume mass density ρ. The total mass M of the ball is $M = 4\pi R^3 \rho/3$. We consider again a mass probe m at a distance r from the ball's center, see the sketch 4.3 below. In order to find the gravitational potential and force, we can use Newton's shell theorem. The ball can be considered as a collection of many spherical mass shells. So for all points outside the ball, the gravitational field behaves as if all the mass were concentrated

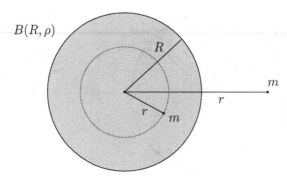

Figure 4.3: Two mass probes residing inside and outside a solid ball

in the center. For points inside the ball, only the spherical mass shells at increased depths toward the center contribute. The other mass shells, at depths toward the outside hull, do not contribute to the gravitational force. Of course, these results can be derived also by direct calculation and this is left to the reader. (*exercise 4.2*) Therefore, for points outside, or on the ball, $r \geq R$, it is again

$$\Phi(r \geq R) = -\frac{G_N M}{r}. \tag{4.88}$$

This justifies why we can treat spherical masses, for instance planets and other celestial bodies, like point masses in certain mechanical problems. For the points inside the ball, $0 \leq r < R$, the potential is given by

$$\Phi(r < R) = -\frac{G_N M}{2R} \left(3 - \left(\frac{r}{R} \right)^2 \right). \tag{4.89}$$

Inside the ball, the potential is proportional to the square of the radius. At the center point, the potential has $3/2$ of its value on the hull. For the gravitational force on a particle m inside the ball we get

$$F(r < R) = -\frac{G_N M m}{R^2} \left(\frac{r}{R} \right), \tag{4.90}$$

i.e. the gravitational force is proportional to the radius. At the points of the hull, $r = R$, the potential $\Phi(r)$ and the force $F(r)$ are continuous functions.

Let us move on to the general case of a *spherical mass distribution*. A spherical mass distribution $B(R, \rho(\boldsymbol{x}))$ can be described by a mass density $\rho(\boldsymbol{x})$ that depends only on the radial distance $x = |\boldsymbol{x}|$ from the center of the spherical geometry, i.e. $\rho(\boldsymbol{x}) = \rho_{\text{rad}}(x)$. The function $\rho_{\text{rad}}(x)$ describes the mass density as a function of the radial coordinate x. The total mass of such a spherical object with radius R is

$$M = 4\pi \int_0^R x^2 \rho_{\text{rad}}(x) \, \mathrm{d}x. \tag{4.91}$$

The gravitational potential for points outside, or on the spherical object, is once again given by $\Phi(r \geq R) = -G_N M/r$. For points inside the spherical object, with $0 \leq r < R$, the potential is given as

$$\Phi(r < R) = -4\pi G_N \int_0^R x \, \rho_{\text{rad}}(x) \, \mathrm{d}x. \tag{4.92}$$

Again, only the spherical layers toward the center contribute to the gravitational force $F(r)$.

In the last part of this section we ask the following question: what is the general form of a potential (or a force), which obeys Newton's shell theorem? It means that we do not start with the specific form of Newtonian gravitation but instead ask which form the potential (or the force) must have in order to fulfill the properties of the shell theorem. In order to solve this problem, we will work with the scalar potential, but we will discuss both, the potential and the force for space regions inside and outside the spherical shell. We start with the situation as depicted in figure 4.2 and aim to determine the potential function $\Phi(r)$, evaluated at the point P. Because we are in the comfortable position to know what the result could be, we make the ansatz

$$\Phi(r) \equiv \frac{f(r)}{r}, \tag{4.93}$$

so that we transform the initial problem into the one of determining the function $f(r)$. We integrate out all shell element contributions $s^{-1}f(s)$ to the potential at the point P over the solid angle $d\Omega = d\varphi \sin\theta \, d\theta$. Let us denote the corresponding quantity $\widetilde{\Phi}(P)$, i.e.

$$\widetilde{\Phi}(P) \equiv \int_{S(R,\rho)} \frac{f(s)}{s} \, d\Omega = 2\pi \int_0^\pi \frac{f(s)}{s} \sin\theta \, d\theta. \tag{4.94}$$

Moreover, we use that $s^2 = r^2 + R^2 - 2rR\cos\theta$ and thus $ds = s^{-1}rR\sin\theta \, d\theta$. Now we need to distinct. For points outside, or on the shell, $r \geq R$, the integral becomes

$$\widetilde{\Phi}(P) = \frac{2\pi}{rR} \int_{r-R}^{r+R} f(s) \, ds. \tag{4.95}$$

At this point we invoke the shell theorem and equate the integral $\widetilde{\Phi}(P)$ with the single value $\Phi(r)$, at distance r to the center, times the total solid angle 4π. Therefore it must be

$$\widetilde{\Phi}(P) = 4\pi\Phi(r) + 4\pi K, \tag{4.96}$$

where the constant $4\pi K$ is conveniently chosen. The two last equations give immediately the condition

$$2Rf(r) + 2rRK = \int_{r-R}^{r+R} f(s) \, ds. \tag{4.97}$$

Differentiating twice with respect to r and once with respect to R yields the condition $2f''(r) = f''(r+R) + f''(r-R)$. Differentiating further twice with respect to R leads to $f^{(4)}(r+R) + f^{(4)}(r-R) = 0$. Thus, in particular it must be

$$f^{(4)}(r) = 0. \tag{4.98}$$

Hence, $f(r)$ has the form

$$f(r) = A + Br + Cr^2 + Dr^3. \tag{4.99}$$

By inserting this expression into the condition 4.97, one obtains readily that it must be $C = 0$ and $D = KR^{-2}$. Thus, the potential $\Phi(r)$ must have the form

$$\Phi(r) = \frac{A}{r} + B + \frac{K}{R^2}r^2. \tag{4.100}$$

This is the sought after result. The first term corresponds to the Newtonian law. The second constant term is an additive normalization of the potential. The third term is interesting and corresponds to the so-called *cosmological constant*, which we will encounter later. The force $F(r)$ on a particle of mass m at the point P is given by the formula

$$F(r) = -\frac{G_{\mathrm{N}}Mm}{r^2} + \lambda r, \tag{4.101}$$

where we have set $A = -G_{\mathrm{N}}M$, in order to obtain an attractive force proportional to the involved mass and $\lambda = -2mKR^{-2}$. The cosmological constant term corresponds to a force λr proportional to the distance. In fact, this particular force term vanishes identically if we invoke the shell theorem for the space region inside the spherical shell. So let us proceed to the case $0 \leq r < R$. Here the shell theorem is expressed by saying that

$$\widetilde{\Phi}(P) = 4\pi K, \tag{4.102}$$

which translates to the condition

$$2rRK = \int_{R-r}^{R+r} f(s)\,\mathrm{d}s. \tag{4.103}$$

Differentiation with respect to r gives $2RK = f(R+r)+f(R-r)$ and therefore the value $K = R^{-1}f(R)$ for the constant. Differentiating the resulting condition $2f(R) = f(R+r)+f(R-r)$ twice with respect to r, finally shows that $f''(R) = 0$. So $f(r)$ must be of the form

$$f(r) = A + Br, \tag{4.104}$$

and therefore it is

$$\Phi(r) = \frac{A}{r} + B. \tag{4.105}$$

So we have proved that requiring the shell theorem for the interior of the shell leads to $\lambda = 0$ and thus exactly to Newton's gravitational law. We can summarize the above results by saying:

> *Newton's law of gravitation is equivalent to the shell theorem.*

In terms of usability, however, for most problems in classical mechanics, Newton's law of gravitation is simpler and more flexible in application than the shell theorem. Here we conclude our discussion of mechanics and gravitation within the Newtonian framework.

Further Reading

In this chapter, we have focused mainly on foundational topics concerning Galileian spacetime, Newton's mechanical laws, and Newtonian gravitation. Classical Newtonian mechanics has many practical and exciting applications that we have not discussed here. For an accessible treatment of problems in celestial mechanics, the reader may turn to the book of Fitzpatrick [29]. For an original discussion of the reverse of Newton's shell theorem, the reader should turn to the articles by Arens [2] and Gurzadyan [37].

5

Lagrangian Methods and Symmetry

Here we introduce the most basic facts about the Lagrangian formalism. We formulate classical mechanics by starting from Hamilton's global variational principle, which employs the action of the system. This approach leads to the Euler-Lagrange equations of motion and their specific invariances. In the following, we express the exact relation between symmetry and conservation within a mechanical system through the famous theorem of Emmy Noether. Finally, we apply Noether's theorem to the complete set of Galilei transformations.

5.1 Applying the Principle of Stationary Action

Two Examples of Variational Calculus

Before we apply the variational principle in classical mechanics, let us consider how the method works for two classical problems. The first example is provided by the question, which curve connecting two points in the Euclidean plane has the minimum length. Of course, we know it is a straight line, but here we will derive this fact by means of the variational principle. Suppose we have two points P_1 and P_2 in the Euclidean plane and consider different curves connecting these points, see the figure 5.1 below. The line element

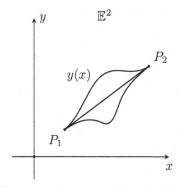

Figure 5.1: Paths connecting two points in the Euclidean plane

$\mathrm{d}s$ is given in Cartesian coordinates by

$$\mathrm{d}s^2 = \mathrm{d}x^2 + \mathrm{d}y^2, \tag{5.1}$$

DOI: 10.1201/9781003087748-5

so that the length s of any curve connecting the points P_1 and P_2 is

$$s = \int_{P_1}^{P_2} \mathrm{d}s = \int_{x_1}^{x_2} \sqrt{1 + (y')^2}\, \mathrm{d}x. \tag{5.2}$$

In the last equation, we treat y as a function of x. Thus, the problem is a variational one, with $s\,[y]$ being the functional. The *Euler-Lagrange equations*, see appendix B.6, lead to

$$y'' = 0, \tag{5.3}$$

with the sought after solution

$$y = ax + b, \tag{5.4}$$

which represents a straight line in the plane. The integration constants are found to be $a = (y_2 - y_1)/(x_2 - x_1)$ and $b = (y_1 x_2 - y_2 x_1)/(x_2 - x_1)$.

The next example is the *brachistochrone problem*, posed by Bernoulli in 1696 (*Johann Bernoulli*). Suppose one has a mass point m under the influence of the constant gravitational field, $g = $ const, near the surface of the Earth and this particle slides frictionless on a surface connecting two given points P_1 and P_2. See figure 5.2 below. Suppose the particle

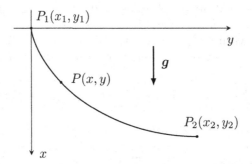

Figure 5.2: Curve connecting two points P_1 and P_2 within a constant gravitational field $\boldsymbol{g}$

is positioned at the point P_1 and left free to roll down the surface to the point P_2. The question is now, which shape of a curve connecting the points will lead to the *shortest time* for the motion between the given points. The velocity of the particle is $v = \mathrm{d}s/\mathrm{d}t$, so the total time for the motion between the points P_1 and P_2 is

$$t_{12} = \int_{P_1}^{P_2} \frac{\mathrm{d}s}{v}. \tag{5.5}$$

The total energy of the particle is conserved, so we have at each instant of time

$$mgx = \frac{m}{2} v^2. \tag{5.6}$$

For the total time we thus obtain

$$t_{12} = \int_{x_1}^{x_2} \sqrt{\frac{1 + (y')^2}{2gx}}\, \mathrm{d}x, \tag{5.7}$$

where we consider the coordinate x as the independent variable and the coordinate y as being dependent on x. The problem is formulated as a variational one and the integrand in the last equation represents the Lagrangian function $L(y, y'; x)$. It is $\partial L / \partial y = 0$ and therefore the Euler-Lagrange equation reads

$$\frac{y'}{\sqrt{2gx(1 + y'^2)}} = \frac{1}{\sqrt{2a}}, \tag{5.8}$$

where $a = \text{const}$ is a positive real parameter. By solving for y, we obtain the definite integral

$$y = \int_0^x \sqrt{\frac{x}{b - x}}\, dx, \tag{5.9}$$

with the new parameter $b = ag^{-1}$. Standard integration leads to

$$y = \left(-\sqrt{(b - x)x} - b \arctan \sqrt{\frac{b - x}{x}} \right) \Big|_0^x. \tag{5.10}$$

In order to simplify, we use the substitution $(b - x)x^{-1} = w^2$ to obtain

$$y = -b \left(\frac{w}{1 + w^2} + \arctan w - \frac{\pi}{2} \right). \tag{5.11}$$

By applying the substitution $w = \tan(-\theta/2)$ with an angle parameter θ, we get

$$\frac{w}{1 + w^2} = -\frac{\sin \theta}{2} \quad \text{and} \quad \arctan w = -\frac{\theta}{2}. \tag{5.12}$$

Hence, the function y is

$$y = \frac{b}{2} [(\theta + \pi) - \sin(\theta + \pi)], \tag{5.13}$$

where we have made use of the identity $\sin(\theta + \pi) = -\sin \theta$. For determining the solution for x, we use $x = b(1 + w^2)^{-1}$ and obtain

$$x = \frac{b}{2} [1 - \cos(\theta + \pi)], \tag{5.14}$$

where we have used $\cos(\theta + \pi) = -\cos \theta$. So the curve leading to the shortest travel time is the one parametrized as

$$y = c(\varphi - \sin \varphi) \quad \text{and} \quad x = c(1 - \cos \varphi), \tag{5.15}$$

with real parameters c and φ. This curve is known as the *cycloid*. The parameter c is the radius of the circle generating the cycloid and φ measures the angle of the rolling circle. From a historical point of view, the solution of the brachistochrone problem was important for sparking the development of the calculus of variations.

Generalized Coordinates

We proceed to the Lagrangian formulation of mechanics according to *Joseph-Louis Lagrange*. The first crucial step is to depart from the usual Cartesian position coordinates $\boldsymbol{x}_\alpha$ and velocities $\boldsymbol{v}_\alpha$ for a system of N particles, and instead use so-called *generalized coordinates* q^j and *generalized velocities* $\dot{q}^j$, with $j = 1, \ldots, n$. The time t parametrizes once again the motion of the system and all its constituents. The generalized coordinates q^j can be positions, distances, angles, or other useful measures that define the state of the system in an n-dimensional manifold, called the *configuration space*. The generalized velocities $\dot{q}^j$ are simply the time derivatives of the generalized coordinates. If the mechanical system contains N particles, then there are $n = 3N$ generalized coordinates and $n = 3N$ generalized velocities. A useful property of this formalism is that one can very effectively implement additional conditions on the mechanical system. We can formulate *constraints* in the form of equations such as

$$f_s(q^1, \ldots, q^n, t) = 0, \quad s = 1, \ldots, m, \tag{5.16}$$

and thus define the dynamics of a mechanical system with $n = 3N - m$ degrees of freedom in the configuration space. We will not dwell into these matters more in detail here, but instead turn to the question of how we can formulate classical mechanics through a variational principle.

Hamilton's Principle of Stationary Action

For a mechanical system with n degrees of freedom, we introduce a *Lagrangian function* $L(q, \dot{q}; t)$ that depends on the generalized coordinates $q = (q^1, \ldots, q^n)$, the generalized velocities $\dot{q} = (\dot{q}^1, \ldots, \dot{q}^n)$, and possibly explicitly on the time t. The detailed functional form of the Lagrangian is specific for each physical system under consideration. Lagrangian functions of mechanical systems, however, have typically a general form, which we will derive below. We define the *action* $S[q]$ for the mechanical system as the functional

$$S[q] \equiv \int_{t_1}^{t_2} L\left(q(t), \dot{q}(t); t\right) \mathrm{d}t \tag{5.17}$$

which depends on the multicomponent function q. The time points t_1 and t_2 represent the start and end points of the dynamical evolution considered. Now we invoke the variational principle, see appendix B.6. The *principle of stationary action* of Hamilton (*Sir William Rowan Hamilton*) demands that the evolution of the system must be such that the condition

$$\delta S[q] = 0 \tag{5.18}$$

is fulfilled for all variations with the boundary conditions $\delta q^j(t_1) = \delta q^j(t_2) = 0$. From this requirement, one can derive the *Euler-Lagrange equations of motion* (*Leonhard Euler*). Exactly as displayed in appendix B.6, with the only difference that now we have multiple components, we can conclude to

$$\frac{\partial L}{\partial q^j} - \frac{\mathrm{d}}{\mathrm{d}t}\frac{\partial L}{\partial \dot{q}^j} = 0 \tag{5.19}$$

as the system of second-order differential equations encoding the dynamics of the system. In order to have equivalence with the Newtonian equations of motion, the Lagrangian function must be of the form

$$L = T - V \tag{5.20}$$

In other words, the Lagrangian $L(q, \dot{q}; t)$ must be equal to the total kinetic energy $T(\dot{q})$ minus the total potential energy $V(q, \dot{q}, t)$ of the system. Here the potential energy can have an explicit dependence on the velocities and the time. In order to make contact with the Newtonian force F_j, we observe that it is given by

$$F_j = -\frac{\partial V}{\partial q^j} + \frac{\mathrm{d}}{\mathrm{d}t}\frac{\partial V}{\partial \dot{q}^j}. \tag{5.21}$$

It is very useful to write the Lagrangian function in the more general form*

$$L(q, \dot{q}; t) = \frac{1}{2} \sum_{j,k=1}^{n} a_{jk}(q, t)\, \dot{q}^j \dot{q}^k - V(q, \dot{q}, t), \tag{5.22}$$

with a symmetric matrix $a_{jk}(q, t)$. Let us also introduce the *generalized* (or *conjugate*) *momenta* p_j as

$$p_j \equiv \frac{\partial L}{\partial \dot{q}^j}, \tag{5.23}$$

for later use. The *total energy*, also called the *Hamiltonian*, E of a Lagrangian mechanical system can be expressed as

$$\boxed{E = \sum_j p_j \dot{q}^j - L} \tag{5.24}$$

If the Lagrangian function is not explicitly dependent on time, the total energy of the system is constant in time. (*exercise 5.1*) Note that the Lagrangian function leading to certain equations of motion is not fully determined. If one transforms a given Lagrangian function $L(q, \dot{q}; t)$ to a new one $\overline{L}(q, \dot{q}; t)$ by

$$\overline{L}(q, \dot{q}; t) \equiv L(q, \dot{q}; t) + \frac{\mathrm{d}M}{\mathrm{d}t}(q, t), \tag{5.25}$$

then the equations of motion stay unchanged. This is most easily seen by using the action, which transforms from S to $\overline{S} = S + \text{const}$, which in turn results to $\delta \overline{S} = \delta S$. In other words, the mechanical system defined by the new Lagrangian exhibits the same dynamical evolution as the original Lagrangian.

Diffeomorphism Invariance

We would like to have the freedom to change the generalized coordinates to a suitable set. This is equivalent to applying a diffeomorphism and transforming from the original coordinates q to the new coordinates q'. Because the action integral is independent of such coordinate transformations, we can assume that the equations of motion will stay unchanged. Let us consider a diffeomorphism expressed as

$$q^j(t) \mapsto q'^j(q, t), \tag{5.26}$$

for the coordinates, and as

$$\dot{q}^j(t) \mapsto \dot{q}'^j(q, \dot{q}, t) = \sum_k \frac{\partial q'^j(q, t)}{\partial q^k}\, \dot{q}^k(t) + \frac{\partial q'^j(q, t)}{\partial t}, \tag{5.27}$$

for their time derivatives. For the inverse diffeomorphism, the analogous formulae apply, where primed and unprimed quantities are simply exchanged. We note also that it is

$$\frac{\partial \dot{q}^k}{\partial \dot{q}'^j} = \frac{\partial q^k}{\partial q'^j}. \tag{5.28}$$

---

*For the sake of clarity, we explicitly write down the sums over the contracted coordinate indices in this section.

Starting from a Lagrangian function $L(q, \dot{q}; t)$, we define a new Lagrangian $L'(q', \dot{q}'; t)$ by

$$L'(q', \dot{q}'; t) \equiv L(q, \dot{q}; t). \tag{5.29}$$

Note that the time parameter is not transformed. We will prove that the Euler-Lagrange derivative transforms as

$$\frac{\delta S'[q']}{\delta q'^j(t)} = \sum_k \left(\frac{\partial q^k}{\partial q'^j} \right) \frac{\delta S[q]}{\delta q^k(t)}, \tag{5.30}$$

where the primed action $S'[q']$ is calculated by employing the primed Lagrangian $L'(q', \dot{q}'; t)$. The equations of motion are valid for the q-coordinates

$$\frac{\delta S}{\delta q^k} = 0, \tag{5.31}$$

exactly if they are valid for the q'-coordinates. Indeed, it is

$$\frac{\partial L'}{\partial q'^j} = \sum_k \left(\frac{\partial L}{\partial q^k} \frac{\partial q^k}{\partial q'^j} + \frac{\partial L}{\partial \dot{q}^k} \frac{\partial \dot{q}^k}{\partial q'^j} \right) \tag{5.32}$$

and

$$\frac{\mathrm{d}}{\mathrm{d}t} \frac{\partial L'}{\partial \dot{q}'^j} = \frac{\mathrm{d}}{\mathrm{d}t} \sum_k \frac{\partial L}{\partial \dot{q}^k} \frac{\partial \dot{q}^k}{\partial \dot{q}'^j}$$
$$= \sum_k \left(\left(\frac{\mathrm{d}}{\mathrm{d}t} \frac{\partial L}{\partial \dot{q}^k} \right) \frac{\partial q^k}{\partial q'^j} + \frac{\partial L}{\partial \dot{q}^k} \frac{\partial \dot{q}^k}{\partial q'^j} \right). \tag{5.33}$$

Subtracting the second equation from the first yields the assertion. For a Lagrangian function of type 5.22, the differentiable coordinate transformation leads to the expression

$$L'(q', \dot{q}'; t) = \frac{1}{2} \sum_{j,k} a'_{jk}(q', t) \, \dot{q}'^j \dot{q}'^k - V'(q', \dot{q}', t), \tag{5.34}$$

with the transformed matrix

$$a'_{jk}(q', t) = \sum_{l,m} \frac{\partial q^l}{\partial q'^j} \frac{\partial q^m}{\partial q'^k} a_{lm}(q, t) \tag{5.35}$$

and the transformed potential function $V'(q', \dot{q}', t) = V(q, \dot{q}, t)$. The transformation of the matrix a_{jk} is exactly the one of a covariant $(0,2)$-tensor. (*exercise 5.2*)

Reparametrization Invariance

Let us now consider a differentiable and invertible transformation of the time parameter t, while the coordinates are not changed. We let the time parameter transform as $t \mapsto t'(t)$ with $\mathrm{d}t'/\mathrm{d}t \neq 0$ and with the inverse transformation $t' \mapsto t(t')$. The action integral is obviously invariant and thus the considered mechanical system, described now by the new time parameter t', must obey the same dynamical equations. Indeed, by using the relation

$$\frac{\mathrm{d}}{\mathrm{d}t} = \frac{\mathrm{d}t'}{\mathrm{d}t} \frac{\mathrm{d}}{\mathrm{d}t'}, \tag{5.36}$$

we see immediately that the Euler-Lagrange equations are satisfied also with the new time parameter t'.

Scale Invariance

We now repeat the analysis we did in the Newtonian framework in Lagrangian mechanics and ask what the conditions are for our theory to be scale invariant. Let us consider a *scale transformation* of the form

$$\left. \begin{array}{rcl} q' & = & a\,q \\ t' & = & b\,t \end{array} \right\}, \tag{5.37}$$

which uses positive real parameters a and b. The velocities transform as

$$\dot{q}' = \frac{a}{b}\,\dot{q}. \tag{5.38}$$

We assume the mechanical system to be described by a Lagrangian

$$L\,(q, \dot{q}; t) = \frac{1}{2} \sum_{j,k=1}^{n} a_{jk}\,\dot{q}^{j}(t)\,\dot{q}^{k}(t) - V(q(t)), \tag{5.39}$$

with a constant matrix a_{jk} and the potential energy $V(q)$ being a homogeneous function of it arguments, i.e.

$$V(a\,q^{1}, \ldots, a\,q^{n}) = a^{-d_V} V(q^{1}, \ldots, q^{n}). \tag{5.40}$$

The number d_V denotes the *scaling dimension* of the function $V(q)$. We see again that scale invariance is reached whenever

$$b = a^{1 + \frac{1}{2} d_V} \tag{5.41}$$

holds, exactly as in the Newtonian formulation. In that case, the Lagrangian is only multiplied by a global constant factor c

$$L\left(aq, \frac{a}{b}\dot{q}; bt\right) = c\,L\,(q, \dot{q}; t) \tag{5.42}$$

and the Euler-Lagrange equations continue to be valid. For the ratios of the time variable and the position variables we obtain the relation

$$\left(\frac{t'}{t}\right) = \left(\frac{q'^{j}}{q^{j}}\right)^{1 + \frac{1}{2} d_V}, \tag{5.43}$$

for each index $j = 1, \ldots, n$. To illustrate the use of the above relation, let us view the Lagrangian function describing small oscillations. In this example, the potential energy $V(q)$ is proportional to q^2. Consequently it is $d_V = -2$ and we obtain that $t' = t$. In other words, the period of the oscillation is independent of the amplitude. By using the property of scale invariance we have gained insight into this physical system without explicitly solving the equations of motion.

5.2 Noether's Theorem in Mechanics

Symmetry Transformations in Mechanics

The Lagrange formulation of mechanics is particularly well-suited for treating questions concerning symmetries and conservation. The core result is the celebrated theorem of Noether (*Amalie Emmy Noether*). The concepts of the Lagrangian formalism and the Noether theorem can be nicely extended to relativistic mechanics and to the framework of classical field theory, as we will see later. But first, let us define what we mean by symmetry in mechanics. Our mechanical theory is described by an action principle with an underlying Lagrangian function. The dynamical variables $q^j(t)$ and the time parameter t enter into the description. We can consider transformations of these dynamical variables. If the action stays unaltered, then we say that the transformation is a *symmetry transformation*. Let us proceed in writing this down. We consider the set of transformations

$$\left.\begin{aligned} t &\mapsto t' = t + \delta t(t) & \equiv t + \epsilon^a \mathrm{T}_a(t) \\ q^j(t) &\mapsto q'^j(t) = q^j(t) + \delta q^j(t) & \equiv q^j(t) + \epsilon^a \mathrm{Q}_a^j(t) \\ q^j(t) &\mapsto q'^j(t') = q^j(t) + \overline{\delta} q^j(t) & \equiv q^j(t) + \epsilon^a \Phi_a^j(t) \end{aligned}\right\}. \tag{5.44}$$

The first row expresses an infinitesimal transformation of the time parameter. The second transformation with $\delta q^j(t)$ corresponds to the *form* (or *function*) *variation* of the dynamical variables. The third transformation with $\overline{\delta} q^j(t)$ represents the *total variation*, taking also into account the transformation of the time parameter. The index j takes the usual values $j = 1, \dots, n$ for the number of degrees of freedom. The infinitesimal transformations are parametrized by the real parameters ϵ^a, $|\epsilon^a| \ll 1$, with the index range $a = 1, \dots, r$. For the repeatedly appearing upper and lower indices a, the summation convention applies. The functions $\mathrm{T}_a(t)$, $\mathrm{Q}_a^j(t)$ and $\Phi_a^j(t)$, defined as

$$\mathrm{T}_a \equiv \left.\frac{\partial(\delta t)}{\partial \epsilon^a}\right|_{\epsilon^a = 0}, \tag{5.45}$$

$$\mathrm{Q}_a^j \equiv \left.\frac{\partial(\delta q^j)}{\partial \epsilon^a}\right|_{\epsilon^a = 0}, \tag{5.46}$$

$$\Phi_a^j \equiv \left.\frac{\partial(\overline{\delta} q^j)}{\partial \epsilon^a}\right|_{\epsilon^a = 0}, \tag{5.47}$$

are called the *generators of the symmetry transformation*. If we know them, we can reconstruct the transformation completely. The formalism might seem a bit heavy now, but as we will see later, these notions apply equally well to the field theory case. Let us recall the basic relation between the form variation

$$\delta q^j(t) \equiv q'^j(t) - q^j(t) \tag{5.48}$$

on one hand, and the total variation

$$\overline{\delta} q^j(t) \equiv q'^j(t') - q^j(t) \tag{5.49}$$

on the other hand. By applying a Taylor expansion, up to first order in the variations, we are led to

$$\overline{\delta} q^j(t) = \delta q^j(t) + \dot{q}^j(t)\, \delta t. \tag{5.50}$$

Per definition, the above transformation is called a *symmetry transformation*, if the action of the theory remains invariant. This means that the condition

$$S[q'; t'] = S[q; t] \tag{5.51}$$

or, written differently,

$$\delta S[q; t] = 0 \tag{5.52}$$

must hold. It should be emphasized that the variation δ appearing above is the one caused by a symmetry transformation 5.44, which includes the dynamic variables q and the time parameter t. It is not the variation of the dynamic variables with special boundary conditions that we have employed for deriving the Euler-Lagrange equations.

Noether's Theorem in Mechanics

The starting point is the symmetry condition 5.52. For simplicity, we assume that our Lagrangian $L(q^j, \dot{q}^j)$ does not depend explicitly on the time parameter t. Let us calculate the symmetry variation $\delta S[q; t]$ step by step. It is

$$\delta S = \delta \int_{t_1}^{t_2} L \, dt = \int_{t_1}^{t_2} (\delta L) \, dt + \int_{t_1}^{t_2} L \, \delta(dt). \tag{5.53}$$

The integrals here are evaluated between two given time points t_1 and t_2. To first order in ϵ^a it is

$$dt' = \frac{dt'}{dt} \, dt = \left[1 + \frac{d}{dt}(\delta t) \right] dt. \tag{5.54}$$

In other words

$$\delta(dt) = \frac{d}{dt}(\delta t) \, dt. \tag{5.55}$$

The second integral on the rhs above is thus

$$\int_{t_1}^{t_2} L \, \delta(dt) = \int_{t_1}^{t_2} \frac{d}{dt}(L \, \delta t) \, dt, \tag{5.56}$$

since L does not depend explicitly on t. For the integrand of the first integral on the rhs above we have

$$\delta L = \frac{\partial L}{\partial q^j} \delta q^j + \frac{\partial L}{\partial \dot{q}^j} \delta \dot{q}^j - \delta \left(\frac{dM}{dt} \right). \tag{5.57}$$

The term $\delta(dM/dt)$ corresponds to the freedom in the definition of a Lagrangian, as expressed in 5.25. The minus sign appears because the dM/dt-term in 5.25 has to be transferred from the rhs to the lhs. In addition, we use the calculation rule for function variations (see appendix B.6)

$$\delta \left(\frac{dM}{dt} \right) = \frac{d}{dt}(\delta M). \tag{5.58}$$

Further, we can express the variation $\delta M(q, t)$ as

$$\delta M(q, t) = \epsilon^a \mathrm{m}_a(q, t), \tag{5.59}$$

or equivalently, by using a generator function $\mathrm{m}_a(q, t)$, as

$$\mathrm{m}_a \equiv \left. \frac{\partial(\delta M)}{\partial \epsilon^a} \right|_{\epsilon^a = 0}, \tag{5.60}$$

in complete analogy to the expressions for symmetry transformations. After these technical comments, let us return to the first integral above. A partial integration leads to

$$\int_{t_1}^{t_2} (\delta L) \, dt = \int_{t_1}^{t_2} \frac{\delta S}{\delta q^j} \delta q^j \, dt + \int_{t_1}^{t_2} \frac{d}{dt} \left(p_j \delta q^j - \delta M \right) dt. \tag{5.61}$$

We assume that the Euler-Lagrange equations of motion are fulfilled, i.e.

$$\frac{\delta S}{\delta q^j} = 0, \tag{5.62}$$

so that the corresponding integral above vanishes. Summarizing all these results, we obtain for the variation of the action

$$\delta S = \int_{t_1}^{t_2} \frac{d}{dt} \left\{ p_j \delta q^j + L \, \delta t - \delta M \right\} dt, \tag{5.63}$$

or equivalently,

$$\delta S = \int_{t_1}^{t_2} \frac{\mathrm{d}}{\mathrm{d}t} \left\{ p_j \left(\overline{\delta} q^j - \dot{q}^j \, \delta t \right) + L \, \delta t - \delta M \right\} \mathrm{d}t, \tag{5.64}$$

and, by using the symmetry generators, we obtain finally

$$\delta S = \int_{t_1}^{t_2} \frac{\mathrm{d}}{\mathrm{d}t} \left\{ \left[p_j \Phi_a^j - \left(p_j \dot{q}^j - L \right) \mathrm{T}_a - \mathrm{m}_a \right] \epsilon^a \right\} \mathrm{d}t. \tag{5.65}$$

By using now the symmetry condition $\delta S = 0$ and the time-independence of the parameters ϵ^a, we deduce that it must be

$$\frac{\mathrm{d}}{\mathrm{d}t} \left\{ p_j \Phi_a^j - \left(p_j \dot{q}^j - L \right) \mathrm{T}_a - \mathrm{m}_a \right\} = 0. \tag{5.66}$$

This is the desired result, which motivates us to define the *Noether charges* $J_a(t)$ as

$$\boxed{ J_a \equiv E \, \mathrm{T}_a - p_j \Phi_a^j + \mathrm{m}_a } \tag{5.67}$$

with $E(t)$ being the *total energy*

$$E = p_j \dot{q}^j - L \tag{5.68}$$

of the system. The indices $a = 1, \dots, r$ take the values according to the number and type of symmetry transformations. Thus, the *Noether conservation equation* reads

$$\boxed{ \frac{\mathrm{d}J_a}{\mathrm{d}t} = 0, \quad a = 1, ..., r } \tag{5.69}$$

Noether's theorem in mechanics expressed in words states:

> *For every global symmetry transformation of the action, there is a temporally conserved Noether charge.*

As mentioned earlier, the Noether theorem does not only apply to classical mechanics and later we will generalize the theorem to the cases of relativistic mechanics and field theory. We should remark that the rather abstract functions $\mathrm{m}_a(q, t)$ can be cast into a more concrete form for each given Lagrangian, and we will see an example of how this works below.

5.3 Galilei Symmetry and Conservation

Constants of Motion

Noether's theorem provides the framework in which we can relate conserved quantities, the so-called *constants of motion*, to corresponding symmetries of the theory. Our classical mechanical theory is expected to be invariant under Galilei transformations, introduced in Section 4.1. Phrased differently, the Galilei transformations are expected to be *symmetry transformations*. In this section we will derive, one by one, the ten constants of motion for each of the ten Galilei transformations. These ten conserved quantities are the total energy E, the total momentum $\boldsymbol{P}$, the total angular momentum $\boldsymbol{L}$, and the center of mass momentum $\boldsymbol{G}$. In order to be specific, let us consider a system of N point particles, with Cartesian position coordinates $\boldsymbol{q}_\alpha$, where the index $\alpha = 1, \dots, N$ enumerates the particles. The Lagrangian L shall be given as

$$L\left(\boldsymbol{q}_\alpha, \dot{\boldsymbol{q}}_\alpha \right) = \frac{1}{2} \sum_\alpha m_\alpha \dot{\boldsymbol{q}}_\alpha - \sum_{\alpha < \beta} V_{\alpha\beta}(|\boldsymbol{q}_\alpha - \boldsymbol{q}_\beta|). \tag{5.70}$$

The potential energy is only a function of the mutual positions of all pairs of particles, compare with Section 4.3. What we will do practically is to apply the single Galilei transformations to this Lagrangian and examine how it transforms.

Homogeneity of Time and Energy Conservation

Homogeneity of time means that the theory must remain unaltered if a translation in time is carried out. The combined (infinitesimal) transformations we need to consider are

$$t' = t + \tau, \tag{5.71}$$

and

$$q'_\alpha = q_\alpha, \tag{5.72}$$

for all $\alpha = 1, \ldots, N$. This overall transformation is parametrized by the real parameter τ, the time shift. As one can immediately see, the Lagrangian 5.70 remains unaltered under such a time translation. The corresponding generators of the transformation are

$$T = \left. \frac{\partial(\delta t)}{\partial \tau} \right|_{\tau=0} = 1, \tag{5.73}$$

and

$$\Phi^k_\alpha = \left. \frac{\partial(\bar\delta q^k_\alpha)}{\partial \tau} \right|_{\tau=0} = 0. \tag{5.74}$$

The index $k = 1, 2, 3$ denotes the Cartesian coordinates. By using the equation 5.67, we can read off that the conserved Noether charge is the total energy of the system,

$$E = T + V. \tag{5.75}$$

The corresponding conservation equation 5.69 reads

$$\boxed{\frac{\mathrm{d}E}{\mathrm{d}t} = 0} \tag{5.76}$$

Expressed in words, we can state:

> *Homogeneity in time implies energy conservation.*

Homogeneity of Space and Linear Momentum Conservation

For studying *homogeneity of space*, we need to consider the combined (infinitesimal) transformations

$$t' = t, \tag{5.77}$$

and

$$q'_\alpha = q_\alpha + a, \tag{5.78}$$

for all $\alpha = 1, \ldots, N$. This specific transformation is parametrized by three real parameters a^j, $j = 1, 2, 3$, which are the Cartesian coordinates of the translation vector a in space. One can see again that the Lagrangian 5.70 keeps its original value. The generators of the transformation are

$$T_j = \left. \frac{\partial(\delta t)}{\partial a^j} \right|_{a^j=0} = 0, \tag{5.79}$$

and

$$\Phi^k_{\alpha,j} = \left. \frac{\partial(\bar\delta q^k_\alpha)}{\partial a^j} \right|_{a^j=0} = \delta^k_j, \tag{5.80}$$

for all $\alpha = 1, \ldots, N$. We deduce that the conserved Noether charge J_j is

$$J_j = -\sum_\alpha (\boldsymbol{p}_\alpha)_k \Phi^k_{\alpha,j} = -\sum_\alpha (\boldsymbol{p}_\alpha)_j = -P_j. \tag{5.81}$$

This means that the total linear momentum of the system

$$\boldsymbol{P} = \sum_\alpha \boldsymbol{p}_\alpha \tag{5.82}$$

is conserved in time:

$$\boxed{\frac{\mathrm{d}\boldsymbol{P}}{\mathrm{d}t} = 0} \tag{5.83}$$

In words, we can state:

> *Homogeneity of space implies linear momentum conservation.*

Isotropy of Space and Angular Momentum Conservation

For expressing *isotropy of space*, we need to consider rotations. Let us summarize some basic facts first. A rotation is given by a *rotation matrix* $R(\boldsymbol{n}, \theta)$, where θ is the *angle parameter*, with values $0 \leq \theta < 2\pi$, and $\boldsymbol{n}$ is the *normal vector*, being normal to the plane of rotation. The condition $|\boldsymbol{n}| = 1$ holds. We must choose three real parameters to define the rotation. The vector $\boldsymbol{\theta} \equiv \theta \boldsymbol{n}$ is called the *rotation vector*. There is a general formula describing the rotation of any space vector $\boldsymbol{q}$ by a rotation $R(\boldsymbol{n}, \theta)$. The formula reads (for a derivation see e.g. [36])

$$R(\boldsymbol{n}, \theta)\, \boldsymbol{q} = (\cos\theta)\boldsymbol{q} + (1 - \cos\theta)(\boldsymbol{n} \cdot \boldsymbol{q})\boldsymbol{n} + (\sin\theta)\boldsymbol{n} \times \boldsymbol{q}. \tag{5.84}$$

For a small angle $\theta \ll 1$, we obtain the linear approximation

$$R(\boldsymbol{n}, \theta)\, \boldsymbol{q} = \boldsymbol{q} + \boldsymbol{\theta} \times \boldsymbol{q}. \tag{5.85}$$

Hence, the variation of the position vector is given by $\delta\boldsymbol{q} = \boldsymbol{\theta} \times \boldsymbol{q}$, see also the sketch 5.3.

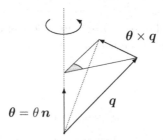

Figure 5.3: Vectors of angle, position and displacement for an infinitesimal rotation

Now we can formulate the transformations for infinitesimal rotations in space, they are given by

$$t' = t, \tag{5.86}$$

and

$$\boldsymbol{q}'_\alpha = \boldsymbol{q}_\alpha + \boldsymbol{\theta} \times \boldsymbol{q}_\alpha, \tag{5.87}$$

for all $\alpha = 1, \ldots, N$. This transformation is parametrized by the three real rotation parameters θ^j, $j = 1, 2, 3$, as described above. Once again, the Lagrangian stays unchanged. The generators of the transformation are

$$\mathrm{T}_j = \left. \frac{\partial(\delta t)}{\partial \theta^j} \right|_{\theta^j = 0} = 0, \tag{5.88}$$

and

$$\Phi^k_{\alpha, j} = \left. \frac{\partial(\bar{\delta} q^k_\alpha)}{\partial \theta^j} \right|_{\varphi^j = 0} = \epsilon^k{}_{jl} (\boldsymbol{q}_\alpha)^l. \tag{5.89}$$

Thus, the conserved Noether charge J_j is found to be

$$J_j = - \sum_\alpha \epsilon_{jl}{}^k (\boldsymbol{q}_\alpha)^l (\boldsymbol{p}_\alpha)_k = -L_j. \tag{5.90}$$

Therefore the total angular momentum of the system

$$\boldsymbol{L} = \sum_\alpha \boldsymbol{q}_\alpha \times \boldsymbol{p}_\alpha \tag{5.91}$$

is conserved in time:

$$\boxed{\frac{\mathrm{d}\boldsymbol{L}}{\mathrm{d}t} = 0} \tag{5.92}$$

Expressed in words:

Isotropy of space implies angular momentum conservation.

Galilei Boost Symmetry and Galilei Momentum Conservation

The last specific Galilei symmetry we are going to consider is the one relating two reference frames moving with constant velocity to each other, the so-called *Galilei boost symmetry*. The (infinitesimal) transformations of time and space are respectively

$$t' = t, \tag{5.93}$$

and

$$\boldsymbol{q}'_\alpha = \boldsymbol{q}_\alpha + \boldsymbol{u} t, \tag{5.94}$$

for all $\alpha = 1, \ldots, N$. The boost transformation is parametrized by the three real parameters u^j, $j = 1, 2, 3$, the Cartesian coordinates of the constant boost velocity $\boldsymbol{u}$. The particle velocities transform like $\dot{\boldsymbol{q}}'_\alpha = \dot{\boldsymbol{q}}_\alpha + \boldsymbol{u}$. Here we encounter a case where the Lagrangian does not keep its original value but instead collects a total time derivative. By inserting the transformed coordinates in the Lagrangian 5.70, we obtain

$$L\left(\boldsymbol{q}'_\alpha, \dot{\boldsymbol{q}}'_\alpha\right) = L\left(\boldsymbol{q}_\alpha, \dot{\boldsymbol{q}}_\alpha\right) + \sum_\alpha \frac{m_\alpha}{2} \boldsymbol{u}^2 + \sum_\alpha m_\alpha \dot{\boldsymbol{q}}_\alpha \cdot \boldsymbol{u}. \tag{5.95}$$

The two last terms in the rhs constitute the total time derivative $\mathrm{d}M(\boldsymbol{q}_\alpha, t)/\mathrm{d}t$. Integration over time gives the function $M(\boldsymbol{q}_\alpha, t)$ itself,

$$M(\boldsymbol{q}_\alpha, t) = \frac{M \boldsymbol{u}^2 t}{2} + \sum_\alpha m_\alpha \boldsymbol{q}_\alpha \cdot \boldsymbol{u}, \tag{5.96}$$

where M in the rhs is the total mass of the system,

$$M = \sum_\alpha m_\alpha, \tag{5.97}$$

and should not be confused with the function $M(\boldsymbol{q}_\alpha, t)$. The part of $M(\boldsymbol{q}_\alpha, t)$ linear in the transformation parameter $\boldsymbol{u}$ is the variation $\delta M(\boldsymbol{q}_\alpha, t)$, i.e.

$$\delta M(\boldsymbol{q}_\alpha, t) = \sum_\alpha m_\alpha \boldsymbol{q}_\alpha \cdot \boldsymbol{u}. \tag{5.98}$$

With this result, we can calculate the generator function $m_j(\boldsymbol{q}_\alpha, t)$ and obtain

$$m_j = \left.\frac{\partial(\delta M)}{\partial u^j}\right|_{u^j=0} = \sum_\alpha m_\alpha (\boldsymbol{q}_\alpha)_j. \tag{5.99}$$

For the symmetry generators we have

$$T_j = \left.\frac{\partial(\delta t)}{\partial u^j}\right|_{u^j=0} = 0, \tag{5.100}$$

and

$$\Phi^k_{\alpha,j} = \left.\frac{\partial(\bar\delta q^k_\alpha)}{\partial u^j}\right|_{u^j=0} = \delta^k_j t, \tag{5.101}$$

for all values of α. Thus, the conserved Noether charge J_j is

$$J_j = -\sum_\alpha t(\boldsymbol{p}_\alpha)_j + \sum_\alpha m_\alpha(\boldsymbol{q}_\alpha)_j = -tP_j + MX_j, \tag{5.102}$$

where we use the center of mass vector

$$\boldsymbol{X} = \frac{1}{M}\sum_\alpha m_\alpha \boldsymbol{q}_\alpha, \tag{5.103}$$

as introduced in Section 4.3. Therefore, we obtain that the Galilei momentum of the system

$$\boldsymbol{G} = t\boldsymbol{P} - M\boldsymbol{X} \tag{5.104}$$

is the quantity conserved in time:

$$\boxed{\frac{\mathrm{d}\boldsymbol{G}}{\mathrm{d}t} = \boldsymbol{0}} \tag{5.105}$$

In words, we can state:

   Galilei boost symmetry implies Galilei momentum conservation.

This completes our derivation of the ten conserved quantities in time, i.e. energy, linear momentum, angular momentum, and center of mass momentum, corresponding to the ten Galilei symmetries of the system.

Further Reading

In this chapter, we introduced the basics of the Lagrangian formalism, with the primary goal of formulating Noether's theorem. A more complete treatment of the Lagrangian formalism and its applications can be found in the classic book of Goldstein, Poole, and Safko [36]. We should note that have completely omitted the Hamiltonian formalism of classical mechanics, which is also treated in [36]. The mathematically inclined reader may turn to the text of Arnold [3].

6

Relativistic Mechanics

In this chapter we introduce the special relativistic version of spacetime and discuss classical relativistic point particles. First, we derive the special relativistic transformations of spacetime coordinates, the so-called Lorentz boosts, which generalize Galilei boosts. This leads us to consider space and time as a unified entity, Minkowski space. We then proceed to the relativistic generalization of the notions of force, momentum, energy, and other physical quantities. Subsequently, we construct a manifestly relativistic version of Lagrangian mechanics. Finally, we discuss how Noether's symmetry theorem is expressed in this relativistic framework, and we derive the main conserved quantities.

6.1 Lorentz Transformations

Relativity and Constancy of Light Speed

In Chapter 4, we have introduced the notions of time, space, and inertial frames of reference to describe mechanical systems. These notions are revisited here. In particular, we invoke again the *principle of relativity*, but we now generalize it to be valid for all physical systems, not only for mechanical ones. In addition, we introduce a second basic principle, which is the *constancy of light speed*. The constancy of light speed is an experimental fact, which we elevate to a principle. The theory of special relativity is concerned with the notion of spacetime, events and the transformations between inertial frames, where the speed of light is a universal constant, independent of any observer. Special relativity provides the arena for all physical theories that are compatible with these spacetime transformations. Within the realm of classical physics, these theories are mechanics and field theory. Gravitational interactions and accelerated frames require a further generalization, which we will discuss later. So let us express the *principle of relativity*, which states:

All physical laws have the same form in all inertial reference systems.

In addition, we introduce the *principle of constant light speed*:

The speed of light, $c = 299.792.458\,\mathrm{m\,s^{-1}}$, is constant for every inertial observer.

Note that the above value for the speed of light is exact and there are no digits after a comma. This definition is valid since 1983, when the unit of meter was redefined to be the distance traveled by light in vacuum in $1/299.792.458$ seconds. This ratio is held fixed by

DOI: 10.1201/9781003087748-6

convention and any change in actually measured light speed results in a change of the actual length of a meter. With the two fundamental principles above, we are able to derive the special relativistic spacetime transformations, to which we now turn.[*]

Lorentz Boosts

Consider two inertial frames labeled S and S', with their respective Cartesian axes pointing to the same direction and their origins coinciding at the time $t = t' = 0$. The frame S' shall move with constant velocity u along the positive direction of the x^1-axis of S. This is exactly the same situation as depicted in figure 4.1. We will derive the relation between the coordinates of the same event, call it P, as measured in the two reference frames. This is the *passive* point of view for transformations. Later we can switch to *active* transformations. The event coordinates for P are (t, x^1, x^2, x^3) in S and (t', x'^1, x'^2, x'^3) in S'. The relation between the coordinates is assumed to be linear, a fact we will actually derive later in the differential geometric framework. It is

$$\left. \begin{aligned} t' &= At + Bx^1 \\ x'^1 &= Ct + Dx^1 \\ x'^2 &= x^2 \\ x'^3 &= x^3 \end{aligned} \right\}. \tag{6.1}$$

Because $x^1 = ut$ corresponds to $x'^1 = 0$ and, inversely, because $x^1 = 0$ corresponds to $x'^1 = -ut'$, we infer that the transformation simplifies to

$$\left. \begin{aligned} t' &= At + Bx^1 \\ x'^1 &= A(x^1 - ut) \\ x'^2 &= x^2 \\ x'^3 &= x^3 \end{aligned} \right\}. \tag{6.2}$$

In the non-relativistic limit, the function A must tend to 1 and the function B must tend to zero. In order to determine these two unknowns, we need another condition for the coordinates. Here the constancy of light speed comes into play. We consider a spherical light wave, which departs at the common origin of the two inertial frames at the time $t = t' = 0$. The points on the wave front obey the equation

$$(ct)^2 - (x^1)^2 - (x^2)^2 - (x^3)^2 = (ct')^2 - (x'^1)^2 - (x'^2)^2 - (x'^3)^2 = 0, \tag{6.3}$$

which expresses the constancy of light speed for the two reference systems. Using this constraint and the system of equations for the coordinates, we can unambiguously determine the unknown functions. The result is (*exercise 6.1*)

$$\left. \begin{aligned} ct' &= \gamma(ct - \beta x^1) \\ x'^1 &= \gamma(x^1 - \beta ct) \\ x'^2 &= x^2 \\ x'^3 &= x^3 \end{aligned} \right\}, \tag{6.4}$$

---

[*]The principle of relativity alone is sufficient to determine almost completely the allowed coordinate transformations, up to a global constant. The constancy of light speed is then only a "small" additional requirement, which then fully characterizes the transformations. For this remarkable fact see [82]. In our derivation of the spacetime transformations, we will follow the easier path by requiring upfront the constancy of light speed.

with the function $\beta(u) \equiv u/c$ being the boost velocity u divided by the speed of light c and the function, called the *gamma factor*, $\gamma(\beta)$ being defined as

$$\gamma \equiv \frac{1}{\sqrt{1 - \beta^2}} \qquad (6.5)$$

We see now that in the non-relativistic limit of small velocities, $u \ll c$, the function β vanishes and the function γ approaches the value one. In this case, the mixing of the space and time coordinates disappears and the Galilei boost transformation is recovered. We note that the above transformation is a special case and does not contain any rotation, which would be still allowed within our basic principles. In addition, we have not included any translations in time and space, which are also allowed within the relativity principle. We will consider these generalizations soon. The specific transformation we have just derived is a *Lorentz boost* of the new reference frame S' in x^1-direction, named after the physicist *Hendrik Antoon Lorentz*. The other Cartesian directions are similar. The inverse transformation is obtained by changing the sign of the relative velocity. This change provides also the value of an active boost in the in x^1-direction. A Lorentz boost reveals that space and time are now intertwined into a unified scheme. From now on we view $x^0 \equiv ct$ as the fourth (or zeroth) coordinate. Also we will use physical units, where the speed of light c is equal to 1, thus simplifying the appearance of equations. By using this new notation, an active *Lorentz boost* takes the form

$$\begin{array}{rcl}
x'^0 &=& \gamma(x^0 + \beta x^1) \\
x'^1 &=& \gamma(x^1 + \beta x^0) \\
x'^2 &=& x^2 \\
x'^3 &=& x^3
\end{array} \qquad (6.6)$$

As mentioned, in addition to the Lorentz boost, we have still the freedom to apply rotations and space and time translations on our reference frames. We defer the full mathematical discussion of these combined transformations, called Lorentz transformations and Poincaré transformations to later. Historically, Hendrik Lorentz and Henri Poincaré had derived the transformations of space and time coordinates in order to explain the absence of an "ether" medium, in which electromagnetic waves were supposed to propagate. Lorentz and Poincaré initially considered these transformations only as a mathematical vehicle, being devoid of a direct physical significance. It was only later, in 1905, when *Albert Einstein* rederived and successfully interpreted these Lorentz boost transformations as the ones capturing how actual physical space and time transform. The fact that space and time are no longer absolute but behave like dynamical entities has profound physical consequences. Let us examine two such physical effects.

Length Contraction
Suppose we have two inertial frames S and S', with parallel axes coinciding at the time $t = t' = 0$ and with the frame S' moving with constant velocity β along the x-axis. We now ask how finite lengths and finite time intervals change, when the reference frame is changed. I.e. we will investigate how the intervals between two events A and B, with the respective coordinates $x_A = (t_A, x_A, y_A, z_A)$ and $x_B = (t_B, x_B, y_B, z_B)$, change. Let us consider the length of a rigid rod lying along the x-axis at rest in the moving frame S', with the length Δl_0 given by

$$\Delta l_0 = x'_B - x'_A. \qquad (6.7)$$

What is the length of this rigid rod as it seen moving by the observer in the frame S? By applying the Lorentz boost in the x-direction, we have for the two events

$$
\begin{aligned}
x'_A &= \gamma(x_A - \beta t_A), \\
x'_B &= \gamma(x_B - \beta t_B).
\end{aligned}
\tag{6.8}
$$

The rod in S has, by definition, the length

$$
\Delta l = x_B - x_A,
\tag{6.9}
$$

as measured at the *same time point* $t_A = t_B$. Thus, the observed length Δl of the rod, as seen in the frame S, is given by

$$
\Delta l = \frac{1}{\gamma}\Delta l_0.
\tag{6.10}
$$

This means that for the observer in S, the moving rod appears *contracted* by the factor $\gamma^{-1} = \sqrt{1 - \beta^2}$. This effect is more complicated if we ask how a three-dimensional object appears when it moves. The crucial point is that the photons arriving at the *same* time at the observer's lens were emitted at *different* times by the different parts of the object. A cube, for instance, will appear rotated. For a discussion of the length contraction appearance of three-dimensional objects, one should turn to the references [64], [82], [83], and [88].

Time Dilatation

We consider again the previous setup of reference frames and the two events A and B. In the frame S', the two events have the time points t'_A and t'_B, with the time interval Δt_0 given by

$$
\Delta t_0 = t'_B - t'_A,
\tag{6.11}
$$

measured at the *same space point* $\boldsymbol{x}'_A = \boldsymbol{x}'_B$. The corresponding time points in the reference frame S are

$$
\begin{aligned}
t_A &= \gamma(t'_A + \beta x'_A), \\
t_B &= \gamma(t'_B + \beta x'_B).
\end{aligned}
\tag{6.12}
$$

Thus, the time interval Δt between the two events, as observed in S and defined by

$$
\Delta t = t_B - t_A,
\tag{6.13}
$$

is given by the relation

$$
\Delta t = \gamma\Delta t_0.
\tag{6.14}
$$

In other words, the time interval for the observer in S appears *stretched* by the factor $\gamma = 1/\sqrt{1 - \beta^2}$. A known physical effect is the extended lifetime of decaying Pions when observed in ultra-relativistic cosmic radiation, compared to the shorter lifetime at rest in the laboratory.

6.2   Minkowski Spacetime

Four-Dimensional Minkowski Space

The form of transformations of space and time urge us to view space and time as a unity, called *spacetime*. Minkowski introduced this view and opened up a new chapter in physics, positioning spacetime as a physical entity with its own dynamics. We call the special relativistic spacetime *Minkowski space* and denote it by $\mathbb{M}_4$. It is a 4-dimensional real affine space with its elements being the points in space and time, called *events*. When we choose an origin and a reference basis in $\mathbb{M}_4$, denoted by $\{e_\mu\}$, we can represent an event as a *four-vector* x and expand it with respect to this basis,

$$x = x^\mu e_\mu. \tag{6.15}$$

We employ the Einstein summation convention again. The *contravariant four-vector* x^μ is written as a column vector

$$x^\mu = \begin{pmatrix} x^0 \\ x^1 \\ x^2 \\ x^3 \end{pmatrix}, \tag{6.16}$$

or, using the three-dimensional quantities t and $\boldsymbol{x}$, as

$$x^\mu = \begin{pmatrix} t \\ \boldsymbol{x} \end{pmatrix}. \tag{6.17}$$

We will use the geometric notation x and the coordinate notation x^μ interchangeably and adopt the so-called *abstract indices notation*, where the symbol x^μ is meant to be the geometric vector with the index showing its rank. The invariance of light speed allows us to define an invariant *norm* $|\cdot|$ for the four-vectors x of Minkowski space, with its square given by

$$|x|^2 \equiv x^2 \equiv (x^0)^2 - (x^1)^2 - (x^2)^2 - (x^3)^2 \tag{6.18}$$

in coordinates. The invariance of the norm is equivalent to the invariance of a *scalar product*, which we define on Minkowski space for any two four vectors x and y to be

$$(x \cdot y) \equiv x^0 y^0 - x^1 y^1 - x^2 y^2 - x^3 y^3. \tag{6.19}$$

The invariance of the scalar product is simply derived by expanding $(x+y)^2$. We can define the scalar product alternatively by using the norm through

$$(x \cdot y) \equiv \frac{1}{2} \left(|x+y|^2 - |x|^2 - |y|^2 \right). \tag{6.20}$$

Furthermore, the scalar product can be expressed also as a bilinear form as

$$(x \cdot y) = (x, \eta y) = \eta_{\mu\nu} x^\mu y^\nu, \tag{6.21}$$

with the *metric tensor* $\eta_{\mu\nu}$ given by

$$\eta_{\mu\nu} = (e_\mu \cdot e_\nu), \tag{6.22}$$

or, written as a matrix η, as

$$\eta \equiv (\eta_{\mu\nu}) = \begin{pmatrix} 1 & 0 & 0 & 0 \\ 0 & -1 & 0 & 0 \\ 0 & 0 & -1 & 0 \\ 0 & 0 & 0 & -1 \end{pmatrix}. \tag{6.23}$$

It is customary to use the notations η and $\eta_{\mu\nu}$ interchangeably. Note that the metric tensor is symmetric in its indices. We can trace back the indefiniteness of the metric to the form of the invariance equation of a spherical light wave. Thus, Minkowski space $\mathbb{M}_4$ can be identified with the pseudo-Euclidean space $\mathbb{E}^{(1,3)}$. The inverse of the metric $\eta_{\mu\nu}$ in components is denoted by $\eta^{\mu\nu}$, and the defining relation is

$$\eta^{\mu\rho} \eta_{\rho\nu} = \delta^\mu_\nu. \tag{6.24}$$

Let us note that the *trace of the metric*, denoted $\eta^\mu{}_\mu$, has the value $\eta^\mu{}_\mu = \delta^\mu_\mu = 4$. The metric tensor mediates between the space $\mathbb{M}_4$ and its dual space $\tilde{\mathbb{M}}_4$, so we can define a *covariant four-vector* (or simply *covector*) $\tilde{x}$ by

$$\tilde{x} \equiv \eta x. \tag{6.25}$$

The covector has the components x_μ, which are calculated as

$$x_\mu = \eta_{\mu\nu} x^\nu, \tag{6.26}$$

and is properly written as a row vector

$$x_\mu = (x_0, x_1, x_2, x_3) = (x^0, -x^1, -x^2, -x^3). \tag{6.27}$$

With the usage of dual vectors, the Minkowski scalar product can be written as

$$(x \cdot y) = x_\mu y^\mu = x^\mu y_\mu = x_0 y^0 + x_1 y^1 + x_2 y^2 + x_3 y^3. \tag{6.28}$$

So we must be careful about the covariant and contravariant position of indices. In terms of the geometric notation introduced in Chapter 2, the metric of Minkowski space $\mathbb{M}_4$ can be expressed by the invariant *line element* ds^2, given as

$$\boxed{ds^2 = \left(dx^0\right)^2 - \left(dx^1\right)^2 - \left(dx^2\right)^2 - \left(dx^3\right)^2} \tag{6.29}$$

We call this metric the *Lorentzian metric*. Let us consider now only two spacetime dimensions, one time dimension $x^0 = t$ and one space dimension $x^1 = x$. The corresponding Lorentzian metric

$$ds^2 = dt^2 - dx^2 \tag{6.30}$$

is not at all like the metric of the two-dimensional Euclidean plane. The distance between any two points in the Lorentzian geometry can be rather counter-intuitive. As an example, consider the distance between the two points P and Q in the spacetime diagram 6.1 below. If the diagram is understood to possess the Euclidean plane geometry, the shortest path

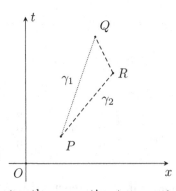

Figure 6.1: Two different paths connecting two events in the Lorentzian plane

between the two points P and Q is the straight line segment γ_1 connecting them. However, if the diagram is understood to possess the Lorentzian plane geometry, then the angled path γ_2, which passes through the point R, is shorter in Lorentzian length, compared to the path γ_1. The reader is encouraged to calculate a concrete numerical example. (*exercise 6.2*)

Not only coordinate vectors, but also differential operators are extended to the four-dimensional Minkowski space. The relativistic extension of the three-dimensional gradient (nabla) operator ∇ is the four-dimensional gradient operator ∂_μ, defined as

$$\partial_\mu \equiv \frac{\partial}{\partial x^\mu} \equiv \left(\frac{\partial}{\partial x^0}, \nabla \right). \tag{6.31}$$

Note that ∂_μ is defined as a covariant row vector. The corresponding contravariant differential operator ∂^μ is given by

$$\partial^\mu \equiv \eta^{\mu\nu} \partial_\nu = \frac{\partial}{\partial x_\mu} = \left(\frac{\partial}{\partial x^0}, -\nabla \right)^T, \tag{6.32}$$

and is a column vector. The square ∂^2 of the four-dimensional gradient operator,

$$\partial^2 \equiv \partial_\mu \partial^\mu = \eta_{\mu\nu} \partial^\mu \partial^\nu = \left(\frac{\partial}{\partial x^0} \right)^2 - \left(\frac{\partial}{\partial x^1} \right)^2 - \left(\frac{\partial}{\partial x^2} \right)^2 - \left(\frac{\partial}{\partial x^3} \right)^2, \tag{6.33}$$

is the relativistic extension of the Laplace operator ∇^2 and includes the time coordinate. The operator ∂^2 is called the *d'Alembert operator (Jean-Baptiste le Rond d'Alembert)*.

Light Cone and Causality

All four-vectors of Minkowski space, v, w, z, etc. are classified according to the sign of their norm into three categories:

$$\left. \begin{array}{ll} v^2 > 0, & \text{timelike vectors} \\ w^2 = 0, & \text{lightlike (or null) vectors} \\ z^2 < 0, & \text{spacelike vectors} \end{array} \right\}. \tag{6.34}$$

The set of all lightlike points forms a three-dimensional hypersurface within the four-dimensional spacetime which we call the *light cone*. The light cone at an event O in Minkowski space is depicted in diagram 6.2.

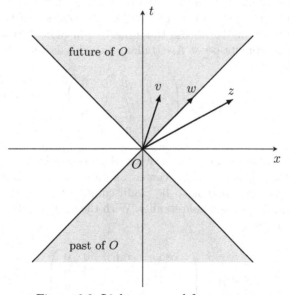

Figure 6.2: Light cone and four-vectors

The light cone consists of two cones, a *backward cone*, expanding into the past, and a *forward cone*, expanding into the future. The tops of the two cones meet at the coordinate origin O. All timelike vectors reside inside the domain engulfed by the light cone and all spacelike vectors reside outside that domain. The coordinate origin, as an event, influences all points within and on the forward light cone, which represents the *future of this event*. All points within and on the backward light cone have influenced the event at the origin, they represent the *past of this event*. All other points outside the cone interior and the cone surfaces cannot influence the event and cannot be influenced by the event, they are the *present of this event*. This is the basic *causality structure* of Minkowski spacetime. This classification can be applied to any pair of events x and y in Minkowski space by considering their invariant distance square $(x - y)^2$, from which we can infer their causal relation.

Lorentz Transformations as Linear Maps

Since Minkowski space is, in particular, a vector space, we can view Lorentz transformations of the spacetime coordinates as linear maps. We remember that the principle of relativity allows also pure space rotations. An active rotation $\Lambda_1(\theta)$ around the x^1-axis by the Euler angle θ will be given as a matrix as

$$\Lambda_1(\theta) = \begin{pmatrix} 1 & 0 & 0 & 0 \\ 0 & 1 & 0 & 0 \\ 0 & 0 & \cos\theta & -\sin\theta \\ 0 & 0 & \sin\theta & \cos\theta \end{pmatrix}, \tag{6.35}$$

mixing the x^2 and x^3 coordinates, but leaving the time coordinate and the rotation axis direction unchanged. A general rotation $\Lambda(\boldsymbol{\theta})$ around a fixed axis by the angle θ has the matrix form

$$\Lambda(\boldsymbol{\theta}) = \left(\begin{array}{c|c} 1 & \mathbf{0}^T \\ \hline \mathbf{0} & R(\boldsymbol{\theta}) \end{array} \right), \tag{6.36}$$

where $R(\boldsymbol{\theta})$ is the element of the group $SO(3)$ representing this rotation. The rotation group is indeed a subgroup of the full Lorentz group. The composition of two rotations is again a pure rotation. We are going to discuss these matters in more detail in the Chapters 10 and 12. An active pure Lorentz boost $\Lambda_1(\beta)$ along the x^1-axis with the boost velocity β is written as a matrix as

$$\Lambda_1(\beta) = \begin{pmatrix} \gamma & \gamma\beta & 0 & 0 \\ \gamma\beta & \gamma & 0 & 0 \\ 0 & 0 & 1 & 0 \\ 0 & 0 & 0 & 1 \end{pmatrix}, \tag{6.37}$$

mixing the x^0 and x^1 coordinates, but leaving the perpendicular coordinate directions unchanged. The Lorentz boosts can be viewed as "rotations" by introducing a different parametrization, the *rapidity* η (not to be confused with the metric tensor), defined by $\cosh\eta \equiv \gamma(\beta)$, or equivalently as $\eta \equiv \text{arctanh}\,\beta$. With the rapidity, the above Lorentz boost takes the form (*exercise 6.3*)

$$\Lambda_1(\eta) = \begin{pmatrix} \cosh\eta & \sinh\eta & 0 & 0 \\ \sinh\eta & \cosh\eta & 0 & 0 \\ 0 & 0 & 1 & 0 \\ 0 & 0 & 0 & 1 \end{pmatrix}. \tag{6.38}$$

One can use either the parameter β or the parameter η to describe a Lorentz boost. A general active Lorentz boost $\Lambda(\boldsymbol{\beta})$ along the velocity 3-vector $\boldsymbol{\beta}$ has the matrix form

$$\Lambda(\boldsymbol{\beta}) = \left(\begin{array}{c|c} \gamma & \gamma \boldsymbol{\beta}^T \\ \hline \gamma \boldsymbol{\beta} & 1 + (\gamma - 1)\hat{\boldsymbol{\beta}} \otimes \hat{\boldsymbol{\beta}} \end{array} \right), \tag{6.39}$$

where $\hat{\boldsymbol{\beta}} \equiv \boldsymbol{\beta}/\beta$ is simply the unit vector in the direction of the boost. For a boost in the x^1-direction there is $\hat{\boldsymbol{\beta}} \equiv (1,0,0)^T$ and we recover the special result for $\Lambda_1(\beta)$. The composition of two Lorentz boosts results in a pure Lorentz boost again only if the two boosts point in the same direction. Otherwise the effect of the resulting transformation is a boost and a rotation.

A nice application of the matrix representation of Lorentz transformations is the relativistic addition of velocities. Suppose we have a two successive boosts in the same direction, with velocities β_1 and β_2. What is the resulting velocity β of the total boost? The matrix multiplication shows immediately that it is

$$\beta = \frac{\beta_1 + \beta_2}{1 + \beta_1 \beta_2}. \tag{6.40}$$

For non-relativistic velocities, $\beta_i \ll 1$, this leads to the Galileian result. At the other extreme, if we have two very high velocities, for example $\beta_1 = \beta_2 = 0.75$ (in units of $c = 1$) where the Galileian result would mean a total velocity greater than 1, the above relativistic formula yields $\beta = 0.96$. Generally, the maximum achievable speed by successive boosts is 1. Later we will see that massive bodies are not able to reach exactly light speed, but can only move below this speed limit.

Relativistic Tensors

A general Lorentz transformation will be denoted symbolically as a $(1,1)$-type tensor $\Lambda^{\mu}{}_{\nu}$, with the indices running from 0 to 3. A Lorentz transformation of a *contravariant* spacetime four-vector x^{μ} is then

$$x'^{\mu} = \Lambda^{\mu}{}_{\nu} x^{\nu}, \tag{6.41}$$

or in matrix notation

$$x' = \Lambda x. \tag{6.42}$$

The invariance of the norm of a four-vector is equivalent to the condition

$$\Lambda^{\rho}{}_{\mu} \Lambda^{\sigma}{}_{\nu} \eta_{\rho\sigma} = \eta_{\mu\nu}, \tag{6.43}$$

or, in matrix notation,

$$\boxed{\Lambda^T \eta \Lambda = \eta} \tag{6.44}$$

The last relation fully characterizes Lorentz transformations. The defining equation 6.44 is the analogue to the defining equation

$$R^T 1 R = 1 \tag{6.45}$$

for rotations. The inverse of a transformation Λ is given by

$$\Lambda^{-1} = \eta \Lambda^T \eta. \tag{6.46}$$

A *covariant* four-vector x_μ in components transforms as

$$x'_\mu = \Lambda_\mu{}^\nu x_\nu, \tag{6.47}$$

since it is

$$\eta_{\mu\rho} x'^\rho = \eta_{\mu\rho} \Lambda^\rho{}_\sigma \eta^{\sigma\nu} \eta_{\nu\tau} x^\tau. \tag{6.48}$$

In matrix notation, the Lorentz transformation of the covector $\tilde{x}$ is given by

$$\tilde{x}' = (\Lambda^{-1})^T \tilde{x}, \tag{6.49}$$

since the relation 6.46 shows that the inverted and transposed of Λ is the transformation matrix for the covector,

$$\Lambda_\mu{}^\nu \equiv (\Lambda^{-1T})_\mu{}^\nu = \eta_{\mu\rho} \Lambda^\rho{}_\sigma \eta^{\sigma\nu}. \tag{6.50}$$

We use the fact that for a matrix A the identity $(A^{-1})^T = (A^T)^{-1}$ holds. In the formula 6.49 we observe a difference between 3-dimensional covectors and 4-dimensional Lorentz covectors. In the 3-dimensional case, the transformation matrix R^i_j is the same for vectors and covectors, as it obeys $R^T = R^{-1}$, so that covectors transform as

$$x'_k = R_k{}^l x_l, \tag{6.51}$$

or, written in matrix notation

$$\tilde{x}' = R\tilde{x}. \tag{6.52}$$

This should be compared with the transformation law 6.49. By generalizing these definitions, a four-component quantity displaying the transformation properties given above, will be considered to be a *contravariant* or a *covariant four-vector*. Examples we will encounter later are the four-momentum, the four-force, and other four-vectors. A general (m, n)-type tensor $T^{\mu_1 \cdots \mu_m}{}_{\nu_1 \ldots \nu_n}$ per definition transforms as

$$T'^{\mu_1 \cdots \mu_m}{}_{\nu_1 \ldots \nu_n} = \Lambda^{\mu_1}{}_{\rho_1} \cdots \Lambda^{\mu_m}{}_{\rho_m} \Lambda_{\nu_1}{}^{\sigma_1} \cdots \Lambda_{\nu_n}{}^{\sigma_n} T^{\rho_1 \cdots \rho_m}{}_{\sigma_1 \ldots \sigma_n}. \tag{6.53}$$

This means each contravariant index and each covariant index transforms with the corresponding matrix element. Such higher-order tensors can be constructed out of lower-order tensors through the tensor product, or they can be build from scratch as multilinear maps. Let us remark that besides tensors, Lorentz symmetry leads to the existence of other objects, the relativistic spinors, very much the same way as for rotation symmetry. Spinors are more fundamental than tensors, since they can be used for constructing tensors, while the opposite is not possible. We will derive spinors within Lorentz symmetry in the group-theoretical discussion later. Finally, let us state why relativistic tensors are so important: any physical law that can be brought into a form of a relativistic tensor equation will be invariant under Lorentz transformations and as such it will be a relativistically invariant law. Therefore, we aim to express physical laws in a *manifestly Lorentz invariant* way as tensorial equations. One speaks also about the *covariant form* of physical laws.

Poincaré Transformations

We have introduced Minkowski space as an affine-linear space, for which the origin is irrelevant. This is exactly for the implementation of the homogeneity of space and time. For any two four-vectors x^μ and y^μ the (squared) norm of their difference,

$$(x - y)^2 \equiv (x^0 - y^0) - (x^1 - y^1)^2 - (x^2 - y^2)^2 - (x^3 - y^3)^2, \tag{6.54}$$

remains invariant under coordinate transformations of the form

$$x'^\mu = \Lambda^\mu{}_\nu x^\nu + a^\mu, \tag{6.55}$$

where $\Lambda^\mu{}_\nu$ is a Lorentz transformation and a^μ is a translation is spacetime. These transformations generalize Lorentz transformations and are called *Poincaré transformations*, or *inhomogeneous Lorentz transformations*. Poincaré transformations are the most general transformations allowed in affine-linear Minkowski space. Our goal is indeed to discover physical theories that are *Poincaré invariant*. The recipe to this end is to seek for Lorentz tensors that are invariant under spacetime translations.

Hypersurfaces and Integration

Let us examine how integration, as discussed generally in Chapter 2, works in the case of Minkowski space. A *hypersurface* Σ in Minkowski space is a 3-dimensional submanifold of $\mathbb{M}_4$. It is described by one scalar equation of the form $f(x) = 0$, or more explicitly in Minkowski coordinates by

$$f(x^0, x^1, x^2, x^3) = 0, \tag{6.56}$$

where $f(x)$ is a scalar function, compare with Section 1.2. The hypersurface Σ can well encompass time and space coordinates, it is not confined to be a surface in three-dimensional space. A special case of a hypersurface is a *hyperplane*, which possesses a flat Euclidean or Lorentzian geometry. In order to describe a hypersurface Σ geometrically, we can consider the vector field being everywhere normal to Σ. Let us consider the differential,

$$\mathrm{d}f = \partial_\mu f \, \mathrm{d}x^\mu = 0, \tag{6.57}$$

of the above defining equation. All displacements $\mathrm{d}x^\mu$ appearing here lie along Σ, i.e. they are tangential to Σ. Thus, we can introduce a *normal four-vector field* $n^\mu(x)$ through

$$n^\mu \equiv \eta^{\mu\nu} \partial_\nu f. \tag{6.58}$$

If $t^\mu(x)$ represents a *tangent four-vector field* on Σ, then it is $n_\mu t^\mu = 0$ for all points x on the hypersurface. The normal four-vector field is not uniquely determined, as for any scalar function $s(x) > 0$ the four-vector $s\eta^{\mu\nu}\partial_\nu f$ represents also a vector field normal to Σ. If the normal vector field $n^\mu(x)$ is (everywhere) non-zero, which effectively means that it is not null, it can be normalized and one can introduce a unique *unit normal four-vector field* $N^\mu(x)$ on the hypersurface, defined as

$$N^\mu \equiv \frac{n^\mu}{\sqrt{|n^2|}}, \tag{6.59}$$

with the normalization $N_\mu N^\mu = \pm 1$. Depending on the type of its normal four-vector field $n^\mu(x)$, we can classify a hypersurface Σ in the following way:

$$\left. \begin{array}{lll} n^\mu \text{ is everywhere timelike} & \Leftrightarrow & \Sigma \text{ is a spacelike hypersurface} \\ n^\mu \text{ is everywhere null} & \Leftrightarrow & \Sigma \text{ is a null hypersurface} \\ n^\mu \text{ is everywhere spacelike} & \Leftrightarrow & \Sigma \text{ is a timelike hypersurface} \end{array} \right\}. \tag{6.60}$$

We note that a null normal vector is also a null tangent vector for a null hypersurface. The *hypersurface element* $\mathrm{d}\Sigma_\mu(x)$ for a 3-dimensional hypersurface Σ is given as (see appendix B.8)

$$\mathrm{d}\Sigma_\mu \equiv N_\mu \sqrt{|\det(h_{jk})|} \, \mathrm{d}y^1 \wedge \mathrm{d}y^2 \wedge \mathrm{d}y^3. \tag{6.61}$$

Here, (y^1, y^2, y^3) denotes a set of coordinates on Σ and the $(0,2)$-tensor $h_{jk}(y)$ represents the induced metric on the hypersurface, see the definition 2.42. By using the results of Section 2.5, we can write the hypersurface element $\mathrm{d}\Sigma_\mu(x)$ in Minkowski coordinates as

$$\mathrm{d}\Sigma_\mu = \frac{1}{3!} \, \epsilon_{\mu\nu\rho\sigma} \, \mathrm{d}x^\nu \wedge \mathrm{d}x^\rho \wedge \mathrm{d}x^\sigma. \tag{6.62}$$

This hypersurface element can be expressed also explicitly as a row covector in the form

$$d\Sigma_\mu = \left(dx^1 \wedge dx^2 \wedge dx^3, \; dx^2 \wedge dx^3 \wedge dx^0, \; dx^3 \wedge dx^0 \wedge dx^1, \; dx^0 \wedge dx^1 \wedge dx^2\right). \quad (6.63)$$

In the special case of a spacelike hypersurface Σ, the element $d\Sigma_\mu(x)$ is timelike for all points of Σ. For such spacelike hypersurfaces, the distance $x - y$ is spacelike for all pairs of points x, y of Σ. A special case of a spacelike hypersurface is a *spacelike hyperplane* with the additional condition $x^0 = $ const. In this case, the hyperplane element is given by $d\Sigma_\mu = N_{(0)\mu}d^3x$, with the constant normal vector $N_{(0)\mu} = (1, 0, 0, 0)$, see the sketch 6.3. Given an arbitrary vector field $V^\mu(x)$ on $\mathbb{M}_4$, possibly defined only within a suitable subset

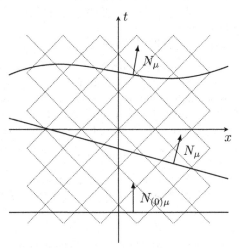

Figure 6.3: A spacelike hypersurface and two spacelike hyperplanes

of $\mathbb{M}_4$, we can consider the integral expression

$$\int_\Sigma V^\mu d\Sigma_\mu, \quad (6.64)$$

where we integrate over an open or closed hypersurface Σ of $\mathbb{M}_4$. In the special case of a spacelike hyperplane Σ, the integral simplifies to the expression

$$\int_\Sigma V^0 d^3x, \quad (6.65)$$

where the integration domain Σ represents a three-dimensional spatial volume within Euclidean space $\mathbb{E}^3$. We want to examine integrals of the divergence $\partial_\mu V^\mu$ of a vector field $V^\mu(x)$. Based on our results in Chapter 2, let us first express the Gauss divergence theorem in Minkowski space as

$$\int_U \partial_\mu V^\mu d^4x = \oint_{\partial U} V^\mu d\Sigma_\mu. \quad (6.66)$$

The lhs integral employs the *Minkowski space volume form* d^4x, given as

$$d^4x = dx^0 \wedge dx^1 \wedge dx^2 \wedge dx^3, \quad (6.67)$$

and represents an integration over the four-dimensional domain $U \subseteq \mathbb{M}_4$. The rhs integral has the three-dimensional closed hypersurface $\Sigma = \partial U$ as integration domain. If the *continuity equation* (also called *divergence equation*)

$$\partial_\mu V^\mu = \partial_0 V^0 + \partial_k V^k = 0 \quad (6.68)$$

holds, some very useful results can be deduced. We consider first the case where the integration domain U is bounded by two spacelike hypersurfaces Σ_a and Σ_b plus a cylindrical boundary Σ_∞ at infinity, i.e. $\partial U = \Sigma_a \cup \Sigma_b \cup \Sigma_\infty$, see the sketch 6.4. We assume now that

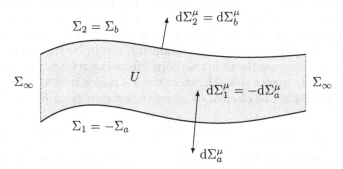

Figure 6.4: Spacetime integration domain, its boundaries and hypersurface elements

the continuity equation 6.68 holds. Concentrating on the rhs of the divergence theorem, we deduce that it is $G(\Sigma_a) + G(\Sigma_b) + G(\Sigma_\infty) = 0$, where we have introduced the integral

$$G(\Sigma_i) \equiv \int_{\Sigma_i} V^\mu d\Sigma_\mu, \tag{6.69}$$

which we call the *generator*, or *charge*. The reason for the naming will be clear at a later stage. Assuming the vector field $V^\mu(x)$ vanishes at the boundary Σ_∞, we obtain the result

$$G(\Sigma_1) = G(\Sigma_2) = \text{const.} \tag{6.70}$$

Here the generator $G(\Sigma_i)$ has to be evaluated on a spacelike hypersurface Σ_i, with a future-oriented surface element $d\Sigma_\mu(x)$, $i = 1, 2$. In other words, the generator stays constant, when evaluated over *any* of these hypersurfaces Σ_i. Choosing specifically a hyperplane with $x^0 = \text{const}$ as the hypersurface, we obtain the expression

$$G(t) = \int_\Sigma V^0(\boldsymbol{x}, t) \, d^3x, \tag{6.71}$$

for which it is $G(t) = \text{const}$. Phrased differently, the equation

$$\frac{dG}{dt} = 0 \tag{6.72}$$

holds, i.e. the generator stays constant in time. Another application of the continuity equation is obtained if we consider from the outset a spacelike hyperplane with $x^0 = \text{const}$ and integrate only over a three-dimensional space volume $R \subseteq \mathbb{E}^3$. Then it is

$$\int_R \partial_\mu V^\mu d^3x = 0. \tag{6.73}$$

By writing out the divergence and by applying the three-dimensional divergence theorem, we obtain

$$\frac{\partial}{\partial x^0} \int_R V^0 d^3x = - \oint_{\partial R} V^k d^2 S_k, \tag{6.74}$$

where $d^2\boldsymbol{S}(\boldsymbol{x})$ is the two-dimensional surface element of the closed spatial surface ∂R. The last equation says that the temporal change of the generator for the three-volume R equals minus the flux of the vector field through the two-dimensional surface ∂R bounding the volume. If the volume is $R = \mathbb{E}^3$ and the vector field $V^k(x)$ falls off sufficiently fast when approaching infinity, one obtains again the constancy equation 6.72.

6.3 Relativistic Particle Mechanics

Proper Time, Four-Velocity and Four-Acceleration

We consider a single point particle and allow relativistic velocities. In order to measure the motion in Minkowski spacetime, we use the *worldline* of the point particle, which is the curve $x^\mu(\tau)$, where τ is a suitable curve parameter. There is no single choice for the real parameter τ. Using the time t as the parameter, like in Newtonian kinematics, is not appropriate anymore, because the time transforms from one reference frame to another. So we need a Lorentz-scalar parameter. The most natural choice is to use the *invariant length* s along the worldline. The line element ds along the curve is given by

$$ds = \sqrt{dx_\mu dx^\mu}, \tag{6.75}$$

where dx^μ is the differential change along the worldline. In the inertial frame in which the particle is *momentarily at rest*, the line element takes the value $ds = dx^0 = dt$. So the invariant length s essentially is the time experienced by the particle, called also the *proper time*. Hence, we define the worldline of the point particle as the curve $x^\mu(s)$, where the curve parameter is the proper time s. It is easily seen that

$$\frac{dt}{ds} = \frac{1}{\sqrt{1 - v^2}} = \gamma. \tag{6.76}$$

Now for any point particle, we define its *four-velocity* u^μ by

$$u^\mu \equiv \frac{dx^\mu}{ds}. \tag{6.77}$$

If the particle has the nonrelativistic three-velocity $\boldsymbol{v}$ in an inertial reference frame, its four-velocity u^μ has the components

$$u^\mu = \begin{pmatrix} \gamma \\ \gamma\boldsymbol{v} \end{pmatrix} \tag{6.78}$$

in this frame. It is always $u_\mu u^\mu = 1$, so the four-velocity is a timelike vector. Furthermore, one defines the *four-acceleration* b^μ of the particle as

$$b^\mu \equiv \frac{du^\mu}{ds} = \frac{d^2 x^\mu}{ds^2}. \tag{6.79}$$

It is always $b_\mu u^\mu = 0$, so the four-acceleration is a spacelike vector. A direct calculation shows that the components of the four-acceleration b^μ are given by

$$b^\mu = \begin{pmatrix} \gamma^4 (\boldsymbol{b} \cdot \boldsymbol{v}) \\ \gamma^2 \boldsymbol{b} + \gamma^4 (\boldsymbol{b} \cdot \boldsymbol{v}) \boldsymbol{v} \end{pmatrix}, \tag{6.80}$$

where $\boldsymbol{b}$ is the nonrelativistic three-acceleration of the particle within the chosen inertial reference frame. It is straightforward to calculate the norm of the four-acceleration, revealing it is a negative quantity. (*exercise 6.4*)

Mass, Momentum, Force and Energy

The mass m of the point particle is an intrinsic property that affects its motion under the influence of external forces. The mass is a positive scalar quantity, $m > 0$, and it is assumed that it does not change in space and time and does not otherwise depend on the state of motion.* The *four-momentum* p^μ of a single particle is defined as

$$p^\mu \equiv m u^\mu, \tag{6.81}$$

where u^μ is the four-velocity of the particle. In components, the four-momentum is given by the formula

$$p^\mu = \begin{pmatrix} \gamma m \\ \gamma m \boldsymbol{v} \end{pmatrix}. \tag{6.82}$$

The time component p^0 is identified with the *relativistic mechanical energy* E_{rel} of the particle,

$$p^0 = \gamma m \equiv E_{\text{rel}}. \tag{6.83}$$

The space components $\boldsymbol{p}$ generalize the nonrelativistic three-momentum to the *relativistic three-momentum* $\boldsymbol{p}_{\text{rel}}$ of the particle,

$$\boldsymbol{p} = \gamma m \boldsymbol{v} \equiv \boldsymbol{p}_{\text{rel}}. \tag{6.84}$$

In the limit of small velocities, $v \ll 1$, one recovers the nonrelativistic three-momentum $m\boldsymbol{v}$. For now we keep the notation E_{rel} and $\boldsymbol{p}_{\text{rel}}$ to make the presence of the relativistic quantities obvious, but later on we will skip the subscripts. The invariant square of the four-momentum is

$$p_\mu p^\mu = E_{\text{rel}}^2 - \boldsymbol{p}_{\text{rel}}^2 = m^2. \tag{6.85}$$

Apparently, the four-momentum is a timelike vector. For measuring relativistic particle dynamics, we define the *four-force* (or *Minkowski force*) f^μ by the dynamical equation

$$\boxed{f^\mu = m b^\mu = \frac{\mathrm{d} p^\mu}{\mathrm{d} s}} \tag{6.86}$$

where once again s is the proper time parameter of the particle and m its scalar mass. The four-force is a spacelike vector. By using the formula $f^\mu = m b^\mu$, we see that the components of the four-force can be written as

$$f^\mu = \begin{pmatrix} \gamma^4 m \left(\boldsymbol{b} \cdot \boldsymbol{v} \right) \\ \gamma^2 m \boldsymbol{b} + \gamma^4 m \left(\boldsymbol{b} \cdot \boldsymbol{v} \right) \boldsymbol{v} \end{pmatrix}, \tag{6.87}$$

with $\boldsymbol{v}$ and $\boldsymbol{b}$ being the usual three-dimensional velocity and acceleration of the particle respectively. Similar to what we did for energy and momentum, we can define a *relativistic three-force* $\boldsymbol{F}_{\text{rel}}$ by

$$\frac{\mathrm{d}}{\mathrm{d}t} \boldsymbol{p}_{\text{rel}} = \frac{\mathrm{d}}{\mathrm{d}t} \left(\gamma m \boldsymbol{v} \right) \equiv \boldsymbol{F}_{\text{rel}}, \tag{6.88}$$

where t is the time in the chosen inertial reference frame. By using the relativistic three-force, the four-force can be written in components as

$$f^\mu = \begin{pmatrix} \gamma \boldsymbol{F}_{\text{rel}} \cdot \boldsymbol{v} \\ \gamma \boldsymbol{F}_{\text{rel}} \end{pmatrix}. \tag{6.89}$$

The time component of the four-force law is

$$\frac{\mathrm{d}}{\mathrm{d}t} \left(\gamma m \right) = \boldsymbol{F}_{\text{rel}} \cdot \boldsymbol{v}. \tag{6.90}$$

---

*Later we will introduce another intrinsic property of particles and fields, the spin.

Assuming that the relativistic three-force is obtained through a potential energy function, i.e. $\boldsymbol{F}_{\text{rel}} = -\boldsymbol{\nabla} V(\boldsymbol{x})$, we can integrate to

$$\gamma m = -V(\boldsymbol{x}) + \text{const.} \tag{6.91}$$

If we define the *total relativistic energy* of the particle E_{total} as

$$E_{\text{total}} \equiv \gamma m + V(\boldsymbol{x}), \tag{6.92}$$

we realize that we have just obtained the conservation law $E_{\text{total}} = \text{const.}$ By using a Taylor expansion with respect to the velocity v, we obtain for the total energy

$$E_{\text{total}} = m + \frac{1}{2}mv^2 + \frac{3}{8}mv^4 + \mathcal{O}(v^6) + V(\boldsymbol{x}). \tag{6.93}$$

The additive constant term m (equal mc^2 in SI units) is the *rest (mass) energy* of the particle and expresses the equivalence of mass and energy. The total energy formula tells us also that for a particle to reach the velocity of light $v = 1$ (which is $v = c$ in SI units), one would need to transfer an infinite amount of energy to the particle. The relativistic kinetic energy T_{rel} of the particle is defined as

$$T_{\text{rel}} \equiv E_{\text{total}} - m - V(\boldsymbol{x}), \tag{6.94}$$

and given as

$$T_{\text{rel}} = \frac{1}{2}mv^2 + \frac{3}{8}mv^4 + \mathcal{O}(v^6), \tag{6.95}$$

as a Taylor expansion. In the limit $v \ll 1$ one recovers the nonrelativistic result $T = mv^2/2$.

Angular Momentum and Center of Energy

Suppose the relativistic point particle has the coordinate x^μ and the momentum p^μ. The four-dimensional generalization of the concept of angular momentum of the particle requires an antisymmetric second rank tensor, see appendix B.1. We define the relativistic *four-angular momentum tensor* $L^{\mu\nu}$ of the particle as

$$L^{\mu\nu} \equiv x^\mu p^\nu - x^\nu p^\mu. \tag{6.96}$$

One can inspect that the tensor $L^{\mu\nu}$ transforms as we would expect under homogeneous Lorentz transformations. Under translations, however, the angular momentum is not invariant. Defining additionally the *four-torque tensor* $N^{\mu\nu}$ by

$$N^{\mu\nu} \equiv x^\mu f^\nu - x^\nu f^\mu, \tag{6.97}$$

one can easily see that the dynamical law

$$\boxed{N^{\mu\nu} = \frac{\mathrm{d}L^{\mu\nu}}{\mathrm{d}s}} \tag{6.98}$$

holds. The pure space components $L^{kl} = x^k p^l - x^l p^k$ of the four-angular momentum are

$$L^{23} = L_1, \qquad L^{31} = L_2, \qquad L^{12} = L_3, \tag{6.99}$$

and identified with the *three-dimensional relativistic angular momentum* vector components

$$L_j = \gamma m \epsilon_{jkl} x^k v^l. \tag{6.100}$$

The mixed time and space components $L^{0k} = x^0 p^k - x^k p^0$ are

$$L^{0k} = t\, p^k - E\, x^k \tag{6.101}$$

and identified with the *three-dimensional relativistic center of mass momentum*

$$G_k = \gamma m \left(t v_k - x_k \right). \tag{6.102}$$

Consequently, the four-angular momentum tensor $L^{\mu\nu}$ has the component form

$$(L^{\mu\nu}) = \begin{pmatrix} 0 & G_1 & G_2 & G_3 \\ -G_1 & 0 & L_3 & -L_2 \\ -G_2 & -L_3 & 0 & L_1 \\ -G_3 & L_2 & -L_1 & 0 \end{pmatrix}. \tag{6.103}$$

For a system consisting of many relativistic point particles the total four-momentum and the total angular momentum tensor are simply the respective sums over all individual particles. However, differentiation with respect to a proper time parameter poses a challenge, since the particles move differently and as such have different proper time values. There are certain physical cases where a generalization of the relativistic equations of motion is possible for multipartite systems. In any case, the relativistic quantities p^μ and $L^{\mu\nu}$ keep their usefulness and occupy a central role in the relativistic symmetry and conservation, as we will show soon. In addition, they can be extended to the case of fields, which actually provide the natural formalism for a relativistic description.

Four-Current Density

As in the nonrelativistic case, it is useful to describe a set of relativistic point particles, call them m_n, with $n = 1, \ldots, N$, through a field function encoding the flow of energy and momentum through spacetime. This is the *mass four-current density* $j^\mu(x)$, defined as the four-vector

$$j^\mu = \begin{pmatrix} \rho \\ \boldsymbol{j} \end{pmatrix}, \tag{6.104}$$

where $\rho(\boldsymbol{x}, t)$ and $\boldsymbol{j}(\boldsymbol{x}, t)$ are the three-dimensional mass density and current density respectively, see Section 4.3. The four-current density $j^\mu(x)$ can be expressed by the formula

$$j^\mu(x) = \sum_n \int_{-\infty}^{\infty} m_n \frac{\mathrm{d} z_n^\mu(s_n)}{\mathrm{d} s_n} \delta \left(x - z_n(s_n) \right) \mathrm{d} s_n, \tag{6.105}$$

where $z_n^\mu(s_n)$ is the four-dimensional trajectory of the nth particle and s_n its associated proper time. (*exercise 6.5*) The proper time parameters s_n are private for each particle. The *current conservation equation* becomes

$$\boxed{\partial_\mu j^\mu = 0} \tag{6.106}$$

This is the relativistic formulation of the analogous conservation equation from Section 4.3. It can be derived also directly by using the properties of the delta distribution. (*exercise 6.6*) This continuity equation once again expresses the local conservation of mass. The general concept of a current density is employed also for describing electric charge distributions, as we will see later. One simply replaces the masses m_n by the respective electric charges q_n.

6.4   Lagrangian Formulation

Covariant Action Principle

In this section we develop mechanics of a relativistic point particle from a formal standpoint, based on the action principle. We carry this out to demonstrate how the variational method works in the relativistic case but also in order to lay down the ground for the subsequent discussion of symmetry and conservation. The *action principle* retains its general form for relativistic physical systems and reads

$$\delta S\left[x^{\mu}\right] = 0, \tag{6.107}$$

where the action $S\left[x^{\mu}\right]$ must be a Lorentz scalar. We choose a general real parameter τ for describing the particle dynamics and consider a Lorentz-scalar Lagrangian

$$L\left(x^{\mu}, \dot{x}^{\mu}; \tau\right). \tag{6.108}$$

Here we use the notation

$$\dot{x}^{\mu}(\tau) \equiv \frac{\mathrm{d}x^{\mu}}{\mathrm{d}\tau}(\tau), \tag{6.109}$$

with the dot denoting the derivation of the trajectory vector $x^{\mu}(\tau)$ with respect to the parameter τ. The corresponding *relativistic particle action* $S\left[x^{\mu}\right]$ is

$$S\left[x^{\mu}\right] \equiv \int_{\tau_1}^{\tau_2} L\left(x^{\mu}(\tau), \dot{x}^{\mu}(\tau); \tau\right) \mathrm{d}\tau. \tag{6.110}$$

We should note that the real parameter τ is chosen without any special constraints. If we would choose specifically the proper time s as the parameter, we would need to respect the additional condition $u^2 = 1$. This condition would actually complicate the variational problem. So in order to avoid that, we let the parameter τ be completely general. We note the relation

$$\dot{x}^{\mu} = u^{\mu}\frac{\mathrm{d}s}{\mathrm{d}\tau}, \tag{6.111}$$

leading to

$$(\dot{x})^2 = \left(\frac{\mathrm{d}s}{\mathrm{d}\tau}\right)^2, \tag{6.112}$$

and therefore

$$\frac{\dot{x}^{\mu}}{|\dot{x}|} = u^{\mu}, \tag{6.113}$$

with the norm $|\dot{x}| = \sqrt{\dot{x}_{\mu}\dot{x}^{\mu}}$. It is desirable that the form of the action 6.110 remains unaltered if a different general parameter $\tau' = \tau'(\tau)$ is used. For a Lagrangian without explicit dependence on the τ parameter, this is achieved if the Lagrangian function is homogeneous in its $\dot{x}^{\mu}$ argument. Now applying the principle of stationary action is straightforward and proceeds in complete analogy to the nonrelativistic case, see also appendix B.6. One obtains the *relativistic Euler-Lagrange equations*

$$\frac{\partial L}{\partial x^{\mu}} - \frac{\mathrm{d}}{\mathrm{d}\tau}\frac{\partial L}{\partial \dot{x}^{\mu}} = 0. \tag{6.114}$$

We introduce also the *generalized momentum* $p_{\mu}(\tau)$, where we use the same symbol as for the mechanical four-momentum,

$$p_{\mu} \equiv \frac{\partial L}{\partial \dot{x}^{\mu}}. \tag{6.115}$$

According to the equations of motion, it is

$$\frac{\partial L}{\partial x^\mu} = \frac{\mathrm{d}p_\mu}{\mathrm{d}\tau}. \tag{6.116}$$

Thus, if the Lagrangian does not depend explicitly on x^μ, which in equivalent to translation invariance, the generalized momentum is conserved.

Relativistic Particle Lagrangians

For a single free relativistic point particle there are many different and equivalent Lagrangians, meaning that they lead to the same free equation of motion. A possible choice for the Lagrangian is

$$L = -m\frac{\mathrm{d}s}{\mathrm{d}\tau} = -m\sqrt{\dot{x}_\mu\dot{x}^\mu}. \tag{6.117}$$

If one chooses an inertial frame and employs the time t as the parameter τ, the above reduces for small velocities, $|v| \ll 1$, to the non-covariant Lagrangian

$$L = -m + \frac{1}{2}mv^2 + \mathcal{O}(v^4). \tag{6.118}$$

This is the nonrelativistic free particle Lagrangian, plus a constant mass term, plus velocity-dependent correction terms. The covariant action $S\left[x^\mu\right]$ for the Lagrangian 6.117 is

$$S\left[x^\mu\right] = -m\int_{s_1}^{s_2}\mathrm{d}s = -m\int_{\tau_1}^{\tau_2}\sqrt{\eta_{\mu\nu}\frac{\mathrm{d}x^\mu}{\mathrm{d}\tau}\frac{\mathrm{d}x^\nu}{\mathrm{d}\tau}}\,\mathrm{d}\tau. \tag{6.119}$$

If we replace in the last expression the rigid Minkowski metric $\eta_{\mu\nu}$ by a general varying metric $g_{\mu\nu}(x)$, we can generalize the action to the case of a particle moving in an arbitrary geometry. The equations of motion for the action above are easily found to be

$$\frac{\mathrm{d}^2x^\mu}{\mathrm{d}\tau^2} = 0, \tag{6.120}$$

and are as expected. Another equivalent Lagrangian for the free particle case is

$$L = -\frac{1}{2}m\dot{x}_\mu\dot{x}^\mu, \tag{6.121}$$

as one can inspect. Introducing an interaction between the relativistic particle and an external relativistic force is achieved by adding an interaction term to the free Lagrangian. If there is a given vector field $A^\mu(x)$ acting on the particle, we can write a scalar interaction Lagrangian L_{int} of the form

$$L_{\mathrm{int}} = -qA_\mu(x)\dot{x}^\mu. \tag{6.122}$$

The real parameter q determines the strength of the coupling between the vector field and the particle. We will encounter this interaction concept again when we discuss fields. The total Lagrangian reads now

$$L = -m\sqrt{\dot{x}_\mu\dot{x}^\mu} - qA_\mu(x)\dot{x}^\mu. \tag{6.123}$$

The variation with respect to the particle trajectory leads to the equation of motion

$$\frac{\mathrm{d}^2x_\mu}{\mathrm{d}\tau^2} = q(\partial_\mu A_\nu - \partial_\nu A_\mu)\frac{\mathrm{d}x^\nu}{\mathrm{d}\tau}. \tag{6.124}$$

This is exactly the equation of the *Lorentz force* of the electromagnetic field acting on a charged particle possessing the electric charge q, see also Section 15.5. The *generalized momentum* $p_\mu(\tau)$ for this theory is

$$p_\mu \equiv mu_\mu + qA_\mu(x) \tag{6.125}$$

revealing that the field $A^\mu(x)$ carries a momentum. Looking more broadly, there are other possible interaction terms that we can consider. The main requirement is that they are Lorentz scalars. If we have a tensor field $\phi^{\mu\nu}(x)$ defined on Minkowski space, an interaction Lagrangian of the form

$$L_{\text{int}} = \phi_{\mu\nu}(x)\dot{x}^\mu\dot{x}^\nu \tag{6.126}$$

is perfectly admissible from the point of view of relativistic invariance. The reader is encouraged to derive the corresponding equations of motion and the generalized momentum. (*exercise 6.7*) Now that we have introduced a relativistic Lagrangian description, we move on to establishing a relativistic version of Noether's theorem.

6.5   Relativistic Symmetry and Conservation

Noether's Theorem in Relativistic Mechanics

Noether's theorem of symmetry and conservation can be extended to relativistic mechanics and is established along the same lines, as in the non-relativistic case. One considers symmetry transformations of the basic variables of the theory, for which the action stays invariant. For each symmetry transformation there is an associated conserved quantity. For demonstrating this theorem, let us consider the most simple case of a single particle, for which the Lagrangian $L(x^\mu, \dot{x}^\mu)$ does not explicitly depend on the evolution parameter τ. The *symmetry transformations* to be considered are

$$\left.\begin{array}{rclcl}
\tau & \mapsto & \tau' = \tau + \delta\tau(\tau) & \equiv \tau + \epsilon^a \mathrm{T}_a(\tau) \\[4pt]
x^\mu(\tau) & \mapsto & x'^\mu(\tau) = x^\mu(\tau) + \delta x^\mu(\tau) & \equiv x^\mu(\tau) + \epsilon^a \mathrm{H}_a^\mu(\tau) \\[4pt]
x^\mu(\tau) & \mapsto & x'^\mu(\tau') = x^\mu(\tau) + \bar{\delta} x^\mu(\tau) & \equiv x^\mu(\tau) + \epsilon^a \Phi_a^\mu(\tau)
\end{array}\right\}. \tag{6.127}$$

We should note immediately that the evolution parameter τ, allowed to transform here, has a different meaning than the time parameter t in the non-relativistic case. The time t belongs to the spacetime variables transforming under Galilei transformations. The evolution parameter τ, however, is beyond the set of spacetime variables, while the later transform under Poincaré transformations. It is the unified variable x^μ which contains time and space. The parametrization of the symmetry transformations is again achieved by the real parameters ϵ^a, with $a = 1, \ldots, r$. The generator functions $\mathrm{T}_a(\tau)$, $\mathrm{H}_a^\mu(\tau)$ and $\Phi_a^\mu(\tau)$ are defined in complete analogy to the non-relativistic case, compare with Section 5.2. For the form variation $\delta x^\mu(\tau)$ and the total variation $\bar{\delta} x^\mu(\tau)$ we have again the fundamental relation

$$\bar{\delta} x^\mu(\tau) = \delta x^\mu(\tau) + \dot{x}^\mu(\tau)\,\delta\tau. \tag{6.128}$$

Now the central requirement is that the action stays invariant under the above symmetry transformations, i.e.

$$S\left[x';\tau'\right] = S\left[x;\tau\right]. \tag{6.129}$$

This condition can be expressed as the variational equation

$$\delta S\left[x;\tau\right] = 0. \tag{6.130}$$

From here onward the derivation of the conserved quantities proceeds exactly as in the non-relativistic case. (*exercise 6.8*) One defines the *Noether charges* $J_a(\tau)$ as

$$J_a \equiv H\,T_a - p_\mu \Phi_a^\mu + \mathrm{m}_a, \tag{6.131}$$

where $H(\tau)$ is the *Hamiltonian function*

$$H \equiv p_\mu \dot{x}^\mu - L \tag{6.132}$$

of the system and $\mathrm{m}_a(\tau)$ is the function that encodes the freedom ones has in the definition of the Lagrangian. The *Noether conservation equation* reads then

$$\frac{\mathrm{d}J_a}{\mathrm{d}\tau} = 0, \tag{6.133}$$

with the index $a = 1, \ldots, r$ enumerating each single symmetry. To give this equation a definite physical meaning, we can choose the parameter τ to be the proper time s of the particle. Then we can express *Noether's theorem* as:

> For every global symmetry transformation of the action, there is a Noether charge that is constant with respect to proper time.

Let us proceed to a few applications of this version of Noether's theorem.

Translation Invariance and Conservation of Four-Momentum

Under a translation of spacetime expressed by

$$\delta\tau = 0 \quad \text{and} \quad \bar{\delta}x^\rho = x^\rho, \tag{6.134}$$

the four-momentum vector p^μ of the particle is conserved

$$\frac{\mathrm{d}p^\mu}{\mathrm{d}s} = 0. \tag{6.135}$$

This equation encompasses the relativistic conservation for the particle's energy and three-momentum.

Lorentz Invariance and Conservation of Angular Momentum Tensor

We consider an infinitesimal Lorentz rotation of spacetime, which is expressed as

$$\delta\tau = 0 \quad \text{and} \quad \bar{\delta}x^\rho = \omega^{\rho\sigma}x_\sigma, \tag{6.136}$$

with the tensor $\omega^{\rho\sigma}$ being antisymmetric,

$$\omega^{\rho\sigma} = -\omega^{\sigma\rho}. \tag{6.137}$$

The antisymmetry of $\omega^{\rho\sigma}$ is a direct consequence of the Lorentz invariance condition

$$x'^2 = x^2. \tag{6.138}$$

By using the matrix equation $x' = (1 + \omega)x$ for the transformed coordinate, we can easily obtain the condition

$$\omega^T \eta + \eta\,\omega = 0, \tag{6.139}$$

which is equivalent to the antisymmetry of ω. So, if the above Lorentz rotation is a symmetry of the action, then the four-angular momentum tensor $L^{\mu\nu} = x^\mu p^\nu - x^\nu p^\mu$ is conserved,

$$\frac{\mathrm{d}L^{\mu\nu}}{\mathrm{d}s} = 0. \tag{6.140}$$

This conservation covers the corresponding three-dimensional relativistic angular momentum and the relativistic center of mass momentum. In the special case of a pure Lorentz boost, without a rotation, the conserved relativistic quantity is again the complete tensor $L^{\mu\nu}$ containing the relativistic center of mass momentum, see also Section 6.3.

More Conserved Quantities

A different type of symmetry transformation can be defined with the infinitesimal changes

$$\delta\tau = \varepsilon \mathrm{T}(\tau) \quad \text{and} \quad \bar{\delta}x^\rho = 0, \tag{6.141}$$

where $\mathrm{T}(\tau)$ is an arbitrary scalar function of τ. This symmetry corresponds to reparametrizations of τ. The conservation equation reads

$$\frac{\mathrm{d}}{\mathrm{d}s}(H\mathrm{T}) = 0. \tag{6.142}$$

For a constant function T, this becomes $\mathrm{d}H/\mathrm{d}s = 0$. It would be tempting to interpret this as energy conservation, but it would not be accurate. The Hamiltonian function H is identically zero for a single relativistic particle. The proper relativistic particle energy is $E_{\text{total}} = \gamma m + V(\boldsymbol{x})$. Nevertheless, the Hamiltonian H has its purpose for the derivation of the canonical form of the equations of motion. For further discussion, see Barut [6].

Finally, let us consider a scale transformation given infinitesimally by

$$\delta\tau = 0 \quad \text{and} \quad \bar{\delta}x^\rho = \alpha x^\rho, \tag{6.143}$$

where α represents a real scaling parameter. The corresponding finite scale transformation of the spacetime coordinate is given by

$$x'^\rho = \exp(\alpha)x^\rho = [1 + \alpha + \mathcal{O}(\alpha^2)]x^\rho. \tag{6.144}$$

It is easy to derive the conservation equation

$$\frac{\mathrm{d}}{\mathrm{d}s}(p_\mu x^\mu) = 0, \tag{6.145}$$

i.e. the contraction $(p \cdot x)$ is conserved with respect to proper time. For example, the theory with the Lagrangian

$$L = -m\sqrt{\dot{x}_\mu \dot{x}^\mu} + \phi_{\mu\nu}(x)\dot{x}^\mu \dot{x}^\nu \tag{6.146}$$

describing a relativistic particle of mass m that interacts with an external tensor field $\phi_{\mu\nu}(x)$ is scale invariant, provided that the tensor field transforms as

$$\phi'_{\mu\nu}(\mathrm{e}^\alpha x) = \mathrm{e}^{-\alpha}\phi_{\mu\nu}(x) \tag{6.147}$$

under such rescalings of spacetime. The scaling behavior of the field is decisive for attaining the overall scale symmetry of the theory and we will discuss more examples later.

In summary, we have seen that there is a deep connection between basic spacetime transformations, on one hand, and conserved quantities, on the other hand. In the following part of this book we will take a closer look at the diverse spacetime symmetries in a more systematic fashion. We will see that these spacetime symmetries fully determine the basic quantities we can use to build theories. This notably achieved by means of group representations.

Further Reading

Accessible treatments of relativistic mechanics can be found in the books of Barut [6], Goldstein, Poole, Safko [36], and Sexl, Urbantke [82]. For an original discussion of the invisibility of Lorentz contraction, one can turn to the original articles of Penrose [64] and Terrel [88]. We have treated the relativistic center of mass momentum as a part of the angular momentum tensor. However, in analogy to the three-dimensional case, it is possible to define a four-vector measuring the relativistic center of mass. The construction of this center of mass quantity is discussed in the book of DeWitt and Christensen [23].

III

Symmetry Groups and Algebras

7

Lie Groups

To grasp the notion of symmetry and explore its possible realizations, we begin with abstract group theory. We consider the basic definitions of groups, subgroups, and the construction of larger groups from existing ones. The next central notion is that of a group representation, which provides a concrete realization of the group structure at hand. Subsequently, we specialize to Lie groups and identify symmetry transformations as the elements of these Lie groups. Initially, we introduce Lie groups in a general manner, but in the following we restrict ourselves to matrix Lie groups. The reason for this focus is twofold: first, matrix Lie groups provide the primary cases for our applications, and second, the central results of Lie theory for this class of groups can be obtained in the simplest way.

7.1   Notion of a Group

Definition of a Group

Let us begin with the definition. A *group* is a set G with a map, called the *(group) product*, from $G \times G$ to G, denoted as $g_1 \cdot g_2$, with the following three properties:

- *Associativity*: for all elements g_1, g_2, $g_3 \in G$ it is

$$g_3 \cdot (g_2 \cdot g_1) = (g_3 \cdot g_2) \cdot g_1. \tag{7.1}$$

- *Existence of identity*: there is an element, $I \in G$, called the *identity* (or *neutral element*), so that for all $g \in G$ it is

$$g \cdot I = I \cdot g = g. \tag{7.2}$$

- *Existence of inverse element*: for each $g \in G$, there is an element denoted by g^{-1}, called the *inverse*, with the property

$$g \cdot g^{-1} = g^{-1} \cdot g = I. \tag{7.3}$$

The identity element is unique and the inverse of each element is unique. (*exercise 7.1*) Implicit in the definition of a group is the *closure property*, which means that the product of any two elements of the group is again an element of the group. The following identities hold for the inversion (*exercise 7.2*)

$$\left(g^{-1}\right)^{-1} = g, \tag{7.4}$$

DOI: 10.1201/9781003087748-7

$$I^{-1} = I, \qquad\qquad\qquad\qquad (7.5)$$

$$(g_2 \cdot g_1)^{-1} = g_1^{-1} \cdot g_2^{-1}. \qquad\qquad\qquad (7.6)$$

Because of the associativity law, we can write a product of three or more group elements without parentheses simply as $g_3 \cdot g_2 \cdot g_1$. Moreover, we will write the product simply without the dot symbol as $g_3\, g_2\, g_1$. If it is

$$g_2 \cdot g_1 = g_1 \cdot g_2, \qquad\qquad\qquad (7.7)$$

or, equivalently, if the commutator vanishes

$$[g_1, g_2] \equiv g_1 \cdot g_2 - g_2 \cdot g_1 = 0, \qquad\qquad (7.8)$$

for all pairs of elements of the group, then the group is called *abelian* or *commutative* (*Niels Henrik Abel*). In all other cases, where in general $g_2 \cdot g_1 \neq g_1 \cdot g_2$ holds, the group is called *non-abelian*. An example of an abelian group is the *additive group* of the real numbers $\mathbb{R}$, where the usual addition is the group "product".

Subgroups and Invariant Subgroups

A subgroup is a subset of a group that maintains the group structure. More precisely, a *subgroup* of a group G is a subset H of G having the following properties:

- *Closure*: for h_1, $h_2 \in H$, the product $h_2 \cdot h_1$ is also an element of H.
- *Existence of identity*: the identity is an element of H.
- *Existence of inverse element*: for each $h \in H$, the inverse h^{-1} is also an element of H.
- *Associativity:* is fulfilled automatically, since all elements of H are also elements of the group G.

Some further definitions are necessary. Assume we have a subgroup H of a group G, with the property that for every element $h \in H$ and for every element $g \in G$ the element $g \cdot h \cdot g^{-1}$ is an element of H. Then the subgroup H is called an *invariant subgroup*. An equivalent terminology used is *normal subgroup*. A group G with no invariant subgroups is called *simple*. A group G with no invariant abelian subgroups is called *semisimple*. A semisimple group can have invariant subgroups, which will be non-abelian however. We define also the *center of a group* G as the subgroup $Cent(G)$, which contains all elements $h \in G$ that commute with every element of G, i.e.

$$Cent(G) \equiv \{h \in G \mid g \cdot h = h \cdot g \text{ for all } g \in G\}. \qquad (7.9)$$

The center of a group, if it exists, is an abelian subgroup and in fact it is an invariant subgroup. (*exercise 7.3*)

Examples of Groups

An example is the commutative group of complex numbers of modulus 1, which have the form $e^{i\theta}$, with the usual multiplication of complex numbers as the composition law. Another example is the commutative group of the vectors in $\mathbb{R}^n$, where the group multiplication is the usual addition of vectors. In fact, every vector space V is also an additive group in terms of the vector addition. The *Möbius transformations* (*August Ferdinand Möbius*), as the maps $m : \mathbb{C} \to \mathbb{C}$, $z \mapsto m(z)$, defined by

$$m(z) = \frac{az + b}{cz + d}, \qquad\qquad\qquad (7.10)$$

with given complex numbers a, b, c, $d \in \mathbb{C}$ satisfying $ad - bc = 1$ form a group, determined by six real parameters. (*exercise 7.4*) We will deal extensively with *linear matrix groups*, whose elements are matrices and the group product is the matrix multiplication. A matrix group is, for instance, $GL(n, \mathbb{R})$, the *general linear group*, defined as the group of all invertible real $n \times n$-matrices M,

$$GL(n, \mathbb{R}) \equiv \{M \in Mat(n, \mathbb{R}) \mid \det M \neq 0\}. \tag{7.11}$$

A subgroup of the general linear group is the group of all matrices with determinant equal one. This group is called the *special linear group*,

$$SL(n, \mathbb{R}) \equiv \{M \in Mat(n, \mathbb{R}) \mid \det M = 1\}. \tag{7.12}$$

Similar are the definitions in the complex valued case for $GL(n, \mathbb{C})$ and $SL(n, \mathbb{C})$. Another important example of a matrix group is the *orthogonal group*, denoted $O(n)$, and defined as the group of real $n \times n$-matrices O, where the transposition operation yields the inverse matrix,

$$O(n) \equiv \{O \in GL(n, \mathbb{R}) \mid O^T O = O O^T = 1\}. \tag{7.13}$$

An example of a matrix group with complex entries is the *unitary group* $U(n)$ of complex $n \times n$-matrices U, where the hermitian conjugation yields the inverse matrix,

$$U(n) \equiv \{U \in GL(n, \mathbb{C}) \mid U^\dagger U = U U^\dagger = 1\}. \tag{7.14}$$

We will examine these matrix groups later in more detail.

Group Homomorphisms

Maps between groups that preserve the group structure are especially useful. If we have two groups G and H and a map $f : G \to H$, with the property

$$f(g_1 \cdot g_2) = f(g_1) \cdot f(g_2), \tag{7.15}$$

for all g_1, $g_2 \in G$, then this map f is called a *group homomorphism* of G and H. If, additionally, this map is 1-to-1 and onto, then it is called a *group isomorphism*. Isomorphic groups are considered to represent the same (group) structure, although their elements might be very different objects. If we have an isomorphism $f : G \to G$ of a group G with itself, then we talk about a *group automorphism*. For a given group G, the set of all its automorphisms, equipped with the map composition as the product, forms itself a group, which we denote as $Aut(G)$. (*exercise 7.5*) A group homomorphism f between two groups G and H maps the identity and the inverses of G to the identity and the corresponding inverses of H, i.e.

$$f(I_G) = I_H, \tag{7.16}$$

and

$$f(g^{-1}) = f(g)^{-1}. \tag{7.17}$$

The proof is straightforward. (*exercise 7.6*) Group isomorphisms help us to identify groups that represent the same structure. Inversely, we can seek to find all groups, which are not isomorphic to each other. Remarkably, this program has been completed by *Élie Joseph Cartan* for groups that are simple (i.e. containing no invariant subgroups) and analytically compact.

Direct and Semidirect Product of Groups

Having two groups G and H at hand, we can construct a new group out of them by using the Cartesian product, $G \times H$, of the two sets, which contains all ordered pairs (g, h), with $g \in G$ and $h \in H$, as elements. The group multiplication is defined as

$$(g_2, h_2) \cdot (g_1, h_1) \equiv (g_2 \cdot g_1, h_2 \cdot h_1). \tag{7.18}$$

This makes the Cartesian product set indeed a group, which we call the *direct product* of G and H and which we denote again as $G \times H$. Obviously, the identity element in $G \times H$ is (I_G, I_H) and the inverse of (g, h) is (g^{-1}, h^{-1}). The group properties of the direct product can be inspected readily. (*exercise 7.7*)

A slightly more delicate construction is the semidirect product. We will use this construction later, e.g. when we discuss the Euclidean group and the Poincaré group. We start with two groups G and H plus a fixed group homomorphism $\chi : H \to Aut(G)$ that assigns to each element of H an automorphism of G. In simpler words, the map χ defines a transformation in which elements of H act on elements of G. The construction again contains the Cartesian product $G \times H$ of the sets but now the group product is defined as

$$(g_2, h_2) \cdot (g_1, h_1) \equiv (g_2 \cdot \chi(h_2)g_1, h_2 \cdot h_1). \tag{7.19}$$

The elements h_1, h_2 of H are combined as usually, but now the element g_1 of G experiences first a transformation through $\chi(h_2)$ before it is multiplied by g_2. This product definition makes the Cartesian product set indeed a group, which we call the *semidirect product* of G and H and denote as $G \rtimes H$. Sometimes it is denoted as $G \rtimes_\chi H$, since χ is essential for the definition. In a semidirect product the inverse of (g, h) is the element $(\chi(h^{-1})g^{-1}, h^{-1})$. The semidirect product is reduced to a direct product if the automorphism $\chi(h)$ is simply the identity map on G, for all $h \in H$. The construction may seem pretty special, but let us see what this construction means is a slightly simpler case. Imagine that the group G is an additive group of vectors (e.g. the group $\mathbb{R}^n$) and H is a group of linear transformations (e.g. the group $GL(n, \mathbb{R})$ or a subgroup of it) acting on the vectors of G. Then the semidirect product will have the form

$$(g_2, h_2) \cdot (g_1, h_1) \equiv (g_2 + h_2(g_1), h_2 \cdot h_1). \tag{7.20}$$

We now see that the first entry on the rhs has the form of an affine-linear transformation. In this way we obtain so called *inhomogeneous*, or *affine groups*. We will encounter this type of groups later when we study space and time transformations. The additive group will be the group of translations, in space or in spacetime.

7.2 Notion of a Group Representation

Definition of Group Representation

The group elements are indeed the transformations we are interested in, and these transformations act on linear vector spaces. The vector spaces can be for instance physical space, physical spacetime or another type of space describing internal degrees of freedom. This means that we are interested to know how the group elements act on the vector space under consideration. To achieve an action of the group of transformations on the space, we need a *group representation*, which assigns to each group element a linear map acting on this space. Let V be an n-dimensional linear vector space and $GL(V)$ the group of invertible linear maps acting on V. Now consider a group G. An *n-dimensional (group) representation* D of G on V is a group homomorphism,

$$
\begin{aligned}
D : \quad G &\to GL(V) \\
g &\mapsto D(g),
\end{aligned}
\tag{7.21}
$$

which assigns to each group element g a linear invertible map $D(g)$ acting on the linear space,

$$D(g) : \begin{array}{ccc} V & \to & V \\ v & \mapsto & D(g)v. \end{array} \tag{7.22}$$

Based on the definition, the homomorphism relation

$$D(g_2 \cdot g_1) = D(g_2) \cdot D(g_1) \tag{7.23}$$

holds. If a representation D is 1-to-1, then it is called *faithful*. By choosing a basis $\{e_i\}$ in the space V, the representation element $D(g)$ takes the form of a matrix multiplication as

$$D(g)v = D(g)^i{}_j v^j e_i. \tag{7.24}$$

Thus, we obtain a so-called *matrix representation*. Of course we can change the basis in V and then the matrix representation changes accordingly. By changing the basis in V from $\{e_i\}$ to $\{e_i'\}$, with the invertible transformation matrix S as

$$e_i' = S_i{}^j e_j, \tag{7.25}$$

the matrix representation will change as (*exercise 7.8*)

$$D(g)' = S^{-1}D(g)S. \tag{7.26}$$

Two matrix representations, which are related by a similarity transformation as above are called *equivalent* and are, in essence, the same representation. We are interested not only in finite-dimensional representations, characterized by discrete vector-space indices, but also in *infinite-dimensional representations* on function spaces, where the "indices" are continuous variables of functions. We will deal with such representations later.

Reducibility and Irreducibility of a Representation

Assume we have a representation D at hand acting on a space V^n. Now if there is a subspace of V^n, let us call it W^m, with dimension $m \leq n$, with the property that $D(g)w$ is again an element of W^m, for all $w \in W^m$ and for all $g \in G$, then this subspace is called *invariant* under the group representation D. There are always the trivial invariant subspaces $\{0\}$ and V^n. A representation, which has no nontrivial invariant subspaces, is called *irreducible*. A *reducible* representation can be brought into a simpler matrix form. Choosing a suitable basis transformation in V^n, the corresponding similarity transformation brings the representation matrices $D(g)$ for all g into the block-form

$$D(g) = \left(\begin{array}{c|c} D_{11}(g) & D_{12}(g) \\ \hline 0 & D_{22}(g) \end{array} \right), \tag{7.27}$$

with $D_{11}(g)$ as $m \times m$-submatrix and $D_{22}(g)$ as $(n-m) \times (n-m)$-submatrix. If the space V^n has not only the invariant subspace W^m but also another invariant subspace U^{n-m}, with the property that $V^n = W^m \oplus U^{n-m}$, then the corresponding representation is called *completely reducible* (or *decomposable*) and, under a suitable basis transformation, it will take the block-diagonal matrix form

$$D(g) = \left(\begin{array}{c|c} D_{11}(g) & 0 \\ \hline 0 & D_{22}(g) \end{array} \right), \tag{7.28}$$

for all elements g. In such a situation, we say that $D(g)$ is decomposed into a direct sum of two representations and write $D(g) = D_{11}(g) \oplus D_{22}(g)$. There are cases where such a direct sum decomposition of the representation can achieved with more than two subspaces. Then we speak again about complete reducibility.

Schur's Lemma

Irreducibility has a surprising consequence, which is phrased in *Schur's lemma* (*Issai Schur*). If we have a group G with an irreducible representation $D(g)$ on a vector space V^n and a matrix A acting on V^n that commutes with all $D(g)$, i.e.

$$AD(g) = D(g)A, \qquad (7.29)$$

for all elements $g \in G$, then the matrix A is proportional to the identity matrix,

$$A = \alpha 1. \qquad (7.30)$$

For the proof we note that A must have at least one eigenvector, call it v_α, i.e. $Av_\alpha = \alpha v_\alpha$, with $\alpha \neq 0$. The vector v_α belongs to the kernel of the operator $A - \alpha 1$, i.e. v_α is in the set of vectors, which are mapped to zero, $v_\alpha \in \ker(A - \alpha 1)$. This kernel is not the trivial space, $\ker(A - \alpha 1) \neq \{0\}$. Because it is also

$$(A - \alpha 1)D(g)v = D(g)(A - \alpha 1)v, \qquad (7.31)$$

for all $g \in G$ and for all $v \in V^n$, we conclude that $\ker(A - \alpha 1)$ is an invariant space under the representation $D(g)$, which means that D is either reducible, or the kernel is equal the whole space V^n. The first option is a contradiction to the assumptions, so the only possibility left is that the extremal case is valid, i.e. that the kernel is the whole space, $\ker(A - \alpha 1) = V^n$. Hence, the underlying operator $A - \alpha 1$ is equal zero and we conclude that $A = \alpha 1$.

Orthogonal and Unitary Representations

In group representations on vector spaces equipped with a scalar product certain representations are especially interesting. Staring with the case of a real vector space with a symmetric positive definite scalar product, a group representation on the space that preserves the scalar product will be very useful. This representation will be one for which all matrices are orthogonal. Per definition, an *orthogonal representation* D of G is one for which

$$D(g)^T D(g) = D(g)D(g)^T = 1 \qquad (7.32)$$

holds, for all $g \in G$. Similarly, if we consider a complex vector space equipped with a hermitian positive definite scalar product, the representations, which are unitary will be special. A *unitary representation* D of G is one for which

$$D(g)^\dagger D(g) = D(g)D(g)^\dagger = 1 \qquad (7.33)$$

holds, for all $g \in G$. Here, $D(g)^\dagger$ denotes the hermitian adjoint of $D(g)$. Very important for our applications are also pseudo-orthogonal and pseudo-unitary representations. We will encounter these later with the Lorentz group and the conformal group.

7.3   Lie Groups and Matrix Groups

Definition of a Lie Group

A *real r-parameter Lie group* is a group, which is also an r-dimensional differentiable manifold. More precisely, the group operations of multiplication and inversion are differentiable maps. As with all differentiable manifolds, the elements can be specified locally by r real parameters, denote them ϵ^a, $a = 1, \ldots, r$, the local coordinates. In the context of Lie groups these parameters are called the *group parameters* and each element of the group can be given as a differentiable real function $g(\epsilon^a) \equiv g(\epsilon^1, \ldots, \epsilon^r)$. For instance, the well-known three Euler angles are a choice for three real parameters for the rotation group. Generally, these parameters need not to be defined globally for the entire manifold and typically there are different coordinate sets needed to describe different parts of the group manifold. The number of parameters r is called also the *dimension of the Lie group*. The definition of a *complex Lie group* is analogous, where the differentiable manifold is complex.

The combination of differentiability and the group properties lead to specific relations between the parameter functions, denoted ϵ, α, β, γ, etc. here. The closure under the group multiplication, $g(\alpha)g(\beta) = g(\epsilon(\alpha, \beta))$, means that the function $\epsilon(\alpha, \beta)$ is differentiable it its arguments. A convenient and possible choice of parametrization is

$$g(0) = g(0, \ldots, 0) = I. \tag{7.34}$$

Associativity leads to the relation

$$\epsilon(\alpha, \epsilon(\beta, \gamma)) = \epsilon(\epsilon(\alpha, \beta), \gamma). \tag{7.35}$$

The neutral element is translated as

$$\epsilon(\alpha, 0) = \epsilon(0, \alpha) = \alpha. \tag{7.36}$$

The existence of inverse means that if we define α^{-1} as the parameter set for which $g(\alpha^{-1}) = g(\alpha)^{-1}$, then it is

$$\epsilon(\alpha, \alpha^{-1}) = \epsilon(\alpha^{-1}, \alpha) = 0. \tag{7.37}$$

The above relations between parameter functions encode the differential and the group structure of a Lie group. A Lie group is called *compact*, if its set of parameters is bounded, i.e. $\lambda^a \leq \epsilon^a \leq \upsilon^a$, for certain values of λ^a and υ^a, for all $a = 1, \ldots, r$ and if it is closed, meaning that the limit of every convergent series of parameters is again a parameter value of the group.

Transformation Groups and Symmetry Groups

For our applications, we are foremost interested in those Lie groups that define symmetry transformations on geometric manifolds. Generally, given a Lie group G, with elements g, h, etc., and a differentiable manifold $\mathcal{M}$, with elements denoted x, the group G is called a *transformation group* acting on $\mathcal{M}$, if the two properties

$$g\left(h(x)\right) = (g \cdot h)\left(x\right) \quad \text{and} \quad I(x) = x \tag{7.38}$$

hold, for all g, $h \in G$ and for all $x \in \mathcal{M}$. In particular, it is then also

$$g^{-1}\left(g(x)\right) = x. \tag{7.39}$$

Each transformation group element g represents a diffeomorphism of the manifold. If we use the parametrization $g(\epsilon)$ for the group elements, then the result of a transformation $g(\epsilon)$ acting on x can be written as

$$x \mapsto g(\epsilon)(x) \equiv x'(x, \epsilon). \tag{7.40}$$

In the special case of a geometric manifold $\mathcal{M}$, if the group elements g are all symmetry transformations of the manifold, the group G is called a *symmetry group of the manifold* $\mathcal{M}$. We should point out that the group elements in general do not operate linearly on the manifold elements. In fact, in most cases the action of a group transformation g on $\mathcal{M}$ is nonlinear. It is a remarkable core result of Lie theory that the entire action of a Lie group is determined by just considering the linearly acting elements near the identity $I \in G$. We will discuss these fundamental facts in the next chapter.

Matrix Groups

We recall that the set of all real (or complex) $n \times n$-matrices $Mat(n, \mathbb{R})$ (or $Mat(n, \mathbb{C})$) can be viewed as the n^2-dimensional manifold $\mathbb{R}^{n^2}$ (or the $2n^2$-dimensional manifold $\mathbb{R}^{2n^2}$) but it is not a group under the matrix multiplication. However, the sets $GL(n, \mathbb{R})$ and $GL(n, \mathbb{C})$ are in fact groups under the multiplication. The case of *linear matrix groups* will be the most interesting for us. These are *closed subgroups* of the real group $GL(n, \mathbb{R})$ and the complex group $GL(n, \mathbb{C})$. These matrix groups are actually more general than one may think. Actually, any Lie group can be embedded into the group $GL(N, \mathbb{R})$ with a suitable dimensionality N. Since the matrix groups are relevant to our applications, let us review some important cases in the following.

Special Linear Groups $SL(n, \mathbb{R})$ and $SL(n, \mathbb{C})$

The *special linear groups* are defined as those for which the determinant is equal to one, i.e.

$$SL(n, \mathbb{R}) \equiv \{M \in GL(n, \mathbb{R}) \mid \det M = 1\}, \tag{7.41}$$

and

$$SL(n, \mathbb{C}) \equiv \{M \in GL(n, \mathbb{C}) \mid \det M = 1\}. \tag{7.42}$$

The concrete case of $SL(2, \mathbb{C})$ provides a (spinor) representation of the Lorentz group, as we will see later.

Orthogonal and Special Orthogonal Groups $O(n)$ and $SO(n)$

The *orthogonal group* $O(n)$ in n dimensions is defined as the group of real $n \times n$-matrices O for which the transposition gives the inverse matrix,

$$O(n) \equiv \{O \in GL(n, \mathbb{R}) \mid O^T O = OO^T = 1\}. \tag{7.43}$$

The matrices O leave the scalar product of vectors in $\mathbb{R}^n$ invariant. The defining property above, $O^T O = OO^T = 1$, has as consequence that $\det O = \pm 1$. An important subgroup of $O(n)$ is the group for which the matrices have a determinant equal one. This is the *special orthogonal group* $SO(n)$,

$$SO(n) \equiv \{O \in GL(n, \mathbb{R}) \mid O^T O = OO^T = 1, \, \det O = 1\}. \tag{7.44}$$

The orthogonal group consists of rotations plus reflections at the origin, whereas the special orthogonal group contains only the pure rotations.

Unitary and Special Unitary Groups $U(n)$ and $SU(n)$

The *unitary group* $U(n)$ in n dimensions is defined as the group of complex $n \times n$-matrices U for which the hermitian conjugation gives the inverse matrix,

$$U(n) \equiv \{U \in GL(n, \mathbb{C}) \mid U^\dagger U = UU^\dagger = 1\}. \tag{7.45}$$

The matrices U leave the hermitian (sesquilinear) scalar product of vectors in $\mathbb{C}^n$ invariant. The defining property above, $U^\dagger U = UU^\dagger = 1$, has as consequence that $|\det U| = 1$. A subgroup of $U(n)$ is the group for which the matrices have a determinant equal one. This is the *special unitary group* $SU(n)$,

$$SU(n) \equiv \{U \in GL(n, \mathbb{C}) \mid U^\dagger U = UU^\dagger = 1, \, \det U = 1\}. \tag{7.46}$$

The group $SU(2)$ provides the spinor representation of 3-dimensional rotations. The unitary groups are of great importance for quantum physics, particle physics, and in the construction of gauge theories of the *Standard Model*.

Pseudo-Orthogonal Groups $O(n, m)$ **and** $SO(n, m)$

We consider the $(n + m)$−dimensional vector space $\mathbb{R}^{(n,m)}$, endowed with the metric

$$\eta \equiv \text{diag}(1, \ldots, 1, -1, \ldots, -1), \tag{7.47}$$

where the $+1$ appears n times and the -1 appears m times. The metric defines naturally a symmetric bilinear scalar product of the vectors v and w by

$$(v \cdot w) \equiv (v, \eta w) \equiv v^1 w^1 + \ldots + v^n w^n - v^{n+1} w^{n+1} - \ldots - v^{n+m} w^{n+m}. \tag{7.48}$$

The *pseudo-orthogonal group* $O(n, m)$ is the matrix group defined as

$$O(n, m) \equiv \{L \in GL(n + m, \mathbb{R}) \mid L^T \eta L = L \eta L^T = \eta\}. \tag{7.49}$$

Similarly, the *special pseudo-orthogonal group* $SO(n, m)$ is defined as the subgroup

$$SO(n, m) \equiv \{L \in GL(n + m, \mathbb{R}) \mid L^T \eta L = L \eta L^T = \eta, \det L = 1\}. \tag{7.50}$$

The matrices of these groups leave the above scalar product invariant, i.e.

$$(Lv \cdot Lw) = (v \cdot w). \tag{7.51}$$

The dimension is similar to the orthogonal case. Examples we will consider in more detail later are $O(1, 3)$ as a realization of the Lorentz group, $SO(1, 3)$ as the proper orthochronous Lorentz group, and $SO(2, 4)$ as the 6-dimensional representation of the conformal group.

Pseudo-Unitary Groups $U(n, m)$ **and** $SU(n, m)$

Similar to the above constructions of the real case, we can view the *pseudo-unitary group* $U(n, m)$ as the complex matrix group defined as

$$U(n, m) \equiv \{U \in GL(n + m, \mathbb{C}) \mid U^\dagger \eta U = U \eta U^\dagger = \eta, \}. \tag{7.52}$$

The *special pseudo-unitary group* $SU(n, m)$ is defined as

$$SU(n, m) \equiv \{U \in GL(n + m, \mathbb{C}) \mid U^\dagger \eta U = U \eta U^\dagger = \eta, \det U = 1\}. \tag{7.53}$$

These groups leave the naturally defined hermitian scalar product invariant. The dimension is analogous to the unitary case. The group $SU(2, 2)$ provides a complex (twistor) representation of the conformal group.

Symplectic Groups $Sp(2n, \mathbb{R})$ **and** $Sp(2n, \mathbb{C})$*

We start with the real case and consider the vector space $\mathbb{R}^{2n}$ with the metric J given by the real $2n \times 2n$-matrix

$$J \equiv \left(\begin{array}{c|c} 0 & 1_n \\ \hline -1_n & 0 \end{array} \right). \tag{7.54}$$

The associated *symplectic bilinear form* on $\mathbb{R}^{2n}$ is

$$(v \cdot w) \equiv (v, Jw) = \sum_{j=1}^{n} (v^j w^{n+j} - v^{n+j} w^j), \tag{7.55}$$

---

*Note that these groups are sometimes denoted as $Sp(n, \mathbb{R})$ and $Sp(n, \mathbb{C})$.

and it is antisymmetric, $(w \cdot v) = -(v \cdot w)$. The set of real $2n$-dimensional matrices S leaving the above bilinear form invariant constitutes the *real symplectic group* $Sp(2n, \mathbb{R})$. It is characterized by the condition

$$S^T J S = S J S^T = J. \tag{7.56}$$

So the definition of the real symplectic group reads

$$Sp(2n, \mathbb{R}) \equiv \{S \in GL(2n, \mathbb{R}) \mid S^T J S = S J S^T = J\}. \tag{7.57}$$

In the complex case, we start with the vector space $\mathbb{C}^{2n}$ and with the same metric J as above. The resulting bilinear form is not hermitian and we do not consider complex conjugation. The matrices which leave this bilinear form unchanged obey again $S^T J S = S J S^T = J$, i.e.

$$Sp(2n, \mathbb{C}) \equiv \{S \in GL(2n, \mathbb{C}) \mid S^T J S = S J S^T = J\}. \tag{7.58}$$

There is another important version of the symplectic group, which is the compact version. Per definition, a compact group is topologically closed and bounded. The *compact symplectic group*, denoted $Sp(2n)$, is given by the intersection of sets

$$Sp(2n) \equiv Sp(2n, \mathbb{C}) \cap U(2n). \tag{7.59}$$

Symplectic groups appear in classical mechanics within the Hamiltonian formalism.

Inhomogeneous Matrix Groups

We noted already that the vector space $\mathbb{R}^n$ (or $\mathbb{C}^n$) can be viewed as an abelian additive group with the vector addition as the group operation. Now the group (or any subgroup of) $GL(n, \mathbb{R})$ (or $GL(n, \mathbb{C})$) defines a group homomorphism from this group to the group of automorphisms of $\mathbb{R}^n$ (or $\mathbb{C}^n$) by means of simple matrix multiplication of vectors. Thus, we can construct the semidirect product $\mathbb{R}^n \rtimes GL(n, \mathbb{R})$ (or $\mathbb{C}^n \rtimes GL(n, \mathbb{C})$). The group composition is given by

$$(a_2, L_2)(a_1, L_1) = (a_2 + L_2 a_1, L_2 L_1), \tag{7.60}$$

and the inversion is

$$(a, L)^{-1} = (-L^{-1}a, L^{-1}), \tag{7.61}$$

with $L_1, L_2 \in GL(n, \mathbb{R})$ (or $GL(n, \mathbb{C})$) and $a \in \mathbb{R}^n$ (or $\mathbb{C}^n$). This semidirect product is called the *inhomogeneous group* (or *affine group*). It is an important construction, which we will encounter as Euclidean group, Galileian group and Poincaré group. The inhomogeneous group of $SO(n)$ is denoted $ISO(n)$, etc. It is interesting to note that an inhomogeneous group element can still be represented as a linear transformation by considering the space $GL(n+1, \mathbb{R})$ (or $GL(n+1, \mathbb{C})$) and by identifying the pair (a, L) with the $(n+1)$-dimensional matrix

$$\left(\begin{array}{c|c} L & a \\ \hline 0 & 1 \end{array} \right), \tag{7.62}$$

which acts on $(n + 1)$-dimensional vectors

$$\begin{pmatrix} v \\ 1 \end{pmatrix}. \tag{7.63}$$

The group composition and inversion of inhomogeneous group elements are calculated as matrix multiplication and matrix inversion.

Summary of the Classical Groups

The matrix Lie groups described above constitute the so-called *classical groups*. In table 7.1 we note their number of real dimensions and their specific defining conditions. The classical groups are in fact part of a more general scheme discovered by Cartan, which we briefly describe below.

Table 7.1: Classical Lie groups and their defining conditions

Classical Group	Real Dimension	Defining Conditions
$GL(n, \mathbb{R})$	n^2	invertible real matrices
$GL(n, \mathbb{C})$	$2n^2$	invertible complex matrices
$SL(n, \mathbb{R})$	$n^2 - 1$	invertible, real, $\det M = 1$
$SL(n, \mathbb{C})$	$2(n^2 - 1)$	invertible, complex, $\det M = 1$
$O(n)$	$\frac{1}{2}n(n - 1)$	real, $O^T O = 1$
$SO(n)$	$\frac{1}{2}n(n - 1)$	real, $O^T O = 1$, $\det O = 1$
$U(n)$	n^2	complex, $U^\dagger U = 1$
$SU(n)$	$n^2 - 1$	complex, $U^\dagger U = 1$, $\det U = 1$
$Sp(2n, \mathbb{R})$	$n(2n + 1)$	real, $S^T J S = J$
$Sp(2n, \mathbb{C})$	$2n(2n + 1)$	complex, $S^T J S = J$

Cartan's Classification of Simple Lie Groups

It was in 1894 when Élie Cartan completed the classification of all simple Lie groups according to their factors, thus identifying the basic Lie groups from which all others can be constructed. Cartan found four distinct classes of infinite series of groups, called A_n, B_n, C_n, and D_n, which belong to the *classical groups*, plus the five so-called *exceptional groups*, E_6, E_7, E_8, F_4, and G_2. The *Cartan classification* is given in table 7.2 below including the compact group of each class. The index n is called the *rank* of the group.

Table 7.2: The Cartan classification

Lie Group Class	Rank Values	Real Dimension	Compact Group
A_n	$n \geq 1$	$n(n + 2)$	$SU(n + 1)$
B_n	$n \geq 2$	$n(2n + 1)$	$SO(2n + 1)$
C_n	$n \geq 3$	$n(2n - 1)$	$Sp(2n)$
D_n	$n \geq 4$	$n(2n - 1)$	$SO(2n)$
G_2	$n = 2$	14	G_2
F_4	$n = 4$	52	F_4
E_n	$n = 6, 7, 8$	78, 133, 248	E_6, E_7, E_8

Exceptional groups are believed to play an important role in fundamental physics. The group E_8, for example, appears prominently in string theory. The exceptional groups reveal surprising symmetry patterns and are the subject of active research. The smallest exceptional group G_2, for instance, is the group of automorphisms of octonions. The largest exceptional group E_8 is the symmetry group of its own Lie algebra.

Further Reading

Our exposition here focused primarily on algebraic properties and less on analytic or topological aspects of Lie groups. A mathematically more complete development of Lie groups can be found in the excellent books of Hall [38] and Rossmann [75]. We have introduced Lie groups on the basis of the notion of subgroups of linear groups. This approach has the advantage that it requires few prerequisites and gives the main results unobscured by technicalities. The reader interested in the foundations of Lie groups within the differential geometric framework can consult the reference by Warner [93]. A gentle introduction to the field of exceptional groups is provided by Ramond [70].

8

Lie Algebras

A Lie algebra is derived naturally from a Lie group as the tangent space of the group manifold at the group identity element. A Lie algebra describes group elements near the identity and encodes the essential properties of the corresponding Lie group. The great advantage of Lie algebras is that they are much easier to handle than Lie groups. We begin with a discussion about the exponential function of matrices and the Baker-Campbell-Hausdorff formula, which are the necessary technical tools to properly treat Lie algebras. The exponential acts as the mediator between a Lie algebra and the associated Lie group. We study the defining properties of Lie algebras and then we provide concrete examples of matrix algebras corresponding to matrix groups.

8.1 Matrix Exponential and the BCH Formula

Exponential of a Matrix

For any $n \times n$-matrix A in $GL(n, \mathbb{R})$, or $GL(n, \mathbb{C})$, we define the *exponential* $\exp A$ through the series expansion

$$\exp(A) \equiv \sum_{k=0}^{\infty} \frac{A^k}{k!}. \tag{8.1}$$

Writing out the first few terms, this is

$$\exp A = 1 + A + \frac{A^2}{2!} + \frac{A^3}{3!} + \frac{A^4}{4!} + \dots . \tag{8.2}$$

For any matrix A, the exponential $\exp A$ is a non-singular matrix and a continuous function of its argument. The convergence of this series is to be proven with a certain matrix norm. A common but not the only choice for a matrix norm $\|A\|$ is

$$\|A\|^2 \equiv \sum_{i,j=1}^{n} |A_{ij}|^2. \tag{8.3}$$

In appendix B.2 we summarize some of the basic properties of the matrix exponential. To note an important one here, for any two matrices A and B that commute, $AB = BA$, it is

$$e^A e^B = e^{A+B} = e^B e^A, \tag{8.4}$$

DOI: 10.1201/9781003087748-8

exactly as for real or complex numbers. For the proof we have simply to multiply the series expansions term by term,

$$
\begin{aligned}
e^A e^B &= \left(\sum_{k=0}^{\infty} \frac{A^k}{k!} \right) \left(\sum_{l=0}^{\infty} \frac{B^l}{l!} \right) \\
&= \sum_{l=0}^{\infty} \sum_{k=0}^{l} \frac{A^k}{k!} \frac{B^{l-k}}{(l-k)!} = \sum_{l=0}^{\infty} \frac{1}{l!} \sum_{k=0}^{l} \frac{l!}{k!(l-k)!} A^k B^{l-k} .
\end{aligned}
\tag{8.5}
$$

Since the matrices commute, the binomial expansion

$$
(A + B)^l = \sum_{k=0}^{l} \frac{l!}{k!(l-k)!} A^k B^{l-k}
\tag{8.6}
$$

holds and hence it is

$$
e^A e^B = e^{A+B} ,
\tag{8.7}
$$

as desired. Another important formula expresses the exponential as a limit of products. For any $n \times n$-matrix A in $GL(n, \mathbb{R})$, or $GL(n, \mathbb{C})$, the formula

$$
\exp A = \lim_{n \to \infty} \left(I + \frac{A}{n} \right)^n
\tag{8.8}
$$

holds, exactly as in the case with numbers. A proof is provided in appendix B.2.

Logarithm of a Matrix

Here we consider the inverse of the exponential. For any $n \times n$-matrix A in $GL(n, \mathbb{R})$, or $GL(n, \mathbb{C})$, with $\|A - I\| < 1$, we define the (*natural*) *logarithm* $\ln A$ through the series expansion

$$
\ln A \equiv \sum_{k=1}^{\infty} (-1)^{k-1} \frac{(A - I)^k}{k} ,
\tag{8.9}
$$

which converges with respect to the norm we have defined above. The logarithm is the inverse of the exponential, in the sense that for any matrix A, with $\|A - I\| < 1$, it is

$$
\exp (\ln A) = A ,
\tag{8.10}
$$

and for a matrix A, with $\|A\| < \ln 2$, it is

$$
\ln (\exp A) = A .
\tag{8.11}
$$

The interested reader can find the proof in Rossmann [75].

Baker-Campbell-Hausdorff Formula

Now that we have defined the exponential e^A of a matrix A, we need to understand how to multiply two such exponentials as $e^A e^B$. As seen above, the formula of two commuting matrices A and B is the same as the one for numbers. This is different when the matrices do not commute, $[A, B] \neq 0$. The *Baker-Campbell-Hausdorff formula* (or *BCH* for short) says that the product of exponentials can be computed as an infinite sum of nested commutators of the original matrices A and B. The important point of the BCH theorem is the very existence of this formula containing nested commutators. Somewhat less important is the explicit algebraic formula. In most applications of the BCH theorem, the mere fact that the formula is completely defined by the nested commutators is a sufficient information. Now the proof of existence is delicate and has many forms. The mathematicians *John E. Campbell*, *Henry F. Baker*, *Felix Hausdorff*, and *Henri Poincaré* had provided proofs for the existence of the formula. Only later, in 1947, *Eugene B. Dynkin* derived the explicit algebraic formula. A modern derivation of the theorem can be found in the book of Rossmann [75]. We restrict ourselves to deriving the formula for two simple cases, once for the first few terms and then by computing the formula in an iterative procedure that yields increasingly higher order terms.

First Terms in the BCH Formula

We have seen that for commuting matrices the formula

$$e^A e^B = e^{A+B} \tag{8.12}$$

holds. For non-commuting matrices A and B, where $AB \neq BA$, there is a generalization needed. We assume now the matrices A and B to be small, in the sense of a suitable matrix norm. For the definition of a matrix norm see e.g. [75]. We will show by direct computation that the generalization of the above formula, up to second order in the matrices, is

$$e^A e^B = e^{A+B+\frac{1}{2}[A,B]+\mathcal{O}(3)}. \tag{8.13}$$

In other words, the commutator appears as the first-order correction. Let us simply carry out the multiplication on the lhs,

$$\left(1 + A + \frac{A^2}{2} + \mathcal{O}(3)\right)\left(1 + B + \frac{B^2}{2} + \mathcal{O}(3)\right), \tag{8.14}$$

and keep terms up to second order. This is exactly the same result if we calculate the rhs up to the same order,

$$1 + \left(A + B + \frac{1}{2}[A,B] + \mathcal{O}(3)\right) + \left(A + B + \frac{1}{2}[A,B] + \mathcal{O}(3)\right)^2. \tag{8.15}$$

We see that the commutator of the original matrices represents the first and largest correction term to the basic formula. The BCH theorem states that all corrections are computed as commutators of the original matrices with certain numerical factors.

Derivation of the Explicit BCH Formula

Here we will derive a method how to compute the BCH formula systematically order by order. In a first step, we prove the useful formula

$$e^{-B} A e^B = A + [A,B] + \frac{1}{2!}[[A,B],B] + \frac{1}{3!}[[[A,B],B],B] + \dots . \tag{8.16}$$

To this end, we define the function $F(\tau)$ of the real parameter τ as

$$F(\tau) \equiv e^{-\tau B} A e^{\tau B}, \tag{8.17}$$

and we write this function as a Taylor series in τ as

$$F(\tau) = \sum_{k=0}^{\infty} \frac{1}{k!} f_k \tau^k. \tag{8.18}$$

Now we compute the derivative of $F(\tau)$ and see that it is

$$\frac{\mathrm{d}}{\mathrm{d}\tau} F(\tau) = [F(\tau), B]. \tag{8.19}$$

By means of the corresponding Taylor expansion,

$$\sum_{k=0}^{\infty} \frac{1}{k!} f_{k+1} \tau^k = \sum_{k=0}^{\infty} \frac{1}{k!} [f_k, B] \tau^k, \tag{8.20}$$

we obtain the recursion formula

$$f_{k+1} = [f_k, B]. \tag{8.21}$$

As it is $f_0 = A$, we obtain, by setting $\tau = 1$, the desired formula 8.16. In the next step we prove for a matrix-valued function $A(\tau)$ depending on the real parameter τ the formula

$$e^{-A(\tau)} \frac{d}{d\tau} e^{A(\tau)} = \int_0^1 e^{-sA} \frac{dA}{d\tau} e^{sA} \, ds. \tag{8.22}$$

First, we consider the expression $\frac{d}{d\tau} e^A$ and insert the exponential series, to obtain

$$\frac{d}{d\tau} e^A = \sum_{k=0}^{\infty} \sum_{l=0}^{\infty} \frac{A^k A' A^l}{(k+l+1)!}, \tag{8.23}$$

where $A' \equiv dA/d\tau$. On the other hand, by using once again the exponential series, the integral expression becomes

$$\int_0^1 e^{(1-s)A} \frac{dA}{d\tau} e^{sA} \, ds = \sum_{k=0}^{\infty} \sum_{l=0}^{\infty} \frac{A^k A' A^l}{k! \, l!} \int_0^1 (1-s)^k s^l \, ds. \tag{8.24}$$

By applying the standard integral expression

$$\int_0^1 (1-s)^k s^l \, ds = \frac{k! \, l!}{(k+l+1)!}, \tag{8.25}$$

we obtain the desired formula 8.22. Combing the formulae 8.16 and 8.22 yields the useful formula

$$e^{-A} \frac{d}{d\tau} e^A = A' + \frac{1}{2} [A', A] + \frac{1}{3!} [[A', A], A] + \dots . \tag{8.26}$$

In the final step we consider the expression $e^{\tau A} e^{\tau B}$, where we use the real parameter τ as the defining element for computing the BCH formula order by order. We will restrict ourselves to a computation up to second order in τ, but the procedure is applicable to arbitrary order. We set

$$e^{\tau A} e^{\tau B} \equiv e^{C(\tau)} \tag{8.27}$$

and seek to compute $C(\tau)$ order by order. This matrix-valued function has the form

$$C(\tau) = c_1 \tau + c_2 \tau^2 + c_3 \tau^3 + \dots . \tag{8.28}$$

We consider the equation

$$e^{-C(\tau)} \frac{d}{d\tau} e^{C(\tau)} = e^{-\tau B} e^{-\tau A} \frac{d}{d\tau} e^{\tau A} e^{\tau B}. \tag{8.29}$$

Simply carrying out the differentiation yields, according to 8.16,

$$e^{-C(\tau)} \frac{d}{d\tau} e^{C(\tau)} = B + A + \tau [A, B] + \frac{\tau^2}{2} [[A, B], B] + \mathcal{O}(\tau^3). \tag{8.30}$$

On the other hand, the expression can be computed also by the formula 8.26 as

$$e^{-C(\tau)} \frac{\mathrm{d}}{\mathrm{d}\tau} e^{C(\tau)} = C' + \frac{1}{2} \left[C', C \right] + \frac{1}{3!} \left[\left[C', C \right], C \right] + \dots \tag{8.31}$$

By using

$$C' = c_1 + 2c_2\tau + 3c_3\tau^2 + \mathcal{O}(\tau^3), \tag{8.32}$$

and

$$[C', C] = - \left[c_1, c_2 \right] \tau^2 + \mathcal{O}(\tau^3), \tag{8.33}$$

we obtain

$$e^{-C(\tau)} \frac{\mathrm{d}}{\mathrm{d}\tau} e^{C(\tau)} = c_1 + 2c_2\tau + \left(3c_3 - \frac{1}{2} \left[c_1, c_2 \right] \right) \tau^2 + \mathcal{O}(\tau^3). \tag{8.34}$$

Now by comparing the two series expansions in powers of τ and by equating the coefficients with the same order, we obtain the first three terms of the BCH formula:

$$\begin{aligned}
c_1 &= A + B, \\
c_2 &= \tfrac{1}{2} \left[A, B \right], \\
c_3 &= \tfrac{1}{12} \left[A, \left[A, B \right] \right] + \tfrac{1}{12} \left[B, \left[B, A \right] \right].
\end{aligned} \tag{8.35}$$

This means that the *BCH formula*, up to third order in the matrices A and B, is given by

$$\exp A \exp B = \exp \left\{ A + B + \frac{1}{2} \left[A, B \right] + \frac{1}{12} \left[A, \left[A, B \right] \right] + \frac{1}{12} \left[B, \left[B, A \right] \right] + \dots \right\}. \tag{8.36}$$

We will not directly use the explicit form of the above formula but only the simpler conclusion that the *product of the exponentials is completely defined by commutators of the original matrices.*

8.2 Lie Algebra of a Lie Group

Lie Algebra Generators

Lie algebras can be introduced abstractly as vector spaces with a commutator product, and only later can the connection between Lie algebras and Lie groups be established. Here we will introduce Lie algebras and their properties as natural conclusions from the properties of the corresponding Lie groups. We define the *Lie algebra* of a Lie group as the tangent vector space attached at the group identity element I. Remember that our Lie group has p real parameters, so that the corresponding Lie algebra is a p-dimensional real vector space. Let us first introduce a basis for the vector space, the *generators of the Lie group*. We can always choose our manifold coordinates ϵ^a to be such that

$$g(0) = I. \tag{8.37}$$

Now we can carry out a Taylor expansion near the group identity with respect to the group parameters ϵ^a, which are supposed to be small, $|\epsilon^a| \ll 1$, and we write

$$\boxed{g(\epsilon) = I - \mathrm{i}\epsilon^a X_a + \mathcal{O}(2)} \tag{8.38}$$

The elements X_a, $a = 1, \dots, r$, are the *infinitesimal generators* of the Lie group and are defined as

$$\boxed{X_a \equiv \mathrm{i} \left. \frac{\partial g}{\partial \epsilon^a}(\epsilon) \right|_{\epsilon=0}} \tag{8.39}$$

The imaginary i is used in the above definitions, so that a hermitian matrix representation X_a corresponds to a unitary matrix representation $g(\epsilon)$.* As a remark on notation, we will use the symbol X_a for the generators of the Lie group G itself and the symbol $\mathcal{G}_a$ for the generators defined within a representation $D(G)$ of the Lie group. If we use a different parametrization of the Lie group with parameters $\alpha^b(\epsilon^a)$, where the group elements are given by $g(\alpha)$, and we require as usual that $g(\alpha = 0) = I$ (or equivalently $\alpha(0) = 0$), then this corresponds to a coordinate transformation of the generators as vector fields living in the tangent space, i.e.

$$X_a = i\frac{\partial \alpha^b}{\partial \epsilon^a}(0)\frac{\partial g}{\partial \alpha^b}(0) = \frac{\partial \alpha^b}{\partial \epsilon^a}(0)\, X'_b\,. \tag{8.40}$$

A group element $g(\epsilon)$ that is near the identity I can be written, to first order in ϵ^a, as

$$g(\epsilon) = I - i\epsilon^a X_a. \tag{8.41}$$

We can divide the parameter values ϵ^a into k equal steps and let k go to infinity. Thus, we can represent an arbitrary group element as an infinite sequence of infinitesimal transformations,

$$g(\epsilon) = \lim_{k \to \infty} \left(I - \frac{i\epsilon^a X_a}{k}\right)^k. \tag{8.42}$$

By using the formula 8.8, we obtain the following representation for the Lie group element:

$$\boxed{g(\epsilon) = \exp\left(-i\epsilon^a X_a\right)} \tag{8.43}$$

We see that by exponentiating Lie algebra elements, we recover Lie group elements. We must note, however, that it is not always guaranteed that all elements of the Lie group are reached by this exponentiation.†

Lie Algebra Commutator

The most basic property, the group closure property, gives

$$g(\alpha)g(\beta) = \exp(-i\alpha^a X_a)\exp(-i\beta^a X_a) = \exp(-i\epsilon^a X_a) = g(\epsilon). \tag{8.44}$$

Now we recall the BCH formula,

$$e^A e^B = e^{A+B+C(A,B)}, \tag{8.45}$$

where the expression $C(A, B)$ in the rhs is a infinite series of nested commutators of the matrices A and B of the form

$$C(A, B) = \frac{1}{2}[A, B] + \frac{1}{12}[A, [A, B]] + \frac{1}{12}[B, [B, A]] + \dots. \tag{8.46}$$

We deduce that the commutator of Lie group generators is itself a generator. Thus, we obtain an entire set of *Lie algebra commutator relations*, which we write as

$$\boxed{[X_a, X_b] = if_{abc}X_c} \tag{8.47}$$

---

*This is done especially to comply with the requirements of operators in quantum physics.
†If the Lie group is compact and connected, it can be completely reconstructed with the exponential mapping.

The *structure constants* f_{abc} above are real numbers and there is summation implied over the repeated index c. What can we read from this commutator formula? It says that the products of Lie group elements are completely determined by the commutator of the Lie algebra generators. More loosely, we can say that the Lie group structure is encoded in the Lie algebra structure. *The Lie algebra is obtained from the group by differentiation, while the group is obtained from the Lie algebra by exponentiation.* This is the so-called *Lie correspondence*. Algebraically, the antisymmetry of the commutator is equivalent to the antisymmetry of the structure constants in their first two indices,

$$f_{abc} = -f_{bac}. \tag{8.48}$$

Furthermore, we can consider the generally valid *Jacobi identity* for linear operators, see appendix B.1, and apply it to the generators,

$$\boxed{[[X_a, X_b], X_c] + [[X_b, X_c], X_a] + [[X_c, X_a], X_b] = 0} \tag{8.49}$$

This is equivalent to the condition

$$f_{abe}f_{ecd} + f_{bce}f_{ead} + f_{cae}f_{ebd} = 0 \tag{8.50}$$

for the structure constants. Note that we have a summation over the repeated index e. The above algebraic properties combined define the Lie algebra of the Lie group. As a final remark let us note that a Lie group is abelian exactly if its corresponding Lie algebra is abelian, meaning that the commutator for all algebra elements vanishes.

8.3 Abstract Lie Algebras and Matrix Algebras

Abstract Lie Algebras

We have introduced Lie algebras as (tangent) vector spaces equipped with a commutator product that obeys antisymmetry and the Jacobi identity. Now we take these algebraic properties as the defining ones for a Lie algebra. More explicitly, a (finite-dimensional) real or complex *Lie algebra* consists of a vector space V over $\mathbb{R}$ or $\mathbb{C}$ with a bilinear product $[X, Y]$, called *Lie commutator*, or *bracket*, with the properties of *antisymmetry*,

$$[X, Y] = -[Y, X], \tag{8.51}$$

and the *Jacobi identity*,

$$[[X, Y], Z] + [[Y, Z], X] + [[Z, X], Y] = 0. \tag{8.52}$$

In the previous section, we have derived these Lie algebra properties from the Lie group structure, now we just postulate them. The Lie algebra of a Lie group G is typically denoted (in fraktur letters) as $\mathfrak{g}$. Two Lie groups G and G' that have the same Lie algebra, $\mathfrak{g} = \mathfrak{g}'$, are isomorphic *near the identity*, $G \cong G'$. A *subalgebra* of a Lie algebra $\mathfrak{g}$ is a subspace $\mathfrak{h}$ of $\mathfrak{g}$, which is closed under the commutator product, i.e. for each $P, Q \in \mathfrak{h}$ also $[P, Q] \in \mathfrak{h}$. Maps between Lie algebras that preserve the structure are of special interest. For two Lie algebras $\mathfrak{g}$ and $\mathfrak{h}$, a linear map $\varphi : \mathfrak{g} \to \mathfrak{h}$, with the property

$$\varphi([X, Y]) = [\varphi(X), \varphi(Y)], \tag{8.53}$$

for all $X, Y \in \mathfrak{g}$, is called a *Lie algebra homomorphism* of $\mathfrak{g}$ and $\mathfrak{h}$. If, additionally, this map is 1-to-1 and onto, then it is called a *Lie algebra isomorphism*. An isomorphism of a Lie algebra with itself is called a *Lie algebra automorphism*. We will discuss representations of Lie algebras (and of Lie groups) in the next chapter.

Semisimple Lie Algebras

Some further definitions are useful. A Lie algebra $\mathfrak{g}$ is called *abelian* if there is $[X, Y] = 0$, for all its elements. A Lie subalgebra $\mathfrak{h}$ of a Lie algebra $\mathfrak{g}$ is called *invariant* if $[P, X] = 0$, for all $P \in \mathfrak{h}$ and for all $X \in \mathfrak{g}$. A Lie subalgebra $\mathfrak{h}$ is called *simple* if it is not abelian and it has no invariant subalgebra. A Lie subalgebra $\mathfrak{h}$ is called *semisimple* if it is not abelian and it has no abelian invariant subalgebra. Semisimple Lie algebras are important for elementary particle theory. For a Lie algebra $\mathfrak{g}$ with the structure constants f_{abc}, we define the *Cartan-Killing metric*, κ_{ab}, as

$$\kappa_{ab} \equiv \sum_{c,d} f_{acd} f_{bdc}. \tag{8.54}$$

The Cartan-Killing metric can be used to determine if a Lie algebra is semisimple. Without proof, we state that semisimplicity is fulfilled whenever κ_{ab} is invertible. Then $\det \kappa \neq 0$ and the inverse metric κ^{ab} exists. In the case of a semisimple Lie algebra $\mathfrak{g}$, we can define further the *Casimir operator* C, according to *Hendrik B. G. Casimir*, as

$$C \equiv \sum_{a,b} \kappa^{ab} X_a X_b. \tag{8.55}$$

Although this definition looks to be dependent of the choice of basis vectors X_a, it can be shown that the Casimir operator is actually independent of this choice. Now we come to the main point: the Casimir operator commutes with every element of the Lie algebra, most easily seen as

$$[C, X_a] = 0. \tag{8.56}$$

Therefore, the Casimir operator commutes also with every group element,

$$[C, \exp(-i\epsilon^a X_a)] = 0. \tag{8.57}$$

In the next chapter we will see that there is a one-to-one correspondence between a Lie algebra representation and a Lie group representation. So if we use the last formula within an irreducible representation of the group, it means, according to Schur's lemma, that the Casimir operator is a multiple of the identity,

$$C = cI. \tag{8.58}$$

The proportionality factor c is characteristic for the specific irreducible representation and can be used for the classification of these representations.

Direct and Semidirect Sum of Lie Algebras

Without proofs we summarize a few important facts. We can ask what the Lie algebra of a direct product of Lie groups is. Suppose we have two Lie groups G and H and we consider the direct product $G \times H$. Then this is again a Lie group and its Lie algebra is the *direct sum* $\mathfrak{g} \oplus \mathfrak{h}$ of the associated Lie algebras $\mathfrak{g}$ and $\mathfrak{h}$. The direct sum is defined as the vector space $\mathfrak{g} \oplus \mathfrak{h} \equiv \{(X, Y) \mid X \in \mathfrak{g}, Y \in \mathfrak{h}\}$ in conjunction with the commutator

$$[(X, Y), (X', Y')] \equiv ([X, X'], [Y, Y']). \tag{8.59}$$

Slightly more involved is the case for a semidirect product of Lie groups. First let us define the *semidirect sum* of algebras. We need two Lie algebras $\mathfrak{g}$ and $\mathfrak{h}$ and an algebra homomorphism $\lambda : \mathfrak{h} \to gl(\mathfrak{g})$, $Y \mapsto \lambda_Y$. The linear map λ_Y acts on the Lie algebra $\mathfrak{g}$, i.e. $\lambda_Y : \mathfrak{g} \to \mathfrak{g}$, $X \mapsto \lambda_Y(X)$. Now the semidirect sum $\mathfrak{g} \oplus_S \mathfrak{h}$ (note that we use the symbol $\oplus_S$)

as a vector space is defined again as in the simpler case, but the commutator must have the form

$$[(X, Y), (X', Y')] \equiv ([X, X'] + \lambda_Y(X') - \lambda_{Y'}(X), [Y, Y']). \tag{8.60}$$

If we have the case of two Lie groups G and H, with their *semidirect product* $G \rtimes H$ defined by means of a group homomorphism $\chi : H \to Aut(G)$, then the corresponding Lie algebra is the *semidirect sum* $\mathfrak{g} \oplus_S \mathfrak{h}$, with the defining algebra homomorphism λ being simply the differential $d\chi$.

One-Parameter Lie Groups and Alternative Definition of Lie Algebras

Let us define a *1-parameter Lie group* as a group homomorphism from the additive group of the real numbers to the multiplicative group of invertible matrices $\gamma : \mathbb{R} \to GL(n, \mathbb{C})$, $t \mapsto \gamma(t)$, with $\gamma(0) = I$. According to the definition, it is $\gamma(s + t) = \gamma(s)\gamma(t)$. We consider the first-order differential equation for matrices

$$\frac{d\gamma}{dt}(t) = -iA\gamma(t), \quad \gamma(0) = I, \tag{8.61}$$

with a constant matrix A. Its unique solution is the matrix exponential

$$\gamma(t) = \exp(-itA). \tag{8.62}$$

The proof is easy, one differentiates the expression $\gamma(t)\exp(itA)$ and notes that it is always equal 1. (*exercise 8.1*) After these technical preparations, let us proceed to the main point. We know that we can define the tangent space of a manifold at a point by considering the class of all smooth curves in the manifold that go through this point. This provides the basis for an alternative definition of the Lie algebra of a Lie group. We consider the class of all 1-parameter groups γ going through the identity I of the Lie group with $\gamma(0) = I$. Consequently, all matrices of the type

$$X = i \left. \frac{d\gamma}{dt}(t) \right|_{t=0} \tag{8.63}$$

are elements of the tangent space, i.e. of the Lie algebra. Inversely, the exponential expressions $\exp(-itX)$ are elements of the Lie group. Thus, our alternative definition of the Lie algebra takes the set of all matrices X, so that the exponential $\exp(-itX)$ is an element of the Lie group for all real parameter values of t. To make the connection to our previous definition with Lie group elements of the form $\exp(-i\epsilon^a X_a)$, we have to identify $tX = \epsilon^a X_a$, which means that for different values of the parameters ϵ^a one obtains different Lie algebra elements X, which in turn correspond to different directions of the tangential curves γ at the point I.

The 1-parameter definition is very well suited in order to derive the defining conditions of the Lie algebra from the defining conditions of the Lie group, because we have to deal with only one differentiation. If we denote the defining conditions of a Lie group elements $g(t)$ by an equation of the form $F(g(t)) = 0$, then the defining conditions for the Lie algebra elements are obtained from the equation

$$\left. \frac{d}{dt} F(g(t)) \right|_{t=0} = 0. \tag{8.64}$$

We can apply this recipe in order to derive the Lie algebras of the classical groups.

Matrix Lie Algebras

As an example of a *matrix Lie algebra*, let us determine the Lie algebra of the group $SU(n)$. Special unitary matrices U are defined by the two conditions

$$U^\dagger = U^{-1}, \quad \det U = 1. \tag{8.65}$$

By writing the matrix U as exponential,

$$U(t) = \exp(-itX), \tag{8.66}$$

we immediately infer that X is a hermitian matrix,

$$X^\dagger = X. \tag{8.67}$$

The determinant condition above means that X has a vanishing trace,

$$\mathrm{Tr}X = 0. \tag{8.68}$$

These two conditions define the Lie algebra of $SU(n)$, which we denote as $su(n)$. In other words, the Lie algebra

$$su(n) \equiv \{X \in GL(n,\mathbb{C}) \mid X^\dagger = X, \mathrm{Tr}X = 0\} \tag{8.69}$$

consists of the n-dimensional hermitian and traceless matrices. This program can be applied to all classical Lie groups and delivers the corresponding Lie algebras. (*exercise 8.2*) The result is summarized at the end of this section.

Lie Algebra of Inhomogeneous Groups

We saw how to build *inhomogeneous* (or *affine*) *groups* as semidirect products of the form $\mathbb{F}^n \rtimes H$, with $\mathbb{F}^n$ being the vector space (over $\mathbb{R}$ or $\mathbb{C}$) and H being a subgroup of $GL(n, \mathbb{F})$. The Lie algebra of this inhomogeneous group is given by the set of pairs

$$\{(a, Y) \mid a \in \mathbb{F}^n, Y \in \mathfrak{h}\}, \tag{8.70}$$

where the vector space $\mathbb{F}^n$ is elevated to a Lie algebra by choosing a commutator, which is identically zero and where $\mathfrak{h}$ is the usual Lie algebra of H. The commutator in the Lie algebra is given by

$$[(a, Y), (a', Y')] = (Ya' - Y'a, [Y, Y']). \tag{8.71}$$

To see why this is the case, we again change our point of view from n to $n + 1$ dimensions and identify the pairs (a, Y) with $(n + 1)$-dimensional matrices of the form

$$\left(\begin{array}{c|c} Y & a \\ \hline 0 & 0 \end{array}\right). \tag{8.72}$$

In the Lie algebra of matrices $Mat(n + 1, \mathbb{F})$, the simple matrix commutator delivers the above formula. (*exercise 8.3*) Apparently, this is a special case of the general definition

$$[(X, Y), (X', Y')] \equiv ([X, X'] + \lambda_Y(X') - \lambda_{Y'}(X), [Y, Y']), \tag{8.73}$$

which we discussed earlier in this section, if we identify $X = a$ and consider $\lambda_Y(a)$ as the simple matrix multiplication Ya.

Summary of the Lie Algebras of the Classical Groups

In the following table 8.1 we list the *Lie algebras of the classical groups*, their dimensionality, and the defining conditions for the Lie algebra elements X.

Table 8.1: Lie algebras and their defining conditions

Lie Algebra	Real Dimension	Defining Conditions
$gl(n, \mathbb{R})$	n^2	All real matrices $Mat(n, \mathbb{R})$
$gl(n, \mathbb{C})$	$2n^2$	All complex matrices $Mat(n, \mathbb{C})$
$sl(n, \mathbb{R})$	$n^2 - 1$	X real, $\mathrm{Tr}X = 0$
$sl(n, \mathbb{C})$	$2(n^2 - 1)$	X complex, $\mathrm{Tr}X = 0$
$o(n) = so(n)$	$\frac{1}{2}n(n-1)$	X real, $X^T = X$
$o(n, m) = so(n, m)$	$\frac{1}{2}(n+m)(n+m-1)$	X real, $\eta X^T \eta = X$
$u(n)$	n^2	X complex, $X^\dagger = X$
$u(n, m)$	$(n+m)^2$	X complex, $\eta X^\dagger \eta = X$
$su(n)$	$n^2 - 1$	X complex, $X^\dagger = X$, $\mathrm{Tr}X = 0$
$su(n, m)$	$(n+m)^2 - 1$	X complex, $\eta X^\dagger \eta = X$, $\mathrm{Tr}X = 0$
$sp(2n, \mathbb{R})$	$n(2n+1)$	X real, $JX^T J = X$
$sp(2n, \mathbb{C})$	$2n(2n+1)$	X complex, $JX^T J = X$

With this summary we conclude our review of the basic facts about Lie groups and Lie algebras and proceed to the question how we can represent these algebraic structures by new quantities.

Further Reading

Rossmann [75] provides a complete proof of the BCH formula and gives a succinct discussion of its meaning. An illuminating mathematical development of the relation between Lie groups and Lie algebras is found in Hall [38] and Rossmann [75]. We have not covered the machinery of Lie algebra roots, weights, and Dynkin diagrams for their classification, which are more advanced topics. The interested reader will find readable treatments in the classics of Georgi [33] and Gilmore [35].

9

Representations

Lie groups and Lie algebras, which capture certain symmetries of interest, are rather abstract mathematical objects. Given a Lie group or algebra, a representation of it yields objects that satisfy that symmetry from the outset. In this sense, a representation makes a symmetry more concrete. In this chapter we derive the basic results of representation theory of Lie groups and Lie algebras that we need for our later applications. We discuss general properties and introduce the adjoint representation. We systematically address the question of how symmetry transformations act on tensors, on functions, and on tensor fields. In the last section we study the Lie algebra spanned by Killing vector fields.

9.1   Representations of Groups and Algebras

Definition of Representations

Representations of groups (or algebras) are maps preserving the structure of the original group (or the original algebra) on a set of operators, which in turn act on a linear space. Let V be a linear vector space and $GL(V)$ be the group of invertible linear maps acting on V. A *Lie group representation* D of a Lie group G on V is a group homomorphism

$$
\begin{aligned}
D: \quad G &\to GL(V) \\
g &\mapsto D(g),
\end{aligned}
\tag{9.1}
$$

which assigns to each group element g a linear invertible map $D(g)$ that acts on the linear space V. The group representation is, by definition, structure preserving, in the sense that

$$
D(gh) = D(g)D(h),
\tag{9.2}
$$

for all elements $g, h \in G$. In the case of a Lie algebra, the structure to be preserved is the commutator. Again, let V be a linear vector space and $gl(V)$ be the Lie algebra of the group $GL(V)$, which consists of linear maps acting on V. A *Lie algebra representation* $\mathcal{D}$ of a Lie algebra $\mathfrak{g}$ on V is an algebra homomorphism

$$
\begin{aligned}
\mathcal{D}: \quad \mathfrak{g} &\to gl(V) \\
X &\mapsto \mathcal{D}(X),
\end{aligned}
\tag{9.3}
$$

preserving the bracket,

$$\mathcal{D}\left([X,Y]\right) = [\mathcal{D}(X), \mathcal{D}(Y)], \tag{9.4}$$

for all elements $X, Y \in \mathfrak{g}$. Specifically for the element $\mathcal{D}(X_a)$ obtained as representation of a generator X_a, $a = 1, \ldots, r$, of the Lie algebra $\mathfrak{g}$, we will use the streamlined notation

$$\mathcal{G}_a \equiv \mathcal{D}(X_a). \tag{9.5}$$

Consequently, the generators $\mathcal{G}_a$, $a = 1, \ldots, r$, of the Lie algebra $gl(V)$ satisfy the fundamental *commutator relations*

$$\boxed{[\mathcal{G}_a, \mathcal{G}_b] = \mathrm{i} f_{abc} \mathcal{G}_c} \tag{9.6}$$

We know that there is a close correspondence between Lie groups and Lie algebras. Is there a similar relationship between representations of Lie groups and representations of Lie algebras? The answer is affirmative, as we will see in the following.

Lie Correspondence

First consider two Lie groups, G and H, and a differentiable Lie group homomorphism $f : G \rightarrow H$. The exponential $\exp(\mathrm{i}tX)$, where X is an element of the Lie algebra $\mathfrak{g}$ and t is a real parameter, describes a curve within the manifold G that passes through the identity I_G of G for $t = 0$. The differential of the map f at the identity I_G is actually a map between the two Lie algebras $\mathfrak{g}$ and $\mathfrak{h}$, let us denote it by $\mathrm{d}f : \mathfrak{g} \rightarrow \mathfrak{h}$. According to our discussions on manifolds, we know that this differential map $\mathrm{d}f$ is given by

$$\mathrm{d}f(X) \equiv \frac{\mathrm{d}}{\mathrm{d}(-\mathrm{i}t)} f(\exp(-\mathrm{i}tX))\Big|_{t=0}. \tag{9.7}$$

We will show that the differential map $\mathrm{d}f$ is in fact a Lie algebra homomorphism between $\mathfrak{g}$ and $\mathfrak{h}$ and that the relation

$$\boxed{f(\exp X) = \exp(\mathrm{d}f(X))} \tag{9.8}$$

holds. In terms of a commutative diagram, this reads as

$$
\begin{array}{ccc}
G & \xrightarrow{\ f\ } & H \\
{\scriptstyle \exp} \uparrow & & \uparrow {\scriptstyle \exp} \\
\mathfrak{g} & \xrightarrow[\ \mathrm{d}f\ ]{} & \mathfrak{h}
\end{array}
\tag{9.9}
$$

It means that a Lie group homomorphism determines a Lie algebra homomorphism, and vice versa. This theorem is called the *Lie correspondence*. For simplicity, let us use the symbol φ for the map $\mathrm{d}f$. We calculate

$$
\begin{aligned}
\frac{\mathrm{d}}{\mathrm{d}s} f(\exp(-\mathrm{i}sX)) &= \frac{\mathrm{d}}{\mathrm{d}t} f(\exp(-\mathrm{i}(s+t)X))\Big|_{t=0} \\
&= \frac{\mathrm{d}}{\mathrm{d}t} f(\exp(-\mathrm{i}sX)\exp(-\mathrm{i}tX))\Big|_{t=0} \\
&= f(\exp(-\mathrm{i}sX)) \frac{\mathrm{d}}{\mathrm{d}t} f(\exp(-\mathrm{i}tX))\Big|_{t=0} \\
&= f(\exp(-\mathrm{i}sX))\,(-\mathrm{i})\,\varphi(X).
\end{aligned}
\tag{9.10}
$$

This means that the function $f(\exp(-\mathrm{i}sX))$ obeys the first-order differential equation of one-parameter groups and thus it is

$$f(\exp(-\mathrm{i}sX)) = \exp(-\mathrm{i}s\varphi(X)). \tag{9.11}$$

By setting $s = \mathrm{i}$, we obtain the desired result. Now we show that the map φ preserves the bracket. According to what we have just proved, it is

$$f(\exp(\mathrm{i}tX)\exp(\mathrm{i}sY)\exp(-\mathrm{i}tX)) = \exp(\mathrm{i}t\varphi(X))\exp(\mathrm{i}s\varphi(Y))\exp(-\mathrm{i}t\varphi(X)). \tag{9.12}$$

By differentiating with respect to s at $s = 0$, we obtain

$$\varphi\left(\exp(\mathrm{i}tX)Y\exp(-\mathrm{i}tX)\right) = \exp(\mathrm{i}t\varphi X)\varphi(Y)\exp(-\mathrm{i}t\varphi X). \tag{9.13}$$

Finally, differentiating with respect to t at $t = 0$ yields the desired result,

$$\varphi\left([X,Y]\right) = [\varphi(X),\varphi(Y)]. \tag{9.14}$$

Correspondence between Lie Group and Lie Algebra Representations

The above theorem about the relation of Lie group homomorphisms and Lie algebra homomorphisms means that there is a correspondence between a Lie group representation D and a Lie algebra representation $\mathcal{D}$ in the form

$$D(\exp X) = \exp(\mathcal{D}(X)). \tag{9.15}$$

Expressing this through a commutative diagram, this reads

$$
\begin{array}{ccc}
G & \xrightarrow{\ D\ } & GL(V) \\
{\scriptstyle\exp}\uparrow & & \uparrow{\scriptstyle\exp} \\
\mathfrak{g} & \xrightarrow[\ \mathcal{D}\ ]{} & gl(V)
\end{array}
\tag{9.16}
$$

Therefore, if we have a group representation, we can immediately determine an algebra representation, and vice versa. The *representation of a Lie group element* has the form

$$\boxed{D(\epsilon) = \exp\left(-\mathrm{i}\epsilon^a \mathcal{G}_a\right)} \tag{9.17}$$

Here we have used the shorthand notation $D(\epsilon) \equiv D(g(\epsilon))$. We will use the formula 9.17 extensively for defining field representations.

9.2 Adjoint Representations

Conjugation Map and Adjoint Map

Let us first define the conjugation operation. For any element a in our Lie group G, define the *conjugation map* $c(a)$ as

$$
\begin{array}{rccc}
c(a): & G & \to & G \\
& g & \mapsto & c(a)g = aga^{-1}.
\end{array}
\tag{9.18}
$$

The conjugation map $c(a)$ is in fact a group automorphism, since it is

$$c(a)(gh) = c(a)(g)\,c(a)(h). \tag{9.19}$$

The Lie algebra map corresponding to $c(a)$ is determined by the usual differentiation operation at the identity,

$$\frac{d}{d(-it)}(a\exp(-itX)a^{-1})\Big|_{t=0} = aXa^{-1}. \tag{9.20}$$

We give this Lie algebra automorphism its own name, the *adjoint map* $\mathrm{Ad}(a)$,

$$\mathrm{Ad}(a): \quad \begin{matrix} \mathfrak{g} & \to & \mathfrak{g} \\ X & \mapsto & \mathrm{Ad}(a)X = aXa^{-1}. \end{matrix} \tag{9.21}$$

Note that the expression aXa^{-1} is always a Lie algebra element. According to the Lie correspondence theorem, it is

$$a\exp(X)a^{-1} = \exp(aXa^{-1}), \tag{9.22}$$

a fact, which is seen easily also by direct computation.

Adjoint Representations of Lie Groups and Lie Algebras

Now let us turn our attention to the map Ad itself. The map is

$$\mathrm{Ad}: \quad \begin{matrix} G & \to & GL(\mathfrak{g}) \\ a & \mapsto & \mathrm{Ad}(a). \end{matrix} \tag{9.23}$$

In fact, this map is a group homomorphism, as it is $\mathrm{Ad}(ab) = \mathrm{Ad}(a)\mathrm{Ad}(b)$ and $\mathrm{Ad}(a^{-1}) = \mathrm{Ad}(a)^{-1}$. (*exercise 9.1*) The Lie algebra map $d(\mathrm{Ad})$ corresponding to Ad, is denoted ad,

$$\mathrm{ad}: \quad \begin{matrix} \mathfrak{g} & \to & gl(\mathfrak{g}) \\ X & \mapsto & \mathrm{ad}(X), \end{matrix} \tag{9.24}$$

and it is determined by the expression

$$\mathrm{ad}(X)Y = \frac{d}{d(-it)}\mathrm{Ad}(\exp(-itX))Y\Big|_{t=0} = XY - YX = [X,Y]. \tag{9.25}$$

So the Lie algebra homomorphism $\mathrm{ad}(X)$ has the explicit form

$$\mathrm{ad}(X): \quad \begin{matrix} \mathfrak{g} & \to & \mathfrak{g} \\ Y & \mapsto & \mathrm{ad}(X)Y = [X,Y], \end{matrix} \tag{9.26}$$

and it respects the algebra structure,

$$\mathrm{ad}\,([X,Y]) = [\mathrm{ad}(X),\mathrm{ad}(Y)], \tag{9.27}$$

which can be proven by using the Jacobi identity. The maps Ad and ad are not only homomorphisms but also special cases of representations where the vector space V of the representation is simply the Lie algebra $\mathfrak{g}$ itself. They are the so-called *adjoint representations* of the Lie group and the Lie algebra. The commutative diagram we saw in the previous section, now with $f = \mathrm{Ad}$ and $df = \mathrm{ad}$, becomes

$$\begin{matrix} G & \xrightarrow{\ \mathrm{Ad}\ } & GL(\mathfrak{g}) \\ {\scriptstyle\exp}\uparrow & & \uparrow{\scriptstyle\exp} \\ \mathfrak{g} & \xrightarrow[\ \mathrm{ad}\ ]{} & gl(\mathfrak{g}) \end{matrix} \tag{9.28}$$

Finally, let us write down once again the Lie correspondence, this time for the case of the adjoint representations,

$$\text{Ad}(\exp X) = \exp(\text{ad}(X)). \tag{9.29}$$

The adjoint representation has a dimension equal to the dimension of the Lie group and Lie algebra. The adjoint representation $\text{ad}(X)$ can be given as a matrix, if we choose a basis $\{X_a\}$ in the Lie algebra. As for any linear map, it is

$$\text{ad}(X_a)X_b = (\text{ad}X_a)_{cb}X_c, \tag{9.30}$$

but it is also

$$\text{ad}(X_a)X_b = [X_a, X_b] = \mathrm{i}f_{abc}X_c. \tag{9.31}$$

So we infer that the matrix representation of the adjoint representation is given by the structure constants,

$$(\text{ad}X_a)_{cb} = \mathrm{i}f_{abc}. \tag{9.32}$$

With the above results we conclude the general theory of Lie groups, Lie algebras, and representations. In the next sections we will proceed to the topics that are central to our purposes, by identifying group elements as the symmetry transformations of interest. We will consider the action of Lie group transformations on the base manifold (or space), on tensorial objects, on scalar functions, and on arbitrary tensor fields, which combine all transformation properties.

9.3 Tensor and Function Representations

Representations on Tensors

We denote by A_i an arbitrary mixed covariant and contravariant tensor, where the multi-index i stands for the collection of all indices. Note that A_i is not a tensor *field* but an element $A^{\mu_1 \cdots \mu_m}{}_{\nu_1 \ldots \nu_n}$ of a tensor space $V^{\otimes m} \otimes \tilde{V}^{\otimes n}$. An infinitesimal transformation of A_i is given by

$$A_i \mapsto A_i' = A_i + \delta A_i. \tag{9.33}$$

On the other hand, we can consider this transformation to be implemented by a transformation group G acting on the tensor space through a representation D. I.e. the group G acts on the tensor space as

$$A_i \mapsto A_i' \equiv D(\epsilon)(A_i). \tag{9.34}$$

We remember that the group elements have the general form 9.17 in the representation, with $\mathcal{G}_a$, $a = 1, \ldots, r$, being the group generators. Thus, the variation δA_i of the tensor can be written as

$$\delta A_i = -\mathrm{i}\,\epsilon^a (\mathcal{G}_a)_{ij} A_j, \tag{9.35}$$

where summation over the repeated multi-index j is implied. The generators $(\mathcal{G}_a)_{ij}$ are called *internal generators* and the reason for this will be clear soon. The generators $(\mathcal{G}_a)_{ij}$ acting on tensors A_i have only the multi-indices ij attached, they do not have any other functional dependence, like a dependence on coordinates. Let us memorize: *tensor representations act on tensors and are characterized by discrete tensorial indices. Tensorial representations correspond to internal generators.*

Representations on Scalar Functions

Let us consider smooth complex-valued functions, denoted $\phi(x)$, $\psi(x)$, etc., depending on the d-dimensional real spacetime coordinate x of a base manifold $\mathcal{M}$. The functions $\phi(x)$, $\psi(x)$ are elements of an infinite-dimensional Hilbert space $\mathcal{H}$, for which a suitable scalar product $\langle \phi, \psi \rangle$ is defined, e.g.

$$\langle \phi, \psi \rangle \equiv \int_{\mathbb{R}^d} \phi^*(x)\, \psi(x)\, \mathrm{d}^d x. \tag{9.36}$$

Typically, it is required that the functions $\phi(x)$ are bounded, in the sense that $\langle \phi, \phi \rangle < \infty$. The coordinate variable x can be viewed as a continuous index. We want the elements of the function space to be *scalar functions*, so that under a symmetry transformation of the coordinates, $x \mapsto x' = D(\epsilon)(x)$, they transform as

$$\phi'(x') = \phi(x). \tag{9.37}$$

We say that the scalar function is invariant under the transformation. This means that the value $\phi'(x')$ of the changed function ϕ' at the transformed point x' is equal the value $\phi(x)$ of the original function ϕ at the original point x. We have actually two measures for capturing the function change, namely the *form variation*, measuring how the function itself changes, and the *total variation*, measuring how the function value changes. Let us make it more precise. Consider again an infinitesimal spacetime transformation $x \mapsto x'(x) = x + \delta x(x)$. Then there are the following induced changes in the function ϕ itself and on its value $\phi(x)$:

$$
\begin{aligned}
\phi &\mapsto \phi' = \phi + \delta\phi, & \text{Form variation} \\
\phi(x) &\mapsto \phi'(x') = \phi(x) + \bar{\delta}\phi(x), & \text{Total variation}
\end{aligned}
\left.\rule{0pt}{4.5ex}\right\} . \tag{9.38}
$$

For a scalar function $\phi(x)$ the form variation $\delta\phi$ is not trivial, while the total variation $\bar{\delta}\phi(x)$ vanishes identically. However, we will consider also non-scalar functions (tensor fields) in the next section, for which the total variation is not zero. The transformation of the scalar function can be again implemented through a transformation group with corresponding generators, which we will call *orbital generators*. Let us memorize already here: *function representations act on scalar functions defined on the coordinate base space, with the coordinates representing continuous indices. Function representations correspond to orbital generators.*

9.4　Symmetry Transformations of Tensor Fields

Symmetry Transformations of Fields

For our applications in physics, we are interested in tensor fields that display a specific transformation behavior under basic symmetry transformations of interest. Technically, *the tensor fields are viewed as functions of discrete tensorial indices and continuous spacetime indices*. In other words, each physical field corresponds to a combined tensorial and functional representation of the symmetry transformation of interest. In this section we will define such symmetry transformations of the base manifold coordinates and tensor fields in detail. We will establish the connection between field transformations and the Lie-algebraic generators of these transformations. The most general *infinitesimal transformations of the base manifold coordinates and tensor fields* we are going to consider in field theory are

$$
\begin{aligned}
x^\mu &\mapsto x'^\mu(x) = x^\mu + \delta x^\mu(x), & \text{Base space variation} \\
\phi_i(x) &\mapsto \phi_i'(x) = \phi_i(x) + \delta\phi_i(x), & \text{Form variation} \\
\phi_i(x) &\mapsto \phi_i'(x') = \phi_i(x) + \bar{\delta}\phi_i(x), & \text{Total variation}
\end{aligned}
\left.\rule{0pt}{7ex}\right\} . \tag{9.39}
$$

We must emphasize that the coordinate functions x^μ must be treated separately, since they are not tensor fields. As we saw already in Section 3.1, there are two quantities for the variation of tensor fields $\phi_i(x)$. One quantity is the *form variation*, $\delta\phi_i$, defined as

$$\delta\phi_i \equiv \phi_i' - \phi_i, \tag{9.40}$$

which captures the *variation of the field function* itself, for all manifold points. The other quantity is the *total variation*, $\overline{\delta}\phi_i(x)$, defined as

$$\overline{\delta}\phi_i(x) \equiv \phi_i'(x') - \phi_i(x), \tag{9.41}$$

which captures the *variation of the field value*, due to the change in the function and the manifold point. We can parametrize the above variations by real parameters ϵ^a, $a = 1, ..., r$, with $|\epsilon^a| \ll 1$. The index a can be a manifold coordinate index, or any other label of the fields. Furthermore, the distinct variations in 9.39 can be written as

$$\left.\begin{aligned}
\delta x^\mu(x) &\equiv \epsilon^a X_a^\mu(x)\,, &\text{Base space variation} \\
\delta\phi_i(x) &\equiv \epsilon^a F_{i,a}(x)\,, &\text{Form variation} \\
\overline{\delta}\phi_i(x) &\equiv \epsilon^a \Phi_{i,a}(x)\,, &\text{Total variation}
\end{aligned}\right\}. \tag{9.42}$$

The functions $X_a^\mu(x)$, $F_{i,a}(x)$ and $\Phi_{i,a}(x)$ are called the *generators of symmetry transformations* and are defined as

$$X_a^\mu \equiv \left.\frac{\partial(\delta x^\mu)}{\partial\epsilon^a}\right|_{\epsilon^a=0}, \tag{9.43}$$

$$F_{i,a} \equiv \left.\frac{\partial(\delta\phi_i)}{\partial\epsilon^a}\right|_{\epsilon^a=0}, \tag{9.44}$$

$$\Phi_{i,a} \equiv \left.\frac{\partial(\overline{\delta}\phi_i)}{\partial\epsilon^a}\right|_{\epsilon^a=0}. \tag{9.45}$$

By knowing the above generators, one can fully reconstruct any symmetry transformation. The total variation and the form variation have a simple and fundamental relation, which we have seen already in Section 3.1. It is

$$\boxed{\overline{\delta}\phi_i = \delta\phi_i + \partial_\mu\phi_i\,\delta x^\mu} \tag{9.46}$$

This means that the total variation $\overline{\delta}\phi_i$ of the field value contains two parts, one stemming from the functional change, $\delta\phi_i$, and another one stemming from the change in the spacetime coordinate, $\delta x^\mu \partial_\mu\phi_i$. This relation translates into an equation that includes all symmetry generator functions and reads

$$\Phi_{i,a} = F_{i,a} + X_a^\mu\,\partial_\mu\phi_i. \tag{9.47}$$

By using the formula 3.25 from Chapter 3, we infer that it is

$$F_{i,a} = -\mathcal{L}_{X_a}\phi_i, \tag{9.48}$$

where $\mathcal{L}_{X_a}\phi_i$ is the Lie derivative of the tensor field $\phi_i(x)$ with respect to the vector field $X_a(x)$. Thus, the relation of the variation-defining functions can be rewritten as

$$\mathcal{L}_{X_a}\phi_i = X_a^\mu\,\partial_\mu\phi_i - \Phi_{i,a}. \tag{9.49}$$

This is once again the formula for the Lie derivative expressed as the directional derivative, $X_a^\mu\,\partial_\mu\phi_i$, minus the tensorial dragging terms, $\Phi_{i,a}$.

Internal and Orbital Generators of a Lie Group

We now consider symmetry transformations of the tensor fields of interest that contain both, a certain tensor indices part and a part involving the base coordinates. The set of these symmetry transformations form a Lie group, whose generators have representations, which are denoted as $\mathcal{G}_a(x)$, $a = 1, ..., r$. We can consider the generators purely for the *tensorial* (or *internal*) *part*, $\mathcal{G}_a^{\text{internal}}(x)$, and the generators purely for the *spacetime* (or *orbital*) *part*, $\mathcal{G}_a^{\text{orbital}}(x)$. Correspondingly, we have the decomposition

$$\boxed{\mathcal{G}_a = \mathcal{G}_a^{\text{internal}} + \mathcal{G}_a^{\text{orbital}}} \tag{9.50}$$

Let us derive the relation between these group generators $\mathcal{G}_a^{\text{internal}}(x)$ and $\mathcal{G}_a^{\text{orbital}}(x)$, on one hand, and the functions $\Phi_{i,a}(x)$ and $X_a^\mu(x)$, on the other hand. The *field function* ϕ_i transforms according to the total generator $\mathcal{G}_a$, since it contains both the internal and the orbital part. We have

$$\boxed{\phi_i'(x) = \exp\left[-i\,\epsilon^a \mathcal{G}_a(x)\right]_{ij}\phi_j(x)} \tag{9.51}$$

or, infinitesimally,

$$\delta\phi_i(x) = -i\,\epsilon^a \mathcal{G}_a(x)_{ij}\,\phi_j(x). \tag{9.52}$$

In contrast, the *field value* $\phi_i(x)$, for a particular argument x, transforms according to the internal generators only. Infinitesimally it is

$$\bar\delta\phi_i(x) = -i\,\epsilon^a \mathcal{G}_a^{\text{internal}}(x)_{ij}\,\phi_j(x), \tag{9.53}$$

or, equivalently, for a finite transformation of the field value,

$$\boxed{\phi_i'(x') = \exp\left[-i\,\epsilon^a \mathcal{G}_a^{\text{internal}}(x)\right]_{ij}\phi_j(x)} \tag{9.54}$$

By using 9.47, we obtain the relation

$$-i\,(\mathcal{G}_a)_{ij}\phi_j = F_{i,a} = \Phi_{i,a} - X_a^\mu\,\partial_\mu\phi_i. \tag{9.55}$$

Thus, for the internal generator we have the formula

$$(\mathcal{G}_a^{\text{internal}})_{ij}\phi_j = i\,\Phi_{i,a}, \tag{9.56}$$

and for the orbital generator the formula

$$(\mathcal{G}_a^{\text{orbital}})_{ij}\phi_j = -i\,X_a^\mu\,\partial_\mu\phi_i. \tag{9.57}$$

These last two formulae are useful for computing $\mathcal{G}_a$ if we know $\Phi_{i,a}$ and X_a^μ, or vice versa. We see that the generator for internal transformations $(\mathcal{G}_a^{\text{internal}})_{ij}$ operates as matrix multiplication with respect to the tensor indices i of the field ϕ_i. In contrast, the generator for spacetime transformations $(\mathcal{G}_a^{\text{orbital}})_{ij}$ acts trivially, simply as δ_{ij}, on the the tensor indices of the field. It is generally

$$(\mathcal{G}_a^{\text{orbital}})_{ij} = -i\,\delta_{ij}X_a^\mu\,\partial_\mu, \tag{9.58}$$

which shows that the generator of orbital transformations is independent of the tensorial character of the field at hand. The orbital part is the same for all tensor fields. Consequently, we can take the simplest field, i.e. the scalar field, in order to derive the orbital generators. We will derive explicit cases of $(\mathcal{G}_a^{\text{orbital}})_{ij}$ later when we look closely on specific spacetime symmetries.

9.5   Induced Representations

Formula for Induced Representations

In this section we will derive how a tensor field representation can be obtained for the semidirect product $T \rtimes G$ of a commutative group T and a symmetry group G. Both groups, the commutative group of translations T and the group G, are supposed to act on tensor fields. Let us denote a translation in T that is parametrized by the vector a as $T(a)$, and a symmetry transformation in G as $\Lambda(\omega)$. The group G is embedded in the semidirect product $T \rtimes G$ in a natural way with the identification $\Lambda(\omega) = \Lambda(0, \omega)$. i.e. the symmetry transformations $\Lambda(\omega) \in G$ are the ones at the particular point $a = 0$. Now let us write down how a tensor field $\phi(x)$, where we suppress the tensorial indices for notational simplicity, changes under these transformations. The following diagram is commutative:

$$
\begin{array}{ccc}
\phi & \xrightarrow{\;\Lambda(0,\omega)\;} & \phi' \\[2pt]
{\scriptstyle T(-a)}\Big\downarrow & & \Big\downarrow{\scriptstyle T(-a)} \\[2pt]
\phi_{(-a)} & \xrightarrow[\;\Lambda(a,\omega)\;]{} & \phi'_{(-a)}
\end{array}
\tag{9.59}
$$

The two transformations $\Lambda(0, \omega)$ and $T(-a)$ are known and we want to find the "translated" transformation $\Lambda(a, \omega)$ at the point a. The transformation $\Lambda(a, \omega)$ belongs to the so-called *induced representation*. Since the diagram is commutative, we obtain the relation

$$
\Lambda(a, \omega) = T(a)^{-1} \Lambda(0, \omega)\, T(a).
\tag{9.60}
$$

Note that $T(a)^{-1} = T(-a)$. In the next step, we take the limit of an infinitesimal parameter ω and consider the linear part of the transformation $\Lambda(a, \omega)$, i.e. we introduce the Lie algebra generators, let us call then $M_{bc}(a, \omega)$, $b, c = 1, \ldots, r$. Then, the above relation becomes

$$
M_{bc}(a, \omega) = T(a)^{-1} M_{bc}(0, \omega)\, T(a).
\tag{9.61}
$$

This is the result we are looking for, because it gives the recipe how to calculate the generator $M_{bc}(a, \omega)$ of the induced representation as a product of known elements. We will use this result later to determine the conformal algebra.

9.6   Lie Algebra of Killing Vector Fields

Killing Vector Fields revisited

In Section 3.3 we have introduced Killing vector fields (KVFs) as the key quantities for describing symmetry transformations of geometric manifolds. So how do KVFs fit in the general scheme of Lie groups and Lie algebras? To answer this question, let us first examine KVFs again. Consider a geometric manifold $(\mathcal{M}, g)$ and a set of Killing vector fields $K_a = K_a^\mu(x)\partial_\mu$, labeled by indices $a = 1, \ldots, r$, thus defining an infinitesimal symmetry transformation of the manifold coordinates through

$$
x'^\mu(x) = x^\mu + \delta x^\mu(x) = x^\mu + \epsilon^a K_a^\mu(x).
\tag{9.62}
$$

In the above transformation, we consider not just one real parameter defining the symmetry transformation, but rather a finite set of real parameters. Each of the Killing vectors $K_a^\mu(x)$, $a = 1, \ldots, r$, is supposed to fulfill the Killing condition

$$
\mathcal{L}_{K_a} g_{\mu\nu} = 0,
\tag{9.63}
$$

thus ensuring that the metric stays invariant, $g'_{\mu\nu}(x) = g_{\mu\nu}(x)$. We can find out the origin of the Killing vector fields by exploiting the parametrization with the r-tuple $\epsilon = (\epsilon^1, \ldots, \epsilon^r)$. As explained in Section 7.3, we can write each of the geometric symmetry transformations as a map $x^\mu \mapsto x'^\mu(x, \epsilon)$. The trivial value is $x^\mu = x'^\mu(x, 0)$. By Taylor-expanding the coordinate functions $x'^\mu(x, \epsilon)$ with respect to the parameters ϵ^a at the point $\epsilon = 0$, we obtain the identification

$$K_a^\mu(x) = \left.\frac{\partial x'^\mu(x, \epsilon)}{\partial \epsilon^a}\right|_{\epsilon^a = 0}. \tag{9.64}$$

In other words, each Killing vector field $K_a^\mu(x)$ can be identified with the linear part of the associated change of the coordinate functions x^μ. The isometry transformations, in turn, constitute a Lie group. In addition, we know that the Lie algebra of a Lie group is the tangent space at the identity element of the group. We can thus anticipate that the KVFs constitute a Lie algebra. This is confirmed in the following.

Lie Algebra of Isometries

When we consider the isometries of a manifold, the following proposition is central:

> *The Killing vector fields of a geometric manifold constitute the Lie algebra of the group of isometries of this manifold.*

Indeed, by infinitesimal transformations as expressed in 9.62, the KVFs generate all isometry transformations, if the latter exist, of a given manifold. The Lie algebra of the KVFs is the real vector space spanned by the linearly independent KVFs with constant real coefficients. The product of this Lie algebra is the usual commutator $[\cdot, \cdot]$ of vector fields. With K and L being KVFs, also their Lie product $[K, L]$ is a KVF, since it is

$$\mathcal{L}_{[K,L]} g_{\mu\nu} = [\mathcal{L}_K, \mathcal{L}_L] \, g_{\mu\nu} = 0, \tag{9.65}$$

where we have applied the identity 3.35. The KVFs constitute a real Lie algebra and the corresponding commutator has the general form

$$[K_a, K_b] = f_{abc} K_c, \tag{9.66}$$

with real structure constants f_{abc}. Later, in Chapter 20, we will derive that the maximum possible number of linearly independent KVFs is $D(D+1)/2$, where D is the dimension of the underlying manifold. Based on the Lie theory, as developed previously, we can infer that an isometry transformation generated by KVFs $K_a(x)$ has the form $\exp(\epsilon^a K_a(x))$. For example, a finite isometry transformation of the 3-dimensional Euclidean space, as discussed in Section 3.4, can be written as

$$\exp\left(c^j \partial_j + \varphi^j \epsilon_{jkl} x^k \partial^l\right), \tag{9.67}$$

where c^j represents the three parameters for the translations and φ^j the three parameters for the rotations. We will rederive this result in the next chapter.

Lie Algebra of Conformal Maps

For conformal maps of a manifold, we have the proposition:

> *The conformal Killing vector fields of a geometric manifold constitute the Lie algebra of the group of conformal maps of this manifold.*

The reasoning in the conformal case is done in the same way as in the isometric case. It is easy to see that if K and L are conformal KVFs, then their commutator $[K, L]$ is also a conformal KVF. (*exercise 9.2*) The commutator relation of the conformal KVF has the same structure as in 9.66. To give a concrete example, a finite scale transformation, as discussed initially in Section 3.4, has the form

$$\exp\left(sx^\mu \partial_\mu\right). \tag{9.68}$$

Here the real parameter s fixes the scale transformation. We will obtain this result again in Chapter 14 by systematically studying the Lie algebra of conformal transformations. At this point we conclude our survey of representations and their basic properties. In the following chapters we will apply these methods to concrete symmetries.

Further Reading

Representation theory is a vast field with many applications, and in this chapter we have covered only the basics necessary for our purposes. A discussion about tensor products of representations and the notion of reducibility is found in the book of Hall [38]. A thorough reference work for representation theory and in particular for the technically involved subject of induced representations is the book of Barut and Raczka [7].

10

Rotations and Euclidean Symmetry

In this chapter we begin our study of the continuous symmetries of space and time. We first introduce these symmetries in their defining representation, acting on the base space of spacetime coordinates. Rotations and translations are the symmetries of nonrelativistic space. The rotation group is, in a sense, the drosophila within linear groups and provides enough structure to illustrate the central notions of Lie group theory. In the first part of this chapter we give a basic overview of the rotation group, realized as $SO(3)$ acting on vectors, and realized as $SU(2)$ acting on spinors. We then derive the rotation algebra. In the second part, we study translations in space and construct the Euclidean group as the semidirect product of translations and rotations. The Euclidean group is the symmetry group of classical affine space. In the last part we study the associated Euclidean algebra.

10.1 Rotation Group

Rotation Group as $SO(3)$

The most natural way to introduce the *rotation group* is by considering the symmetry transformations of the linear space $\mathbb{R}^3$ equipped with the standard scalar product and the metric

$$\delta_{kl} = \mathrm{diag}(1,1,1). \tag{10.1}$$

A *rotation*, denoted R, is a linear transformation of the vectors $\boldsymbol{x}$,

$$\boldsymbol{x} \mapsto \boldsymbol{x}' = R\boldsymbol{x}, \tag{10.2}$$

where we require that the lengths are kept invariant,

$$x'^2 = x^2. \tag{10.3}$$

This is equivalent to the requirement

$$\boxed{R^T R = 1} \tag{10.4}$$

which fully defines the rotation group as the group of orthogonal transformations, $O(3)$. A rotation needs 3 real parameters to be defined, since the condition of orthogonality fixes already 6 real parameters. The orthogonal group $O(3)$ also contains the *inversions*, which

DOI: 10.1201/9781003087748-10

we will, however, not consider further here. In other words, form now on we restrict ourselves to the group $SO(3)$ of *proper rotations*. When we choose a Cartesian coordinate system, the active *counter-clockwise** rotations $R_k(\theta)$ around the axes numbered 1, 2, and 3 by an angle θ are given by

$$R_1(\theta) = \begin{pmatrix} 1 & 0 & 0 \\ 0 & \cos\theta & -\sin\theta \\ 0 & \sin\theta & \cos\theta \end{pmatrix}, \tag{10.5}$$

for the 1-axis,

$$R_2(\theta) = \begin{pmatrix} \cos\theta & 0 & \sin\theta \\ 0 & 1 & 0 \\ -\sin\theta & 0 & \cos\theta \end{pmatrix}, \tag{10.6}$$

for the 2-axis, and

$$R_3(\theta) = \begin{pmatrix} \cos\theta & -\sin\theta & 0 \\ \sin\theta & \cos\theta & 0 \\ 0 & 0 & 1 \end{pmatrix}, \tag{10.7}$$

for the 3-axis. One might ask why the rotation $R_2(\theta)$ has the opposite angle sign than the other two rotations. Let us take a look on the diagram 10.1 below. It shows a right-handed

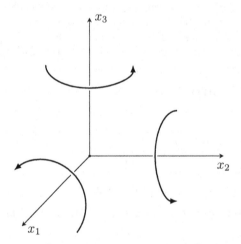

Figure 10.1: Positive rotations around Cartesian axes

Cartesian system and the counter-clockwise rotations around each of the axes. Note that the rotation $R_1(\theta)$ happens in the (2,3)-plane, the rotation $R_3(\theta)$ happens in the (1,2)-plane, but the rotation $R_2(\theta)$ happens in the (3,1)-plane and the order of the directions is important, i.e. a counter-clockwise rotation in the (3,1)-plane turns from the 3-axis to the 1-axis.

---

*Note that the signs resulting in concrete formulae for the generators and their commutators depend on the definitions of the parameter sets and the metric signature.

Rotation Group as $SU(2)$

The rotation group does not need to act necessarily on 3-vectors. We can assign to every 3-vector $\boldsymbol{x}$ a hermitian (i.e. self-adjoint) 2×2-matrix X with the map

$$\boldsymbol{x} = \begin{pmatrix} x^1 \\ x^2 \\ x^3 \end{pmatrix} \mapsto X = \begin{pmatrix} x^3 & x^1 - \mathrm{i}x^2 \\ x^1 + \mathrm{i}x^2 & -x^3 \end{pmatrix}. \tag{10.8}$$

The resulting matrix X is hermitian, $X^\dagger = X$, and traceless, $\mathrm{Tr}\,X = 0$. It is also

$$\det X = -(x^1)^2 - (x^2)^2 - (x^3)^2 = -\boldsymbol{x}^2. \tag{10.9}$$

So we seek for transformations of the matrices X that leave the determinant, or equivalently, the 3-vector norm, invariant. This is achieved by using unitary transformations U belonging to the group $SU(2)$. Indeed, the transformation

$$X \mapsto X' = UXU^\dagger \tag{10.10}$$

yields a matrix X', which is again hermitian and traceless, and has the property of leaving the determinant unchanged,

$$\det X' = \det X, \tag{10.11}$$

exactly as needed. So we have found that unitary transformations $U \in SU(2)$ correspond to rotations $R \in SO(3)$, and that transformations in the space of hermitian traceless matrices correspond to rotations in the base space. In formulae this reads

$$\boldsymbol{x}' = R\boldsymbol{x} \quad \Leftrightarrow \quad X' = UXU^\dagger, \tag{10.12}$$

and

$$\boldsymbol{x}'^2 = \boldsymbol{x}^2 \quad \Leftrightarrow \quad \det X' = \det X. \tag{10.13}$$

The two unitary matrices U and $-U$ lead to the same rotation R. Hence, there is a 1:2-homomorphism

$$SO(3) \rightarrow SU(2), \tag{10.14}$$

which expresses the fact that $SU(2)$ is topologically the double cover of $SO(3)$. The unitary transformations of $SU(2)$ naturally act also on the 2-complex-dimensional space $\mathbb{C}^2$, the so-called *spinor space*, which contains complex vectors with two components, the *spinors*. We will examine spinors in detail in the chapter on Lorentz symmetry. We introduce the three *Pauli matrices*, $\{\sigma_1, \sigma_2, \sigma_3\}$, which are defined as

$$\sigma_1 \equiv \begin{pmatrix} 0 & 1 \\ 1 & 0 \end{pmatrix}, \quad \sigma_2 \equiv \begin{pmatrix} 0 & -\mathrm{i} \\ \mathrm{i} & 0 \end{pmatrix}, \quad \sigma_3 \equiv \begin{pmatrix} 1 & 0 \\ 0 & -1 \end{pmatrix}, \tag{10.15}$$

and represent a basis for traceless hermitian matrices (*Wolfgang Ernst Pauli*). By using this matrix basis, the matrix X can be written as

$$X = x^k \sigma_k = x^1 \sigma_1 + x^2 \sigma_2 + x^3 \sigma_3. \tag{10.16}$$

The inverse relation is given by

$$x_k = \frac{1}{2} \mathrm{Tr}(X \sigma_k). \tag{10.17}$$

For the two transformations R and U, the relation can be made also explicit. Note that it is

$$R^l{}_k \sigma_l = U \sigma_k U^\dagger. \tag{10.18}$$

By multiplying with σ_m and by taking the trace, we obtain the formula

$$R_{kl} = \frac{1}{2}\text{Tr}(U^\dagger \sigma_k U \sigma_l). \tag{10.19}$$

Later, we will study the Lorentz group $L = SO(1,3)$ as the relativistic analogue of the rotation group $R = SO(3)$. The above analysis and the above results can indeed be readily generalized to the relativistic case.

10.2   Rotation Algebra

Infinitesimal Rotations

Before we proceed to the derivation of the *rotation algebra*, let us first examine infinitesimal rotations in a direct way. Writing for simplicity all indices as lower indices, an infinitesimal rotation is given by

$$R_{kl} = \delta_{kl} + \omega_{kl}, \tag{10.20}$$

with $|\omega_{kl}| \ll 1$. The defining condition for rotations 10.4 is thus equivalent to

$$\left(1 + \omega^T\right)\left(1 + \omega\right) = 1, \tag{10.21}$$

which means that the parameter matrix ω_{kl} is antisymmetric,

$$\omega_{kl} = -\omega_{lk}. \tag{10.22}$$

We can write ω_{kl} as a matrix as

$$(\omega_{kl}) \equiv \begin{pmatrix} 0 & -\theta_3 & \theta_2 \\ \theta_3 & 0 & -\theta_1 \\ -\theta_2 & \theta_1 & 0 \end{pmatrix}, \tag{10.23}$$

where the rotation parameters ω_{kl} and θ_k have the relation

$$\omega_{kl} = -\epsilon_{klm}\theta_m. \tag{10.24}$$

So an infinitesimal rotation of a vector V_k takes the form

$$\delta V_k = \omega_{kl}V_l = -\epsilon_{klm}\theta_m V_l, \tag{10.25}$$

a result, which will be used below when we derive the vector representation of the generators of rotations.

Generators and Commutators of Rotation Algebra

In the previous section we have identified the three independent rotations, $R_1(\theta)$, $R_2(\theta)$, and $R_3(\theta)$. The *generators of rotations*, denoted J_k, as elements of the Lie algebra of rotations are defined as

$$J_k \equiv \mathrm{i}\,\frac{\partial R_k}{\partial\theta}(\theta)\Big|_{\theta=0}. \tag{10.26}$$

A direct calculation leads to the hermitian matrix representation

$$J_1 = \begin{pmatrix} 0 & 0 & 0 \\ 0 & 0 & -\mathrm{i} \\ 0 & \mathrm{i} & 0 \end{pmatrix}, \quad J_2 = \begin{pmatrix} 0 & 0 & \mathrm{i} \\ 0 & 0 & 0 \\ -\mathrm{i} & 0 & 0 \end{pmatrix}, \quad J_3 = \begin{pmatrix} 0 & -\mathrm{i} & 0 \\ \mathrm{i} & 0 & 0 \\ 0 & 0 & 0 \end{pmatrix}. \tag{10.27}$$

Thus, an infinitesimal rotation about an angle vector $\delta\boldsymbol{\theta}$ can be written as

$$R(\delta\boldsymbol{\theta}) = 1 - \mathrm{i}\,\delta\boldsymbol{\theta} \cdot \boldsymbol{J}. \tag{10.28}$$

Equivalently, a finite rotation about an angle $\boldsymbol{\theta}$ is given by the exponential expression

$$R(\boldsymbol{\theta}) = \exp(-\mathrm{i}\,\boldsymbol{\theta} \cdot \boldsymbol{J}). \tag{10.29}$$

Having a concrete matrix representation of the generators at hand, it is easy to obtain the *commutator relations of the rotation algebra. (exercise 10.1)*

$$\boxed{[J_k, J_l] = \mathrm{i}\,\epsilon_{klm} J_m} \tag{10.30}$$

Here a summation over the repeated index m is implied. These relations define fully the rotation algebra, independently of any concrete representation. They are valid for the Lie algebra representation $so(3)$ as for the representation $su(2)$. We see that the structure constants f_{klm} for the rotation group are given by the Levi-Civita antisymmetric tensor ϵ_{klm}. From our general considerations for field representations, we know that the generators of rotations J_k possess an *internal part*, denoted S_k, and an *orbital part*, denoted L_k, with $k = 1, 2, 3$. This means that the generators of rotations J_k have the decomposition

$$\boxed{J_k = S_k + L_k} \tag{10.31}$$

The two distinct representations, S_k and L_k, are derived in the following.

Vector Representation of Rotations

First, we derive the representation of the generators of rotations, when these act on vectors. This is an example of an internal (or spin) representation. As mentioned before, we denote the generators acting on tensorial indices by the symbol S_k, with $k = 1, 2, 3$. The corresponding representation of a rotation, $D_V(R(\boldsymbol{\theta}))$, can be written as an exponential as

$$D_V(R(\boldsymbol{\theta})) = \exp(-\mathrm{i}\,\boldsymbol{\theta} \cdot \boldsymbol{S}). \tag{10.32}$$

Therefore, an infinitesimal transformation δV_k of a vector V_k is expressed as

$$\delta V_k = -\mathrm{i}\,\theta_m \,(S_m)_{kl}\, V_l. \tag{10.33}$$

On the other hand, the infinitesimal transformation of the vector is also given by

$$\delta V_k = \epsilon_{kml}\theta_m V_l. \tag{10.34}$$

Thus, we obtain the *vectorial (spin) representation of the generators of rotations*,

$$\boxed{(S_m)_{kl} = -\mathrm{i}\,\epsilon_{mkl}} \tag{10.35}$$

acting on vector indices. By using the identity $\epsilon_{mkl} = \epsilon_{mrs}\delta_{kr}\delta_{ls}$, we can express the internal generators of rotations alternatively as antisymmetric tensors S_{kl} as

$$\boxed{(S_{kl})_{rs} = -\mathrm{i}\,(\delta_{kr}\,\delta_{ls} - \delta_{lr}\,\delta_{ks})} \tag{10.36}$$

We will encounter such internal generators later also in their 4-dimensional relativistic generalization. Higher order tensorial representations can be derived in the same way as shown here.

Orbital Representation of Rotations

In order to derive the orbital representation, we start with the general formula that we have developed in the previous chapter,

$$(\mathcal{G}_a^{\text{orbital}})_{ij} = -\mathrm{i}\,\delta_{ij} \mathrm{X}_a^\mu \partial_\mu. \tag{10.37}$$

In the present case, a is a space index, say $m = 1, 2, 3$, the i, j are the field indices describing the internal (spin) character, and the orbital index μ is a space index too, say $k = 1, 2, 3$. Thus, the above formula is translated to

$$L_m = -\mathrm{i}\,X_m^k \partial_k. \tag{10.38}$$

Here L_m, $m = 1, 2, 3$, denotes the *orbital representation of the generators of rotations*. By noting that

$$X_m^k = \frac{\partial}{\partial \omega^m} \delta x^k \tag{10.39}$$

and that

$$\delta x^k = -\mathrm{i}\,\omega^l \, (S_l)^k{}_n x^n, \tag{10.40}$$

with S_l given as previously, we reach at the result

$$\boxed{L_m = -\mathrm{i}\,\epsilon_{mkl} x_k \partial_l} \tag{10.41}$$

The formula can also be written as $\boldsymbol{L} = -\mathrm{i}\,\boldsymbol{x} \times \boldsymbol{\nabla}$ by using the cross product. However, the antisymmetric tensor formula

$$\boxed{L_{kl} = -\mathrm{i}\,(x_k \partial_l - x_l \partial_k)} \tag{10.42}$$

is more useful, since it is valid for spaces with dimensions higher than three. Later, when we discuss the Lorentz group, we will see that this three-dimensional result can be embedded in the four-dimensional Lorentzian framework.

10.3   Translations and the Euclidean Group

Pure Translations

The base space we have considered so far in this chapter has been the linear space $\mathbb{R}^3$. However, we would like the origin of the space to be irrelevant, which means that we actually need to consider the affine-linear Euclidean space $\mathbb{E}^3$, with the metric $\delta_{kl} = \text{diag}(1, 1, 1)$. Within Euclidean space, the position of the coordinate origin can be chosen freely. This corresponds to the passive point of view. In the active point of view, this freedom means that we can freely translate the vectors of $\mathbb{E}^3$ by a constant vector, say $\boldsymbol{a}$. This *translation in space* is

$$\boldsymbol{x} \mapsto \boldsymbol{x}' = \boldsymbol{x} + \boldsymbol{a}. \tag{10.43}$$

Obviously, the translations constitute an abelian group, the *group of space translations*, which we denote as T. It is clear that the length of vectors is not invariant anymore. What is invariant, however, is the difference between vectors, i.e. the *distance* between points. Having two vectors $\boldsymbol{x}$ and $\boldsymbol{y}$ being translated, we see that the norm of their difference remains invariant,

$$(\boldsymbol{x}' - \boldsymbol{y}')^2 = (\boldsymbol{x} - \boldsymbol{y})^2. \tag{10.44}$$

An important point is that we require (physical) tensor fields of any valence, $\phi_i(\boldsymbol{x})$, to remain unchanged under a translation,

$$\phi_i'(\boldsymbol{x}') = \phi_i(\boldsymbol{x}). \tag{10.45}$$

That is, the components of tensor fields of arbitrary valence should transform as scalar functions under translations.

Motions
In Euclidean space $\mathbb{E}^3$, we can combine translations and rotations and thus obtain affine-linear transformations called *Euclidean transformations*, or *motions*, which have the form

$$\boldsymbol{x} \mapsto \boldsymbol{x}' = R\boldsymbol{x} + \boldsymbol{a}. \tag{10.46}$$

As explained above, the invariant quantity is now the difference between vectors,

$$\left(\boldsymbol{x}' - \boldsymbol{y}'\right)^2 = \left(\boldsymbol{x} - \boldsymbol{y}\right)^2. \tag{10.47}$$

Conversely, we can ask what are the maximally allowed transformations that leave the Euclidean space structure, i.e., the metric and the distances, invariant. It is straightforward to show that these are the affine-linear transformations above. We can state:

> The Euclidean motions are exactly the symmetry transformations of Euclidean space $\mathbb{E}^3$.

Geometrically, motions change the position and orientation of a 3-dimensional object, but leave its size and shape unchanged.

Euclidean Group, Semidirect Product
The motions form a group, called the *Euclidean group E*. In three dimensions, the Euclidean group has 6 parameters, namely 3 for translations and 3 for rotations. The composition of two successive motions,

$$\boldsymbol{x}' = R_1\boldsymbol{x} + \boldsymbol{a}_1 \quad \text{and} \quad \boldsymbol{x}'' = R_2\boldsymbol{x}' + \boldsymbol{a}_2, \tag{10.48}$$

leads to the combined motion

$$\boldsymbol{x}'' = R_2R_1\boldsymbol{x} + R_2\boldsymbol{a}_1 + \boldsymbol{a}_2. \tag{10.49}$$

If we denote a motion by an ordered pair $(\boldsymbol{a}, R)$, then the composition law is given by

$$(\boldsymbol{a}_2, R_2)(\boldsymbol{a}_1, R_1) = (\boldsymbol{a}_2 + R_2\boldsymbol{a}_1, R_2R_1). \tag{10.50}$$

The inverse transformation is given by

$$(\boldsymbol{a}, R)^{-1} = (-R^{-1}\boldsymbol{a}, R^{-1}), \tag{10.51}$$

and the neutral transformation is the element $(0, 1)$. Associativity can be shown in a straightforward way. Every motion can be written as a composition of a translation and a rotation as $(\boldsymbol{a}, R) = (\boldsymbol{a}, 1)(0, R)$, or, in the reverse order as $(\boldsymbol{a}, R) = (0, R)(R^{-1}\boldsymbol{a}, 1)$. The formula 10.50 reveals that the Euclidean group is the semidirect product of the translation group T with the rotation group R, i.e. $E = T \rtimes R$. As mentioned before, we restrict ourselves here to the proper rotations group $R = SO(3)$ and omit the space reflections. The Euclidean group is occasionally written as $ISO(3)$ and called the *inhomogeneous rotation group*.

Linearization of Euclidean Transformations

Euclidean transformations can be made linear, when we view them in a 4-dimensional space with the vector representation

$$\begin{pmatrix} \boldsymbol{x} \\ 1 \end{pmatrix}, \tag{10.52}$$

where the column entry 1 remains unchanged under all transformations. A Euclidean transformation

$$\boldsymbol{x}' = R\,\boldsymbol{x} + \boldsymbol{a} \tag{10.53}$$

is then translated to a linear matrix multiplication in 4 dimensions,

$$\begin{pmatrix} \boldsymbol{x}' \\ 1 \end{pmatrix} = \left(\begin{array}{c|c} R & \boldsymbol{a} \\ \hline \mathbf{0} & 1 \end{array} \right) \begin{pmatrix} \boldsymbol{x} \\ 1 \end{pmatrix} = \begin{pmatrix} R\,\boldsymbol{x} + \boldsymbol{a} \\ 1 \end{pmatrix}. \tag{10.54}$$

This 4-dimensional representation can be used also to derive the Lie algebra, to which we now turn.

10.4   Euclidean Algebra

Generator of Space Translations

One way to derive the *Euclidean algebra* is to employ the previous 4-dimensional matrix representation of Euclidean transformations. Here we will proceed in a different way and focus first on a special case, the case of a scalar field, and then generalize the result. Under a translation

$$\boldsymbol{x}' = \boldsymbol{x} + \boldsymbol{a}, \tag{10.55}$$

a scalar field $\phi(\boldsymbol{x})$ is invariant,

$$\phi'(\boldsymbol{x}') = \phi(\boldsymbol{x}). \tag{10.56}$$

Equivalently, we can write $\phi'(\boldsymbol{x}) = \phi(\boldsymbol{x} - \boldsymbol{a})$. On the other hand, according to the formula 9.51 for the form variation of a field, $\phi'(\boldsymbol{x})$ can be written also as

$$\phi'(\boldsymbol{x}) = \exp(-\mathrm{i}\,a^k P_k)\phi(\boldsymbol{x}), \tag{10.57}$$

where the 3-covector P_k represents the three *generators of space translations*. Now let us consider a small translation $\boldsymbol{a}$, with $|\boldsymbol{a}| \ll 1$. The above relations yield respectively

$$\phi(\boldsymbol{x} - \boldsymbol{a}) = \phi(\boldsymbol{x}) - a^k \partial_k \phi(\boldsymbol{x}) + \mathcal{O}(2), \tag{10.58}$$

and

$$\exp(-\mathrm{i}\,\boldsymbol{a} \cdot \boldsymbol{P})\phi(\boldsymbol{x}) = \phi(\boldsymbol{x}) - \mathrm{i}\,a^k P_k \phi(\boldsymbol{x}) + \mathcal{O}(2). \tag{10.59}$$

By comparing terms, we arrive at the explicit formula for the generator of space translations:

$$\boxed{P_k = -\mathrm{i}\,\partial_k} \tag{10.60}$$

We have derived this result as the orbital part of the generator of translations for a scalar field. However, this is already the most general representation of the generator of translations acting on *any* type of tensor field. The reason is because *we require all tensor fields, carrying any type of indices, to transform under translations like a scalar field*, i.e. internal indices are not affected by translations. Consequently, there is no internal (spin-specific) part for the generator of translations, only the orbital part exists.

Commutators of the Euclidean Algebra

Now we can derive the commutator relations of the Euclidean algebra, which is easily done by employing the representations derived above. The commutator of two translations is trivial, encoding the fact that the order of translations is irrelevant,

$$[P_k, P_l] = 0. \tag{10.61}$$

For the commutator of J_k with P_k, we use the complete representation

$$J_k = S_k - i\,\epsilon_{klm} x_l \partial_m. \tag{10.62}$$

The spin part S_k is independent of the coordinates $\boldsymbol{x}$ and as such it does not contribute to the commutator with P_k. A straight calculation yields the commutator (*exercise 10.2*)

$$[J_k, P_l] = i\,\epsilon_{klm} P_m. \tag{10.63}$$

We summarize the *commutator relations of the Euclidean algebra* here:

$$
\boxed{
\begin{aligned}
{[J_k, J_l]} &= i\,\epsilon_{klm} J_m \\
{[J_k, P_l]} &= i\,\epsilon_{klm} P_m \\
{[P_k, P_l]} &= 0
\end{aligned}
}
\tag{10.64}
$$

It is interesting to note that the commutator of a rotation and a translation corresponds to a pure translation. This corresponds to the fact that the generator of translations behaves as a vector under rotations. A finite Euclidean transformation is given by the exponential expression

$$\exp\left(-i\,\boldsymbol{\theta} \cdot \boldsymbol{J} - i\,\boldsymbol{a} \cdot \boldsymbol{P}\right). \tag{10.65}$$

Here we conclude our analysis of the symmetries of Euclidean space. We will repeat the group-theoretic combination of translations and rotations that we have considered here for the Euclidean space later when we discuss translations and Lorentz rotations in the four-dimensional Minkowski space.

Further Reading

In our presentation of the rotation group here, we have omitted the discrete space reflections. The full rotation group, including the reflections, is treated in Costa and Fogli [16] and in somewhat more detail in Sexl and Urbantke [82]. The Euclidean group as the symmetry group of Euclidean space is typically found in books on geometry, for example in Dubrovin, Fomenko, and Novikov [25].

11

Boosts and Galilei Symmetry

Galilei symmetry is the symmetry of nonrelativistic spacetime. It corresponds to the maximally admissible set of transformations between Galileian inertial reference frames. Beyond this global definition, it is enlightening to construct the Galilei group G step by step from the ground up. We start by introducing the time parameter and the group of velocity boosts, B. Then we move on to the construction the homogeneous Galilei group as the semidirect product of velocity boosts and rotations, $B \rtimes R$. In a final step, we reach to the full Galilei group by considering the semidirect product of translations in space and time, T, with the homogeneous Galilei group, $G = T \rtimes B \rtimes R$. In the last section we derive the Galilei algebra.

11.1 Group of Boosts

Time Parameter and Galilei Boosts

So far we have considered transformations of space only, the rotations and translations, comprising the 6-parameter Euclidean group. We now introduce the *time parameter* t, which can be viewed as a fourth coordinate in addition to the x_k, $k = 1, 2, 3$. Furthermore, we consider transformations between Galileian inertial reference frames by boosts, where the *boost velocity* $\boldsymbol{u}$ is nonrelativistic, $|\boldsymbol{u}| \ll 1$. Alternatively to this passive view, we can adopt the active view and transform the coordinates according to a *Galilei boost*, which is defined as the transformation

$$\left. \begin{array}{rcl} \boldsymbol{x} & \mapsto & \boldsymbol{x}' = \boldsymbol{x} + \boldsymbol{u}\,t \\ t & \mapsto & t' = t \end{array} \right\}. \tag{11.1}$$

The Galilei boost is parametrized by the three parameters u_k, $k = 1, 2, 3$. Under such pure boosts, the time coordinate t remains untransformed and therefore it is $\ddot{\boldsymbol{x}}' = \ddot{\boldsymbol{x}}$, so that Newtonian forces remain unchanged. Conversely, if we assume the invariance of Newtonian forces, we conclude that the admissible transformations are the complete set of *Galilei transformations*,

$$\left. \begin{array}{rcl} \boldsymbol{x} & \mapsto & \boldsymbol{x}' = R\,\boldsymbol{x} + \boldsymbol{u}\,t + \boldsymbol{a} \\ t & \mapsto & t' = t + \tau \end{array} \right\}, \tag{11.2}$$

DOI: 10.1201/9781003087748-11

parametrized by R, $\boldsymbol{u}$, $\boldsymbol{a}$, and τ, compare with Chapter 4. The *Galilei group* G is a 10-parameter group that takes a central spot in classical nonrelativistic mechanics. We have stated already in our discussion of Newtonian mechanics the main result:

The Galilei group G is exactly the symmetry group of Galileian spacetime $\mathbb{G}_4$.

Turning our attention again to pure Galilei boosts 11.1, we can deduce the group properties immediately. The *group of Galilei boosts* is denoted B and is an abelian group. As in the case of Euclidean transformations, it is possible to render Galilei boosts as linear transformations, by letting them act on a 4-dimensional *spacetime* with the vector representation

$$\begin{pmatrix} \boldsymbol{x} \\ t \end{pmatrix}. \tag{11.3}$$

A Galilei boost becomes a linear matrix multiplication in 4 dimensions then,

$$\begin{pmatrix} \boldsymbol{x}' \\ t' \end{pmatrix} = \begin{pmatrix} 1_3 & \boldsymbol{u} \\ \boldsymbol{0} & 1 \end{pmatrix} \begin{pmatrix} \boldsymbol{x} \\ t \end{pmatrix} = \begin{pmatrix} \boldsymbol{x} + \boldsymbol{u}\,t \\ t \end{pmatrix}. \tag{11.4}$$

We will continue with this concept of linearization, while progressing to the full Galilei group.

Algebra of Galilei Boosts

In order to derive the Lie algebra of the group of Galilei boosts, we start with the orbital representation and consider scalar functions $\phi(\boldsymbol{x}, t)$ of $\boldsymbol{x}$ and t, for which

$$\phi'(\boldsymbol{x}', t') = \phi(\boldsymbol{x}, t) \tag{11.5}$$

is assumed to hold. In the case of a boost transformation, the above invariance reads

$$\phi'(\boldsymbol{x}, t) = \phi(\boldsymbol{x} - \boldsymbol{u}\,t, t). \tag{11.6}$$

By considering a small boost parameter, $|\boldsymbol{u}| \ll 1$, we obtain

$$\phi'(\boldsymbol{x}, t) = \phi(\boldsymbol{x}, t) - \boldsymbol{u}\,t \cdot \frac{\partial}{\partial \boldsymbol{x}} \phi(\boldsymbol{x}, t) + \mathcal{O}(2). \tag{11.7}$$

On the other hand, we can express the variation $\delta\phi(\boldsymbol{x}, t)$ of the scalar field as

$$\delta\phi(\boldsymbol{x}, t) = -\mathrm{i}\,\boldsymbol{u} \cdot \boldsymbol{G}\phi(\boldsymbol{x}, t), \tag{11.8}$$

where $\boldsymbol{G}$ is the *orbital representation of the generator of Galilei boosts*. From the two last equations, we conclude that it is

$$\boldsymbol{G} = -\mathrm{i}\,t\frac{\partial}{\partial \boldsymbol{x}}, \tag{11.9}$$

or, in components,

$$\boxed{G_k = -\mathrm{i}\,t\partial_k} \tag{11.10}$$

This generator contains a term proportional to $t\partial_k$, but it does not contain a term of the form $x_k\partial_t$, the latter actually appearing in Lorentz boosts that mix space and time coordinates.*

---

*The orbital generator of Lorentz boosts contains among others the components $L_{0k} = -\mathrm{i}\,(x_0\partial_k - x_k\partial_0)$.

Galilei boosts do not mix space and time. The *commutator relations of Galilei boosts* are obtained through a straightforward calculation and read

$$[G_k, G_l] = 0 \qquad (11.11)$$

Now let us review the space part of a Galilei boost 11.1 again. It can be viewed as a parametrized translation $\boldsymbol{x}' = \boldsymbol{x} + \boldsymbol{y}(t)$. For each t, there is a translation by $\boldsymbol{y}(t)$. We saw that translations do not change the tensorial indices. This means that Galilei boots indeed do not transform the tensorial indices. At first glance, one might conclude that no internal degrees of freedom of a particle or a system are changed under a Galilei boost, but this is not the complete picture, as we explain below.

Momentum and Mass
Under Galilei boosts, there is another internal quantity, which is changed, and this is the momentum. For a system of particles the change of the total momentum $\boldsymbol{P}$ is

$$\boldsymbol{P} \mapsto \boldsymbol{P}' = \boldsymbol{P} + M\boldsymbol{u}, \qquad (11.12)$$

with M being the total mass of the system. So we need to take into account how the momentum changes and how this change is implemented with a corresponding boost generator. The above equation reveals that in *momentum space*, a Galilei boost is a translation. We know already from the previous chapter how translations are generated. In this way, we conclude that the Galilei boost generator in momentum space, denoted $\widetilde{\boldsymbol{G}}$, is given by

$$\widetilde{\boldsymbol{G}} = -\mathrm{i}\, M \frac{\partial}{\partial \boldsymbol{P}}. \qquad (11.13)$$

This result can be derived also directly if we consider a scalar function $\psi(\boldsymbol{p}, \omega)$ on momentum space. The variable $\boldsymbol{p}$ is conjugate to $\boldsymbol{x}$, while the variable ω is conjugate to t. Because of

$$\psi'(\boldsymbol{P}', \omega') = \psi(\boldsymbol{P}, \omega), \qquad (11.14)$$

which in the case of a boost translates to

$$\psi'(\boldsymbol{P}, \omega) = \psi(\boldsymbol{P} - M\boldsymbol{u}, \omega), \qquad (11.15)$$

we obtain again the formula 11.13, for $|\boldsymbol{u}| \ll 1$. The generator 11.13 can be transformed to coordinate space by a Fourier transformation, see the definitions in appendix B.4. The result is (*exercise 11.1*)

$$\boldsymbol{G} = -M\boldsymbol{X}. \qquad (11.16)$$

In a sense, this is the *internal representation of the generator of Galilei Boosts*. In fact, we used the same symbol for the orbital and the internal generator of Galilei boosts. Let us combine the two parts in one single formula for the *complete generator of Galilei boosts*, acting on functions of space and time coordinates:

$$G_k = -\mathrm{i}\, t\partial_k - MX_k \qquad (11.17)$$

Remembering that the generator of translations is $P_k = -\mathrm{i}\partial_k$, we arrive at the familiar expression

$$\boldsymbol{G} = t\boldsymbol{P} - M\boldsymbol{X}. \qquad (11.18)$$

It is reassuring to see that the complete generator of Galilei boosts has exactly the form of the conserved Galilei momentum, as known from Newtonian mechanics. It can be inspected that the commutator relations 11.11 are also valid for the expression in 11.17. Let us interpret the above result. Not only is the *spin* an intrinsic property of a particle, related to rotations, but also its *mass*, which is related to boosts. Exactly as the total angular momentum $J = L + S$ contains the orbital part L and the spin part S, so does the center of mass momentum $G = tP - MX$ contain the orbital part tP and the internal part $-MX$. We also note that the mass M can be elevated to a generator within an extension of the Galilei algebra, the so-called *Bargmann algebra*.

11.2 Group of Boosts and Rotations

Semidirect Product of Boosts and Rotations, $B \rtimes R$

Let us move on with the construction of the full Galilei group. Our next stage is the semidirect product of boosts and rotations, $B \rtimes R$. This is the group consisting of the transformations

$$\left. \begin{array}{rcl} x & \mapsto & x' = R\,x + u\,t \\ t & \mapsto & t' = t \end{array} \right\}. \tag{11.19}$$

The reason for the semidirect product is that a repeated application of the above transformation leads to an action of rotations on the boost velocity. This 6-parameter group is interesting, as it represents the nonrelativistic limit of the Lorentz group. The group

$$HG \equiv B \rtimes R \tag{11.20}$$

is also called the *homogeneous Galilei group*. The transformations in 11.19 can be linearized by moving to a 4-dimensional spacetime representation and by writing the transformations as

$$\begin{pmatrix} x' \\ t' \end{pmatrix} = \begin{pmatrix} R & u \\ 0 & 1 \end{pmatrix} \begin{pmatrix} x \\ t \end{pmatrix} = \begin{pmatrix} R\,x + u\,t \\ t \end{pmatrix}. \tag{11.21}$$

The relativistic Lorentz group consists of linear transformations from the outset, and it acts on a 4-dimensional spacetime. However, the Lorentz transformations have a crucial difference compared to the matrix transformation above. Within the above matrix, where the zero 3-vector entry is placed, Lorentz transformations have a non-zero entry that accounts for the change in the time coordinate due to the influence of the spacial coordinates.

Lie Algebra of $B \rtimes R$

We know already the single generators of the homogeneous Galilei group, it is the pair of J and G. Their mutual commutator is

$$[J_k, G_l] = i\,\epsilon_{klm} G_m. \tag{11.22}$$

Thus, the full algebra of $B \rtimes R$ is:

$$\boxed{\begin{array}{rcl} [J_k, J_l] & = & i\,\epsilon_{klm} J_m \\ [J_k, G_l] & = & i\,\epsilon_{klm} G_m \\ [G_k, G_l] & = & 0 \end{array}} \tag{11.23}$$

Let us proceed to the final step in the construction of the complete Galilei group.

11.3   Galilei Group

Galilei Group as Semidirect Product

Now we add the space and time translations to the homogeneous group $B \rtimes R$. The complete Galilei transformations 11.2 are elements of the semidirect product $T \rtimes (B \rtimes R)$, where T now represents the 4-dimensional abelian group of translations in space and time. One can inspect that the full Galilei group G is equally well given by the semidirect product $(T \rtimes B) \rtimes R$. Hence, we can leave the brackets and simply write down the *Galilei group* as

$$G = T \rtimes B \rtimes R \tag{11.24}$$

Note, however, that G is *not* equal to $B \rtimes (T \rtimes R)$. (*exercise 11.2*)

Linearization of Galilei Transformations

As before with the Euclidean group and the subgroups of the Galilei group, it is possible to represent the complete Galilei transformations as linear transformations if we view them in a 5-dimensional space that uses the vector representation

$$\begin{pmatrix} \boldsymbol{x} \\ t \\ 1 \end{pmatrix}. \tag{11.25}$$

Once again, the column entry 1 remains unchanged under all transformations. A complete Galilei transformation becomes a linear matrix multiplication in 5 dimensions then,

$$\begin{pmatrix} \boldsymbol{x}' \\ t' \\ 1 \end{pmatrix} = \left(\begin{array}{cc|c} R & \boldsymbol{u} & \boldsymbol{a} \\ \boldsymbol{0} & 1 & \tau \\ \hline 0 & 0 & 1 \end{array} \right) \begin{pmatrix} \boldsymbol{x} \\ t \\ 1 \end{pmatrix} = \begin{pmatrix} R\boldsymbol{x} + \boldsymbol{u}\,t + \boldsymbol{a} \\ t + \tau \\ 1 \end{pmatrix}. \tag{11.26}$$

The linear transformation matrix above contains the ten parameters, R, $\boldsymbol{u}$, $\boldsymbol{a}$, and τ, of a complete Galilei transformation. In the following chapters we will study Lorentz, Poincaré, and conformal transformations. Lorentz transformations are linear from the outset. Poincaré transformations, which are affine-linear, are rendered linear by considering them in one additional dimension. Conformal transformations are nonlinear, but remarkably, they can also be made linear by adding two additional dimensions.

11.4   Galilei Algebra

Generator of Time Translations

We have identified almost all generators of the full Galilei group. The only one missing is the *generator of time translations*, call it H, so that a *translation in time* is given by the expression

$$\exp\left(-\mathrm{i}\,\tau H\right). \tag{11.27}$$

An analysis of the transformation behavior of a scalar field reveals that the (orbital) generator of time translations is given by (*exercise 11.3*)

$$\boxed{H = -\mathrm{i}\frac{\partial}{\partial t}} \tag{11.28}$$

Exactly as with translations in space, translations in time have no influence on the internal degrees of freedom, such as tensorial indices or the momentum. There is no internal generator of time translations. Thus, a complete finite Galilei transformation can be written as an exponential in the form

$$\exp\left\{-\mathrm{i}\left(\boldsymbol{\theta}\cdot\boldsymbol{J}+\boldsymbol{u}\cdot\boldsymbol{G}+\boldsymbol{a}\cdot\boldsymbol{P}+\tau H\right)\right\}. \tag{11.29}$$

We can now finally move on to the commutator relations.

Galilei Algebra Commutators

In order to derive all mutual commutators of the generators, it is convenient to use the orbital parts only. In this way we obtain the following set of *commutator relations of the Galilei algebra*: (*exercise 11.4*)

$$
\begin{aligned}
\left[J_k, J_l\right] &= \mathrm{i}\,\epsilon_{klm}J_m \\
\left[J_k, G_l\right] &= \mathrm{i}\,\epsilon_{klm}G_m \\
\left[J_k, P_l\right] &= \mathrm{i}\,\epsilon_{klm}P_m \\
\left[J_k, H\right] &= 0 \\
\left[G_k, G_l\right] &= 0 \\
\left[G_k, P_l\right] &= 0 \\
\left[G_k, H\right] &= \mathrm{i}\,P_k \\
\left[P_k, P_l\right] &= 0 \\
\left[P_k, H\right] &= 0 \\
\left[H, H\right] &= 0
\end{aligned}
\tag{11.30}
$$

The above table completely defines the *Galilei algebra*. There is one subtlety for the commutator $[G_k, P_l]$, however, that we need to mention. This commutator actually does not vanish in general. If we use the complete expression for the generator of Galilei boosts, $G_k = -\mathrm{i}\,t\partial_k - MX_k$, we arrive at the commutator relation

$$[G_k, P_l] = -\mathrm{i}\,M\delta_{kl}, \tag{11.31}$$

where M is the total mass of the system. The mass is not an element of the Galilei algebra. However, we can elevate the mass to an operator $M \cdot I$, where I is the identity operator commuting with all other generators of the Galilei group. The resulting 11-parameter group containing also the generator $M \cdot I$ is called the *Bargmann group* (*Valentine Bargmann*). A further discussion of the Bargmann group is beyond our scope, and the interested reader may consult Sundermeyer's book [85] for more.

Further Reading

The Galilei transformations are typically treated in books on classical mechanics and quantum mechanics. A discussion of the Galilei group from a symmetry point of view is found in the book of Sundermeyer [85]. The reader will find there also a description on the Bargmann group and its role for the Schrödinger equation.

12

Lorentz Symmetry

We have already seen that Lorentz transformations are the allowed spacetime transformations which satisfy the physical requirements of special relativity. Lorentz transformations contain the usual space rotations and Lorentz boosts and intertwine them into a unified relativistic formalism. In this chapter, we derive the Lorentz transformations group-theoretically and study the corresponding Lie group. Our focus is on the restricted Lorentz group which excludes space and time inversions. For both the Lorentz group and the Lorentz algebra, we derive the existence of spinors. We then derive the finite-dimensional tensorial representations for scalars, vectors, and tensors, as well as the representations for Weyl and Dirac spinors. Finally, we derive the infinite-dimensional orbital representations where the Lorentz group acts on functions of spacetime coordinates.

12.1 Lorentz Group

Definition of the Lorentz Group

Our starting point is Minkowski space $\mathbb{M}_4 \equiv (\mathbb{E}^4, \eta)$, i.e. the affine-linear space of spacetime events, endowed with the constant metric tensor

$$\eta_{\mu\nu} = \mathrm{diag}(1, -1, -1, -1). \tag{12.1}$$

A *Lorentz transformation* relating the coordinates of two inertial frames has the linear form

$$x^\mu \mapsto x'^\mu = \Lambda^\mu{}_\nu x^\nu, \tag{12.2}$$

where we require that 4-lengths are preserved,

$$\eta_{\mu\nu} x'^\mu x'^\nu = \eta_{\mu\nu} x^\mu x^\nu. \tag{12.3}$$

This is equivalent to the tensor transformation relation

$$\boxed{\Lambda^\rho{}_\mu \Lambda^\sigma{}_\nu \eta_{\rho\sigma} = \eta_{\mu\nu}} \tag{12.4}$$

DOI: 10.1201/9781003087748-12

which in turn is equivalent to the constancy of the metric tensor,

$$\eta'_{\mu\nu} = \eta_{\mu\nu}. \tag{12.5}$$

In matrix notation, the characterization of Lorentz transformations reads

$$\boxed{\Lambda^T \eta \Lambda = \eta} \tag{12.6}$$

The above equivalent relations 12.3 to 12.6 fully characterize the *Lorentz group*, which we denote by L. The group properties can be inspected readily. (*exercise 12.1*) A Lorentz transformation is characterized by 6 real parameters. Of the 16 possible real parameters of a Lorentz transformation, only 6 are required to determine the transformation, since the condition 12.6 fixes the remaining 10 parameters. The inverse Λ^{-1} of a transformation Λ is given by

$$\Lambda^{-1} = \eta \Lambda^T \eta. \tag{12.7}$$

The Lorentz group L is isomorphic to the group $O(1,3)$ of orthogonal transformations acting on the pseudo-Euclidean space $\mathbb{E}^{(1,3)}$. The Lorentz group consequently contains also the *space inversions*, *time inversions*, and *spacetime inversions* as possible Lorentz transformations. In coordinates, these are given respectively by

$$x^\mu \mapsto x'^\mu = (x^0, -x^1, -x^2, -x^3), \tag{12.8}$$

$$x^\mu \mapsto x'^\mu = (-x^0, x^1, x^2, x^3), \tag{12.9}$$

$$x^\mu \mapsto x'^\mu = (-x^0, -x^1, -x^2, -x^3). \tag{12.10}$$

The full Lorentz group is indeed the union of four different subsets, which are disconnected in the sense that any transformation in each part cannot be connected smoothly to a transformation in another part by varying the group parameters. From now on, we will focus on that part of the Lorentz group, which does *not* contain the above discrete transformations. Let us characterize this part of the Lorentz group. Consider the group defining equation 12.4 above, which includes the case

$$(\Lambda^0{}_0)^2 - (\Lambda^1{}_0)^2 - (\Lambda^2{}_0)^2 - (\Lambda^3{}_0)^2 = 1, \tag{12.11}$$

which means that either $\Lambda^0{}_0 \geq 1$ or $\Lambda^0{}_0 \leq -1$. We will focus on the case $\Lambda^0{}_0 \geq 1$ from now on, which ensures that transformations do not reverse the sign of time. The defining relation 12.4 leads also to the condition

$$\det \Lambda = \pm 1 \tag{12.12}$$

for the determinant of Λ. From now on we will focus on the case $\det \Lambda = 1$, which means that we exclude inversions of space. These two conditions,

$$\Lambda^0{}_0 \geq 1 \quad \text{and} \quad \det \Lambda = 1 \tag{12.13}$$

combined define actually a subgroup of the full Lorentz group, the so-called *restricted* or *proper orthochronous Lorentz group*, denoted by $L^\uparrow_+$. It can be shown that this is indeed a group. (*exercise 12.2*) The transformations of $L^\uparrow_+$ do not change the direction of time nor do they invert space. All transformations of $L^\uparrow_+$ can be connected to the identity in a smooth way. The proper orthochronous Lorentz group is isomorphic to the group $SO(1,3)$. From the 6 parameters needed to define a Lorentz transformation, 3 parameters are needed for a space rotation and 3 parameters for a Lorentz boost. Because of this combination, Lorentz transformations are also called *Lorentz rotations*.

Lorentz Boosts

Let us derive here the formula for a pure *Lorentz boost*. Consider again the setup of two inertial frames labeled S and S', with their respective axes pointing to the same direction and their origins coinciding at the time point $x^0 = x'^0 = 0$. We take the passive view and change from the frame S to the frame S', where the latter moves with constant velocity β along the positive direction of the x^1-axis of S. For the coordinates the relation $x'^\mu = \Lambda(\beta)^\mu{}_\nu x^\nu$ holds and the Lorentz boost matrix $\Lambda(\beta)$ has the special form

$$\Lambda(\beta) = \begin{pmatrix} \Lambda^0{}_0 & \Lambda^0{}_1 & 0 & 0 \\ \Lambda^1{}_0 & \Lambda^1{}_1 & 0 & 0 \\ 0 & 0 & 1 & 0 \\ 0 & 0 & 0 & 1 \end{pmatrix}. \tag{12.14}$$

We must determine four matrix elements. The defining equation 12.6 gives us three relations

$$\begin{aligned} \left(\Lambda^0{}_0\right)^2 - \left(\Lambda^1{}_0\right)^2 &= 1, \\ \left(\Lambda^0{}_1\right)^2 - \left(\Lambda^1{}_1\right)^2 &= -1, \\ \Lambda^0{}_0\Lambda^0{}_1 - \Lambda^1{}_0\Lambda^1{}_1 &= 0. \end{aligned} \tag{12.15}$$

The fourth relation is obtained from the condition that the space origin of S' as measured with the coordinates $(x'^0, 0, 0, 0)$ in S' stems from the Lorentz boost of the coordinates $(x^0, \beta x^0, 0, 0)$ in S. This leads to $\Lambda^1{}_0 x^0 + \Lambda^1{}_1 \beta x^0 = 0$. Since the time coordinate x^0 is arbitrary, the relation holds for all values and we obtain

$$\Lambda^1{}_0 + \Lambda^1{}_1 \beta = 0. \tag{12.16}$$

The solution of this system of equations leads to the known result for the Lorentz boost matrix,

$$\Lambda(\beta) = \begin{pmatrix} \gamma & -\gamma\beta & 0 & 0 \\ -\gamma\beta & \gamma & 0 & 0 \\ 0 & 0 & 1 & 0 \\ 0 & 0 & 0 & 1 \end{pmatrix}, \tag{12.17}$$

with the *gamma factor*

$$\gamma \equiv \frac{1}{\sqrt{1 - \beta^2}}. \tag{12.18}$$

The boosts for the other Cartesian directions are similar. This repeats the results from our previous considerations in Section 6.2. We note here that the pure Lorentz boosts *do not* constitute a subgroup of the Lorentz group. In fact two Lorentz boosts in different directions correspond to a Lorentz boost and a spatial rotation. This is in contrast to the Galilei boosts which *do* constitute a subgroup of the Galilei group. However, Lorentz boosts *in the same direction* do constitute a subgroup of the Lorentz group. (*exercise 12.3*)

12.2 Spinor Representation of the Lorentz Group

Lorentz Transformations of Hermitian Matrices

We have seen that representations are a means of identifying new quantities that obey the symmetry under consideration. So far we have worked solely in the base space of Lorentz symmetry, i.e. in Minkowski space. Now we will construct a representation of the Lorentz group that acts on a different space. We assign to every four-vector x^μ in Minkowski space a hermitian (i.e. self-adjoint) 2×2-matrix X via the map

$$\mathbb{M}_4 \to Herm(2, \mathbb{C}), \tag{12.19}$$

through the assignment

$$x^\mu = \begin{pmatrix} x^0 \\ x^1 \\ x^2 \\ x^3 \end{pmatrix} \mapsto X = \begin{pmatrix} x^0 + x^3 & x^1 - \mathrm{i}x^2 \\ x^1 + \mathrm{i}x^2 & x^0 - x^3 \end{pmatrix}. \tag{12.20}$$

The matrix X is hermitian, $X^\dagger = X$, and its determinant yields

$$\det X = (x^0)^2 - (x^1)^2 - (x^2)^2 - (x^3)^2 = x^2. \tag{12.21}$$

Therefore, we want to find transformations for the hermitian matrices X such that for the transformed matrix X' the invariance $\det X' = \det X$ holds. This is equivalent to $x'^2 = x^2$ then. This means we seek for similarity transformations that preserve hermiticity and the value of the determinant. The transformation formula that keeps hermiticity is simply

$$X \mapsto X' = LXL^\dagger, \tag{12.22}$$

where L is an arbitrary complex 2×2-matrix. If we more specifically consider a transformation $L \in SL(2, \mathbb{C}) \equiv \{L \in GL(2, \mathbb{C}) \mid \det L = 1\}$, then we have in addition

$$\det X' = \det X, \tag{12.23}$$

exactly as desired. Thus, we have found that for every transformation $L \in SL(2, \mathbb{C})$ a Lorentz transformation $\Lambda(L)$ is assigned, leading to the equivalence relations

$$x' = \Lambda(L)x \quad \Leftrightarrow \quad X' = LXL^\dagger, \tag{12.24}$$

and

$$x'^2 = x^2 \quad \Leftrightarrow \quad \det X' = \det X. \tag{12.25}$$

The two matrix transformations L and $-L$ lead to the same Lorentz transformation $\Lambda(L)$. In other words, there is a 1:2-homomorphism

$$SO(1,3) \rightarrow SL(2, \mathbb{C}), \tag{12.26}$$

with the assignment

$$\Lambda \mapsto L(\Lambda). \tag{12.27}$$

The Lorentz transformation Λ acts on Minkowski space $\mathbb{M}_4$, whereas its representation $L(\Lambda)$ acts on $Herm(2, \mathbb{C})$. However, $L(\Lambda)$ naturally acts also on the 2-complex-dimensional linear space $\mathbb{C}^2$. This simply corresponds to the 2-complex-dimensional fundamental representation of $SL(2, \mathbb{C})$. The linear space $\mathbb{C}^2$ is the so-called *spinor space* and contains two-component complex vectors, the *spinors*. But before we move on to these 2-dimensional quantities, let us make the relation between the two different representations of the Lorentz transformations clearer. The four *Pauli matrices* $\{\sigma_0, \sigma_1, \sigma_2, \sigma_3\}$, defined as

$$\sigma_0 \equiv \begin{pmatrix} 1 & 0 \\ 0 & 1 \end{pmatrix}, \quad \sigma_1 \equiv \begin{pmatrix} 0 & 1 \\ 1 & 0 \end{pmatrix}, \quad \sigma_2 \equiv \begin{pmatrix} 0 & -\mathrm{i} \\ \mathrm{i} & 0 \end{pmatrix}, \quad \sigma_3 \equiv \begin{pmatrix} 1 & 0 \\ 0 & -1 \end{pmatrix}, \tag{12.28}$$

constitute a basis for hermitian matrices so that the matrix X can be written as

$$X = x^\mu \sigma_\mu = x^0 \sigma_0 + x^1 \sigma_1 + x^2 \sigma_2 + x^3 \sigma_3. \tag{12.29}$$

The inversion is given by

$$x^\mu = \frac{1}{2}\mathrm{Tr}(X\,\sigma^\mu). \tag{12.30}$$

We define also the notation, see also appendix B.3,

$$\overline{\sigma}^\mu \equiv \sigma_\mu \equiv (1, -\boldsymbol{\sigma}) \quad \text{and} \quad \overline{\sigma}_\mu \equiv \sigma^\mu \equiv \begin{pmatrix} 1 \\ \boldsymbol{\sigma} \end{pmatrix}, \tag{12.31}$$

in order to ensure correct placement of upper and lower indices in equations. We remark that $\overline{\sigma}^\mu$ and $\overline{\sigma}_\mu$ are not obtained through simple contractions with $\eta^{\mu\nu}$ and $\eta_{\mu\nu}$. Further, we note that it is

$$\Lambda^\mu{}_\nu \sigma_\mu = L\sigma_\nu L^\dagger. \tag{12.32}$$

By multiplying with $\overline{\sigma}^\rho$ and by taking the trace, we obtain the explicit formula

$$\Lambda^\mu{}_\nu(L) = \frac{1}{2}\mathrm{Tr}(L^\dagger \overline{\sigma}^\mu L\sigma_\nu). \tag{12.33}$$

We still need to show that a matrix $\Lambda^\mu{}_\nu(L)$ defined as above is a proper orthochronous Lorentz transformation. Consider a timelike vector $(1,0,0,0)$, or $X = 1$. Then it is $X' = LL^\dagger$ and this matrix has only positive diagonal elements and thus it is $x'^0 \geq 0$. This means that $\Lambda^\mu{}_\nu(L)$ is orthochronous. Further, we note that both $L = \pm 1$ are mapped to $\Lambda = 1$ and we conclude that $\det \Lambda = 1$, as desired.

Relativistic Spinors

The representation $SL(2,\mathbb{C})$ of the Lorentz group acts naturally on the spinor space $\mathbb{C}^2$ consisting of the 2-complex-dimensional spinors. Let us examine here some basic properties of these spinors. In terms of terminology, we talk about *relativistic spinors* φ, χ, etc., which we consider to be 2-dimensional contravariant vectors, with complex components φ^A, χ^B, etc. that use the *spinor indices* $A, B = 1, 2$. We write

$$\varphi = \begin{pmatrix} \varphi^1 \\ \varphi^2 \end{pmatrix}. \tag{12.34}$$

The spinors transform under a Lorentz transformation $L(\Lambda)$ as

$$\varphi' = L\varphi, \tag{12.35}$$

or, in component notation,

$$\varphi'^A = L^A{}_B \varphi^B. \tag{12.36}$$

The 2-dimensional spinors are the more fundamental objects, compared to the 4-dimensional Lorentz vectors. We have to define also a scalar product on spinor space that should be invariant under Lorentz transformations. In analogy to four-vectors, we consider for spinors a bilinear form

$$(\varphi, C\chi) = \varphi^A C_{AB} \chi^B, \tag{12.37}$$

Here we use the summation convention for the spinor indices $A, B = 1, 2$. This scalar product is invariant iff

$$L^T C L = C. \tag{12.38}$$

This condition is fulfilled if C is chosen to be real and antisymmetric, as can be readily inspected. The standard form chosen for C is that of the totally antisymmetric epsilon tensor in two dimensions, $C_{AB} = \epsilon_{AB}$. So, remarkably, the antisymmetric epsilon tensor

$$\epsilon \equiv (\epsilon_{AB}) = \begin{pmatrix} 0 & 1 \\ -1 & 0 \end{pmatrix} \tag{12.39}$$

operates as the *metric tensor of spinor space*. Some important properties are

$$\epsilon = i\sigma_2 = \epsilon^* = -\epsilon^{-1} = -\epsilon^T. \tag{12.40}$$

We use the notation ϵ^{AB} with upper indices for the inverse of ϵ_{AB}, i.e. $\epsilon_{AB} = -\epsilon^{AB}$. The following relations are useful in calculations (*exercise 12.4*)

$$\epsilon^{AC}\epsilon_{BC} = \delta^A_B \tag{12.41}$$

and

$$\epsilon_{AB}\epsilon^{CD} = \delta^C_A\delta^D_B - \delta^D_A\delta^C_B. \tag{12.42}$$

We can ask if we could have defined the scalar product of spinors as a sesquilinear form $(\varphi, C\chi) = \varphi^{A*}C_{AB}\chi^B$, which would result to a condition of the form $L^\dagger CL = C$. However, the last equation has no solution for C. This is the reason why we implement the scalar product on spinor space as a bilinear form, despite the fact that the spinors are complex-valued. Having the metric ϵ_{AB} of spinor space at hand, we can define covariant spinors as

$$\varphi_A = \epsilon_{AB}\varphi^B. \tag{12.43}$$

The inversion needs some care concerning the placement of signs and the order of factors,

$$\varphi^A = -\epsilon^{AB}\varphi_B = \varphi_B\epsilon^{BA}. \tag{12.44}$$

The transformation of covariant spinor components under Lorentz transformations is given by

$$\varphi'_A = L_A{}^B\varphi_B, \tag{12.45}$$

with the usual notation

$$L_A{}^B \equiv ((L^{-1})^T)_A{}^B \tag{12.46}$$

for the contragredient transformation, see appendix B.1. With the usage of covariant spinors, the spinorial scalar product can be calculated as

$$(\varphi, \epsilon\chi) = \varphi^A\chi_A = -\varphi_A\chi^A = -(\chi, \epsilon\varphi), \tag{12.47}$$

which shows that it is *antisymmetric*. The two representations based on L and on $(L^T)^{-1}$ are interchangeable, since they are equivalent in the sense

$$(L^T)^{-1} = \epsilon L \epsilon^{-1}. \tag{12.48}$$

This allows us to consider contravariant and covariant spinors as equivalent. There exist two more 2-dimensional spinor representations that leave the spinorial scalar product invariant. These are L^* and $(L^\dagger)^{-1}$, i.e. the complex conjugated versions of the previous two representations. Since the complex conjugation cannot be expressed as an equivalence relation, we have here two entirely new representations, which are nevertheless equivalent to each other,

$$(L^\dagger)^{-1} = \epsilon L^* \epsilon^{-1}. \tag{12.49}$$

The spinors specifically transforming under L^* and $(L^\dagger)^{-1}$ are denoted with dotted indices $\dot{A}$, $\dot{B}$, etc., to make the distinction visible. Sometimes barred symbols are used for the complex conjugated spinors, but we will not follow this practice as it does not add any new information. Hence, we will simply write for the complex conjugated spinors

$$\chi^{\dot{A}} \equiv (\chi^A)^* \quad \text{and} \quad \chi_{\dot{A}} \equiv (\chi_A)^*. \tag{12.50}$$

Thus, the contravariant components transform as

$$\chi'^{\dot{A}} = L^{*\dot{A}}{}_{\dot{B}}\chi^{\dot{B}}, \tag{12.51}$$

while the covariant components transform with $(L^\dagger)^{-1}$. For the associated spinorial metric tensor $\epsilon_{\dot{A}\dot{B}}$ of these new representations, we require simply

$$\epsilon_{\dot{A}\dot{B}} = \epsilon_{AB}. \tag{12.52}$$

In summary, we have four 2-complex-dimensional spinor representations of the Lorentz transformations acting on spinor indices in the following way:

$$\left.\begin{array}{rcc} L & \text{transforms} & \varphi^A \\ (L^T)^{-1} & \text{transforms} & \varphi_A \\ L^* & \text{transforms} & \chi^{\dot{A}} \\ (L^\dagger)^{-1} & \text{transforms} & \chi_{\dot{A}} \end{array}\right\}. \tag{12.53}$$

This classification is called the *van der Waerden notation (Bartel L. van der Waerden)*. Later we will combine the two spinor types φ_A and $\chi^{\dot{A}}$ to a 4-complex-dimensional object, the *Dirac spinor*. But first let us express how a general spinor of general rank transforms under Lorentz transformations. A possible way to construct spinors of higher rank is through tensor products of lower rank spinors. Let us denote a general spinor as

$$\varphi^{AB...\dot{C}\dot{D}...}_{EF...\dot{G}\dot{H}...}, \tag{12.54}$$

where the position of dotted indices relative to undotted indices is irrelevant and only a matter of convention. Such a general spinor transforms according to the rule

$$\varphi'^{AB...\dot{C}\dot{D}...}_{EF...\dot{G}\dot{H}...} = L^A{}_a L^B{}_b \cdots L^{*\dot{C}}{}_{\dot{c}} L^{*\dot{D}}{}_{\dot{d}} \cdots L_E{}^e L_F{}^f \cdots L^*{}_{\dot{G}}{}^{\dot{g}} L^*{}_{\dot{H}}{}^{\dot{h}} \cdots \varphi^{ab...\dot{c}\dot{d}...}_{ef...\dot{g}\dot{h}...}, \tag{12.55}$$

which includes the Lorentz transformation of all four types of spinor indices.

Connection between Spinors and Tensors

The spinor representation of the Lorentz group cannot be obtained from a tensor representation. This fact was proved by É. Cartan in 1923 for the general case of $SO(n, \mathbb{C})$. The inverse is true however, tensor representations can be constructed from spinor representations. In this sense, spinor representations are more "fundamental" than tensor representations. Algebraically, we can relate tensors and spinors to each other. The connection between a four-vector and a two-index spinor is already visible in the definition formula

$$X = x^\mu \sigma_\mu, \tag{12.56}$$

which leads to the transformation law

$$X' = LXL^\dagger, \tag{12.57}$$

or, in components,

$$X'^{A\dot{A}} = L^A{}_B L^{*\dot{A}}{}_{\dot{B}} X^{B\dot{B}}. \tag{12.58}$$

We see that X is a spinor with two indices, one dotted and one undotted. This means that generally *a real four-vector corresponds to a hermitian spinor of rank two*. If we write 12.56 in spinor components, it attains the form

$$X^{A\dot{A}} = x^\mu \sigma_\mu^{A\dot{A}}, \tag{12.59}$$

with $\sigma_\mu^{A\dot{A}}$ being simply the matrix elements of the Pauli matrices. Let us continue with some more index gymnastics. There is a wealth of relations between spinorial objects and tensorial objects and we list here only the most basic of these. (*exercise 12.5*) The lowering and raising of spinorial indices happens by contraction with the corresponding spinorial metric,

$$\bar{\sigma}_{\mu\dot{A}A} \equiv \epsilon_{\dot{A}\dot{B}}\epsilon_{AB}\sigma_\mu^{B\dot{B}}, \tag{12.60}$$

and

$$\sigma_\mu^{A\dot{A}} \equiv \epsilon^{AB}\epsilon^{\dot{A}\dot{B}}\bar{\sigma}_{\mu\dot{B}B}. \tag{12.61}$$

The reader should note the placement of dotted and undotted indices in these definitions. A very useful relation is

$$\bar{\sigma}_{\mu\dot{A}A}\sigma_\nu^{B\dot{A}} + \bar{\sigma}_{\nu\dot{A}A}\sigma_\mu^{B\dot{A}} = 2\eta_{\mu\nu}\epsilon_A{}^B, \tag{12.62}$$

from which also

$$\bar{\sigma}_{\mu\dot{A}A}\sigma_\nu^{A\dot{A}} = 2\eta_{\mu\nu} \tag{12.63}$$

follows. We note also the relation

$$\bar{\sigma}_{\mu\dot{A}A}\sigma^\mu_{B\dot{B}} = 2\delta_{AB}\delta_{\dot{A}\dot{B}}. \tag{12.64}$$

The inversion of 12.59 is thus

$$x^\mu = \frac{1}{2}\bar{\sigma}^\mu_{\dot{A}A}X^{A\dot{A}}. \tag{12.65}$$

Generalizing the fundamental equation 12.59 to, let's say, a $(1,1)$-tensor T^μ_ν the relation to the corresponding spinor $T^{A\dot{A}}_{B\dot{B}}$ is

$$T^{A\dot{A}}_{B\dot{B}} = \sigma_\mu^{A\dot{A}}\sigma^\nu_{B\dot{B}}T^\mu_\nu, \tag{12.66}$$

and inversely

$$T^\mu_\nu = \frac{1}{4}\bar{\sigma}^\mu_{\dot{A}A}\bar{\sigma}_\nu^{\dot{B}B}T^{A\dot{A}}_{B\dot{B}}. \tag{12.67}$$

In particular, we can transfer the Minkowski metric $\eta_{\mu\nu}$ to spinor space by defining the spinorial tensor $\eta_{A\dot{A}B\dot{B}}$ naturally as

$$\eta_{A\dot{A}B\dot{B}} \equiv \sigma^\mu_{A\dot{A}}\sigma^\nu_{B\dot{B}}\eta_{\mu\nu}. \tag{12.68}$$

This tensor is the product of two spinorial metric tensors in the form

$$\eta_{A\dot{A}B\dot{B}} \equiv \epsilon_{AB}\epsilon_{\dot{A}\dot{B}}. \tag{12.69}$$

Now if we start from a spinor φ^A and its complex conjugate $\varphi^{\dot{A}} \equiv (\varphi^A)^*$, we can construct the tensor product $\Phi^{A\dot{A}} \equiv \varphi^A\varphi^{\dot{A}}$, which is a hermitian spinor of second rank corresponding to a four-vector. Further, its Minkowskian scalar product is

$$\eta_{A\dot{A}B\dot{B}}\Phi^{A\dot{A}}\Phi^{B\dot{B}} = (\epsilon_{AB}\varphi^A\varphi^B)(\epsilon_{\dot{A}\dot{B}}\varphi^{\dot{A}}\varphi^{\dot{B}}) = 0, \tag{12.70}$$

i.e. the specially constructed spinor $\Phi^{A\dot{A}} \equiv \varphi^A\varphi^{\dot{A}}$ actually corresponds to a lightlike four-vector. Loosely we say that a spinor is the "square root" of a lightlike four-vector.

Let us summarize. We have deduced the existence of relativistic spinors and their general properties by considering the structure of the Lorentz group. There is another way to introduce spinors, namely by considering the structure of the Lie algebra. Later we will see that spinors can be "left-handed" or "right-handed" and we will group them into a single object, the Dirac spinor. Spinors can be used to reformulate tensorial theories, such as Maxwell's electromagnetism or the Einstein equations, and they can be a powerful tool for deriving new results. However, the most important application of spinors is their role as wave functions describing relativistic particles.

12.3   Lorentz Algebra

Overview of the Program

In this section we will deal extensively with the Lie algebra of the restricted Lorentz group $L_+^\uparrow$ and study it from different points of view. We first derive various forms of the *Lorentz algebra* and examine the algebraic commutator relations. In the following sections, we derive the most important representations of the Lorentz algebra. These are the finite-dimensional representations on tensorial and spinorial indices (i.e. the internal / spin representations), and subsequently the infinite-dimensional representations on spacetime coordinates (i.e. the orbital representations). Finally, we combine internal and orbital parts to establish how physical fields transform under Lorentz transformations.

Infinitesimal Lorentz Transformations

We begin with a direct treatment of infinitesimal Lorentz transformations and proceed later to the full machinery of generators and the exponential map between algebra and group. An infinitesimal Lorentz transformation has the form

$$\Lambda^\mu{}_\nu = \delta^\mu{}_\nu + \omega^\mu{}_\nu, \tag{12.71}$$

with the parameter matrix $\omega^\mu{}_\nu$ being small, $|\omega^\mu{}_\nu| \ll 1$. In matrix form, the transformation is written as

$$\Lambda = 1 + \omega. \tag{12.72}$$

The defining equation of Lorentz transformations 12.6 imposes the condition

$$\left(1 + \omega^T\right) \eta \left(1 + \omega\right) = \eta, \tag{12.73}$$

which means simply that

$$\omega^{\mu\nu} = -\omega^{\nu\mu}, \tag{12.74}$$

i.e. infinitesimal Lorentz transformations are parametrized by an antisymmetric tensor $\omega^{\mu\nu}$ that needs 6 real parameters to be defined. We write the mixed indices version $\omega^\mu{}_\nu$ explicitly as matrix* in the form

$$(\omega^\mu{}_\nu) \equiv \begin{pmatrix} 0 & \eta^1 & \eta^2 & \eta^3 \\ \eta^1 & 0 & -\theta^3 & \theta^2 \\ \eta^2 & \theta^3 & 0 & -\theta^1 \\ \eta^3 & -\theta^2 & \theta^1 & 0 \end{pmatrix}. \tag{12.75}$$

Equivalently, by multiplying with $(\eta^{\nu\rho})$ from the left, the antisymmetric upper indices version $\omega^{\mu\rho}$ is gained, given by

$$(\omega^{\mu\rho}) = \begin{pmatrix} 0 & -\eta^1 & -\eta^2 & -\eta^3 \\ \eta^1 & 0 & \theta^3 & -\theta^2 \\ \eta^2 & -\theta^3 & 0 & \theta^1 \\ \eta^3 & \theta^2 & -\theta^1 & 0 \end{pmatrix}. \tag{12.76}$$

Connecting the two definitions of the parametrization, we have

$$\omega^{0k} = -\eta^k \quad \text{and} \quad \omega^{kl} = \epsilon^{klm}\theta^m, \tag{12.77}$$

---

*The choice of the signs of the parameters η^k and θ^k is convention. Choosing different signs will lead to different signs in the algebra commutators.

or inversely

$$\eta^k = -\omega^{0k} \quad \text{and} \quad \theta^k = \frac{1}{2}\epsilon^{klm}\omega^{lm}. \tag{12.78}$$

The η^k parametrize the Lorentz boosts, while the θ^k parametrize the pure rotations.

Generators and Commutators of the Lorentz Algebra

First, we treat rotations and boosts separately and combine them afterward to a unified relativistic view. The infinitesimal parameter matrix relevant for the 3-dimensional treatment here is the one in equation 12.75. The pure rotations around the three directions are $\Lambda_1(\theta)$, $\Lambda_2(\theta)$, and $\Lambda_3(\theta)$. The corresponding *generators of rotations*, J_k , are defined as

$$J_k \equiv i \frac{\partial \Lambda_k}{\partial \theta}(\theta)\Big|_{\theta=0} , \tag{12.79}$$

leading to the hermitian matrix representation

$$J_1 = \begin{pmatrix} 0 & 0 & 0 & 0 \\ 0 & 0 & 0 & 0 \\ 0 & 0 & 0 & -i \\ 0 & 0 & i & 0 \end{pmatrix}, \quad J_2 = \begin{pmatrix} 0 & 0 & 0 & 0 \\ 0 & 0 & 0 & i \\ 0 & 0 & 0 & 0 \\ 0 & -i & 0 & 0 \end{pmatrix}, \quad J_3 = \begin{pmatrix} 0 & 0 & 0 & 0 \\ 0 & 0 & -i & 0 \\ 0 & i & 0 & 0 \\ 0 & 0 & 0 & 0 \end{pmatrix}. \tag{12.80}$$

An infinitesimal rotation about an angle vector $\delta\boldsymbol{\theta}$ can be written as

$$\Lambda(\delta\boldsymbol{\theta}) = 1 - i\,\delta\boldsymbol{\theta} \cdot \boldsymbol{J}. \tag{12.81}$$

Equivalently, a finite rotation about an angle $\boldsymbol{\theta}$ is given by the exponential map

$$\Lambda(\boldsymbol{\theta}) = \exp(-i\,\boldsymbol{\theta} \cdot \boldsymbol{J}). \tag{12.82}$$

in complete analogy to the case of $SO(3)$ rotations. Let us move on to the three Lorentz boosts, $\Lambda_1(\eta)$, $\Lambda_2(\eta)$, and $\Lambda_3(\eta)$. The *generators of Lorentz boosts*, K_k, are defined as

$$K_k \equiv i \frac{\partial \Lambda_k}{\partial \eta}(\eta)\Big|_{\eta=0} , \tag{12.83}$$

leading to the antihermitian matrix representation

$$K_1 = \begin{pmatrix} 0 & i & 0 & 0 \\ i & 0 & 0 & 0 \\ 0 & 0 & 0 & 0 \\ 0 & 0 & 0 & 0 \end{pmatrix}, \quad K_2 = \begin{pmatrix} 0 & 0 & i & 0 \\ 0 & 0 & 0 & 0 \\ i & 0 & 0 & 0 \\ 0 & 0 & 0 & 0 \end{pmatrix}, \quad K_3 = \begin{pmatrix} 0 & 0 & 0 & i \\ 0 & 0 & 0 & 0 \\ 0 & 0 & 0 & 0 \\ i & 0 & 0 & 0 \end{pmatrix}. \tag{12.84}$$

An infinitesimal boost along the rapidity vector $\delta\boldsymbol{\eta}$ can be written as

$$\Lambda(\delta\boldsymbol{\eta}) = 1 - i\,\delta\boldsymbol{\eta} \cdot \boldsymbol{K}. \tag{12.85}$$

Equivalently, a finite boost along the rapidity vector $\boldsymbol{\eta}$ is given by the exponential map

$$\Lambda(\boldsymbol{\eta}) = \exp(-i\,\boldsymbol{\eta} \cdot \boldsymbol{K}). \tag{12.86}$$

A general Lorentz transformation will contain both a rotation and a boost, and as such can be obtained through the exponential map as

$$\Lambda(\boldsymbol{\theta},\boldsymbol{\eta}) = \exp(-i\,\boldsymbol{\theta} \cdot \boldsymbol{J} - i\,\boldsymbol{\eta} \cdot \boldsymbol{K}). \tag{12.87}$$

The product of a rotation $\Lambda(\boldsymbol{\theta})$ and a boost $\Lambda(\boldsymbol{\eta})$ is a Lorentz transformation, but we have to note that in general it is

$$\exp(-\mathrm{i}\,\boldsymbol{\theta}\cdot\boldsymbol{J})\exp(-\mathrm{i}\,\boldsymbol{\eta}\cdot\boldsymbol{K})\neq\exp(-\mathrm{i}\,\boldsymbol{\theta}\cdot\boldsymbol{J}-\mathrm{i}\,\boldsymbol{\eta}\cdot\boldsymbol{K}), \tag{12.88}$$

due to the BCH formula. Conversely, every Lorentz transformation $\Lambda\in L_+^\uparrow$ can be uniquely written as a product

$$\Lambda=\Lambda(\boldsymbol{\theta})\,\Lambda(\boldsymbol{\eta}) \tag{12.89}$$

of a pure rotation $\Lambda(\boldsymbol{\theta})$ and a pure Lorentz boost $\Lambda(\boldsymbol{\eta})$, with certain parameters $\boldsymbol{\theta}$ and $\boldsymbol{\eta}$. This is the *decomposition theorem*. This decomposition into a unitary and hermitian matrix is similar to the decomposition of a complex number as $z=\mathrm{e}^{\mathrm{i}\arg(z)}|z|$. For the proof of the decomposition theorem, recall that an active pure boost $\Lambda(\boldsymbol{\beta})$ is given by

$$\Lambda(\boldsymbol{\beta})=\left(\begin{array}{c|c} \gamma & \gamma\boldsymbol{\beta}^T \\ \hline \gamma\boldsymbol{\beta} & 1+(\gamma-1)\hat{\boldsymbol{\beta}}\otimes\hat{\boldsymbol{\beta}} \end{array}\right), \tag{12.90}$$

where $\hat{\boldsymbol{\beta}}\equiv\boldsymbol{\beta}/\beta$ is the boost unit vector, and a pure rotation $\Lambda(\boldsymbol{\theta})$ is given by

$$\Lambda(\boldsymbol{\theta})=\left(\begin{array}{c|c} 1 & \mathbf{0}^T \\ \hline \mathbf{0} & R(\boldsymbol{\theta}) \end{array}\right), \tag{12.91}$$

where $R(\boldsymbol{\theta})$ is an orthogonal matrix. A general Lorentz transformation has the form

$$\Lambda=\left(\begin{array}{cc} \Lambda^0{}_0 & \Lambda^0{}_j \\ \Lambda^i{}_0 & \Lambda^i{}_j \end{array}\right). \tag{12.92}$$

Now if we compute the product $\Lambda\,\Lambda(\boldsymbol{\beta})^{-1}$ by using the parametrization $\beta_j=\Lambda^0{}_j/\Lambda^0{}_0$ and $\gamma=\Lambda^0{}_0$, we conclude that the product is a pure rotation matrix, i.e. $\Lambda\,\Lambda(\boldsymbol{\beta})^{-1}=\Lambda(\boldsymbol{\theta})$. This proves the theorem.

Since we have a concrete matrix representation of the generators now, it is straightforward to obtain the *commutator relations of the Lorentz algebra*. One finds the relations (*exercise 12.6*)

$$\boxed{\begin{aligned} {[J_k,J_l]} &= \mathrm{i}\,\epsilon_{klm}J_m \\ {[J_k,K_l]} &= \mathrm{i}\,\epsilon_{klm}K_m \\ {[K_k,K_l]} &= -\mathrm{i}\,\epsilon_{klm}J_m \end{aligned}} \tag{12.93}$$

These relations define fully the Lorentz algebra, independently of any concrete representation. They are valid for the 4-dimensional $so(1,3)$ representation as for the 2-complex-dimensional $sl(2,\mathbb{C})$ representation. We can see that the commutator of two boosts results into a rotation, which corresponds to the fact that the set of boosts does not constitute a subgroup of the Lorentz group.

It is possible to achieve a relativistic formulation for the Lorentz algebra by introducing the antisymmetric matrix $M_{\mu\nu}$ of *generators of Lorentz transformations* by

$$(M_{\mu\nu})\equiv\left(\begin{array}{cccc} 0 & -K_1 & -K_2 & -K_3 \\ K_1 & 0 & J_3 & -J_2 \\ K_2 & -J_3 & 0 & J_1 \\ K_3 & J_2 & -J_1 & 0 \end{array}\right). \tag{12.94}$$

This is complementary to the definition of the Lorentz group parameters in 12.76. In terms of 3-dimensional indices, this correspondence is expressed as

$$M_{0k} = -K_k \quad \text{and} \quad M_{kl} = \epsilon_{klm} J_m, \tag{12.95}$$

or, inversely, by

$$K_k = -M_{0k} \quad \text{and} \quad J_k = \frac{1}{2}\epsilon_{klm} M_{lm}. \tag{12.96}$$

A general Lorentz transformation attains the form

$$\Lambda(\omega) = \exp\left(-\frac{\mathrm{i}}{2}\omega^{\mu\nu} M_{\mu\nu}\right). \tag{12.97}$$

The factor $1/2$ is needed because we sum twice over the relevant $\mu\nu$-indices. The 3-dimensional commutators 12.93 can now be combined to a 4-dimensional relativistic relation: (*exercise 12.7*)

$$\boxed{[M_{\mu\nu}, M_{\rho\sigma}] = -\mathrm{i}\left(\eta_{\mu\rho} M_{\nu\sigma} - \eta_{\nu\rho} M_{\mu\sigma} + \eta_{\nu\sigma} M_{\mu\rho} - \eta_{\mu\sigma} M_{\nu\rho}\right)} \tag{12.98}$$

These are the *commutator relations of the Lorentz algebra* in their unified relativistic form.

Lorentz Algebra as a Direct Sum

Within the Lie algebra, the basis elements can be freely chosen. Instead of the set of the generators $\boldsymbol{J}$ and $\boldsymbol{K}$ and instead of the $M_{\mu\nu}$ we can choose another linear combination. A useful linear combination is

$$\boldsymbol{J}^{\pm} \equiv \frac{1}{2}(\boldsymbol{J} \pm \mathrm{i}\boldsymbol{K}), \tag{12.99}$$

since the commutator relations of the Lorentz algebra take the simple form (*exercise 12.8*)

$$\boxed{\begin{aligned}
\left[J_k^+, J_l^+\right] &= \mathrm{i}\,\epsilon_{klm} J_m^+ \\
\left[J_k^-, J_l^-\right] &= \mathrm{i}\,\epsilon_{klm} J_m^- \\
\left[J_k^+, J_l^-\right] &= 0
\end{aligned}} \tag{12.100}$$

It means that the Lorentz algebra is decomposed into a direct sum of two independent (i.e. commuting) spin algebras $su(2)$. In other words, there is a Lie algebra homomorphism $so(1,3) \cong sl(2,\mathbb{C}) \cong su(2) \oplus su(2)$. It is a fascinating result that the Lorentz algebra is equivalent to two copies of the spin algebra. Once again we are led to the existence of relativistic spinors, this time from a Lie-algebraic point of view. The inversion of 12.99 is

$$\boldsymbol{J} = \boldsymbol{J}^+ + \boldsymbol{J}^- \quad \text{and} \quad \boldsymbol{K} = -\mathrm{i}(\boldsymbol{J}^+ - \boldsymbol{J}^-). \tag{12.101}$$

By using this, we see that a general Lorentz transformation now obtains the form

$$\Lambda(\boldsymbol{\theta}, \boldsymbol{\eta}) = \exp\left\{(-\mathrm{i}\boldsymbol{\theta} - \boldsymbol{\eta}) \cdot \boldsymbol{J}^+ + (-\mathrm{i}\boldsymbol{\theta} + \boldsymbol{\eta}) \cdot \boldsymbol{J}^-\right\}. \tag{12.102}$$

The just derived direct sum decomposition of the Lorentz algebra is especially useful for the classification of various representations, as we will see below.

Casimir Operators and Classification of Representations

By using the relativistic generators $M_{\mu\nu}$, the *Casimir operators* are expressed as the contractions

$$\frac{1}{4}M_{\mu\nu}M^{\mu\nu} = \frac{1}{2}\left(\boldsymbol{J}^2 - \boldsymbol{K}^2\right) = (\boldsymbol{J}^+)^2 + (\boldsymbol{J}^-)^2, \tag{12.103}$$

and

$$\frac{1}{8}\epsilon^{\mu\nu\rho\sigma}M_{\mu\nu}M_{\rho\sigma} = \boldsymbol{J}\cdot\boldsymbol{K} = -\mathrm{i}\left((\boldsymbol{J}^+)^2 - (\boldsymbol{J}^-)^2\right). \tag{12.104}$$

These quantities commute with the generators $M_{\mu\nu}$. In the framework of the direct sum decomposition of the Lorentz algebra, one can use the independent quantities $(\boldsymbol{J}^+)^2$ and $(\boldsymbol{J}^-)^2$ as the Casimir operators. We are let to a scheme of how to classify the representations of the Lorentz algebra. We can use the set of hermitian operators $(\boldsymbol{J}^+)^2$, J_3^+ and $(\boldsymbol{J}^-)^2$, J_3^- to organize all the finite-dimensional irreducible representations. Each of those irreducible representations, denoted by $D^{(j^+,j^-)}$, is determined by a pair (j^+,j^-) of two independent indices, j^+ and j^-, the so-called *weights*, which can be either zero, a positive integer, or a positive half-integer number. The dimension of the representation $D^{(j^+,j^-)}$ is equal $(2j^+ + 1)(2j^- + 1)$. The representations where $j^+ + j^-$ is an integer, are the so-called *tensorial representations*, whereas the representations where $j^+ + j^-$ is a half-integer, are the so-called *spinorial representations*. In the representation $D^{(j^+,j^-)}$, the J_3-component of the operator $\boldsymbol{J} = \boldsymbol{J}^+ + \boldsymbol{J}^-$ can take all integer steps between $|j^+ - j^-|$ and $(j^+ + j^-)$ as values. To give some concrete examples: the representation $D^{(0,0)}$ is the scalar representation; the representation $D^{(\frac{1}{2},\frac{1}{2})}$ is the 4-dimensional vector representation; the representation $D^{(\frac{3}{2},\frac{3}{2})}$ is a 16-dimensional representation on tensors of order two. We will develop in great detail the 2-dimensional representations $D^{(\frac{1}{2},0)}$, the *left-handed spinor representation*, and $D^{(0,\frac{1}{2})}$, the *right-handed spinor representation*. The terminology of "handedness" will be explained very soon. These spinorial representations are the most fundamental nontrivial ones and can be used to build all higher-valued representations.

12.4   Representation on Scalars, Vectors and Tensors

Tensor Representations

So far we have viewed the generators of the Lorentz algebra either as abstract entities, solely defined through the commutator relations, or as four-dimensional matrices acting on the natural base space. Now we are going to give concrete representations for the relativistic generators $M_{\mu\nu}$ when these act on scalars, vectors, and tensors. These are so-called *tensor* (or *internal*) *representations* of the Lorentz transformations acting on tensorial indices. We will denote the generators acting on tensorial indices with the symbol $S_{\mu\nu}$, and leave the symbol $M_{\mu\nu}$ for the more general case when tensorial indices and spacetime indices are present.

Representation on Scalars

This corresponds to the representation $D^{(0,0)}$. A scalar quantity $c \in \mathbb{C}$ by definition does not change under Lorentz transformations. Infinitesimally, this means

$$\delta c = 0. \tag{12.105}$$

On the other hand, according to 12.97 the change must be

$$\delta c = \left(-\frac{\mathrm{i}}{2}\omega^{\mu\nu}S_{\mu\nu}\right)c, \tag{12.106}$$

from which we deduce that for scalar quantities the generator vanishes identically,

$$S_{\mu\nu} = 0. \tag{12.107}$$

This is the trivial case.

Representation on Vectors

This corresponds to the representation $D^{(\frac{1}{2},\frac{1}{2})}$. A four-vector V^α transforms infinitesimally as

$$\delta V^\alpha = \omega^\alpha{}_\beta V^\beta. \tag{12.108}$$

The general formula 12.97 requires however

$$\delta V^\alpha = -\frac{i}{2}\omega^{\mu\nu}(S_{\mu\nu})^\alpha{}_\beta V^\beta. \tag{12.109}$$

The unique solution is the following value of the generator (*exercise 12.9*)

$$\boxed{(S_{\mu\nu})^\alpha{}_\beta = -i\,(\eta_{\mu\beta}\,\eta^\alpha{}_\nu - \eta_{\nu\beta}\,\eta^\alpha{}_\mu)} \tag{12.110}$$

The reader is encouraged to inspect that the 3-dimensional result 10.36 is included as special case. (*exercise 12.10*) By using the concrete vectorial representation, one can deduce again the Lorentz algebra commutator relations 12.98. A Lorentz transformation acting on four-vector indices has the representation $D_V(\Lambda(\omega))$ given by the exponential

$$D_V(\Lambda(\omega)) = \exp\left(-\frac{i}{2}\omega^{\mu\nu}S_{\mu\nu}\right), \tag{12.111}$$

where the spin generator $S_{\mu\nu}$ is given by 12.110.

Representation on Tensors

We consider here only the simple case of a second order tensor $T^{\alpha\kappa}$. Its transformation is infinitesimally, i.e. to first order in the transformation parameters, given by

$$\delta T^{\alpha\kappa} = (\omega^\alpha{}_\beta\eta^\kappa{}_\lambda + \omega^\kappa{}_\lambda\eta^\alpha{}_\beta)\,T^{\beta\lambda}. \tag{12.112}$$

On the other hand it is

$$\delta T^{\alpha\kappa} = -\frac{i}{2}\omega^{\mu\nu}(S^{\text{tensor}}_{\mu\nu})^{\alpha\kappa}{}_{\beta\lambda}T^{\beta\lambda}. \tag{12.113}$$

The solution is the following representation of the generator (*exercise 12.11*)

$$(S^{\text{tensor}}_{\mu\nu})^{\alpha\kappa}{}_{\beta\lambda} = -i\,((S_{\mu\nu})^\alpha{}_\beta\eta^\kappa{}_\lambda - (S_{\mu\nu})^\kappa{}_\lambda\eta^\alpha{}_\beta). \tag{12.114}$$

Here, $(S_{\mu\nu})^\alpha{}_\beta$ is again the vector representation of the generator 12.110. Higher order tensors are similar, although the explicit formulae can become rather complicated. Now as we have seen the most important integer representations, we proceed to the half-integer cases.

12.5 Representation on Weyl and Dirac Spinors

Spinor Representations

In our discussion of the Lorentz group, we saw that spinors arise naturally as fundamental relativistic objects. We classified them according to their transformation and saw that there are *two non-equivalent types of spinors*. In the Lie-algebraic discussion of Lorentz transformations, we again deduced the existence of spinors and classified them according to their weights (j^+, j^-). In this section we will derive the basic *spinor representations* of the Lorentz algebra. In doing so, we will make the connection between the group-theoretical and the Lie-algebraic notions of spinors, the latter being the so-called *Weyl spinors*. There are *left-handed* and *right-handed Weyl spinors*. In order to have a unified description of both handedness states, we will introduce *Dirac spinors* and the associated *Dirac matrices*.

Left-Handed Weyl Spinors

This case corresponds to the representation $D^{(\frac{1}{2}, 0)}$. It is the 2-dimensional *left-handed Weyl spinor representation*, where the weights take the values $(j^+, j^-) = (\frac{1}{2}, 0)$, which is equivalent to the generators

$$\boxed{J^+ = \frac{\sigma}{2}, \quad J^- = 0}$$

(12.115)

since the Pauli matrices, divided by 2, provide a representation of $su(2)$,

$$\left[\frac{\sigma_k}{2}, \frac{\sigma_l}{2} \right] = i \, \epsilon_{klm} \frac{\sigma_m}{2}.$$

(12.116)

Consequently, the representation of a Lorentz transformation takes the form

$$D_L(\Lambda(\boldsymbol{\theta}, \boldsymbol{\eta})) = \exp \left\{ (-i\boldsymbol{\theta} - \boldsymbol{\eta}) \cdot \frac{\sigma}{2} \right\}.$$

(12.117)

This Lorentz transformation $D_L(\Lambda)$ acts on *left-handed Weyl spinors*, which we will denote by ψ_L, where the subscript L stands for left-handed,

$$\psi_L' = D_L(\Lambda(\boldsymbol{\theta}, \boldsymbol{\eta}))\psi_L .$$

(12.118)

What is the connection between the just introduced left-handed Weyl spinors and the definitions from our group-theoretical analysis? The identification is that left-handed Weyl spinors ψ_L are the covariant spinors φ_A, with the latter transforming as

$$\varphi_A' = L_A{}^B \varphi_B.$$

(12.119)

Thus, the identification of the objects from the group-theoretical and the algebraic view is

$$\left. \begin{array}{rcl} \varphi_A &=& (\psi_L)_A \\ L_A{}^B &=& (D_L(\Lambda))_A{}^B \end{array} \right\},$$

(12.120)

where A, B are the spinor indices. The reason for choosing spinors of covariant type for the left-handed case is pure convention. In the case of right-handed spinors, we will choose contravariant indices, which is also pure convention.

We would like to bring the spinor representation into a relativistic form, like

$$D_L(\Lambda(\omega)) = \exp \left(-\frac{i}{2} \omega_{\mu\nu} S_L^{\mu\nu} \right).$$

(12.121)

The question is if we can find a suitable generator $S_L^{\mu\nu}$ for the left-handed Weyl spinor representation. For this purpose, let us introduce the following relativistic combinations of Pauli matrices, having two antisymmetric Lorentz indices:

$$\sigma^{\mu\nu} \equiv \frac{i}{4} \left(\sigma^\mu \overline{\sigma}^\nu - \sigma^\nu \overline{\sigma}^\mu \right) \quad \text{as} \quad (\sigma^{\mu\nu})_A{}^B,$$

(12.122)

$$\overline{\sigma}^{\mu\nu} \equiv \frac{i}{4} \left(\overline{\sigma}^\mu \sigma^\nu - \overline{\sigma}^\nu \sigma^\mu \right) \quad \text{as} \quad (\overline{\sigma}^{\mu\nu})^{\dot{A}}{}_{\dot{B}}.$$

(12.123)

These combinations look complicated at first sight, but are simple expressions when viewed in 3-dimensional notation. A direct calculation shows that

$$\sigma^{0k} = -\overline{\sigma}^{0k} = -\frac{\mathrm{i}}{2}\sigma_k \quad \text{and} \quad \sigma^{kl} = \overline{\sigma}^{kl} = \frac{1}{2}\epsilon^{klm}\sigma_m. \tag{12.124}$$

Let us note also the property $(\sigma^{\mu\nu})^\dagger = \overline{\sigma}^{\mu\nu}$, which follows from the relation $(\sigma_k)^\dagger = \sigma_k$ of Pauli matrices. Each of the set of matrices $\sigma^{\mu\nu}$ and $\overline{\sigma}^{\mu\nu}$ is a basis for 2-dimensional traceless matrices and as such constitutes a basis of the algebra $sl(2,\mathbb{C})$. A direct calculation reaffirms this, as each of this set of matrices obeys the commutator relations of the Lorentz algebra,

$$[\sigma^{\mu\nu}, \sigma^{\rho\tau}] = -\mathrm{i}\left(\eta^{\mu\rho}\sigma^{\rho\tau} - \eta^{\nu\rho}\sigma^{\mu\tau} + \eta^{\nu\tau}\sigma^{\mu\rho} - \eta^{\mu\tau}\sigma^{\nu\rho}\right), \tag{12.125}$$

and

$$[\overline{\sigma}^{\mu\nu}, \overline{\sigma}^{\rho\tau}] = -\mathrm{i}\left(\eta^{\mu\rho}\overline{\sigma}^{\rho\tau} - \eta^{\nu\rho}\overline{\sigma}^{\mu\tau} + \eta^{\nu\tau}\overline{\sigma}^{\mu\rho} - \eta^{\mu\tau}\overline{\sigma}^{\nu\rho}\right). \tag{12.126}$$

These algebraic considerations indicate that a multiple of $\sigma^{\mu\nu}$ must be the desired relativistic form of the generator $S^{\mu\nu}$. Let us prove this. First, we note that according to 12.95, it is generally for $S^{\mu\nu}$

$$S^{0k} = K_k \quad \text{and} \quad S^{kl} = \epsilon^{klm}J_m. \tag{12.127}$$

Now we translate the generators $\boldsymbol{J}^+ = \frac{\boldsymbol{\sigma}}{2}$ and $\boldsymbol{J}^- = \boldsymbol{0}$ into the equivalent generators

$$\boldsymbol{J} = \frac{\boldsymbol{\sigma}}{2}, \quad \boldsymbol{K} = -\mathrm{i}\frac{\boldsymbol{\sigma}}{2}, \tag{12.128}$$

and we obtain the result

$$S_L^{0k} = -\frac{\mathrm{i}}{2}\sigma_k = \sigma^{0k} \quad \text{and} \quad S_L^{kl} = \frac{1}{2}\epsilon^{klm}\sigma_m = \sigma^{kl}. \tag{12.129}$$

So we have found the desired *generators of Lorentz transformations for left-handed Weyl spinors*, $S_L^{\mu\nu}$. The result is

$$\boxed{(S_L^{\mu\nu})_A{}^B = (\sigma^{\mu\nu})_A{}^B} \tag{12.130}$$

Consequently, the representation of Lorentz transformations on left-handed Weyl spinors is given by

$$D_L(\Lambda(\omega)) = \exp\left(-\frac{\mathrm{i}}{2}\omega_{\mu\nu}\sigma^{\mu\nu}\right). \tag{12.131}$$

The reasoning for the right-handed case is exactly the same, as we explain below.

Right-Handed Weyl Spinors

This case corresponds to the representation $D^{(0,\frac{1}{2})}$. It is the 2-dimensional *right-handed Weyl spinor representation*, where the weights take the values $(j^+, j^-) = (0, \frac{1}{2})$, which corresponds to the generators

$$\boxed{\boldsymbol{J}^+ = 0, \quad \boldsymbol{J}^- = \frac{\boldsymbol{\sigma}}{2}} \tag{12.132}$$

The representation of a Lorentz transformation takes the form

$$D_R(\Lambda(\boldsymbol{\theta}, \boldsymbol{\eta})) = \exp\left\{(-\mathrm{i}\boldsymbol{\theta} + \boldsymbol{\eta}) \cdot \frac{\boldsymbol{\sigma}}{2}\right\}. \tag{12.133}$$

This Lorentz transformation $D_R(\Lambda)$ acts on *right-handed Weyl spinors*, which we will denote by ψ_R, where the subscript R stands for right-handed,

$$\psi'_R = D_R(\Lambda(\boldsymbol{\theta}, \boldsymbol{\eta}))\psi_R. \tag{12.134}$$

Similar to the left-handed case, we identify right-handed Weyl spinors ψ_R with the contravariant spinors $\chi^{\dot{A}}$, the latter transforming as

$$\chi'^{\dot{A}} = L^{*\dot{A}}{}_{\dot{B}}\chi^{\dot{B}}. \tag{12.135}$$

Thus, the identification of the objects from the group-theoretical and the algebraic view is

$$\left.\begin{array}{rcl} \chi^{\dot{A}} & = & (\psi_R)^{\dot{A}} \\ L^{*\dot{A}}{}_{\dot{B}} & = & (D_R(\Lambda))^{\dot{A}}{}_{\dot{B}} \end{array}\right\}, \tag{12.136}$$

where $\dot{A}$, $\dot{B}$ are the spinor indices.

Again, the spinor representation can be brought into the relativistic form

$$D_R(\Lambda(\omega)) = \exp\left(-\frac{\mathrm{i}}{2}\omega_{\mu\nu}S_R^{\mu\nu}\right). \tag{12.137}$$

We translate the generators $\boldsymbol{J}^+ = \boldsymbol{0}$ and $\boldsymbol{J}^- = \frac{\boldsymbol{\sigma}}{2}$ into the equivalent generators

$$\boldsymbol{J} = \frac{\boldsymbol{\sigma}}{2}, \quad \boldsymbol{K} = \mathrm{i}\frac{\boldsymbol{\sigma}}{2}, \tag{12.138}$$

and obtain the result

$$S_R^{0k} = \overline{\sigma}^{0k} \quad \text{and} \quad S_R^{kl} = \overline{\sigma}^{kl}. \tag{12.139}$$

This means that the *generators of Lorentz transformations for right-handed Weyl spinors*, $S_R^{\mu\nu}$, is

$$\boxed{(S_R^{\mu\nu})^{\dot{A}}{}_{\dot{B}} = (\overline{\sigma}^{\mu\nu})^{\dot{A}}{}_{\dot{B}}} \tag{12.140}$$

Consequently, the representation of Lorentz transformations on right-handed Weyl spinors is given by

$$D_R(\Lambda(\omega)) = \exp\left(-\frac{\mathrm{i}}{2}\omega_{\mu\nu}\overline{\sigma}^{\mu\nu}\right). \tag{12.141}$$

In the following we proceed to the construction of the Dirac spinor, which combines the two cases of left-handed and right-handed Weyl spinors into one object.

Dirac Spinors

This case corresponds to the representation $D^{(\frac{1}{2},0)} \oplus D^{(0,\frac{1}{2})}$. The independent Weyl spinors $\psi_L = \varphi_A$ and $\psi_R = \chi^{\dot{A}}$ (and their complex conjugates $\varphi_{\dot{A}}$ and χ^A) are combined to a *Dirac spinor*, sometimes also called *Dirac bispinor*, Ψ and its conjugate $\overline{\Psi}$ respectively (*Paul Adrien Maurice Dirac*). We define

$$\Psi \equiv \left(\begin{array}{c} \psi_L \\ \psi_R \end{array}\right) \equiv \left(\begin{array}{c} \varphi_A \\ \chi^{\dot{A}} \end{array}\right) \tag{12.142}$$

and

$$\overline{\Psi} = (\chi^A, \varphi_{\dot{A}}). \tag{12.143}$$

The general formula for the *conjugate Dirac spinor* $\overline{\Psi}$ is

$$\overline{\Psi} \equiv \Psi^{\dagger} \begin{pmatrix} 0 & 1 \\ 1 & 0 \end{pmatrix}, \tag{12.144}$$

and we note that $\Psi^{\dagger} = (\varphi_{\dot{A}}, \chi^{A})$. The components of Dirac spinors Ψ_{a}, $\overline{\Psi}_{b}$, etc. have the Dirac spinor indices $a, b = 1, 2, 3, 4$. Typically, we will write the Dirac spinor indices only as lower, covariant indices and we will assume summation over repeated indices, even if they are not placed in upper and lower positions. Dirac spinors take values in $\mathbb{C}^{4}$. Any complex 4×4-matrix $U = (U_{ab})$ acting on Dirac spinors must have a certain placement of Weyl spinor indices in order to multiply correctly. The general form is

$$(U_{ab}) = \begin{pmatrix} V_{A}{}^{B} & W_{A\dot{B}} \\ X^{\dot{A}B} & Y^{\dot{A}}{}_{\dot{B}} \end{pmatrix}, \tag{12.145}$$

with the complex 2×2-matrices V, W, X, Y. The correct multiplication is seen by inspecting the result of a multiplication

$$(U_{ab}\Psi_{b}) = \begin{pmatrix} V_{A}{}^{B}\varphi_{B} + W_{A\dot{B}}\chi^{\dot{B}} \\ X^{\dot{A}B}\varphi_{B} + Y^{\dot{A}}{}_{\dot{B}}\chi^{\dot{B}} \end{pmatrix}, \tag{12.146}$$

showing a consistent index placement, as desired. Under Lorentz transformations, Dirac spinors change according to the corresponding 4-complex-dimensional representation of the Lorentz group, $D_{\Psi}(\Lambda)$,

$$\Psi'_{a} = D_{\Psi}(\Lambda)_{ab}\Psi_{b}, \tag{12.147}$$

which is fully defined by the transformation of the two Weyl components. We recall that the Weyl spinors transform according to

$$\varphi'_{A} = L_{A}{}^{B}\varphi_{B}, \tag{12.148}$$

and

$$\chi'^{\dot{A}} = L^{*\dot{A}}{}_{\dot{B}}\chi^{\dot{B}}, \tag{12.149}$$

so that the Lorentz transformation of a Dirac spinor is given explicitly by

$$\Psi' = \begin{pmatrix} \varphi'_{A} \\ \chi'^{\dot{A}} \end{pmatrix} = \begin{pmatrix} L_{A}{}^{B} & 0 \\ 0 & L^{*\dot{A}}{}_{\dot{B}} \end{pmatrix} \begin{pmatrix} \varphi_{B} \\ \chi^{\dot{B}} \end{pmatrix}. \tag{12.150}$$

i.e. the matrix of the Lorentz transformation of a Dirac spinor has the block diagonal form

$$D_{\Psi}(\Lambda) = \begin{pmatrix} L_{A}{}^{B} & 0 \\ 0 & L^{*\dot{A}}{}_{\dot{B}} \end{pmatrix} = \begin{pmatrix} D_{L}(\Lambda) & 0 \\ 0 & D_{R}(\Lambda) \end{pmatrix}. \tag{12.151}$$

The question now is if we can write the Dirac representation $D_{\Psi}(\Lambda)$ also in the form of an exponential expression in a manifestly relativistic manner. To this end, we introduce the 4-complex-dimensional *Dirac matrices*, also called *gamma matrices*, denoted γ^{μ} and shown here in the so-called *Weyl* (or *chiral*) *representation*

$$\gamma^{\mu} \equiv \begin{pmatrix} 0 & \sigma^{\mu} \\ \overline{\sigma}^{\mu} & 0 \end{pmatrix}. \tag{12.152}$$

In addition, we define the 4×4 *spin matrix* $\gamma^{\mu\nu}$ acting on Dirac spinor indices,

$$\gamma^{\mu\nu} \equiv \frac{i}{4}[\gamma^\mu, \gamma^\nu], \qquad (12.153)$$

where the bracket is the usual commutator. It can be easily seen that

$$\gamma^{\mu\nu} = \begin{pmatrix} \sigma^{\mu\nu} & 0 \\ 0 & \bar{\sigma}^{\mu\nu} \end{pmatrix}. \qquad (12.154)$$

By comparing with the results 12.130 and 12.140 for Weyl spinors, we deduce that the spin matrices $\gamma^{\mu\nu}$ in fact constitute the *generators of Lorentz transformations for Dirac spinors*, denoted $S_\Psi^{\mu\nu}$. This means that

$$\boxed{S_\Psi^{\mu\nu} = \gamma^{\mu\nu} = \frac{i}{4}[\gamma^\mu, \gamma^\nu]} \qquad (12.155)$$

The generators $S_\Psi^{\mu\nu} = \gamma^{\mu\nu}$ provide a unified representation for left-handed and right-handed spinors in the form

$$D_\Psi(\Lambda(\omega)) = \exp\left(-\frac{i}{2}\omega_{\mu\nu}\gamma^{\mu\nu}\right). \qquad (12.156)$$

An infinitesimal Lorentz transformation of a Dirac spinor Ψ has the form

$$\delta\Psi = -\frac{i}{2}\omega_{\mu\nu}S_\Psi^{\mu\nu}. \qquad (12.157)$$

For the conjugated Dirac spinor $\overline{\Psi} = \Psi^\dagger\gamma^0$ the Lorentz group generators, denoted $S_{\overline{\Psi}}^{\mu\nu}$, are different though. The general recipe for an infinitesimal transformation,

$$\delta\overline{\Psi} = -\frac{i}{2}\omega_{\mu\nu}S_{\overline{\Psi}}^{\mu\nu}, \qquad (12.158)$$

still holds. On the other hand, it is

$$\delta\overline{\Psi} = (\delta\Psi)^\dagger\gamma^0. \qquad (12.159)$$

by using additionally the result

$$\overline{S_\Psi^{\mu\nu}} = \gamma^0(S_\Psi^{\mu\nu})^\dagger\gamma^0 = S_\Psi^{\mu\nu}, \qquad (12.160)$$

as described in appendix B.3, we readily obtain the formula for the generators $S_{\overline{\Psi}}^{\mu\nu}$, which reads

$$S_{\overline{\Psi}}^{\mu\nu} = -S_\Psi^{\mu\nu} = -\gamma^{\mu\nu}. \qquad (12.161)$$

We will use the above results in later chapters when we construct various field quantities. Now let us look at the Dirac gamma matrices from a more abstract algebraic standpoint.

Dirac Matrices

The 4-complex-dimensional extension of the Pauli matrices are the 4×4 *Dirac matrices*, introduced previously in the so-called Weyl basis representation. The Dirac matrices obey the following *anticommutation relations*, see appendix B.3,

$$\boxed{\{\gamma^\mu, \gamma^\nu\} \equiv \gamma^\mu\gamma^\nu + \gamma^\nu\gamma^\mu = 2\eta^{\mu\nu}} \qquad (12.162)$$

These anticommutation relations are a special case of defining relations of a *Clifford algebra* (*William Kingdon Clifford*). These relations are valid independently of the concrete representation of the Dirac matrices. Indeed, if γ^μ obeys the Clifford condition 12.162, then the transformed matrix $\gamma^{\mu\prime}$ obtained via a similarity transformation

$$\gamma^{\mu\prime} = U\gamma^\mu U^{-1} \tag{12.163}$$

with an invertible matrix U also obeys the Clifford condition 12.162. In this way, various representations other than the initial definition 12.152 are obtained. These are, for instance, the Dirac and the Majorana representations, as used in quantum field theory. In addition to the four matrices γ^μ, another fifth matrix, denoted γ^5, is important, which we define as

$$\gamma^5 \equiv \gamma_5 \equiv i\gamma^0\gamma^1\gamma^2\gamma^3. \tag{12.164}$$

An equivalent formula for γ^5 is

$$\gamma^5 = -\frac{i}{4!}\epsilon_{\mu\nu\rho\sigma}\gamma^\mu\gamma^\nu\gamma^\rho\gamma^\sigma. \tag{12.165}$$

Using the matrix γ^5 we can construct projection operators in the direction of the left-handed and right-handed spinors, specifically

$$\frac{1}{2}\left(1-\gamma^5\right)\begin{pmatrix}\psi_L\\\psi_R\end{pmatrix} = \begin{pmatrix}\psi_L\\0\end{pmatrix}, \tag{12.166}$$

and

$$\frac{1}{2}\left(1+\gamma^5\right)\begin{pmatrix}\psi_L\\\psi_R\end{pmatrix} = \begin{pmatrix}0\\\psi_R\end{pmatrix}. \tag{12.167}$$

In appendix B.3 we list some of the most important properties of the Dirac gamma matrices we will make use of. Let us note here that the set of 16 complex 4×4-matrices $\{1_4, \gamma^\mu, \gamma^{\mu\nu}, \gamma^5, \gamma^\mu\gamma^5\}$ represents a basis for the corresponding Clifford algebra. In d spacetime dimensions, the Clifford algebra of Dirac matrices is 2^d-dimensional. With γ^μ satisfying the relation 12.162, also the matrices $\Lambda^\mu{}_\nu\gamma^\nu$ satisfy this anticommutation relation. Hence, they must be of the form $\Lambda^\mu{}_\nu\gamma^\nu = U\gamma^\mu U^{-1}$, with a suitable matrix U being a representation of Lorentz transformations. This type of transformation of the Dirac matrices renders the Dirac wave equation relativistically invariant.

12.6   Representation on Fields

Generators for Tensor Fields

We recall that the generators of transformations acting on (physical) tensor fields contain a spin part and an independent orbital part,

$$\mathcal{G}_a = \mathcal{G}_a^{\text{internal}} + \mathcal{G}_a^{\text{orbital}}. \tag{12.168}$$

The indices a are here the Lorentz indices pairs $\mu\nu$, $\alpha\beta$, etc. The above relation is now translated to

$$\boxed{M_{\mu\nu} = S_{\mu\nu} + L_{\mu\nu}} \tag{12.169}$$

where $S_{\mu\nu}$ is the spin part, which we have derived in the sections before, and $L_{\mu\nu}$ is the orbital part, which we will derive in the following.

Orbital Representation of Lorentz Transformations

Our general formula for the orbital generator,

$$(\mathcal{G}_a^{\text{orbital}})_{ij} = -\mathrm{i}\,\delta_{ij} \mathrm{X}_a^\alpha \partial_\alpha, \tag{12.170}$$

is translated to

$$L_{\mu\nu} = -\mathrm{i}\,\mathrm{X}_{\mu\nu}^\alpha \partial_\alpha, \tag{12.171}$$

where we do not show the trivially acting ij-indices. To proceed, we note the formulae

$$\mathrm{X}_{\mu\nu}^\alpha = \frac{\partial}{\partial \omega^{\mu\nu}} \delta x^\alpha \tag{12.172}$$

and

$$\delta x^\alpha = -\frac{\mathrm{i}}{2} \omega^{\rho\sigma} (S_{\rho\sigma})^\alpha{}_\beta x^\beta, \tag{12.173}$$

with $(S_{\rho\sigma})^\alpha{}_\beta$ given by the expression 12.110. These formulae lead to the desired result for the *orbital representation of the generators of Lorentz transformations*:

$$\boxed{L_{\mu\nu} = \mathrm{i}\,(x_\mu \partial_\nu - x_\nu \partial_\mu)} \tag{12.174}$$

This orbital generator is the same for all types of tensor fields. Let us note also that this generator contains the three-dimensional components, as given in formula 10.42.

Now we are able to write down the complete set of *generators of Lorentz transformations* $M_{\mu\nu}$ for the various tensor and spinor fields of interest:

$$
\begin{array}{ll}
M_{\mu\nu} = L_{\mu\nu}\,, & \text{for scalar fields} \\[4pt]
M_{\mu\nu} = L_{\mu\nu} + S_{\mu\nu}^{\text{vector}}, & \text{for vector fields} \\[4pt]
M_{\mu\nu} = L_{\mu\nu} + S_{\mu\nu}^{\text{tensor}}, & \text{for tensor fields} \\[4pt]
M_{\mu\nu} = L_{\mu\nu} + \sigma_{\mu\nu}\,, & \text{for left-handed Weyl fields} \\[4pt]
M_{\mu\nu} = L_{\mu\nu} + \overline{\sigma}_{\mu\nu}\,, & \text{for right-handed Weyl fields} \\[4pt]
M_{\mu\nu} = L_{\mu\nu} + \gamma_{\mu\nu}\,, & \text{for Dirac fields}
\end{array}
\tag{12.175}
$$

Finite Transformations of Fields

We can express how physical tensor and spinor fields transform under finite Lorentz transformations. The fields, considered as functions of both indices types, transform according to

$$\phi_i'(x) = \exp\left[-\frac{\mathrm{i}}{2} \omega^{\mu\nu} \left(L_{\mu\nu}(x) + S_{\mu\nu}\right)\right]_{ij} \phi_j(x). \tag{12.176}$$

If instead of the full functional change we view only the change of the field value at a specific point, the transformations are the ones for the corresponding tensorial or spinorial objects. This total change of the field value is then given by

$$\phi_i'(x') = \exp\left(-\frac{\mathrm{i}}{2} \omega^{\mu\nu} S_{\mu\nu}\right)_{ij} \phi_j(x). \tag{12.177}$$

The scalar field value $\phi(x)$ transforms as

$$\phi'(x') = \phi(x). \tag{12.178}$$

The vector field value $V^\mu(x)$ transforms as

$$V'^\mu(x') = \Lambda^\mu{}_\nu V^\nu(x). \tag{12.179}$$

Equivalently, the covector field value $V_\alpha(x)$ transforms as

$$V'_\alpha(x') = \Lambda_\alpha{}^\beta V_\beta(x). \tag{12.180}$$

A general (m,n)-type tensor field value $\phi^{\mu_1\cdots\mu_m}{}_{\nu_1\ldots\nu_n}(x)$ transforms as

$$\phi'^{\mu_1\cdots\mu_m}{}_{\nu_1\ldots\nu_n}(x') = \Lambda^{\mu_1}{}_{\rho_1}\cdots\Lambda^{\mu_m}{}_{\rho_m}\Lambda_{\nu_1}{}^{\sigma_1}\cdots\Lambda_{\nu_n}{}^{\sigma_n}\,\phi^{\rho_1\cdots\rho_m}{}_{\sigma_1\ldots\sigma_n}(x), \tag{12.181}$$

i.e. each contravariant and each covariant index transforms with the corresponding Lorentz matrix element. Finally, the field values for left-handed, right-handed, and Dirac spinors transform as

$$\psi'_{L,R}(x') = D_{L,R}(\Lambda)\psi_{L,R}(x), \tag{12.182}$$

and

$$\Psi'(x') = D_\Psi(\Lambda)\Psi(x). \tag{12.183}$$

At this point we conclude our study of the Lorentz group, its main representations, and its action on fields.

Further Reading

In this chapter we have concentrated on the proper orthochronous Lorenz group. A development of the full Lorentz group, including reflections, can be found in Sexl and Urbantke [82]. For a discussion of discrete transformations and the subject of spinor handedness, the interested reader may consult Costa and Fogli [16]. The Pauli matrices and the Dirac gamma matrices generate algebras which are special cases of Clifford algebras. The notion of a Clifford algebra and the related quaternion numbers are discussed in Lounesto [53].

13

Poincaré Symmetry

Minkowski space is an affine-linear space, which means in particular that the origin of inertial reference frames can be shifted. Therefore, in addition to the Lorentz transformations considered so far, we must now consider spacetime translations. These two transformations are combined to the Poincaré transformations, which are the maximally allowed symmetry transformations of Minkowski space. The importance of Poincaré transformations arises from the fact that we require classical physical laws to have the same form in all inertial reference frames. Equivalently, we require that the physical laws be invariant under Poincaré transformations. In this chapter we study the foundations of the Poincaré group and its Lie algebra.

13.1 Meaning of Poincaré Transformations

Major Properties of Poincaré Transformations
Before we delve into technical details, let us compile some basic statements to remember when dealing with Poincaré transformations:

- *Poincaré transformations are the affine-linear transformations consisting of Lorentz transformations and spacetime translations.*
- *Inertial reference frames are connected through Poincaré transformations.*
- *The Minkowski space metric is invariant under Poincaré transformations.*
- *Four-distances are invariant under Poincaré transformations.*
- *Physical laws are invariant under the change of inertial reference frames and thus under Poincaré transformations.*

In the following we will examine the meaning of these statements and develop the group-theoretical aspects.

13.2 Poincaré Group

Poincaré Transformations and Minkowski Space
First, let us consider a *spacetime translation* in Minkowski space $\mathbb{M}_4 \equiv (\mathbb{E}^4, \eta)$, written as

$$x'^{\mu} = x^{\mu} + a^{\mu}. \tag{13.1}$$

DOI: 10.1201/9781003087748-13

These translations constitute an abelian group T, the *group of spacetime translations*. We can combine a translation and a Lorentz transformation,

$$x'^{\mu} = \Lambda^{\mu}{}_{\nu}x^{\nu} + a^{\mu}, \tag{13.2}$$

and thus define a *Poincaré transformation* in Minkowski space. These transformations mediate between inertial reference frames defined in Minkowski space. Based on the general transformation rule of a metric 2.17, we can write for Poincaré transformations

$$\eta'_{\mu\nu}(x') = \frac{\partial x^{\rho}}{\partial x'^{\mu}} \frac{\partial x^{\sigma}}{\partial x'^{\nu}} \eta_{\rho\sigma}. \tag{13.3}$$

By using the definition 13.2, we see that the constant translations do not contribute to the derivations and we end up with the invariance condition $\eta'_{\mu\nu}(x') = \eta_{\mu\nu}$, as known already from Lorentz transformations. Thus, we conclude that Poincaré transformations satisfy the condition

$$\boxed{\frac{\partial x'^{\rho}}{\partial x^{\mu}}(x)\frac{\partial x'^{\sigma}}{\partial x^{\nu}}(x)\,\eta_{\rho\sigma} = \eta_{\mu\nu}} \tag{13.4}$$

i.e. the metric $\eta_{\mu\nu}$ of Minkowski space remains invariant under Poincaré transformations. This is the isometry condition 3.41 applied to the present case. Conversely, if we require our Minkowski metric to be invariant as in 13.4, we conclude that the maximally allowed transformations are affine-linear transformations. Indeed, a general infinitesimal coordinate transformation is given by

$$x'^{\mu}(x) = x^{\mu} + \delta x^{\mu}(x). \tag{13.5}$$

The associated transformation of the metric is to first order in $\delta x^{\mu}(x)$ given by

$$\begin{aligned}
\eta'_{\mu\nu}(x') &= \frac{\partial x^{\rho}}{\partial x'^{\mu}} \frac{\partial x^{\sigma}}{\partial x'^{\nu}} \eta_{\rho\sigma} \\
&= (\delta^{\rho}_{\mu} - \partial_{\mu}\delta x^{\rho})(\delta^{\sigma}_{\nu} - \partial_{\nu}\delta x^{\sigma})\eta_{\rho\sigma} \\
&= \eta_{\mu\nu} - (\partial_{\mu}\delta x_{\nu}(x) + \partial_{\nu}\delta x_{\mu}(x)).
\end{aligned} \tag{13.6}$$

So, if we require that the metric is invariant, $\eta'_{\mu\nu}(x') = \eta_{\mu\nu}$, we obtain the *Killing equation*

$$\partial_{\mu}\delta x_{\nu}(x) + \partial_{\nu}\delta x_{\mu}(x) = 0. \tag{13.7}$$

Now let us Taylor-expand the unknown $\delta x^{\mu}(x)$ with respect to the variable x as

$$\delta x^{\mu}(x) = a^{\mu} + b^{\mu}{}_{\nu}x^{\nu} + c^{\mu}{}_{\nu\rho}x^{\nu}x^{\rho} + \dots, \tag{13.8}$$

with the constant factors a^{μ}, $b^{\mu}{}_{\nu}$ and so on. The Killing equation leads to the condition $b_{\mu\nu} = -b_{\nu\mu}$, while a^{μ} remains a free parameter. All other factors in the Taylor expansion vanish. We are thus led to an infinitesimal affine-linear transformation,

$$x'^{\mu} = x^{\mu} + b^{\mu}{}_{\nu}x^{\nu} + a^{\mu}, \tag{13.9}$$

where the factors $b^{\mu}{}_{\nu}$ are nothing else than the infinitesimal parameters of a Lorentz transformation, which we have previously denoted $\omega^{\mu}{}_{\nu}$. Hence, we have proved the equivalence between the invariance of the Minkowski metric and the specific form of the Poincaré transformations. The invariance of four-distances under Poincaré transformations can be shown in a straightforward way. (*exercise 13.1*) We can summarize:

The Poincaré group P is exactly the symmetry group of Minkowski space $\mathbb{M}_4$.

Poincaré Group Properties

We investigate now the basic group properties of Poincaré transformations. Each Poincaré transformation requires 10 parameters to be defined, which are 6 parameters for the Lorentz rotation and 4 parameters for the translation. The composition of two successive Poincaré transformations,

$$x' = \Lambda_1 x + a_1 \quad \text{and} \quad x'' = \Lambda_2 x' + a_2, \tag{13.10}$$

leads to the combined transformation

$$x'' = \Lambda_2 \Lambda_1 x + \Lambda_2 a_1 + a_2. \tag{13.11}$$

If we denote a Poincaré transformation simply by the ordered pair (a, Λ), then the composition law is given by

$$(a_2, \Lambda_2)(a_1, \Lambda_1) = (a_2 + \Lambda_2 a_1, \Lambda_2 \Lambda_1). \tag{13.12}$$

The inverse is given by

$$(a, \Lambda)^{-1} = (-\Lambda^{-1} a, \Lambda^{-1}) \tag{13.13}$$

and the neutral element is $(0, 1)$. Associativity can be shown easily. Every Poincaré transformation can be written as a composition of a translation and a Lorentz transformation as $(a, \Lambda) = (a, 1)(0, \Lambda)$, or, in the reverse order as $(a, \Lambda) = (0, \Lambda)(\Lambda^{-1} a, 1)$. The formula 13.12 reveals that the Poincaré group is the semidirect product of the translation group T with the Lorentz group L, i.e. $P = T \rtimes L$. By comparing with the case of rotations and Euclidean transformations, we can conclude that the Poincaré group results from the Lorentz group in the same manner as the Euclidean group results from the rotation group. Usually, we concentrate on the proper orthochronous Lorentz group $L_+^\uparrow$ and thus on the part $P_+^\uparrow = T \rtimes L_+^\uparrow$, which is indeed a subgroup of the full Poincaré group. The full Poincaré group is sometimes symbolized by $ISO(1, 3)$ and called the *inhomogeneous Lorentz group*.

Linearization of Poincaré Transformations

Spacetime translations can be rendered to be linear transformations, when we view them in a 5-dimensional space. A translation of the form

$$x'^\mu = x^\mu + a^\mu \tag{13.14}$$

can be made linear by using the 5-dimensional vector representation

$$\begin{pmatrix} x^\mu \\ 1 \end{pmatrix}, \tag{13.15}$$

where the column entry 1 stays unchanged under all transformations. The translation is then given by a matrix multiplication,

$$\begin{pmatrix} x'^\mu \\ 1 \end{pmatrix} = \left(\begin{array}{c|c} \eta^\mu{}_\nu & a^\mu \\ \hline 0 & 1 \end{array} \right) \begin{pmatrix} x^\nu \\ 1 \end{pmatrix} = \begin{pmatrix} x^\mu + a^\mu \\ 1 \end{pmatrix}. \tag{13.16}$$

Similarly, a Poincaré transformation

$$x'^\mu = \Lambda^\mu{}_\nu x^\nu + a^\mu \tag{13.17}$$

is linear in five dimensions, if we write it as

$$\begin{pmatrix} x'^\mu \\ 1 \end{pmatrix} = \left(\begin{array}{c|c} \Lambda^\mu{}_\nu & a^\mu \\ \hline 0 & 1 \end{array} \right) \begin{pmatrix} x^\nu \\ 1 \end{pmatrix} = \begin{pmatrix} \Lambda^\mu{}_\nu x^\nu + a^\mu \\ 1 \end{pmatrix}. \tag{13.18}$$

This 5-dimensional matrix representation can be employed to derive the Poincaré algebra.

13.3   Poincaré Algebra and Field Representations

Generator of Translations

There are multiple ways to derive the *Poincaré algebra*. One way is to employ the above 5-dimensional matrix representation of the Poincaré group. Another method is the use of unitary representations of the Poincaré group, as shown in [82]. Here we will first focus on the special case of a scalar field and then generalize the results. Let us first derive the field representation of the generators of translations. Under a translation, $x' = x + a$, a scalar field $\phi(x)$ by definition transforms as

$$\phi'(x') = \phi(x). \tag{13.19}$$

Equivalently, we can write $\phi'(x) = \phi(x - a)$. According to the form variation of a field 9.51, this is also given by

$$\phi'(x) = \exp(-i\, a^\mu P_\mu)\phi(x), \tag{13.20}$$

where the four-covector P_μ represents the four *generators of spacetime translations*. Let us view the two previous relations for a small translation a, with $|a| \ll 1$. We have

$$\phi(x - a) = \phi(x) - a^\mu \partial_\mu \phi(x) + \mathcal{O}(2), \tag{13.21}$$

and

$$\exp(-i\, a \cdot P)\phi(x) = \phi(x) - i\, a^\mu P_\mu \phi(x) + \mathcal{O}(2). \tag{13.22}$$

Thus, we can infer the explicit formula for the generator of spacetime translations:

$$\boxed{P_\mu = -i\, \partial_\mu} \tag{13.23}$$

We have derived this result as the orbital part of the generator of translations for a scalar field. In fact, it is the most general representation of the generator of translations acting on *any* type of tensor field. This is because *we require all tensor fields, carrying any type of indices, to transform under translations like a scalar field.* Internal indices or spin indices are not affected by translations. As a formula, this reads

$$\boxed{\phi_i'(x + a) = \phi_i(x)} \tag{13.24}$$

There is no internal (spin-specific) part for the generator of translations, only an orbital part exists. All tensor fields $\phi_i(x)$ possess the same generator of spacetime translations.

Commutators of the Poincaré Algebra

Now can we derive the commutator relations of the Poincaré algebra, which is best done by using the just obtained field representation. The commutator of two translations is trivial, encoding the fact that the order of translations is immaterial,

$$[P_\mu, P_\nu] = 0. \tag{13.25}$$

For the commutator of $M_{\mu\nu}$ with P_μ, we use the complete field representation

$$M_{\mu\nu} = i\, (x_\mu \partial_\nu - x_\nu \partial_\mu) + S_{\mu\nu}. \tag{13.26}$$

The spin part $S_{\mu\nu}$ is independent of the coordinate x^μ and thus it does not contribute to the commutator with P_μ. The orbital part, however, does contribute and the result is the commutator (*exercise 13.2*)

$$[M_{\mu\nu}, P_\rho] = -i\, (\eta_{\mu\rho} P_\nu - \eta_{\nu\rho} P_\mu). \tag{13.27}$$

Now we only need to add the commutators between the generators of Lorentz transformations, thus obtaining all *commutator relations of the Poincaré algebra*:

$$
\begin{aligned}
[M_{\mu\nu}, M_{\rho\sigma}] &= -\mathrm{i}\left(\eta_{\mu\rho}M_{\nu\sigma} - \eta_{\nu\rho}M_{\mu\sigma} + \eta_{\nu\sigma}M_{\mu\rho} - \eta_{\mu\sigma}M_{\nu\rho}\right) \\
[M_{\mu\nu}, P_{\rho}] &= -\mathrm{i}\left(\eta_{\mu\rho}P_{\nu} - \eta_{\nu\rho}P_{\mu}\right) \\
[P_{\mu}, P_{\nu}] &= 0
\end{aligned}
\tag{13.28}
$$

Let us also write down the form of a finite Poincaré transformation given as an exponential expression:

$$
\exp\left(-\frac{\mathrm{i}}{2}\omega^{\mu\nu}M_{\mu\nu} - \mathrm{i}\,a^{\mu}P_{\mu}\right).
\tag{13.29}
$$

Tensor and Spinor Fields under Poincaré Transformations

In the previous chapter we have seen how tensor fields transform under Lorentz transformations. The transformation law of Lorentz tensor fields is given by 12.181. So the question arises what is the transformation law of tensor fields under Poincaré transformations. As we have stated above, the tensor fields relevant to us do not change under translations. Under translations, these tensor fields behave like scalar fields. Thus, the transformation law of tensor fields under Poincaré transformations is exactly the same as in the formula 12.181. This can also be seen by direct calculation from the general transformation law 1.96 of tensor fields on manifolds. This line of thought applies equally well to spinor fields. Translations do not affect spinor field components and hence the spinor field transformation laws known from the Lorentz group are the exactly same in the case of the Poincaré group.

Casimir Operators, Mass and Helicity

In Section 8.3 we have shown that Casimir operators can be used to classify the irreducible representations of a Lie algebra. In the case of the Poincaré algebra, there are two *Casimir operators* used for this classification, the momentum square, $P^2 = P_{\mu}P^{\mu}$, and the square $W^2 = W_{\mu}W^{\mu}$ of the *Pauli-Lubanski vector (Josef K. Lubanski)* W_{μ}, where the latter is defined as

$$
W_{\mu} \equiv \frac{1}{2}\epsilon_{\mu\nu\rho\sigma}M^{\nu\rho}P^{\sigma}.
\tag{13.30}
$$

The Casimir operators P^2 and W^2 commute with all the generators of the Poincaré group, i.e.

$$
[M_{\mu\nu}, P^2] = [M_{\mu\nu}, W^2] = 0,
\tag{13.31}
$$

and

$$
[P_{\mu}, P^2] = [P_{\mu}, W^2] = 0.
\tag{13.32}
$$

For irreducible representations, the momentum square P^2 can be simplified to the form

$$
P^2 = m^2 I,
\tag{13.33}
$$

with the mass square m^2 being the eigenvalue. By noting that the field representation is $P_{\mu} = -\mathrm{i}\partial_{\mu}$, we can derive the *Klein-Gordon equation*,

$$
(\partial^2 + m^2)\phi = 0,
\tag{13.34}
$$

for a scalar field purely by group-theoretical considerations, compare with Section 15.3. The invariance of P^2 means that the mass m is invariant under Poincaré transformations. The Pauli-Lubanski vector is always perpendicular to the momentum vector,

$$
W_{\mu}P^{\mu} = 0,
\tag{13.35}
$$

and we can derive the following set of commutator relations: (*exercise 13.3*)

$$
\begin{aligned}
[P_\mu, W_\nu] &= 0, \\
[M_{\mu\nu}, W_\rho] &= -\mathrm{i}\,(\eta_{\mu\rho} W_\nu - \eta_{\nu\rho} W_\mu), \\
[W_\mu, W_\nu] &= -\mathrm{i}\,\epsilon_{\mu\nu\rho\sigma} W^\rho P^\sigma.
\end{aligned}
\tag{13.36}
$$

For the square W^2, one obtains the formula

$$
W^2 = -\frac{1}{2} M_{\mu\nu} M^{\mu\nu} P^2 + M_{\mu\rho} M^{\nu\rho} P^\mu P_\nu .
\tag{13.37}
$$

By using the three-dimensional generators $\boldsymbol{J}$ and $\boldsymbol{K}$ of the Lorentz group, the Pauli-Lubanski vector can be expressed in components as

$$
W^\mu = \left(\begin{array}{c} \boldsymbol{J} \cdot \boldsymbol{P} \\ J P^0 + \boldsymbol{K} \times \boldsymbol{P} \end{array} \right),
\tag{13.38}
$$

with $P^2 = (P^0)^2 - \boldsymbol{P}^2 = m^2$, as usual. Depending on the rest mass m we can distinguish between two cases. For a massive particle with $m^2 > 0$, we can transform to the momentary rest frame of the particle and obtain

$$
W^\mu = \left(\begin{array}{c} 0 \\ m\boldsymbol{J} \end{array} \right).
\tag{13.39}
$$

Within the rest frame, the Pauli-Lubanski vector is essentially the angular momentum, or spin, of the particle. Generally, one defines the *helicity* h of a particle as

$$
h \equiv \boldsymbol{J} \cdot \frac{\boldsymbol{P}}{|\boldsymbol{P}|}.
\tag{13.40}
$$

It is the projection of the angular momentum $\boldsymbol{J}$ of the particle in the direction $\boldsymbol{P}/|\boldsymbol{P}|$ of motion. Because the total angular momentum is $\boldsymbol{J} = \boldsymbol{L} + \boldsymbol{S}$ and the orbital angular momentum $\boldsymbol{L} = \boldsymbol{x} \times \boldsymbol{P}$ is always orthogonal to $\boldsymbol{P}$, we have that the helicity h is equal the spin projection along the direction of motion,

$$
h \equiv \boldsymbol{S} \cdot \frac{\boldsymbol{P}}{|\boldsymbol{P}|}.
\tag{13.41}
$$

In the case of a massless particle $m^2 = 0$, it is $P^2 = 0$ and $W^2 = 0$. Because it also $W_\mu P^\mu = 0$, the proportionality

$$
W^\mu = \lambda P^\mu
\tag{13.42}
$$

must hold for some real number λ. Since it is

$$
W^0 = \boldsymbol{J} \cdot \boldsymbol{P} = \lambda P^0 = \pm\lambda |\boldsymbol{P}|,
\tag{13.43}
$$

we obtain that the constant of proportionality is $\lambda = \pm h$. For a further discussion of the helicity of a particle the reader should turn to [23] and [85].

13.4 Correspondence of Spacetime Symmetries

Relations of Spacetime Symmetries

The spacetime symmetry groups considered so far can be summarized in a compact and useful way. To this end, we use a three-dimensional commutative diagram as shown here:

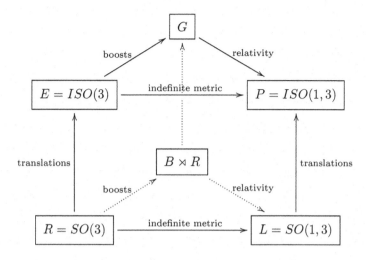

Figure 13.1: Correspondence of symmetry groups

This three-dimensional prism-shaped diagram contains in each of its corners one of the symmetry groups we have studied. The symmetry groups appear naturally in pairs consisting of a base group of transformations and the associated group obtained by translating the base group. The first pair is the rotation group $R = SO(3)$ and the Euclidean group $E = ISO(3)$. The second pair is the homogeneous Galilei group $B \rtimes R$ and the full Galilei group $G = T \rtimes B \rtimes R$. The third pair is the Lorentz group $L = SO(1,3)$ and the Poincaré group $P = ISO(1,3)$. From the pure space symmetries we can reach to the Galilei symmetries by including velocity boosts. From the Galilei symmetries we can reach to the relativistic symmetries by taking into account the finite speed of light. From the pure space symmetries we can arrive to the relativistic spacetime symmetries by increasing the dimension by one and by adopting the indefinite Lorentzian metric.

Further Reading

A succinct discussion of Casimir operators of the Poincaré group and the massive and massless representations can be found in the book by Costa and Fogli [16]. A reference providing insights into the Poincaré group from diverse physical and mathematical viewpoints is the book of Sexl and Urbantke [82].

14

Conformal Symmetry

In this chapter we proceed to the next step in the generalization of spacetime symmetries and discuss conformal symmetry. We consider conformal transformations that either act on Minkowski space or act as structure-preserving transformations on conformal space. We systematically derive the conformal group and the single transformations of which it is comprised. Two new types of transformations appear, the dilatations and the special conformal transformations. Then we move on to the derivation of the Lie algebra for tensor fields in complete generality. We introduce the notion of primary fields and their relation to tensor densities. A special feature of the special conformal transformations is their nonlinearity. The conformal group can act linearly if we employ a representation on a six-dimensional pseudo-Euclidean space or a representation on a four-complex-dimensional space.

14.1 Conformal Group

Minkowski Spacetime vs. Conformal Spacetime

Let us consider a d-dimensional Minkowski space $\mathbb{M}_d \equiv (\mathbb{E}^d, \eta)$. The spacetime dimensionality is $d = 1 + D$, with one time dimension and D spacial dimensions. The metric $\eta_{\mu\nu}$ is given by the constant metric tensor

$$\eta_{\mu\nu} = \text{diag}(1, -1, \ldots, -1), \tag{14.1}$$

where the entry -1 appears D times on the matrix diagonal. We can apply conformal coordinate transformations on this Minkowski space and treat them as symmetry transformations, as discussed in Section 3.2, where the metric stays unchanged. Equally well, we can re-interpret the conformal transformations as structure-preserving transformations of a more general space, the conformal spacetime. The d-dimensional *conformal spacetime*, denoted by $\mathcal{C}_{(1,D)}$, or simply $\mathcal{C}_d$, is defined as

$$\mathcal{C}_d \equiv (\mathbb{E}^d, [\eta]), \tag{14.2}$$

where the equivalence class $[\eta]$ of conformal metrics is given by

$$[\eta] \equiv \left\{ \Omega^2 \eta, \text{ for all real scalar functions } \Omega, \text{ with } \Omega > 0 \right\}. \tag{14.3}$$

DOI: 10.1201/9781003087748-14

The conformal spacetime $\mathcal{C}_d$ encompasses all allowed conformal geometries, compare with Section 3.2. The conformal transformations, as defined by the condition 3.43, represent symmetry transformations of the above spaces and lead naturally to the conformal group.

Conformal Transformations and Conformal Group

Let us first consider Minkowski space $\mathbb{M}_d$. According to the definition 3.43, the *conformal transformations* of the coordinate variables x^μ of Minkowski space are given by

$$\frac{\partial x'^\rho}{\partial x^\mu}(x)\frac{\partial x'^\sigma}{\partial x^\nu}(x)\,\eta_{\rho\sigma} = \Psi^2(x)\,\eta_{\mu\nu} \qquad (14.4)$$

where the *conformal factor* $\Psi(x)$ is a positive scalar function. The above condition is an implicit definition of the conformal transformations of the coordinates x^μ. By interpreting this condition instead as a transformation of the metric, it can be rewritten in the form

$$g'_{\mu\nu}(x') = \Phi^2(x)\,\eta_{\mu\nu}, \qquad (14.5)$$

where $\Phi^2(x) = \Psi^{-2}(x)$ and $\eta_{\mu\nu}$ is the initial metric. The conformal transformations constitute a group called the *conformal group*, which we denote by $C(1, D)$, or simply by C. Obviously, the Poincaré group P is a subgroup of the conformal group C and corresponds to the special case of constant $\Psi = 1$. On the other hand, the conformal group is the largest finite-dimensional subgroup of the group of diffeomorphisms. In order to prove the group property for the set of conformal transformations, we first note that the composition of any two conformal transformations is also a conformal transformation. More specifically, consider the two iterated conformal transformations F and G given by

$$\begin{aligned} F &: x \mapsto x' = F(x), \\ G &: x' \mapsto x'' = G(x'), \end{aligned} \qquad (14.6)$$

with the respective conformal factors $\Psi_F^2(x)$ and $\Psi_G^2(x)$. Then the combined transformation

$$G \circ F : x \mapsto x'' = G \circ F(x) \qquad (14.7)$$

is also a conformal transformation and its conformal factor $\Psi_{G\circ F}^2(x)$ is given by

$$\Psi_{G\circ F}^2(x) = \Psi_G^2(x')\,\Psi_F^2(x). \qquad (14.8)$$

The proof is easy and left to the reader. (*exercise 14.1*) The neutral element is the identity transformation and the inverse conformal transformation is given by the inverse conformal factor. Due to the last formula, the associativity of conformal transformations reduces to the associativity of real numbers. Thus, the conformal transformations on Minkowski space $\mathbb{M}_d$ constitute a group. Geometrically, the conformal transformations leave the angles between d-vectors unchanged. In general, they do not preserve lengths or distances. However, the sign of distances and the light cone, defined by $ds^2 = 0$, remain invariant. (*exercise 14.2*) Now instead of Minkowski space $\mathbb{M}_d$, we can consider the conformal spacetime $\mathcal{C}_d$ and the conformal transformations on it. Then we can state:

> The conformal group C is exactly the symmetry group of conformal space $\mathcal{C}_d$.

The transformation 14.4 contains the special case where the conformal factor $\Psi(x)$ is a constant, $\Psi(x) = \text{const} = \omega$. This is the case of a global, or rigid, *scale transformation*, which is also called a *dilatation*. Hence, we see that *general conformal transformations can be viewed as local scale transformations.* The set of transformations consisting of translations, Lorentz rotations and dilatations constitute a subgroup of the conformal group. The defining equation for this subgroup is

$$\frac{\partial x'^\rho}{\partial x^\mu}(x)\frac{\partial x'^\sigma}{\partial x^\nu}(x)\,\eta_{\rho\sigma} = \omega^2\,\eta_{\mu\nu}, \tag{14.9}$$

with a constant real parameter $\omega > 0$. It is common to write $\omega = \exp(\alpha)$, with the real parameter $\alpha \in \mathbb{R}$. The transformations of this subgroup are written explicitly as

$$x'^\mu = \omega\Lambda^\mu{}_\nu x^\nu + a^\mu. \tag{14.10}$$

In contrast to Poincaré symmetry, which is exactly the allowed symmetry of Minkowski space, a conformal symmetry of Minkowski space is in general *not* realized physically. Still, conformal symmetry is physically relevant for idealized systems or as an approximation when masses vanish, which is an approximation for ultra-high-energy particle reactions, for example. The study of conformal symmetry leads also to surprising insights into known theories, such as Maxwell electromagnetism, which is conformally invariant exactly in four spacetime dimensions, a fact we will prove later.

Example of a Conformal Transformation

It is instructive to give an example of a conformal coordinate transformation before we delve into the derivation of the conformal group. Consider the two-dimensional manifold $\mathbb{R}^2 \setminus \{0\}$ equipped with the usual Euclidean metric $\delta = \mathrm{diag}(1, 1)$ and with Cartesian coordinates (x, y). We view the coordinate transformation from $\mathbb{R}^2 \setminus \{0\}$ to the plane $\mathbb{R}^2$ given by

$$\begin{pmatrix} x \\ y \end{pmatrix} \mapsto \begin{pmatrix} u \\ v \end{pmatrix} = \begin{pmatrix} \left(x^2 - y^2\right)^2 - 4x^2y^2 \\ 4xy\left(x^2 - y^2\right) \end{pmatrix}. \tag{14.11}$$

The question is if this coordinate transformation represents a conformal transformation of the metric. For this to be the case, the transformation must satisfy the defining equation 14.4. This is fulfilled with the conformal factor $\Psi^2(x, y) = 16\left(x^2 + y^2\right)^3$. In order to visualize the effect of this conformal transformation, we apply it to the regular Cartesian coordinate grid, as illustrated in the diagram 14.1 below. The 2×2 sized coordinate patch in the

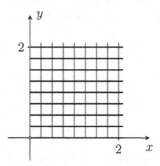

Figure 14.1: Cartesian coordinate patch

Cartesian (x, y)-plane is distorted under the conformal transformation and produces a new coordinate grid in the Cartesian (u, v)-plane. The resulting coordinate grid in the diagram 14.2 visualizes how the new conformal geometry appears from our usual Cartesian point of view. We observe that the one coordinate quadrant of the (x, y)-plane is sufficient to reach a coverage of all four quadrants of the (u, v)-plane. One should note the different diagram

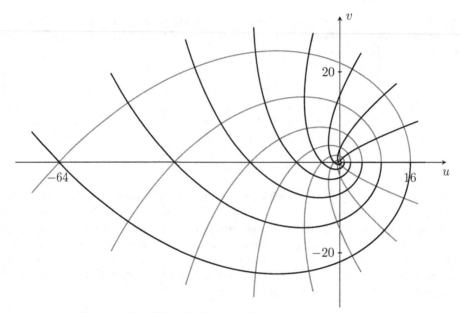

Figure 14.2: The resulting conformal coordinate patch

scales. The important point to remember is that the new conformal coordinate grid is such that all coordinate lines meet still at right angles, as for the initial coordinate grid. This demonstrates the angle-preserving property of conformal transformations.

Conformal Killing Equation

Following the above considerations, let us now derive how the spacetime variables change under conformal transformations. We consider a general d-dimensional conformal spacetime, as introduced above. We will first derive the infinitesimal and subsequently the finite transformations. Generally, an active infinitesimal coordinate transformation is

$$x^\mu \mapsto x'^\mu(x) = x^\mu + \delta x^\mu(x). \tag{14.12}$$

The changes in the coordinates induce a corresponding change in the metric components $g_{\mu\nu}(x)$. This change, to first order in $\delta x^\mu(x)$, is calculated as

$$
\begin{aligned}
g'_{\mu\nu}(x') &= \frac{\partial x^\rho}{\partial x'^\mu} \frac{\partial x^\sigma}{\partial x'^\nu} g_{\rho\sigma}(x) \\
&= (\delta^\rho_\mu - \partial_\mu \delta x^\rho)(\delta^\sigma_\nu - \partial_\nu \delta x^\sigma) g_{\rho\sigma}(x) \\
&= g_{\mu\nu}(x) - (\partial_\mu \delta x_\nu(x) + \partial_\nu \delta x_\mu(x)).
\end{aligned}
\tag{14.13}
$$

For the total change of the metric, $\bar\delta g_{\mu\nu}(x)$, which is defined as

$$\bar\delta g_{\mu\nu}(x) \equiv g'_{\mu\nu}(x') - g_{\mu\nu}(x), \tag{14.14}$$

we have correspondingly, as always,

$$\bar\delta g_{\mu\nu}(x) = -(\partial_\mu \delta x_\nu(x) + \partial_\nu \delta x_\mu(x)). \tag{14.15}$$

If we have specifically a conformal coordinate transformation at hand, it is

$$\bar\delta g_{\mu\nu}(x) = \left(\Phi^2(x) - 1\right) \eta_{\mu\nu}. \tag{14.16}$$

Thus, we are led to the conformal Killing equation,

$$\partial_\mu \delta x_\nu(x) + \partial_\nu \delta x_\mu(x) = \left(1 - \Phi^2(x)\right)\eta_{\mu\nu} \equiv f(x)\,\eta_{\mu\nu}, \tag{14.17}$$

which should be understood as the condition for a conformal transformation. By taking the trace of both sides above, we obtain

$$1 - \Phi^2 = \frac{2}{d}(\partial_\rho \delta x^\rho). \tag{14.18}$$

Thus, the *conformal Killing equation* takes the form

$$\partial_\mu \delta x_\nu + \partial_\nu \delta x_\mu = \frac{2}{d}(\partial_\rho \delta x^\rho)\,\eta_{\mu\nu}. \tag{14.19}$$

This result is in line with the general conformal Killing equation 3.53. In the following we will devise a formula for the variations $\delta x^\mu(x)$ in conformal transformations by using the conformal Killing equation. To this end, we need to derive some additional identities. We apply ∂_μ to the Killing equation 14.17 and change indices cyclically to obtain

$$2\partial_\mu \partial_\nu \delta x_\rho = (\partial_\mu f)\eta_{\nu\rho} + (\partial_\nu f)\eta_{\rho\mu} - (\partial_\rho f)\eta_{\mu\nu}. \tag{14.20}$$

A contraction with $\eta^{\mu\nu}$ yields

$$2\partial^2 \delta x^\rho = (2-d)\partial_\rho f. \tag{14.21}$$

Further we apply ∂_μ to 14.21 to get

$$2\partial^2 \partial_\mu \delta x_\nu = (2-d)\partial_\mu \partial_\nu f. \tag{14.22}$$

From 14.17, by applying ∂^2, we obtain

$$\partial^2 \partial_\mu \delta x_\nu + \partial^2 \partial_\nu \delta x_\mu = (\partial^2 f)\eta_{\mu\nu}. \tag{14.23}$$

From the last two equations, 14.22 and 14.23, we get

$$(2-d)\partial_\mu \partial_\nu f = (\partial^2 f)\eta_{\mu\nu}. \tag{14.24}$$

A contraction with $\eta^{\mu\nu}$ gives finally

$$(d-1)\partial^2 f = 0. \tag{14.25}$$

Now we have to consider different values for the dimension d. In the case $d = 1$, the equation 14.25 is trivially fulfilled and does not impose any specific condition on $f(x)$, i.e. any smooth transformation is conformal in this case. In the case $d = 2$, the equation 14.25 reads $\partial^2 f = 0$ and leads actually to an infinite-dimensional algebra. We will not study this case any further, despite it is an important ingredient for some developments in statistical mechanics and quantum field theory. For the two-dimensional conformal symmetry, the interested reader should consider the reference [24]. In the case $d \geq 3$, from equations 14.24 and 14.25, we see that it is

$$\partial_\mu \partial_\nu f = 0. \tag{14.26}$$

This means that $f(x)$ is at most linear in x, i.e.

$$f(x) = s + t_\mu x^\mu, \tag{14.27}$$

with s and t_μ being constants. Substituting $f(x)$ into equation 14.20, we see that $\partial_\mu \partial_\nu \delta x_\rho$ is a constant and thus we can conclude that $\delta x^\mu(x)$ is at most quadratic in x,

$$\delta x_\mu(x) = a_\mu + b_{\mu\nu}x^\nu + c_{\mu\nu\rho}x^\nu x^\rho, \tag{14.28}$$

with $c_{\mu\nu\rho} = c_{\mu\rho\nu}$. In order to identify the constants a_μ, $b_{\mu\nu}$, and $c_{\mu\nu\rho}$, we investigate the equations 14.17 to 14.20 for each power in x. For the Killing vector $\delta x_\mu = a_\mu$, representing the 0th order, there is no constraint and a_μ represents the known spacetime translation vector. For the Killing vector $\delta x_\mu = b_{\mu\nu} x^\nu$, representing the 1st order in x, the equations 14.17 and 14.18 lead to

$$b_{\mu\nu} + b_{\nu\mu} = \frac{2}{d} b^\sigma{}_\sigma \eta_{\mu\nu}, \tag{14.29}$$

which means that $b_{\mu\nu}$ is the sum of an antisymmetric tensor plus a multiple of $\eta_{\mu\nu}$,

$$b_{\mu\nu} \equiv \omega_{\mu\nu} + \alpha \eta_{\mu\nu}, \tag{14.30}$$

with $\omega_{\mu\nu} = -\omega_{\nu\mu}$. The constant factor α is identified to be $\alpha = d^{-1} b^\sigma{}_\sigma$. The antisymmetric part $\omega_{\mu\nu}$ is the known infinitesimal form of a Lorentz rotation. The trace part $\alpha \eta_{\mu\nu}$ corresponds to the *infinitesimal form of a dilatation*,

$$x'^\mu = x^\mu + \alpha x^\mu. \tag{14.31}$$

For the Killing vector $\delta x_\mu = c_{\mu\nu\rho} x^\nu x^\rho$, representing the 2nd order in x, the equation 14.20 yields

$$4 c_{\rho\mu\nu} = (\partial_\mu f) \eta_{\nu\rho} + (\partial_\nu f) \eta_{\rho\mu} - (\partial_\rho f) \eta_{\mu\nu}. \tag{14.32}$$

At this order it is

$$f(x) = \frac{4}{d} c^\sigma{}_{\sigma\rho} x^\rho, \tag{14.33}$$

and thus

$$\partial_\mu f = \frac{4}{d} c^\sigma{}_{\sigma\mu}. \tag{14.34}$$

This allows us to determine the constant $c_{\rho\mu\nu}$ as

$$c_{\rho\mu\nu} = c_\mu \eta_{\nu\rho} + c_\nu \eta_{\rho\mu} - c_\rho \eta_{\mu\nu}, \tag{14.35}$$

with the constant vector $c_\mu = d^{-1} c^\sigma{}_{\sigma\mu}$. The last part of the infinitesimal coordinate transformations is therefore

$$x'^\mu = x^\mu + 2(c \cdot x) x^\mu - c^\mu x^2. \tag{14.36}$$

This is the *infinitesimal form of a special conformal transformation*, the latter also abbreviated as *SCT*. In summary, the single *infinitesimal conformal coordinate transformations* are given in the following table:

$$
\begin{array}{ll}
x'^\mu = x^\mu + \omega^\mu{}_\nu x^\nu, & \text{Lorentz rotation} \\
x'^\mu = x^\mu + a^\mu, & \text{Translation} \\
x'^\mu = x^\mu + \alpha x^\mu, & \text{Dilatation} \\
x'^\mu = x^\mu + 2(c \cdot x) x^\mu - c^\mu x^2, & \text{SCT}
\end{array}
\tag{14.37}
$$

A complete conformal coordinate transformation that uses all the available parameter freedom has the infinitesimal form

$$\delta x^\mu = \omega^\mu{}_\nu x^\nu + a^\mu + \alpha x^\mu + 2(c \cdot x) x^\mu - c^\mu x^2. \tag{14.38}$$

Finite Conformal Transformations

Let us derive the finite conformal coordinate transformations. The translations and Lorentz rotations are already known to us, so we focus here on the dilatations and the SCTs. First, let us parametrize the spacetime coordinate by a real parameter p, with $0 \leq p \leq 1$, as

$$x^\mu(p) \equiv x^\mu + p\delta x^\mu, \tag{14.39}$$

so that the parameter turns on the transformation of the coordinates in a smooth way, with $x^\mu(1) = x'^\mu$ and $x^\mu(0) = x^\mu$. Now the task is to solve the system of ODEs

$$\frac{\mathrm{d}x^\mu}{\mathrm{d}p}(p) = \delta x^\mu, \tag{14.40}$$

with the known infinitesimal δx^μ and thus determine the finite transformed coordinate $x'^\mu = x^\mu(1)$. For dilatations, the ODE has the form

$$\frac{\mathrm{d}x^\mu}{\mathrm{d}p} = \alpha x^\mu, \tag{14.41}$$

which is integrated by

$$x^\mu(p) = \mathrm{e}^{\alpha p} x^\mu. \tag{14.42}$$

Thus, we conclude

$$x'^\mu = \mathrm{e}^{\alpha} x^\mu. \tag{14.43}$$

This is a *finite dilatation* parametrized by α. For the SCTs, the corresponding ODE is

$$\frac{\mathrm{d}x^\mu}{\mathrm{d}p} = \left(2x^\mu x_\nu - x^2 \delta^\mu_\nu\right) c^\nu. \tag{14.44}$$

Inverting the matrix in the bracket yields

$$\left(\frac{2x^\mu x_\nu - x^2 \delta^\mu_\nu}{x^4}\right) \frac{\mathrm{d}x^\nu}{\mathrm{d}p} = c^\mu, \tag{14.45}$$

or, equivalently,

$$\frac{\mathrm{d}}{\mathrm{d}p}\left(\frac{x^\mu}{x^2}\right) = -c^\mu. \tag{14.46}$$

The integration is easy and leads to

$$\frac{x^\mu(1)}{x^2(1)} - \frac{x^\mu(0)}{x^2(0)} = -c^\mu, \tag{14.47}$$

which is

$$\frac{x'^\mu}{x'^2} = \frac{x^\mu}{x^2} - c^\mu. \tag{14.48}$$

We can see here that the SCT is equivalent to an inversion, followed by a translation by a constant vector, and a subsequent inversion. We can write this symbolically as

$$\mathrm{SCT}\,[c] = \mathrm{Inversion} \circ \mathrm{Translation}\,[-c] \circ \mathrm{Inversion}. \tag{14.49}$$

Two inversions are equivalent to the identity transformation. This is the reason why SCTs are also called *conformal translations*. From the last equation, the square x'^2 can be calculated, which leads finally to the explicit form of a *finite special conformal transformation*,

$$x'^\mu = \frac{x^\mu - c^\mu x^2}{1 - 2(c \cdot x) + c^2 x^2}. \tag{14.50}$$

Conversely, for small c^μ, the infinitesimal form of the SCT can be derived again. (*exercise 14.3*) Note that for certain values of x and c, a SCT can become singular. In summary, *finite conformal coordinate transformations* contain the following single transformations:

$$
\begin{array}{ll}
x'^\mu = \Lambda^\mu{}_\nu x^\nu, & \text{Lorentz rotation} \\[2mm]
x'^\mu = x^\mu + a^\mu, & \text{Translation} \\[2mm]
x'^\mu = e^\alpha x^\mu, & \text{Dilatation} \\[2mm]
x'^\mu = \dfrac{x^\mu - c^\mu x^2}{1 - 2(c \cdot x) + c^2 x^2}, & \text{SCT}
\end{array}
\tag{14.51}
$$

In $d = 1 + D$ spacetime dimensions, the translations combined with the Lorentz rotations form the Poincaré group P. By adding the dilatations and the special conformal transformations to the Poincaré transformations, we obtain the full conformal group C. The full conformal group in $d \geq 3$ dimensions is determined by

$$
\frac{d(d-1)}{2} + d + 1 + d = \frac{(d+1)(d+2)}{2}
\tag{14.52}
$$

parameters. For $d = 4$ this amounts to 15 parameters, where 10 parameters fix the Poincaré group, 1 parameter defines the dilatations, and 4 parameters define the special conformal transformations. Finally, it should be noted that instead of using a pseudo-Euclidean metric, one can start with a proper Euclidean metric, and in a similar way to the above, achieve a generalization of Euclidean space to a conformal space.

Conformal Factors

For the just derived single transformations constituting the conformal group, it is useful to derive the corresponding conformal factors. We leave the details to the reader and list the results here. (*exercise 14.4*) For Lorentz rotations and translations, the conformal factor is trivially constant one. For dilatations, the conformal factor $\Psi_{\mathrm{Dil}}(x)$ is

$$
\Psi_{\mathrm{Dil}}(x) = e^\alpha = \text{const.}
\tag{14.53}
$$

For inversions, the conformal factor $\Psi_{\mathrm{Inv}}(x)$ is

$$
\Psi_{\mathrm{Inv}}(x) = \frac{1}{x^2}.
\tag{14.54}
$$

Finally, for SCTs, the conformal factor $\Psi_{\mathrm{SCT}}(x)$ is

$$
\Psi_{\mathrm{SCT}}(x) = \frac{1}{1 - 2(c \cdot x) + c^2 x^2}.
\tag{14.55}
$$

We will use these results later when we study how fields transform.

14.2 Conformal Algebra

Orbital Part of Conformal Generators

We now proceed to derive the *conformal algebra* defining the conformal group locally, near the identity. We will also derive the complete field representation of the generators of the conformal group. In this section, we consider the case of four-dimensional Minkowski space $\mathbb{M}_4$. Let us start with the orbital portion of the generators of conformal transformations. The *generator of dilatations* is denoted by D and is parametrized with a real scalar parameter α. The corresponding finite dilatation is generally given as the exponential

$$\exp\left(-\mathrm{i}\,\alpha D\right). \tag{14.56}$$

In order to find the orbital portion of the generator D acting on the spacetime coordinates, we use the generating function

$$\mathrm{X}^\mu = \frac{\partial}{\partial\alpha}\delta x^\mu = x^\mu, \tag{14.57}$$

so that we have the relation

$$\mathrm{i}\,D\phi = x^\mu\partial_\mu\phi, \tag{14.58}$$

from which we can simply read off the desired result,

$$D = -\mathrm{i}\,x^\mu\partial_\mu. \tag{14.59}$$

Further, the *generator of SCTs* is denoted by K_μ and is parametrized with a real four-vector parameter c^μ. The corresponding finite SCT is given as the exponential expression

$$\exp\left(-\mathrm{i}\,c^\mu K_\mu\right). \tag{14.60}$$

In order to find the orbital portion of the generator K_μ, we use the generating function

$$\mathrm{X}^\nu_\mu = \frac{\partial}{\partial c^\mu}\delta x^\nu = 2x_\mu x^\nu - \delta^\nu_\mu x^2, \tag{14.61}$$

so that we obtain the relation

$$\mathrm{i}\,K_\mu\phi = \left(2x_\mu x^\nu - \delta^\nu_\mu x^2\right)\partial_\nu\phi, \tag{14.62}$$

from which the desired formula follows,

$$K_\mu = -\mathrm{i}\left(2x_\mu x^\nu\partial_\nu - x^2\partial_\mu\right). \tag{14.63}$$

Thus, we have the generators of dilatations and SCTs in their orbital representation. The conformal algebra contains also the generators of Lorentz rotations and spacetime translations. In table below we summarize all *orbital generators of conformal transformations*:

$$
\begin{array}{ll}
M_{\mu\nu} = \mathrm{i}\left(x_\mu\partial_\nu - x_\nu\partial_\mu\right), & \text{Lorentz rotations, orbital part} \\[4pt]
P_\mu = -\mathrm{i}\,\partial_\mu, & \text{Translations} \\[4pt]
D = -\mathrm{i}\,x^\mu\partial_\mu, & \text{Dilatations, orbital part} \\[4pt]
K_\mu = -\mathrm{i}\left(2x_\mu x^\nu\partial_\nu - x^2\partial_\mu\right), & \text{SCTs, orbital part}
\end{array}
\tag{14.64}
$$

Commutators of the Conformal Algebra

Although the above representation of the generators is special, we can still use it to derive the universally valid commutator relations of the conformal algebra, much the same way as we did with the Poincaré algebra. Some of the commutator relations of the pure Poincaré part can be taken over unchanged, the rest of the relations can be computed in a straightforward way. (*exercise 14.5*) The full set of *commutator relations of the conformal algebra* is given in the following table:

$$
\begin{aligned}
[M_{\mu\nu}, M_{\rho\sigma}] &= -\mathrm{i}\left(\eta_{\mu\rho}M_{\nu\sigma} - \eta_{\nu\rho}M_{\mu\sigma} + \eta_{\nu\sigma}M_{\mu\rho} - \eta_{\mu\sigma}M_{\nu\rho}\right) \\
[M_{\mu\nu}, P_{\rho}] &= -\mathrm{i}\left(\eta_{\mu\rho}P_{\nu} - \eta_{\nu\rho}P_{\mu}\right) \\
[P_{\mu}, P_{\nu}] &= 0 \\
[M_{\mu\nu}, D] &= 0 \\
[P_{\mu}, D] &= -\mathrm{i}\, P_{\mu} \\
[D, D] &= 0 \\
[M_{\mu\nu}, K_{\rho}] &= -\mathrm{i}\left(\eta_{\mu\rho}K_{\nu} - \eta_{\nu\rho}K_{\mu}\right) \\
[P_{\mu}, K_{\nu}] &= -2\mathrm{i}\left(\eta_{\mu\nu}D + M_{\mu\nu}\right) \\
[D, K_{\mu}] &= -\mathrm{i}\, K_{\mu} \\
[K_{\mu}, K_{\nu}] &= 0
\end{aligned}
\tag{14.65}
$$

The first commutator is the one that fully defines the Lorentz algebra. The first three commutators define the Poincaré algebra. The first six commutators determine the algebra of the group containing the Poincaré transformations and the dilatations. By taking all commutators above into account, the full conformal algebra is defined. Lorentz rotations are the only transformations that do not commute among themselves, while the latter is true for translations, for dilatations, and even for SCTs, despite their complicated algebraic appearance. Note also that the commutator of Lorentz rotations and translations has the same structure as the commutator of Lorentz rotations and SCTs, which again justifies why SCTs are also called *conformal translations*.

Internal Part of Conformal Generators

Now we would like to derive the internal generators for dilatations and SCTs acting on tensor indices. However, the results from Section 9.4 about symmetry transformations of fields cannot be applied immediately, since both $\mathcal{G}_a^{\mathrm{internal}}(x)$ and $\Phi_{i,a}(x)$ are unknown for now. So we turn to the method of induced representations as described in Section 9.5. We saw there that induced representations extend a given representation of a subgroup to a larger group containing the subgroup. The subgroup we are going to consider here is the subgroup of the conformal group that leaves the coordinate origin $x = 0$ unchanged. This group contains all conformal transformations except the spacetime translations, i.e. it contains Lorentz transformations, dilatations, and SCTs. The corresponding generators of this subgroup acting on the tensorial multi-indices i, j, etc. of a tensor field $\phi_i(x)$ are

$$
\begin{aligned}
(M_{\mu\nu}(0))_{ij}\, \phi_j(0) &= (S_{\mu\nu})_{ij}\, \phi_j(0), \\
(D(0))_{ij}\, \phi(0) &= \Delta_{ij}\phi_j(0), \\
(K_{\mu}(0))_{ij}\, \phi_j(0) &= (\kappa_{\mu})_{ij}\, \phi_j(0),
\end{aligned}
\tag{14.66}
$$

where $S_{\mu\nu}$ is the known spin tensor of Lorentz symmetry, while the internal parts of dilatations and SCTs, $D(0) \equiv \Delta$ and $K_{\mu}(0) \equiv \kappa_{\mu}$, are still to be determined. The generators

$S_{\mu\nu}$, Δ and κ_μ of the reduced group must satisfy the following commutator relations:

$$\begin{aligned}
[S_{\mu\nu}, S_{\rho\sigma}] &= -\mathrm{i}\left(\eta_{\mu\rho}S_{\nu\sigma} - \eta_{\nu\rho}S_{\mu\sigma} + \eta_{\nu\sigma}S_{\mu\rho} - \eta_{\mu\sigma}S_{\nu\rho}\right), \\
[S_{\mu\nu}, \Delta] &= 0, \\
[\Delta, \Delta] &= 0, \\
[S_{\mu\nu}, \kappa_\rho] &= -\mathrm{i}\left(\eta_{\mu\rho}\kappa_\nu - \eta_{\nu\rho}\kappa_\mu\right), \\
[\Delta, \kappa_\mu] &= -\mathrm{i}\,\kappa_\mu, \\
[\kappa_\mu, \kappa_\nu] &= 0.
\end{aligned} \tag{14.67}$$

So the task now is to start with the representation of the generators at the origin $x = 0$ and then generalize to an arbitrary point $x = a$. In other words, we will add the translations to the reduced group and thus extend the representation to include these translations. For the sought after generators $D(a)$ and $K_\mu(a)$, the formulae are

$$D(a) = \mathrm{e}^{\mathrm{i}a\cdot P}D(0)\mathrm{e}^{-\mathrm{i}a\cdot P}, \tag{14.68}$$

and

$$K_\mu(a) = \mathrm{e}^{\mathrm{i}a\cdot P}K_\mu(0)\mathrm{e}^{-\mathrm{i}a\cdot P}. \tag{14.69}$$

The products in the above rhs have the form $\mathrm{e}^{-B}A\mathrm{e}^{B}$, so that we can use the formula 8.16 from our study of the exponential map,

$$\mathrm{e}^{-B}A\mathrm{e}^{B} = A + [A, B] + \frac{1}{2!}\left[[A, B], B\right] + \frac{1}{3!}\left[[[A, B], B], B\right] + \dots. \tag{14.70}$$

The algebra of the full group satisfies the commutator relations 14.65 and we will use them now. For the dilatation generator $D(a)$, the application of the BCH formula reveals that only the first two terms contribute, while all other terms vanish due to the commutator among the momentum generators. The result is (*exercise 14.6*)

$$D(a) = D(0) + a^\mu P_\mu. \tag{14.71}$$

If we use x instead of a and write the tensor indices explicitly, we have

$$D_{ij}\phi_j = \left(\Delta + x^\mu P_\mu\right)_{ij}\phi_j, \tag{14.72}$$

where it is $(P_\mu)_{ij} = P_\mu \delta_{ij}$, while the tensorial character of Δ_{ij} will be determined shortly. Observe that the orbital part $x^\mu P_\mu$ of a dilatation appears again due to the action of the translation. For the SCT generator $K_\mu(a)$, only the first three terms of the BCH expansion contribute. Remarkably, once again all higher terms vanish due to the momentum commutators. The first two terms, $A + [A, B]$, constitute the internal part of the SCT generator, whereas the third term, $(1/2!)\left[[A, B], B\right]$, delivers the orbital part. The result is (*exercise 14.7*)

$$K_\mu(a) = \underbrace{K_\mu(0) + 2a_\mu D(0) + 2a^\nu S_{\nu\mu}}_{\text{Internal Part}} + \underbrace{2a_\mu(a \cdot P) - a^2 P_\mu}_{\text{Orbital Part}}. \tag{14.73}$$

If we use the coordinate x instead of a and write the tensor indices explicitly, we obtain

$$\left(K_\mu\right)_{ij}\phi_j = \left(\kappa_\mu + 2x_\mu\Delta - 2x^\nu S_{\mu\nu} + 2x_\mu(x \cdot P) - x^2 P_\mu\right)_{ij}\phi_j, \tag{14.74}$$

where $(S_{\mu\nu})_{ij}$ is the well-known spin tensor and the tensorial character of $(\kappa_\mu)_{ij}$ is determined in the following. We concentrate on the cases where the tensor field $\phi_i(x)$ corresponds to an irreducible representation of the Lorentz group. Due to $[S_{\mu\nu}, \Delta] = 0$, and by using Schur's lemma, see equation 7.30, we conclude that Δ is proportional to the identity. The factor is set according to the scaling of the field. We use

$$\Delta_{ij} = -\mathrm{i}\, d_\phi \delta_{ij}, \tag{14.75}$$

where d_ϕ is a positive real number, called the *scaling dimension* of the field $\phi_i(x)$. We assume that the scaling dimension d_ϕ is the same for all components of $\phi_i(x)$. The scaling dimension can take different values for different tensor fields and essentially describes how the value of a field scales under dilatations. From the commutator $[\Delta, \kappa_\mu] = -\mathrm{i}\,\kappa_\mu$ we can conclude that it is $\kappa_\mu = 0$. Interestingly, the SCT generator K_μ has no own, specific internal part. Its internal part stems only from the Lorentz spin and the scaling dimension. Thus, we have derived the complete generators for dilatations and SCTs containing both the orbital and the internal part. In the following table we summarize the entire set of *generators of conformal transformations in their field representation*:

$$
\begin{array}{|ll|}
\hline
M_{\mu\nu} = \mathrm{i}\,(x_\mu \partial_\nu - x_\nu \partial_\mu) + S_{\mu\nu}\,, & \text{Lorentz rotations} \\[4pt]
P_\mu = -\mathrm{i}\,\partial_\mu\,, & \text{Translations} \\[4pt]
D = -\mathrm{i}\,x^\mu \partial_\mu - \mathrm{i}\,d_\phi\,, & \text{Dilatations} \\[4pt]
K_\mu = -\mathrm{i}\left(2x_\mu x^\nu \partial_\nu - x^2 \partial_\mu\right) - 2\mathrm{i}\,x_\mu d_\phi - 2x^\nu S_{\mu\nu}\,, & \text{SCTs} \\
\hline
\end{array}
\tag{14.76}
$$

A finite conformal transformation, given as an exponential expression of the generators, is then

$$\exp\left(-\frac{\mathrm{i}}{2}\omega^{\mu\nu}M_{\mu\nu} - \mathrm{i}\,a^\mu P_\mu - \mathrm{i}\,\alpha D - \mathrm{i}\,c^\mu K_\mu\right). \tag{14.77}$$

We can now move on to the transformation behavior of physical fields.

14.3   Field Transformations

Jacobian Determinant vs. Conformal Factor

We consider the d-dimensional Minkowski space $\mathbb{M}_d$ here. Let us ask what the relation is between the Jacobian determinant $J(x)$,

$$J(x) \equiv \left|\frac{\partial x'}{\partial x}\right|, \tag{14.78}$$

for a general conformal transformation $x \mapsto x'$, as defined in 14.4, on one hand, and the associated conformal factor $\Psi(x)$, on the other hand. In the case of dilatations, $x' = \omega x$, it is simply

$$\left|\frac{\partial x'}{\partial x}\right| = \omega^d = \text{const}, \tag{14.79}$$

where d is the spacetime dimension. Thus, we obtain the relation

$$\left|\frac{\partial x'}{\partial x}\right| = \Psi^d(x). \tag{14.80}$$

This relation is in fact valid for arbitrary conformal transformations. By taking the determinant of both sides of the definition 14.4, we can readily obtain this general result.

Finite Transformation of Fields

Now let us investigate how finite transformations of fields look like under conformal transformations. As we know, under Lorentz rotations, $x' = \Lambda x$, the field value $\phi_i(x)$ transforms as

$$\phi_i'(x') = \phi_i'(\Lambda x) = \exp\left(-\frac{i}{2}\omega^{\mu\nu}\left(S_{\mu\nu}\right)_{ij}\right)\phi_j(x), \tag{14.81}$$

where $S_{\mu\nu}$ is the spin matrix representation for the field type considered. In the case of a Lorentz-scalar field $\phi(x)$, the spin matrix vanishes and the field transforms simply as $\phi'(x') = \phi(x)$. Under translations, $x' = x + a$, we require for *any* field $\phi_i(x)$ that every component transforms like a scalar,

$$\phi_i'(x') = \phi_i'(x + a) = \phi_i(x). \tag{14.82}$$

To proceed to finite transformations of fields under dilatations and SCTs, let us restrict ourselves to the case of a *scalar field* $\phi(x)$. It is by no means guaranteed that a Lorentz-scalar field will behave under dilatations or SCTs as simply as described above. First, let us look at dilatations given by $x' = e^\alpha x$, where α is a real number. If we insist that the scalar field value $\phi(x)$ transforms in the simplest way as $\phi'(x') = \phi(x)$, then the corresponding theory is scale-invariant only for $d = 2$, as we will show later when we review possible Lagrangians. But we would like to consider scale-invariant theories in $d = 4$ dimensions, or more generally, in d dimensions. According to the principles of how field values transform, we have in the case of dilatations the formula

$$\phi'(x') = \phi'(e^\alpha x) = e^{-i\alpha\Delta}\phi(x) = e^{-\alpha d_\phi}\phi(x). \tag{14.83}$$

In the last equality we have employed

$$\Delta = -i\,d_\phi \tag{14.84}$$

for the internal part Δ of the generator of dilatations. The real number d_ϕ is the *scaling dimension* of the field. Actually, the above transformation is exactly what we would expect for a homogeneous function transforming under a dilatation. By using the scale factor $\Psi(x)$ and the Jacobian determinant $J(x)$, the transformation can be written as

$$\phi'(x') = \Psi(x)^{-d_\phi}\phi(x) = J(x)^{-d_\phi/d}\phi(x). \tag{14.85}$$

Now we elevate this type of transformation to a desired property. By definition, a *primary field* $\phi(x)$ is one for which the total change $\phi'(x')$ is given as

$$\phi'(x') = \left|\frac{\partial x'}{\partial x}\right|^{-d_\phi/d}\phi(x). \tag{14.86}$$

We recognize that this is the behavior of a scalar density $\phi(x)$ with the weight $-d_\phi/d$. By considering scalar fields which transform as densities as in 14.86, we are able to construct field theories with dilatation symmetry.

Let us proceed to finite SCTs, whose Jacobian determinant is generally given by

$$\left|\frac{\partial x'}{\partial x}\right| = \frac{1}{(1 - 2(c \cdot x) + c^2 x^2)^d}. \tag{14.87}$$

In the special case where we have a *primary scalar field* $\phi(x)$ at hand, its total transformation under a SCT is given by

$$\phi'(x') = \left(1 - 2(c \cdot x) + c^2 x^2\right)^{d_\phi}\phi(x). \tag{14.88}$$

We observe that the SCTs represent conformal transformations for which the multiplying factor in the transformation formula is coordinate-dependent. In addition, we should emphasize that the most general finite transformation of a scalar field under a SCT is given by the formula

$$\phi'(x') = \exp\left(-i\, c^\mu K_\mu\right)\phi(x),\tag{14.89}$$

where

$$K_\mu = -2i\, x_\mu d_\phi - 2x^\nu S_{\mu\nu}\tag{14.90}$$

stands for the internal part of the generator of the SCT. We recognize that a primary scalar field transforms in a simpler way under a SCT compared to the general transformation rule.

After the above technical results let us express the main point. A conformal transformation of spacetime will induce a change in the fields $\phi_i(x)$, which we expect to transform as in 14.86. If we want the field theory to be *conformally invariant*, we have the scaling dimensions d_{ϕ_i} at hand, which we can tune it in such a way that the Lagrangian remains invariant. In Section 16.4 we will develop such examples of conformally invariant theories. But already here a major property should be pointed out, which is common to all conformally invariant theories.

Scale Invariance of Massless Theories
The rest mass of a particle is not invariant under scale transformations. Indeed, if we consider the momentum square, $P^2 = P_\mu P^\mu = m^2 I$, as the defining quantity for the rest mass of a particle, we see that under a scale transformation D it transforms as

$$e^{i\alpha D}P^2 e^{-i\alpha D} = e^{-2\alpha}P^2,\tag{14.91}$$

where we have used again the formula 8.16. If we require our relativistic theory to be additionally scale invariant, with the dilatation parameter taking all real values and not being restricted to be discrete, then either all masses have to be zero, or the mass spectrum has to be continuous. As we know from particle physics, the rest masses of elementary particles have certain, discrete values, so the mass spectrum is not continuous. So we are left with the requirement that all rest masses must vanish. On the other hand, the limit of zero masses is reached if we consider physics at very high energies, when particle masses can be neglected. Then, scale invariance is reached approximately. We can summarize that scale symmetry and conformal symmetry are not satisfied by theories with non-zero masses. Or, in other words, that a necessary condition for conformal symmetry is that all rest masses vanish.

14.4 Linearization of the Conformal Group

Six-Dimensional Coordinates and $SO(2,4)$
To be definite, we consider again the four-dimensional case $\mathbb{M}_4$. Let us restrict ourselves to the subgroup of the conformal group which does not contain the reflections and denote it by $C_+^\uparrow(1,3)$, or simply by $C_+^\uparrow$, and call it the *restricted conformal group*. As we have seen, conformal transformations are in general nonlinear. We can now ask whether it is possible to express the restricted conformal group in such a way that it consists of linear transformations. The answer is affirmative, and actually there are several ways to achieve this. First, we sketch a method how to linearize the restricted conformal group by considering its action on a higher-dimensional space. We work first on the level of the Lie algebra and construct new suited linear combinations of the known generators. Instead of using a 4-dimensional representation acting on the usual Minkowski space, we let the new generators

act on a 6-dimensional pseudo-Euclidean space. Let us proceed with this construction. Define the new generators, denoted by $\mathbf{M}_{AB}$, with the indices* $A, B = 0, 1, 2, 3, 4, 5$, as

$$
\left.\begin{aligned}
M_{\mu\nu} &= \mathbf{M}_{\mu\nu} \\
P_\mu &= \mathbf{M}_{4\mu} + \mathbf{M}_{5\mu} \\
K_\mu &= \mathbf{M}_{4\mu} - \mathbf{M}_{5\mu} \\
D &= \mathbf{M}_{45}
\end{aligned}\right\}, \tag{14.92}
$$

with $\mu, \nu = 0, 1, 2, 3$. The new 6-dimensional generators $\mathbf{M}_{AB}$ are defined to be antisymmetric, $\mathbf{M}_{AB} = -\mathbf{M}_{BA}$, and are viewed in a representation where they act on the pseudo-Euclidean base space $\mathbb{E}^{(2,4)}$ with the signature

$$
\eta_{AB} = \text{diag}(1, -1, -1, -1, -1, 1). \tag{14.93}
$$

In matrix form, the new generators can be written as

$$
\mathbf{M}_{AB} = \begin{pmatrix}
M_{\mu\nu} & -\frac{1}{2}(P_\mu + K_\mu) & -\frac{1}{2}(P_\mu - K_\mu) \\
\frac{1}{2}(P_\mu + K_\mu) & 0 & D \\
\frac{1}{2}(P_\mu - K_\mu) & -D & 0
\end{pmatrix}. \tag{14.94}
$$

A direct calculation shows that the commutator relations 14.65 of the conformal algebra, when translated for the new generators, lead to a single unified relation,

$$
[\mathbf{M}_{AB}, \mathbf{M}_{CD}] = -i(\eta_{AC}\mathbf{M}_{BD} - \eta_{BC}\mathbf{M}_{AD} + \eta_{BD}\mathbf{M}_{AC} - \eta_{AD}\mathbf{M}_{BC}), \tag{14.95}
$$

which in turn is nothing else than the commutator relation for the pseudo-orthogonal group $SO(2, 4)$. (*exercise 14.8*) We conclude: *the restricted conformal group* $C_+^\uparrow(1, 3)$ *is locally isomorphic to* $SO(2, 4)$, *which is exactly the pseudo-orthogonal group acting on* $\mathbb{E}^{(2,4)}$. Globally, the group $SO(2, 4)$ is the *double cover* of $C_+^\uparrow(1, 3)$, although we will not prove this here. The group $SO(2, 4)$ acts, as desired, linearly on the base space $\mathbb{E}^{(2,4)}$. When we restrict $SO(2, 4)$ to act on the lower-dimensional base space $\mathbb{E}^{(1,3)} = \mathbb{M}_4$, then the conformal transformations are forced to become nonlinear. The number of parameters needed to specify a transformation in $SO(2, 4)$ is equal 15, exactly as for the conformal group. In order to understand how $SO(2, 4)$ acts on the 6-dimensional space $\mathbb{E}^{(2,4)}$, we are going to construct the 6-dimensional coordinates explicitly. We denote the 6-dimensional base space coordinates by y^A, with $A = 0, 1, 2, 3, 4, 5$, or in column vector form as

$$
y^A = \begin{pmatrix} y^\mu \\ y^4 \\ y^5 \end{pmatrix}. \tag{14.96}
$$

A coordinate transformation under the group $SO(2, 4)$ is simply

$$
y'^A = \mathbf{M}^A{}_B y^B, \tag{14.97}
$$

with $\mathbf{M}^A{}_B = \eta^{AC}\mathbf{M}_{CB}$. The generators satisfy the defining equations

$$
\mathbf{M}^A{}_C \mathbf{M}^B{}_D \eta_{AB} = \eta_{CD} \quad \text{and} \quad \det \mathbf{M} = 1. \tag{14.98}
$$

---

*The indices here should not be confused with spinor indices.

The 6-dimensional cone, which is the hypersurface defined by the condition

$$\eta_{AB} y^A y^B = 0, \tag{14.99}$$

remains preserved under the $SO(2,4)$ transformations and this constraint reduces the freedom of the y^A coordinates. In addition to the four coordinates x^μ, we need two additional coordinate variables, which we denote by κ and λ and define through

$$\left. \begin{array}{rcl} y^\mu & \equiv & \kappa x^\mu \\ \lambda & \equiv & \kappa x^2 \end{array} \right\}. \tag{14.100}$$

We can interpret the additional coordinates κ and λ as parameters describing the scale and the local change of scale in our new base space. From the invariance of the cone 14.99, we obtain

$$\kappa \lambda - \left(y^4 + y^5 \right) \left(y^4 - y^5 \right) = 0. \tag{14.101}$$

This motivates us to make the identification

$$\left. \begin{array}{rcl} \kappa & = & y^4 + y^5 \\ \lambda & = & y^4 - y^5 \end{array} \right\}, \tag{14.102}$$

which is, however, not unique. Hence, the new 6-dimensional coordinate vector has the form

$$y^A = \begin{pmatrix} \kappa x^\mu \\ \frac{\kappa}{2} \left(1 + x^2 \right) \\ \frac{\kappa}{2} \left(1 - x^2 \right) \end{pmatrix}, \tag{14.103}$$

which shows the dependence on x^μ and κ. Equally well, we could express the vector through x^μ and λ. Now we are ready to see how all the different transformations of the restricted conformal group become linear transformations of the 6-dimensional coordinates. The basic relation we use is the second equation in 14.100, which defines the ratio of κ and λ. Then the single linear transformations described below are easily derived. (*exercise 14.9*)

Transformations with $SO(2,4)$

Let us write down how the 6-dimensional coordinates transform under the various transformations of the conformal group. Under finite Lorentz rotations, the coordinates transform as

$$\left. \begin{array}{rcl} y'^\mu & = & \Lambda^\mu{}_\nu y^\nu \\ \kappa' & = & \kappa \\ \lambda' & = & \lambda \end{array} \right\}. \tag{14.104}$$

Under translations, the coordinates transform as

$$\left. \begin{array}{rcl} y'^\mu & = & y^\mu + \kappa a^\mu \\ \kappa' & = & \kappa \\ \lambda' & = & \lambda + 2(a \cdot y) + \kappa a^2 \end{array} \right\}. \tag{14.105}$$

Under finite dilatations, the coordinates transform as

$$\left. \begin{array}{rcl} y'^\mu & = & y^\mu \\ \kappa' & = & e^{-\alpha} \kappa \\ \lambda' & = & e^\alpha \lambda \end{array} \right\}. \tag{14.106}$$

Under finite conformal translations, the coordinates transform according to

$$\left. \begin{array}{rcl} y'^\mu & = & y^\mu - \lambda c^\mu \\ \kappa' & = & \kappa - 2(c \cdot y) + \lambda c^2 \\ \lambda' & = & \lambda \end{array} \right\}, \tag{14.107}$$

which is a linear transformation, as desired initially. The 6-dimensional coordinates lead to remarkably similar formulae for the (usual) translations and the conformal translations.

More generally, in the case of $d = 1 + D$ spacetime dimensions, the restricted conformal group $C_+^\uparrow(1, D)$ finds a similar representation with the group $SO(2, d)$. One obtains again the result that the conformal group in d spacetime dimensions has $(d+1)(d+2)/2$ degrees of freedom.

Twistors and $SU(2, 2)$

The previous construction of a linearized conformal group may appear somewhat ad hoc. There is another more elegant method to represent conformal transformations in a linear fashion developed by *Sir Roger Penrose* in 1967, see [66]. The original goal of Penrose was the development of a natural formalism for conformally invariant spaces and the usage of the set of complex numbers. The formalism of Penrose extends the notion of a spinor representation of the Lorentz group to a *twistor representation* of the conformal group. Let us describe the analogy. Remember that the rotation group $R \cong SO(3)$ has a 2-complex-component spinor representation as double cover,

$$R \cong SO(3) \cong SU(2). \tag{14.108}$$

The restricted Lorentz group $L_+^\uparrow \cong SO(1, 3)$ has a 2-complex-component Weyl spinor representation as double cover,

$$L_+^\uparrow \cong SO(1, 3) \cong SL(2, \mathbb{C}). \tag{14.109}$$

The extension of this chain is that the restricted conformal group $C_+^\uparrow$ has a double cover by the group $SO(2, 4)$, which in turn has a double cover by the group $SU(2, 2)$, the latter acting on a 4-complex-dimensional space,

$$C_+^\uparrow \cong SO(2, 4) \cong SU(2, 2). \tag{14.110}$$

Thus, the conformal group acts linearly on the 4-complex-dimensional coordinate space $\mathbb{C}^4$ endowed with a metric of signature $(-, -, +, +)$, the *twistor coordinate space*. The group $SU(2, 2)$ is the 4-fold cover of $C_+^\uparrow$, so the four transformations $\pm A$, $\pm iA$ of $SU(2, 2)$ lead to the same conformal transformation on Minkowski space $\mathbb{M}_4$. The conformal algebra in four dimensions has 15 generators and in the representation as $su(2, 2)$ these generators are complex-valued 4×4-matrices. We can choose these generators to be a linear combinations of our known basis consisting of the 15 elements $\{\gamma^\mu, \gamma^{\mu\nu}, \gamma^5, \gamma^\mu\gamma^5\}$, see Section 12.5. For an explicit construction of twistors, the interested reader should consult the article [66] of Penrose. It should be noted here that twistors have provided powerful methods for solving field equations of massless theories, and at the same time are considered an important mathematical tool in the development of a quantum theory of gravity.

Further Reading

In this chapter we have developed the main results on conformal transformations in dimensions $d \geq 3$. However, we have not treated the two-dimensional conformal group and its applications in quantum field theory. For this topic, the interested reader should consult the monograph of Di Francesco, Mathieu, and Senechal [24]. Realizations of the linear action of the conformal group on six-dimensional coordinates are described, for example, in Kosyakov [49]. The introduction of the notion of twistor space can be found in the original article [66] of Penrose. A gentle introduction to twistors and their role in geometry and field theory is provided by Ward and Wells [92].

IV

Classical Fields

15

Lagrangians and Noether's Theorem

In this chapter we introduce and develop field theory as a powerful framework for building relativistically invariant theories. The focus is entirely on Lagrangian methods. We start with Hamilton's principle of stationary action and derive the Euler-Lagrange equations as the dynamical field equations. We illustrate the methods by providing basic examples of scalar, spinor, and vector fields, with the Maxwell field representing the most important classical case. Then we turn to the formulation of symmetry in a relativistic field theory. This is accomplished by the famous Noether theorem, which we derive and discuss.

15.1 Introducing Fields

Coulomb's Law and Faraday's Notion of Fields

The *electrostatic force* $\boldsymbol{F}_{12}$ on a motionless and electrically charged body carrying the *electric charge* q_1, as generated by another electrically charged body q_2, is described by *Coulomb's law* (*Charles Augustine de Coulomb*),

$$\boxed{\boldsymbol{F}_{12} = \frac{q_1 q_2}{|\boldsymbol{x}_1 - \boldsymbol{x}_2|^2}\, \boldsymbol{e}_{12}} \tag{15.1}$$

where we use the unit vector

$$\boldsymbol{e}_{12} \equiv \frac{\boldsymbol{x}_1 - \boldsymbol{x}_2}{|\boldsymbol{x}_1 - \boldsymbol{x}_2|}. \tag{15.2}$$

It is an inverse square law, exactly as Newton's law of gravitation. Two bodies possessing electric charges of the same sign repel each other, whereas two bodies with electric charges of opposite sign attract each other. In this book, we use the system of *Gaussian units* for formulating all electromagnetic equations, the reader should turn to appendix A.1 for a description of these units. Coulomb's law is, like Newton's law, a force law modeling an instantaneous interaction between the charged bodies. It does not describe what may exist or happen between the two charged bodies. In contrast, in *field theory* the interaction between the two charged bodies is attributed to an *electric field*, which is generated by the first charged body and which influences the second charged body. In the same manner, the

DOI: 10.1201/9781003087748-15

second charged body generates an electric field, which in turn influences the first charged body. In this concept, *an electrically charged body acts as a source of the electric field* and the electric field in turn influences a charged body *only locally*. This shifted point of view, away from the charged body and toward the field and its specific actions was introduced by *Michael Faraday*. By using the electric field three-vector $\boldsymbol{E}(\boldsymbol{x})$, the Coulomb force $\boldsymbol{F}_{12}$ on the charged particle q_1 is

$$\boldsymbol{F}_{12} = q_1 \boldsymbol{E}(\boldsymbol{x}_1), \tag{15.3}$$

with the expression

$$\boldsymbol{E}(\boldsymbol{x}_1) = \frac{q_2}{|\boldsymbol{x}_1 - \boldsymbol{x}_2|^2}\, \boldsymbol{e}_{12} \tag{15.4}$$

for the electric field generated by the charge q_2. Typically, the electric field $\boldsymbol{E}(\boldsymbol{x})$ is thought to be defined over all space $\boldsymbol{x} \in \mathbb{E}^3$. In a more complex setup of electrically charged bodies, where one has not just Coulomb point charges, the electric field $\boldsymbol{E}(\boldsymbol{x})$ is defined at each point $\boldsymbol{x}$ in space as the quotient

$$\boldsymbol{E}(\boldsymbol{x}) \equiv \lim_{q \to 0} \frac{\boldsymbol{F}}{q}, \tag{15.5}$$

where $\boldsymbol{F}$ is the electrostatic force which the *test charge* q experiences when it is located at the point $\boldsymbol{x}$, see figure 15.1. The limit of a very small test charge is needed in order to avoid the effects of the additional electric field generated by this test charge. The electric field has the property of being additive, i.e. the total electric field of two charge distributions is the sum of the single electric fields of these charge distributions.

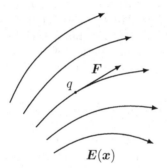

Figure 15.1: Electric field measured by electrostatic forces on a test charge

A physical field contains energy and momentum. The Coulomb field generated by N discrete point charges q_n, $n = 1, \ldots, N$, has the *potential energy* V given by

$$V = \sum_{\substack{n,m=1 \\ n<m}}^{N} \frac{q_n q_m}{|\boldsymbol{x}_n - \boldsymbol{x}_m|}. \tag{15.6}$$

The above sum runs over all distinct pairs of charges. This electrostatic potential energy is completely analogous to the potential energy of the the Newtonian gravitational field. Furthermore, we can introduce the *electric potential* $\varphi(\boldsymbol{x})$ generated by a set of electric point charges by

$$\varphi(\boldsymbol{x}) = \sum_{m} \frac{q_m}{|\boldsymbol{x} - \boldsymbol{x}_m|}. \tag{15.7}$$

So we see that fields, like the electric field here, possess mechanical properties like energy. Conversely, a discrete mechanical system can be described by using the concept of a field, as we show in the following.

Particle System and its Energy-Momentum Tensor

Before proceeding to a general discussion of fields, we would like to show that a discrete set of relativistic particles can be treated as a field. We consider the case of a set of N relativistic point particles with masses m_n and trajectories $z_n^\mu(s_n)$, $n = 1, \ldots, N$. Each particle measures its own proper time s_n. The particle set can be well described by the mass current density, which we have encountered in Section 6.3. However, it is possible to use a different quantity in order to describe the flow of energy and momentum through spacetime. We define the (*mechanical*) *energy-momentum tensor* $T^{\mu\nu}(x)$ as the field

$$T^{\mu\nu}(x) = \sum_n \int_{-\infty}^{\infty} m_n \frac{\mathrm{d}z_n^\mu(s_n)}{\mathrm{d}s_n} \frac{\mathrm{d}z_n^\nu(s_n)}{\mathrm{d}s_n} \delta\left(x - z_n(s_n)\right) \mathrm{d}s_n. \tag{15.8}$$

The meaning of the energy-momentum tensor will become clear shortly. First, one can see immediately that the energy-momentum tensor of the particle system is symmetric,

$$T^{\mu\nu} = T^{\nu\mu}. \tag{15.9}$$

We note this as one of the crucial properties of the energy-momentum tensor. For the trace $T^\mu{}_\mu$ it is

$$T^\mu{}_\mu = \sum_n m_n. \tag{15.10}$$

I.e. the trace vanishes exactly when all particles have vanishing masses. $T^{\mu\nu}(x)$ describes the flow of energy and momentum through spacetime. For the four-divergence of the energy-momentum tensor we can calculate (*exercise 15.1*)

$$\partial_\mu T^{\mu\nu}(x) = \sum_n \int_{-\infty}^{\infty} m_n \frac{\mathrm{d}^2 z_n^\nu(s_n)}{\mathrm{d}s_n^2} \delta\left(x - z_n(s_n)\right) \mathrm{d}s_n. \tag{15.11}$$

Hence, for a system of free, non-interacting particles, the local conservation equation

$$\partial_\mu T^{\mu\nu} = 0 \tag{15.12}$$

holds. For a system in which the particles interact with an external field, the mechanical energy-momentum tensor is not conserved. Instead, only the total energy-momentum tensor, being comprised of the mechanical part and the field part, is locally conserved. We discuss this in Section 16.2 for the electromagnetic field. The energy-momentum tensor has 16 components, out of which 10 are unique. To understand their physical meaning, let us inspect them in detail for the simple case of a single particle with mass m. By writing the particle's four-velocity $u^\mu(s)$ in components as

$$u^\mu(s) = \begin{pmatrix} \gamma\left(u(s)\right) \\ \gamma\left(u(s)\right) \boldsymbol{v}(s) \end{pmatrix}, \tag{15.13}$$

the expression for the energy-momentum tensor

$$T^{\mu\nu}(x) = \int_{-\infty}^{\infty} m\, u^\mu(s)\, u^\nu(s)\, \delta^4(x - z(s))\, \mathrm{d}s \tag{15.14}$$

can be integrated out with respect to the variable $\mathrm{d}z^0(s) = u^0(s)\,\mathrm{d}s$ and yields

$$T^{\mu\nu}(x) = m \frac{u^\mu(s) u^\nu(s)}{u^0(s)} \delta^3(\boldsymbol{x} - \boldsymbol{z}(s)). \tag{15.15}$$

Let us look at each component of $T^{\mu\nu}(x)$ separately. The $(0,0)$ component is

$$T^{00}(x) = \gamma(s)\, m\, \delta^3(\boldsymbol{x} - \boldsymbol{z}(s)), \tag{15.16}$$

which is the relativistic energy density at the point where the particle is positioned. The $(0, j)$ and $(j, 0)$ components are

$$T^{0j}(x) = T^{j0}(x) = \gamma(s)\, m\, v^j(s)\, \delta^3(\boldsymbol{x} - \boldsymbol{z}(s)), \tag{15.17}$$

representing the relativistic three-momentum density, or equivalently, the relativistic energy flux density (= energy per time per perpendicular surface element). Finally, the (j, k) components are

$$T^{jk}(x) = T^{kj}(x) = \gamma(s)\, m\, v^j(s)\, v^k(s)\, \delta^3(\boldsymbol{x} - \boldsymbol{z}(s)), \tag{15.18}$$

representing the jth momentum flux density in the kth direction. In the case of an N-particle system, one uses the sum over all particles but the interpretation of the single components of $T^{\mu\nu}(x)$ remains the same. Written as a matrix, the energy-momentum tensor has the components

$$(T^{\mu\nu}) = \begin{pmatrix} \begin{array}{c|c} \begin{array}{c} T^{00} = \\ \text{energy density} \end{array} & \begin{array}{c} T^{0j} = \\ \text{energy flux density} \end{array} \\ \hline \begin{array}{c} T^{j0} = \\ \text{momentum} \\ \text{density} \end{array} & \begin{array}{c} T^{jk} = \\ \text{momentum} \\ \text{flux density} \end{array} \end{array} \end{pmatrix}. \tag{15.19}$$

The above representation of the energy-momentum tensor was first proposed by Minkowski. The mechanical energy-momentum tensor, with its basic properties, is a raw model for other types of energy-momentum tensors, which we will encounter later. Generally, a physical field possesses energy and momentum, and the most efficient way to capture these quantities in a unified concept is through the notion of the energy-momentum tensor.

15.2 Action Principle for Fields

Hamilton's Principle and Euler-Lagrange Field Equations

A fundamental way to introduce physical fields is to consider them as *field representations of the Poincaré group*. Then, the corresponding tensor field functions automatically respect the Poincaré symmetry, which we want to establish as the most basic spacetime symmetry of our physical theory. The remaining task is then to construct the dynamics while using the field functions as basic elements. This is achieved in the most effective way by applying the Lagrangian formalism.

In Lagrangian mechanics, we use generalized coordinates $q = (q^1, \ldots, q^n)$ and generalized velocities $\dot{q} = (\dot{q}^1, \ldots, \dot{q}^n)$, where the evolution of the physical system is described by a time parameter. In contrast, in the case of Lagrangian field theory, we use fields of the form $\phi_i(x)$ defined on the d-dimensional* Minkowski space $\mathbb{M}_d \equiv (\mathbb{E}^d, \eta)$. The index i on the field

---

*We work as far as possible in general d dimensions, although physically we are foremost interested in the case $d = 4$.

ction $\phi_i(x)$ is a multi-index describing the specific tensorial behavior of the field under
imensional Poincaré transformations. The continuous coordinate variable x^μ plays the
e of what the discrete index $k = 1, \ldots, n$ played in mechanics. In mechanics, one has a
te number of degrees of freedom, whereas in field theory one has a continuously infinite
nber of degrees of freedom. This makes field theory mathematically more demanding but
o very flexible as a framework for describing physical systems.

Each field theory has its own specific *Lagrangian density*, or simply *Lagrangian*, denoted
$\mathscr{L}$, which is a function of the form

$$\mathscr{L} = \mathscr{L}(\phi_i, \partial_\mu \phi_i). \tag{15.20}$$

simplicity, we will leave out the case of an explicit dependence of the Lagrangian density
the coordinate variable. Also, we will exclude a dependence on field derivatives higher
n first order. In analogy to mechanics, the most interesting cases contain a *kinetic term*
1 a *potential term* and we will encounter examples soon. The *action* in field theory is the
ctional $S[\phi_i]$ of the fields $\phi_i(x)$ defined as

$$S[\phi_i] \equiv \int_U \mathscr{L}(\phi_i(x), \partial_\mu \phi_i(x)) \, \mathrm{d}^d x \tag{15.21}$$

ere the integral contains the *Minkowski space volume form* $\mathrm{d}^d x$, which is given as

$$\mathrm{d}^d x = \mathrm{d}x^0 \wedge \mathrm{d}x^1 \wedge \cdots \wedge \mathrm{d}x^D. \tag{15.22}$$

e domain of integration U is either the entire Minkowski space or a suitable finite domain
hin it. We want to have the simplest transformation behavior of the action $S[\cdot]$ under
incaré transformations, so it should be a scalar. Consequently, the Lagrangian density
·) must be a scalar density. Under arbitrary coordinate transformations $x \mapsto x'(x)$, the
ume measure $\mathrm{d}^d x$ transforms as

$$\mathrm{d}^d x' = J(x) \, \mathrm{d}^d x, \tag{15.23}$$

ere $J(x)$ is the Jacobian determinant. Thus, the Lagrangian density must transform like

$$\mathscr{L}\left(\phi_i'(x'), \partial_\mu' \phi_i'(x')\right) = J(x)^{-1} \mathscr{L}(\phi_i(x), \partial_\mu \phi_i(x)) \tag{15.24}$$

the action $S[\cdot]$ to be a proper scalar. In order to define the dynamics of field theory, we
oke the *action principle*. We let the fields vary in the form

$$\phi_i(x) \mapsto \phi_i(x) + \delta\phi_i(x), \tag{15.25}$$

h the additional condition that the variations $\delta\phi_i(x)$ vanish at infinity, or at the boundary
of the integration domain U. Note that there is no variation of the coordinates here.
e action principle in field theory states that under such variations of the fields, the action
ys invariant. This is written as

$$\delta S[\phi_i] = 0 \tag{15.26}$$

in the form of a functional differential equation, see also appendix B.6,

$$\frac{\delta S[\phi_i]}{\delta \phi_i(x)} = 0. \tag{15.27}$$

These equations select the correct field configurations fulfilling the action principle. Let us derive the corresponding equations of motion. The variation of the action $\delta S\,[\phi_i]$ is

$$\delta S = \delta \int_U \mathscr{L}\, \mathrm{d}^d x = \int_U (\delta \mathscr{L})\, \mathrm{d}^d x. \tag{15.28}$$

For the variation $\delta \mathscr{L}$ of the Lagrangian density, we calculate that

$$\delta \mathscr{L} = \frac{\partial \mathscr{L}}{\partial \phi_i}\delta\phi_i + \frac{\partial \mathscr{L}}{\partial(\partial_\mu \phi_i)}\delta(\partial_\mu \phi_i), \tag{15.29}$$

or, written differently,

$$\delta \mathscr{L} = \left(\frac{\partial \mathscr{L}}{\partial \phi_i} - \partial_\mu \frac{\partial \mathscr{L}}{\partial(\partial_\mu \phi_i)} \right)\delta\phi_i + \partial_\mu \left(\frac{\partial \mathscr{L}}{\partial(\partial_\mu \phi_i)}\delta\phi_i \right). \tag{15.30}$$

The last term on the rhs is a divergence, which under the volume integral becomes a surface integral over the boundary, or at infinity. Because the variations $\delta\phi_i(x)$ vanish there, the integral vanishes too. Thus, the action principle leads to the *Euler-Lagrange equations* in field theory:

$$\boxed{\frac{\partial \mathscr{L}}{\partial \phi_i} - \partial_\mu \frac{\partial \mathscr{L}}{\partial(\partial_\mu \phi_i)} = 0} \tag{15.31}$$

These are the sought after *field equations*. A condition or an equation that is valid only if the field equations are satisfied is said to be valid *on-shell*. We emphasize the important fact that a Lagrangian leading to certain equations of motion is not unique. Adding a divergence term to the Lagrangian leaves the resulting field equations invariant. Indeed, replacing the original Lagrangian $\mathscr{L}$ by the equivalent Lagrangian $\overline{\mathscr{L}}$, given by

$$\overline{\mathscr{L}}\,(\phi_i, \partial_\mu \phi_i) = \mathscr{L}\,(\phi_i, \partial_\mu \phi_i) + \partial_\mu M^\mu\,(\phi_i), \tag{15.32}$$

does not alter the equations of motion. The four-vector quantity $M^\mu(\phi_i)$ must depend only on the fields but not on their derivatives and in addition it must vanish on the boundary of the spacetime integration domain.

Let us remark that we will be making use of the definition

$$\Pi_i^\mu \equiv \frac{\partial \mathscr{L}}{\partial(\partial_\mu \phi_i)} \tag{15.33}$$

for the so-called *conjugate momentum* $\Pi_i^\mu(x)$ of the field $\phi_i(x)$. The conjugate momentum is a formal quantity within the canonical formalism. We will discuss about the actual momentum contained in a field in the next chapter. In full analogy to mechanics, we consider the *Hamiltonian density* $\mathscr{H}(x)$ in field theory, which is defined as

$$\mathscr{H} \equiv \Pi_i^0(\partial_0 \phi_i) - \mathscr{L}. \tag{15.34}$$

Note that there is a summation over the repeated field multi-index. In most physically relevant field theories, the Hamiltonian density coincides with the physical *energy density* of the field at hand. By integrating over the entire 3-space $\mathbb{E}^3$,

$$E(t) = \int_{\mathbb{E}^3} \mathscr{H}(\boldsymbol{x}, t)\, \mathrm{d}^3 x, \tag{15.35}$$

one obtains the *total energy* $E(t)$ of the field. After these general constructions, we are now ready to consider concrete examples of field theories in four spacetime dimensions.

5.3   Scalar Fields

al Scalar Field

e *real scalar field* $\phi(x)$ is the simplest field with respect to the tensorial behavior. There different examples of scalar field theories described by different Lagrangians. We would to note that scalar field theories become particularly relevant in the quantum regime, when they are coupled to gravity. But already on the classical level scalar fields contain ery rich structure.

The most basic scalar field theory is described by the *Klein-Gordon Lagrangian* (*Oskar Klein* and *Walter Gordon*),

$$\mathcal{L} = \frac{1}{2}\partial_\mu\phi\,\partial^\mu\phi - \frac{m^2}{2}\phi^2 \tag{15.36}$$

ich leads to the *Klein-Gordon equation*

$$(\partial^2 + m^2)\phi = 0 \tag{15.37}$$

is is exactly the field equation we have derived in Section 13.3 with group-theoretical siderations. Originally, Klein and Gordon introduced this equation in 1926 as a relati-tic wave equation for describing electrons but it turned out that the electron spin had l to be incorporated, which was done later by Dirac. The Klein-Gordon equation de-ibes a spinless and electrically neutral free particle with the rest mass m. In fact, the gss particle, whose existence was experimentally confirmed in 2012, is currently the only own elementary particle being described by a real scalar field. A simple solution of the in-Gordon equation is a plane wave solution

$$\phi(x) = \exp(-ip_\mu x^\mu), \tag{15.38}$$

vided the momentum p_μ obeys the dispersion relation $p_\mu p^\mu = m^2$. The dispersion relation es the energy E of the plane wave solution as

$$E = \sqrt{\boldsymbol{p}^2 + m^2}. \tag{15.39}$$

m the above relation, we obtain the group velocity $\boldsymbol{v}$ of the plane wave as

$$\boldsymbol{v} = \frac{\partial E}{\partial \boldsymbol{p}} = \frac{\boldsymbol{p}}{\sqrt{\boldsymbol{p}^2 + m^2}}. \tag{15.40}$$

other simple solution is a static, spherically symmetric solution. In that case, the scalar d can be described by a function $\phi(r)$ of the radial distance $r = |\boldsymbol{x}|$. By writing the place operator ∇^2 in spherical coordinates

$$\nabla^2\phi(\boldsymbol{x}) = \frac{1}{r}\frac{\mathrm{d}^2}{\mathrm{d}r^2}(r\phi(r)), \tag{15.41}$$

Klein-Gordon equation reduces to the ODE

$$\frac{\mathrm{d}^2}{\mathrm{d}r^2}(r\phi(r)) = m^2 r\phi(r). \tag{15.42}$$

e physical solution reads

$$\phi(r) = \frac{A}{r}\exp(-\alpha mr). \tag{15.43}$$

This $\phi(r)$ is the famous *Yukawa potential* (*Hideki Yukawa*). The overall factor A defines the strength of the potential, while the positive parameter α determines the exponential damping. Obviously, the Yukawa potential contains the Coulomb potential as a special case when the mass vanishes, $m = 0$. Yukawa introduced this potential in 1935 to describe the strong nuclear force between protons and neutrons in the nucleus.

An extension of the Klein-Gordon theory is the ϕ^4-*theory* with the Lagrangian

$$\mathscr{L} = \frac{1}{2}\partial_\mu\phi\,\partial^\mu\phi - \frac{m^2}{2}\phi^2 - \frac{\lambda}{4!}\phi^4, \tag{15.44}$$

where the real parameter λ is assumed to be positive. The corresponding field equation is the nonlinear equation

$$(\partial^2 + m^2)\phi = -\frac{\lambda}{3!}\phi^3. \tag{15.45}$$

In the quantized theory, this describes a spinless and electrically neutral particle with self-interaction. One of the interesting features of the ϕ^4-theory is that its massless version is conformally invariant in four dimensions, a fact which we prove in Section 16.4.

Complex Scalar Fields

We start with two distinct real scalar fields $\phi_1(x)$ and $\phi_2(x)$, where each one is described by the Klein-Gordon Lagrangian 15.36 and define a *complex scalar field* $\phi(x)$ and its *complex conjugate* $\phi^*(x)$ by

$$\phi \equiv \frac{1}{\sqrt{2}}(\phi_1 + i\phi_2) \quad \text{and} \quad \phi^* \equiv \frac{1}{\sqrt{2}}(\phi_1 - i\phi_2). \tag{15.46}$$

One can deal with the two independent real fields ϕ_1 and ϕ_2, or, one can regard ϕ and its complex conjugate ϕ^* as the two independent fields. The sum of the two real Klein-Gordon Lagrangians is equal to the total Lagrangian of the complex scalar fields ϕ and ϕ^*,

$$\boxed{\mathscr{L} = (\partial_\mu\phi)^*(\partial^\mu\phi) - m^2\phi^*\phi} \tag{15.47}$$

This Lagrangian has the property of being real-valued and leads to the two field equations

$$\boxed{(\partial^2 + m^2)\phi = 0 \quad \text{and} \quad (\partial^2 + m^2)\phi^* = 0} \tag{15.48}$$

which are considered as independent equations. As we will see later, the complex scalar field describes two electrically charged particles with opposite electric charges.

The Lagrangian for complex scalar fields with a ϕ^4-type self-interaction has the form

$$\mathscr{L} = (\partial_\mu\phi)^*(\partial^\mu\phi) - m^2\phi^*\phi - \mu\,(\phi^*\phi)^2, \tag{15.49}$$

with μ being a real parameter. This Lagrangian leads to the nonlinear field equation

$$(\partial^2 + m^2)\phi = -2\mu\phi^*\phi^2 \tag{15.50}$$

and its complex conjugate.

Newtonian Gravitational Field

As an example of a non-relativistic field, we can consider the *Newtonian gravitational potential* $\Phi(\boldsymbol{x})$ generated by a mass density $\rho(\boldsymbol{x})$ in 3-space. By employing the possible (but not unique) Lagrangian

$$\mathscr{L} = -\frac{1}{8\pi G_N}\left(\boldsymbol{\nabla}\Phi\right)^2 - \rho\,\Phi, \tag{15.51}$$

can derive the gravitational potential equation 4.77 as the Euler-Lagrange equation well as the total potential energy of the field. (*exercise 15.2*) This is a field theory ich we have introduced on physical grounds, based on Newton's gravitation. A relativistic eralization of gravitation requires radical new concepts, which we will discuss later.

.4 Spinor Fields

eyl Spinor Fields

t us write down a Lagrangian for a *left-handed Weyl spinor field* $\psi_L(x)$. We are guided the requirement of Poincaré invariance for spinorial and vectorial indices and the wish have a real-valued Lagrangian. The most basic *left-handed Weyl Lagrangian* is

$$\mathcal{L} = i\psi_L^\dagger \overline{\sigma}^\mu \partial_\mu \psi_L. \tag{15.52}$$

re, we have left out the mass term, i.e. this particular Lagrangian describes particles th zero mass. By writing out the spinor indices explicitly, the Lagrangian becomes (see ventions in Section 12.5)

$$\mathcal{L} = i(\psi_L^\dagger)_{\dot{A}} (\overline{\sigma}^\mu)^{\dot{A}B} \partial_\mu (\psi_L)_B. \tag{15.53}$$

e corresponding field equation is the *left-handed Weyl equation*

$$\overline{\sigma}^\mu \partial_\mu \psi_L = 0. \tag{15.54}$$

writing out the vector indices, the equation reads

$$\partial_0 \psi_L = \sigma^k \partial_k \psi_L, \tag{15.55}$$

d by using the identity

$$\sigma_k \sigma_l = \delta_{kl} + i\epsilon_{klm}\sigma_m, \tag{15.56}$$

Pauli matrices, this leads to

$$\partial^2 \psi_L = 0. \tag{15.57}$$

other words, each component of the massless Weyl spinor obeys the four-dimensional wave uation. The treatment of a *right-handed Weyl spinor field* $\psi_R(x)$ is totally analogous and can write down the corresponding *right-handed Weyl Lagrangian* as

$$\mathcal{L} = i\psi_R^\dagger \sigma^\mu \partial_\mu \psi_R. \tag{15.58}$$

th the spinor indices written explicitly, this reads

$$\mathcal{L} = i(\psi_R^\dagger)^{\dot{A}} (\sigma^\mu)_{\dot{A}B} \partial_\mu (\psi_L)^B. \tag{15.59}$$

e corresponding field equation is the *right-handed Weyl equation*

$$\sigma^\mu \partial_\mu \psi_R = 0. \tag{15.60}$$

practice, Weyl spinor fields have long been used to describe neutrino particles because it s thought that neutrinos were massless. Experimental evidence in recent years indicates, wever, that neutrinos have a very small, but non-vanishing mass. The lightest neutrino, ich is the electron neutrino, has a mass of less than 2.0 eV, according to the latest Particle ta Group report from 2018.

Dirac Spinor Field

As shown in Section 12.5, we can combine a left-handed and a right-handed Weyl spinor to a Dirac spinor. Here we combine the Weyl spinor fields $\psi_L(x)$ and $\psi_R(x)$ to a *Dirac spinor field* $\Psi(x)$ in the form

$$\Psi \equiv \begin{pmatrix} \psi_L \\ \psi_R \end{pmatrix}, \tag{15.61}$$

being comprised of four complex components $\Psi_a(x)$, with the Dirac indices $a = 1, 2, 3, 4$. We should also consider the *conjugate Dirac spinor field* $\overline{\Psi}(x) \equiv \Psi^\dagger(x)\gamma^0$ and the algebra of gamma matrices. The simplest *Dirac Lagrangian* is

$$\boxed{\mathscr{L} = \overline{\Psi}(i\gamma^\mu\partial_\mu - m)\Psi} \tag{15.62}$$

where the Dirac fields $\Psi(x)$ and $\overline{\Psi}(x)$ are treated as two independent fields. We are led to the *Dirac equation* and its conjugated form,

$$\boxed{i\gamma^\mu(\partial_\mu\Psi) - m\Psi = 0 \quad \text{and} \quad i(\partial_\mu\overline{\Psi})\gamma^\mu + m\overline{\Psi} = 0} \tag{15.63}$$

These equations describe an electron, or more generally, a fermion particle with mass m and spin $1/2$ without external interactions. A Dirac spinor field carries an electric charge, as we will see later. By applying the operator $(i\gamma^\mu\partial_\mu + m)$ to the Dirac equation 15.63 from the left and by using the property 12.162 of gamma matrices, one obtains the Klein-Gordon equation 15.37. So the Dirac equation contains the Klein-Gordon equation. One can say that the Dirac equation is the "square root" of the Klein-Gordon equation.

Instead of the Lagrangian 15.62, one can start with the alternative Dirac Lagrangian

$$\boxed{\mathscr{L} = \frac{i}{2}\left(\overline{\Psi}\gamma^\mu(\partial_\mu\Psi) - (\partial_\mu\overline{\Psi})\gamma^\mu\Psi\right) - m\overline{\Psi}\Psi} \tag{15.64}$$

This alternative Dirac Lagrangian is equal the original Lagrangian plus the divergence term

$$-\frac{i}{2}\partial_\mu(\overline{\Psi}\gamma^\mu\Psi). \tag{15.65}$$

It leads to the same two Dirac equations 15.63. The alternative Lagrangian 15.64 treats the field variables Ψ and $\overline{\Psi}$ in a more symmetric way. Note that the value of both Dirac Lagrangians 15.62 and 15.64 is zero *on-shell*.

In Section 12.5 we saw that we can use different representations of the gamma matrices and the Dirac spinors. By allowing a combined transformation of the Dirac spinors like

$$\Psi' = U\Psi \quad \text{and} \quad \overline{\Psi}' = \Psi'^\dagger\gamma^{0\prime}, \tag{15.66}$$

and the gamma matrices like

$$\gamma^{\mu\prime} = U\gamma^\mu U^{-1}, \tag{15.67}$$

the Dirac Lagrangian 15.62 becomes

$$\mathscr{L}' = \Psi^\dagger U^\dagger U\gamma^0 U^{-1}(iU\gamma^\mu U^{-1}\partial_\mu - m)U\Psi. \tag{15.68}$$

Therefore, the Dirac Lagrangian stays invariant, $\mathscr{L}' = \mathscr{L}$, iff the matrix U of the representation transformation is unitary. Consequently, we are especially interested in unitary transformations between different representations of the gamma matrices. The Lagrangians above describe free spinor fields. In Chapter 17 we will derive how the Dirac spinor field interacts with the electromagnetic field.

15.5   Maxwell Vector Field

Electromagnetic Field

The classical theory of *Maxwell electromagnetism* (*James Clerk Maxwell*) unifies the theoretical descriptions of electric and magnetic phenomena within one elegant physical theory. The complete set of the four *Maxwell equations* in three-dimensional vector notation is comprised of the pair of the homogeneous equations

$$\boxed{\boldsymbol{\nabla} \cdot \boldsymbol{B} = 0 \quad \text{and} \quad \boldsymbol{\nabla} \times \boldsymbol{E} + \partial_t \boldsymbol{B} = \boldsymbol{0}} \tag{15.69}$$

and the pair of the inhomogeneous equations

$$\boxed{\boldsymbol{\nabla} \cdot \boldsymbol{E} = 4\pi\rho \quad \text{and} \quad \boldsymbol{\nabla} \times \boldsymbol{B} - \partial_t \boldsymbol{E} = 4\pi\boldsymbol{j}} \tag{15.70}$$

The homogeneous equations express the *non-existence of magnetic charges* and *Faraday's law of induction* respectively. The inhomogeneous equations express *Coulomb's electric charge law* and *Ampere's current law* (*Andre-Marie Ampere*). The Maxwell equations assume that the sources are given, and are described by the *electric charge density* $\rho(\boldsymbol{x}, t)$ and the three-dimensional *electric current density* $\boldsymbol{j}(\boldsymbol{x}, t)$. The Maxwell theory leads to the *continuity equation*

$$\boxed{\partial_t \rho + \boldsymbol{\nabla} \cdot \boldsymbol{j} = 0} \tag{15.71}$$

expressing the *local conservation of electric current*. The three-dimensional *electric* and *magnetic fields* $\boldsymbol{E}(\boldsymbol{x}, t)$ and $\boldsymbol{B}(\boldsymbol{x}, t)$ can be derived from a *scalar potential* $\varphi(\boldsymbol{x}, t)$ and a *vector potential* $\boldsymbol{A}(\boldsymbol{x}, t)$ through the equations

$$\boldsymbol{E} = -\boldsymbol{\nabla}\varphi - \partial_t \boldsymbol{A} \quad \text{and} \quad \boldsymbol{B} = \boldsymbol{\nabla} \times \boldsymbol{A}. \tag{15.72}$$

Conversely, the electromagnetic field influences the motion of any electrically charged particle or current. The corresponding force $\boldsymbol{F}$ is called the *Lorentz force*. In this constellation, the Maxwell field is given and is acting on a charged particle with the electric charge q and the momentary velocity $\boldsymbol{v}$. The Lorentz force formula reads

$$\boxed{\boldsymbol{F} = q(\boldsymbol{E} + \boldsymbol{v} \times \boldsymbol{B})} \tag{15.73}$$

The fields $\boldsymbol{E}$ and $\boldsymbol{B}$ are applied at the momentary position of the particle. The name of this force goes back to *Hendrik A. Lorentz*, however *Oliver Heaviside* and *James C. Maxwell* are credited with having discovered this force earlier.

The reader should note that we have made a clear distinction between two different situations. One situation is that the currents are given and the resulting Maxwell field is determined. The other situation is that the Maxwell field is given and its influence on charged particles is calculated. We refrain to discuss here the intricate question about the back-reaction of the emitted radiation of a moving particle to its own motion. This analysis touches the conceptual limits of classical electrodynamics and poses a nontrivial problem. The interested reader should turn to Barut [6] and Kosyakov [49] for more.

From a relativistic field theory point of view, the *Maxwell field* is the four-vector field $A^\mu(x)$ written in components as

$$A^\mu = \begin{pmatrix} \varphi \\ \boldsymbol{A} \end{pmatrix}. \tag{15.74}$$

From this vector field $A^\mu(x)$ one derives the *Faraday tensor field*, also called the *electromagnetic tensor field*, $F^{\mu\nu}(x)$, as

$$F^{\mu\nu} = \partial^\mu A^\nu - \partial^\nu A^\mu. \tag{15.75}$$

The Faraday tensor field is obviously antisymmetric,

$$F^{\mu\nu} = -F^{\nu\mu}. \tag{15.76}$$

The contact to the three-dimensional electric and magnetic fields is achieved through the component representation

$$(F^{\mu\nu}) = \begin{pmatrix} 0 & -E_1 & -E_2 & -E_3 \\ E_1 & 0 & -B_3 & B_2 \\ E_2 & B_3 & 0 & -B_1 \\ E_3 & -B_2 & B_1 & 0 \end{pmatrix}. \tag{15.77}$$

The flow of electric charges through spacetime is described by the four-vector *electric current density* $j^\mu(x)$, given in components as

$$j^\mu = \begin{pmatrix} \rho \\ \boldsymbol{j} \end{pmatrix}. \tag{15.78}$$

For an N-particle system, where each particle carries the electric charge q_n, $n = 1, \ldots, N$, the electric current density $j^\mu(x)$ can be expressed by the formula

$$j^\mu(x) = \sum_n \int_{-\infty}^{\infty} q_n \frac{\mathrm{d}z_n^\mu(s_n)}{\mathrm{d}s_n} \delta\left(x - z_n(s_n)\right) \mathrm{d}s_n, \tag{15.79}$$

in full analogy to the mass current density, as described in Section 6.3. One can derive a local current conservation for the electric charge, either by re-expressing the three-dimensional formula 15.71, or, by using the above explicit current for a particle system. The *local current conservation* reads

$$\boxed{\partial_\mu j^\mu = 0} \tag{15.80}$$

The reader is encouraged to carry out both calculations. (*exercise 15.3*) Building on the results from Section 6.2, we can integrate the current conservation equation over a finite space volume of a closed system and derive the constancy of the total electric charge contained in the volume. The definition formula 15.75 leads to the equivalent *integrability condition*

$$\boxed{\partial^\mu F^{\nu\rho} + \partial^\nu F^{\rho\mu} + \partial^\rho F^{\mu\nu} = 0} \tag{15.81}$$

This condition corresponds exactly to the two homogeneous Maxwell equations 15.69. The other two inhomogeneous equations 15.70 are relativistically combined to the single (*inhomogeneous*) *Maxwell equation*

$$\boxed{\partial_\mu F^{\mu\nu} = 4\pi j^\nu} \tag{15.82}$$

From a mathematical standpoint, the relativistic Maxwell equation 15.82 represents a set of four linear, first-order PDEs. Expressed through the vector field $A^\mu(x)$, the inhomogeneous Maxwell equation reads

$$\partial^2 A^\nu - \partial_\mu \partial^\nu A^\mu = 4\pi j^\nu. \tag{15.83}$$

This version of the Maxwell equation can be nicely simplified, as we will see very soon. Let us also note that in the relativistic setup, the *Lorentz force* takes the form of a four-vector f^μ given as

$$\boxed{f^\mu = qF^{\mu\nu}u_\nu} \tag{15.84}$$

The particle possesses the electric charge q and the four-velocity u^μ. In four-vector components, the Lorentz force is explicitly given as (*exercise 15.4*)

$$f^\mu = \gamma q \begin{pmatrix} \boldsymbol{E} \cdot \boldsymbol{v} \\ \boldsymbol{E} + \boldsymbol{v} \times \boldsymbol{B} \end{pmatrix}. \tag{15.85}$$

The reader should also compare the Lorentz force here with the equation of motion we have obtained in Section 6.4 for a relativistic particle under the influence of a vector field $A^\mu(x)$.

Gauge Freedom

Maxwell's theory contains a *gauge freedom* which persists in the four-dimensional relativistic formulation. The field tensor $F^{\mu\nu}(x)$ represents a measurable quantity, but the underlying vector field $A^\mu(x)$ is not completely fixed and can undergo a *gauge transformation*,

$$A^\mu(x) \mapsto A'^\mu(x) = A^\mu(x) - \partial^\mu \theta(x), \tag{15.86}$$

with the electromagnetic field tensor remaining invariant,

$$F^{\mu\nu}(x) \mapsto F'^{\mu\nu}(x) = F^{\mu\nu}(x). \tag{15.87}$$

The *gauge function* $\theta(x)$ in 15.86 can be any well-behaving Lorentz-scalar function. Practically, the gauge freedom can be used to adapt the Maxwell equation to a certain situation. A very common choice of a gauge, i.e. the choice of a condition for $A^\mu(x)$, is the *Lorenz gauge* (*Ludvig Valentin Lorenz*), which is the specific condition

$$\partial_\mu A^\mu = 0. \tag{15.88}$$

By adopting the Lorenz gauge condition, the Maxwell equation 15.83 simplifies to

$$\partial^2 A^\mu = 4\pi j^\mu, \tag{15.89}$$

which is a standard *wave equation* for $A^\mu(x)$ with the external source $j^\mu(x)$. In order to choose the Lorenz gauge, one needs a certain gauge function. If $A^\mu_{\text{init}}(x)$ is the initial vector field, then a gauge function $\theta(x)$ fulfilling the PDE

$$\partial^2 \theta = \partial_\mu A^\mu_{\text{init}} \tag{15.90}$$

leads to a new vector field $A'^\mu(x)$ that fulfills the Lorentz condition. But even then the gauge function $\theta(x)$ is not fully fixed, as one can still add a scalar function $\xi(x)$ if this function satisfies $\partial^2 \xi = 0$. Principally, one can use also other suitable gauges for the vector field. The main point to keep in mind is that when we construct tensorial quantities from the vector field $A^\mu(x)$, we must ensure that the fundamental physical relations besides Poincaré invariance also exhibit *gauge invariance*.

Degrees of Freedom and Propagator Solutions

The previous discussion shows that using all four components of $A^\mu(x)$ leads to a description of the electromagnetic field with an inherent redundancy. How many components are then actually needed and are physically relevant? The answer is that only two *transverse polarization components* matter. The two transverse polarization components represent the physical degrees of freedom of the electromagnetic field. Let us see how this works. We assume a free field in the Lorenz gauge, so that the two equations

$$\partial^2 A^\mu = 0 \quad \text{and} \quad \partial_\mu A^\mu = 0 \tag{15.91}$$

are fulfilled. We consider a monochromatic plane wave given as

$$A^\mu(x) = \varepsilon^\mu \exp(-ik_\nu x^\nu). \tag{15.92}$$

The four-vector ε^μ is the *polarization vector*, whereas the four-vector k^μ represents the *wave vector* in the direction of the plane wave. The free Maxwell equation leads to the condition

$$k_\mu k^\mu = 0, \tag{15.93}$$

which expresses the *masslessness of the electromagnetic field*. The Lorenz condition requires

$$k_\mu \varepsilon^\mu = 0, \tag{15.94}$$

which means that the polarization vector is always transverse to the wave vector. Now we let the gauge freedom come into play and consider the gauge function

$$\theta(x) = i\xi \exp(-ik_\nu x^\nu), \tag{15.95}$$

where ξ is a real parameter. The gauge transformation 15.86 leads to the transformation

$$\varepsilon^\mu \mapsto \varepsilon'^\mu = \varepsilon^\mu - \xi k^\mu \tag{15.96}$$

for the corresponding polarization vector. To be definite, let us consider the plane wave traveling along the positive x^1-direction. The wave vector is then

$$k^\mu = \begin{pmatrix} \omega \\ k \\ 0 \\ 0 \end{pmatrix}, \tag{15.97}$$

with $k = \omega$. By using the condition $k_\mu \varepsilon^\mu = 0$, we are led to $\varepsilon^0 = \varepsilon^1$ for the polarization vector. Now we can choose the real parameter ξ to be equal ε^0/ω and obtain the simplified polarization vector

$$\varepsilon^\mu = \begin{pmatrix} 0 \\ 0 \\ \varepsilon^2 \\ \varepsilon^3 \end{pmatrix}, \tag{15.98}$$

as desired. Due on the gauge freedom, the scalar component ε^0 and the longitudinal component ε^1 have been made to vanish. These components are not physical and correspond to the gauge freedom. In contrast, the other two transverse polarization components, ε^2 and ε^3 is this example, represent the physical degrees of freedom. Of course, each transverse polarization component is not just a real number but a real function. In summary, we can say that electromagnetic waves are transverse waves with their polarization vector residing always in the plane perpendicular to the direction of wave propagation.

It is instructive to write down general solutions for the Maxwell wave equation 15.89 based on the technique of *propagators*, called also *Green's functions* (*George Green*). The problem of solving the equation 15.89 is reduced to the problem of solving the generalized PDE

$$\partial^2 G(x - y) = \delta(x - y), \tag{15.99}$$

which defines the propagator function $G(x-y)$. The defining equation 15.99 expresses that the propagator is essentially the inverse of the d'Alembert wave operator ∂^2. If we have a solution $A^\mu_{\text{hom}}(x)$ of the homogeneous wave equation

$$\partial^2 A^\mu_{\text{hom}}(x) = 0 \tag{15.100}$$

at hand, then a solution $A^\mu(x)$ of the inhomogeneous equation 15.89 can be written as

$$A^\mu(x) = A^\mu_{\text{hom}}(x) + 4\pi \int_{\mathbb{M}_4} G(x-y)\, j^\mu(y)\, \mathrm{d}^4 y. \tag{15.101}$$

In appendix B.5 we derive in detail the two resulting and physically allowed propagators. One solution is the *retarded propagator* $G_{\text{ret}}(x-y)$, given by

$$G_{\text{ret}}(x-y) = \frac{1}{2\pi}\delta\left[(x-y)^2\right]\theta(x^0 - y^0). \tag{15.102}$$

It corresponds physically to *retarded wave* solutions. The retarded propagator contributes only for lightlike signals traveling from the source point at $\boldsymbol{y}$ at the time y^0 to the field point at $\boldsymbol{x}$ at a *later time* x^0. This captures the usual conception that past events can influence

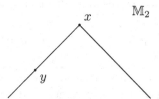

Figure 15.2: Field point x influenced by a source point y with $y^0 < x^0$

future events. In contrast, the *advanced propagator* $G_{\text{adv}}(x-y)$ is expressed by the formula

$$G_{\text{adv}}(x-y) = \frac{1}{2\pi}\delta\left[(x-y)^2\right]\theta(y^0 - x^0), \tag{15.103}$$

and corresponds physically to *advanced wave* solutions. The advanced propagator contributes only for lightlike signals traveling from the source point at $\boldsymbol{y}$ at the time y^0 to the field point at $\boldsymbol{x}$ at an *earlier time* x^0.

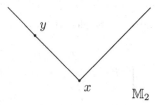

Figure 15.3: Field point x influenced by a source point y with $y^0 > x^0$

Maxwell Lagrangian

The electromagnetic field without external sources, i.e. for $j^\mu = 0$, is governed by the *Maxwell free-field Lagrangian*, discovered by *Joseph Larmor*:

$$\boxed{\mathscr{L} = -\frac{1}{16\pi}F_{\mu\nu}F^{\mu\nu}} \tag{15.104}$$

This Lagrangian is a Lorentz invariant and gauge invariant expression from the outset. One can deduce readily the *Maxwell free-field equation*

$$\partial_\mu F^{\mu\nu} = 0, \tag{15.105}$$

and, equivalently, for the vector field in the Lorenz gauge the equation

$$\partial^2 A^\mu = 0. \tag{15.106}$$

The last equation is again the free wave equation for each component of the vector field. With a non-vanishing, *given external current* $j^\mu(x)$, the *Maxwell Lagrangian with external source* is

$$\boxed{\mathscr{L} = -\frac{1}{16\pi} F_{\mu\nu} F^{\mu\nu} - j_\mu A^\mu} \tag{15.107}$$

From this Lagrangian, the Maxwell equation 15.82 is deduced and correspondingly also the local current conservation 15.80. The last term $-j^\mu A_\mu$ does not appear to be gauge invariant. Indeed, under a gauge transformation, it produces an additional term $j^\mu \partial_\mu \theta$, with the gauge function $\theta(x)$. So, unfortunately, the Lagrangian 15.107 is not gauge invariant. The corresponding action $S[A_\mu]$ receives an additional term of the form

$$\int_U j^\mu \partial_\mu \theta \, \mathrm{d}^4 x = \int_U \partial_\mu(j^\mu \theta) \, \mathrm{d}^4 x. \tag{15.108}$$

The last equality uses the local conservation of the current $j^\mu(x)$. Apparently, the last expression is a volume integral of a total divergence and as such it vanishes for all fields vanishing sufficiently quickly at infinity, or at the boundary ∂U of the system under consideration. In other words, the local current conservation comes to our rescue and assures the gauge invariance of the theory. In Chapter 17 we will investigate the opposite direction and see that local gauge invariance leads to local conservation of the total current.

Duality Symmetry

Starting from the Faraday tensor $F^{\rho\sigma}(x)$, one can define the *dual tensor* $\widetilde{F}_{\mu\nu}(x)$ as

$$\widetilde{F}_{\mu\nu} \equiv \frac{1}{2}\epsilon_{\mu\nu\rho\sigma} F^{\rho\sigma}. \tag{15.109}$$

The dual tensor $\widetilde{F}^{\mu\nu}(x)$ interchanges the electric and magnetic field components in the following way:

$$(\widetilde{F}^{\mu\nu}) = \begin{pmatrix} 0 & -B_1 & -B_2 & -B_3 \\ B_1 & 0 & E_3 & -E_2 \\ B_2 & -E_3 & 0 & E_1 \\ B_3 & E_2 & -E_1 & 0 \end{pmatrix}. \tag{15.110}$$

In other words, the dual tensor is obtained by applying the changes $\boldsymbol{E} \mapsto \boldsymbol{B}$ and $\boldsymbol{B} \mapsto -\boldsymbol{E}$. By using the dual tensor, the homogeneous Maxwell equation 15.81 can be written as

$$\partial_\mu \widetilde{F}^{\mu\nu} = 0. \tag{15.111}$$

For a source-free theory, the above equation has the same form as the dynamical Maxwell equation 15.82, which in this case reads

$$\partial_\mu F^{\mu\nu} = 0. \tag{15.112}$$

This means that for a source-free theory the Faraday tensor and its dual exhibit the same dynamics. This is the so-called *duality symmetry*. The duality symmetry can be generalized by allowing the Faraday tensor and its dual tensor to mix in the following way:

$$\left.\begin{aligned} F'^{\mu\nu} &= F^{\mu\nu}\cos\vartheta - \widetilde{F}^{\mu\nu}\sin\vartheta \\ \widetilde{F}'^{\mu\nu} &= F^{\mu\nu}\sin\vartheta + \widetilde{F}^{\mu\nu}\cos\vartheta \end{aligned}\right\}. \tag{15.113}$$

Here ϑ is a real angle parameter. This *duality transformation* mixes electric and magnetic fields. Despite this mixing, the two above Maxwell equations remain form invariant. However, the Maxwell free-field action does change under duality transformations.

Maxwell Equations with Differential Forms

By using differential forms, as introduced in Section 2.4, it is possible to express Maxwell's equations in an elegant and very general way. Starting from the vector potential $A_\mu(x)$ in its covariant component representation, we can assign to it the 1-form A defined as

$$A \equiv A_\mu \, dx^\mu, \tag{15.114}$$

where $\{dx^\mu\}$ is the covector basis of spacetime. By taking the exterior derivative of A we obtain

$$F = dA, \tag{15.115}$$

where the 2-form F is related to the Faraday tensor by

$$F \equiv F_{\mu\nu} \, dx^\mu \otimes dx^\nu = \frac{1}{2} F_{\mu\nu} \, dx^\mu \wedge dx^\nu. \tag{15.116}$$

The inherent gauge freedom is now expressed by the transformation $A \mapsto A' = A - d\theta$, where $\theta(x)$ is a scalar function. Under this gauge transformation, the Faraday 2-form remains invariant, $F = dA = dA'$. Conversely, if F is expressed by two different 1-forms A and A', then according to Poincaré's lemma it must be $A' = A - d\theta$ within a suitable region of spacetime. Taking the exterior derivative of F, we obtain the 3-form dF given by

$$dF = \frac{1}{3!}(\partial_\mu F_{\nu\rho} + \partial_\nu F_{\rho\mu} + \partial_\rho F_{\mu\nu}) \, dx^\mu \wedge dx^\nu \wedge dx^\rho. \tag{15.117}$$

We see that the integrability condition 15.81 is equivalent to the 3-form equation

$$dF = 0. \tag{15.118}$$

Let us now introduce the dual 2-form $*F$ by

$$*F \equiv \frac{1}{2} \widetilde{F}_{\mu\nu} \, dx^\mu \wedge dx^\nu,$$

with the components $\widetilde{F}_{\mu\nu}(x)$ defined as in 15.109. Moreover, we introduce the 1-form J related to the current density $j_\mu(x)$ by

$$J \equiv j_\mu \, dx^\mu, \tag{15.119}$$

and its dual $*J$ as the 3-form

$$*J \equiv \frac{1}{3!} \widetilde{J}_{\mu\nu\rho} \, dx^\mu \wedge dx^\nu \wedge dx^\rho, \tag{15.120}$$

with the components $\widetilde{J}_{\mu\nu\rho}(x)$ defined by $\widetilde{J}_{\mu\nu\rho} \equiv \epsilon_{\mu\nu\rho\sigma} j^\sigma$. The exterior derivative of $*F$ is the 3-form $\mathrm{d} * F$ given by

$$\mathrm{d} * F = \frac{1}{3!}\epsilon_{\mu\nu\rho\sigma}\partial_\alpha F^{\alpha\sigma}\,\mathrm{d}x^\mu \wedge \mathrm{d}x^\nu \wedge \mathrm{d}x^\rho, \tag{15.121}$$

which can be readily obtained using the Schouten identity. (*exercise 15.5*) We see that the inhomogeneous Maxwell equation 15.82 is equivalent to the 3-form equation

$$\mathrm{d} * F = 4\pi * J. \tag{15.122}$$

Furthermore, by using that it is $\mathrm{d}(\mathrm{d}\alpha) = 0$ for any n-form α, we obtain the 4-form equation

$$\mathrm{d} * J = 0. \tag{15.123}$$

This corresponds to the local current conservation 15.80. The great advantage of formulating Maxwell's equations with differential forms is that these equations are coordinate independent and valid on more general spacetimes than Minkowski space.

Massive Vector Field
The Maxwell field is massless, as we have seen by examining the wave vector. However, one can consider a massive vector field, the *Proca field* (*Alexandru Proca*), which is defined by the *Wentzel-Pauli Lagrangian* (*Gregor Wentzel*)

$$\mathscr{L} = -\frac{1}{16\pi}F_{\mu\nu}F^{\mu\nu} + \frac{m^2}{8\pi}A_\mu A^\mu - j_\mu A^\mu. \tag{15.124}$$

Here we have included also an external source $j^\mu(x)$. The vector field $A^\mu(x)$ possesses the mass m, while the field tensor $F^{\mu\nu}(x)$ is defined as in 15.75. The corresponding EL-Eqs are

$$\partial_\mu F^{\mu\nu} + m^2 A^\nu = 4\pi j^\nu. \tag{15.125}$$

By taking the divergence, we obtain

$$m^2\partial_\nu A^\nu = 4\pi\partial_\nu j^\nu. \tag{15.126}$$

So, whenever the external current is conserved or vanishing, the Lorentz condition $\partial_\nu A^\nu = 0$ is fulfilled, provided $m \neq 0$. Assuming this is the case, the field equations read

$$(\partial^2 + m^2)A^\nu = 4\pi j^\nu \quad \text{and} \quad \partial_\nu A^\nu = 0. \tag{15.127}$$

These equations are called *Proca equations*. The massive Proca field is a model to describe the W and Z bosons mediating the weak nuclear force within the *Standard Model* of particle physics. In the Standard Model, however, the masses are not static but dynamically generated through the so-called spontaneous symmetry breaking mechanism. An accessible discussion of the classical aspects of the symmetry breaking mechanism is provided by Huang [43].

15.6 Noether's Theorem in Field Theory

Symmetry Transformations in Field Theory
In this section we will derive *Noether's theorem* within the framework of field theory. First, let us recall the basic definitions and facts regarding symmetry transformations of spacetime coordinates and fields. The reader should compare with Section 9.4. We work in general

d spacetime dimensions with a Lorentzian metric. The most general *transformations of spacetime coordinates and fields* we are going to consider are:

$$\left. \begin{aligned} x'^{\mu}(x) &= x^{\mu} + \delta x^{\mu}(x) &\equiv x^{\mu} + \epsilon^{a} \mathrm{X}_{a}^{\mu}(x) \\ \phi'_{i}(x) &= \phi_{i}(x) + \delta\phi_{i}(x) &\equiv \phi_{i}(x) + \epsilon^{a} \mathrm{F}_{i,a}(x) \\ \phi'_{i}(x') &= \phi_{i}(x) + \overline{\delta}\phi_{i}(x) &\equiv \phi_{i}(x) + \epsilon^{a} \Phi_{i,a}(x) \end{aligned} \right\}. \tag{15.128}$$

As explained before, we must distinguish between the form variation $\delta\phi_{i}(x) \equiv \phi'_{i}(x) - \phi_{i}(x)$ and the total variation $\overline{\delta}\phi_{i}(x) \equiv \phi'_{i}(x') - \phi_{i}(x)$. We parametrize the variations by using real parameters ϵ^{a}, with the index $a = 1, ..., r$ labeling each transformation. The *generators of symmetry transformations*, $\mathrm{X}_{a}^{\mu}(x)$, $\mathrm{F}_{i,a}(x)$ and $\Phi_{i,a}(x)$, are listed in Section 9.4. We also recall the fundamental relation between total variation and form variation,

$$\overline{\delta}\phi_{i} = \delta\phi_{i} + \partial_{\mu}\phi_{i}\,\delta x^{\mu}. \tag{15.129}$$

According to this relation, we have for the symmetry generators the equivalent relation

$$\Phi_{i,a} = \mathrm{F}_{i,a} + \mathrm{X}_{a}^{\mu}\,\partial_{\mu}\phi_{i}. \tag{15.130}$$

Now we need to define what we mean by a *symmetry transformation of a field theory*. A Lagrangian field theory is defined through its action functional $S[\phi_{i}]$, which depends on the fields $\phi_{i}(x)$ and the spacetime integration domain. A transformation as in 15.128 is a symmetry transformation of the field theory iff the action stays invariant,

$$\boxed{S[\phi'_{i}; x'] = S[\phi_{i}; x]} \tag{15.131}$$

For infinitesimal transformations, this can be written as

$$S[\phi_{i} + \delta\phi_{i};\, x + \delta x] = S[\phi_{i}; x], \tag{15.132}$$

or, by using the variation of the action functional, as

$$\boxed{\delta S[\phi_{i}; x] = 0} \tag{15.133}$$

This is the symmetry condition, expressed in the form we are going to use in the following. We should emphasize that the variation symbol here is meant for the variational changes as defined by the symmetry transformations 15.128. It is not the variation with boundary conditions as used for the Euler-Lagrange equations. Let us also remark that the symmetry condition 15.131 can be expressed at the level of the Lagrangian density and reads

$$\mathscr{L}\left(\phi'_{i}(x'), \partial_{\mu}\phi'_{i}(x')\right) \mathrm{d}^{d}x' = \mathscr{L}\left(\phi_{i}(x), \partial_{\mu}\phi_{i}(x)\right) \mathrm{d}^{d}x + \partial_{\mu}M^{\mu}(x)\,\mathrm{d}^{d}x. \tag{15.134}$$

If we integrate over the whole spacetime $\mathbb{M}_{d}$, the last term on the rhs vanishes, provided the vector field $M^{\mu}(x)$ vanishes sufficiently fast at infinity.

Noether's Theorem in Field Theory

In mechanics we have deduced that for every continuous global symmetry there is an associated time-conserved *charge*. In field theory we work with spacetime densities, so the corresponding Noether theorem states that for every continuous global symmetry there is an associated locally conserved *current*. The following derivation of Noether's theorem is conceptually very similar to the derivation in the mechanical case. The starting point is the symmetry condition 15.133, in which we calculate the variation $\delta S\,[\phi_i; x]$. In the definition of the action, we can consider a bounded spacetime domain $U \subset \mathbb{M}_d$, or the total Minkowski space $\mathbb{M}_d$. However, it is not necessary to specify the integration domain in more detail. For the symmetry variation $\delta S\,[\phi_i; x]$ we have

$$\delta S = \delta \int_U \mathscr{L}\, \mathrm{d}^d x = \int_U (\delta \mathscr{L})\, \mathrm{d}^d x + \int_U \mathscr{L}\, \delta(\mathrm{d}^d x). \qquad (15.135)$$

The first integral on the rhs containing $\delta \mathscr{L}$ is known to us already from the action principle. For the second integral, we calculate the variation in the volume integration measure $\delta(\mathrm{d}^d x)$. It is, to first order in δx^μ,

$$\mathrm{d}^d x' = \left| \det\left(\frac{\partial x'}{\partial x} \right) \right| \mathrm{d}^d x = \left[1 + \partial_\mu(\delta x^\mu) \right] \mathrm{d}^d x, \qquad (15.136)$$

for any orientation-preserving coordinate transformation, see appendix B.6. Hence, we have

$$\delta(\mathrm{d}^d x) = \partial_\mu(\delta x^\mu)\, \mathrm{d}^d x. \qquad (15.137)$$

The second integral is accordingly

$$\int_U \mathscr{L}\, \delta(\mathrm{d}^d x) = \int_U \partial_\mu(\mathscr{L}\, \delta x^\mu)\, \mathrm{d}^d x, \qquad (15.138)$$

because the considered Lagrangian does not depend explicitly on the spacetime coordinate x^μ. The integrand of the first integral is given as

$$\delta \mathscr{L} = \frac{\partial \mathscr{L}}{\partial \phi_i}\, \delta \phi_i + \frac{\partial \mathscr{L}}{\partial(\partial_\mu \phi_i)}\, \delta(\partial_\mu \phi_i) - \delta(\partial_\mu M^\mu). \qquad (15.139)$$

The term $\delta(\partial_\mu M^\mu)$ is a consequence of the freedom 15.32 in a Lagrangian density. The minus sign stems from the transfer of the $\partial_\mu M^\mu$-term from the rhs to the lhs in 15.32. It is further

$$\delta(\partial_\mu M^\mu) = \partial_\mu(\delta M^\mu). \qquad (15.140)$$

The variation $\delta M^\mu(x)$ can be written as

$$\delta M^\mu(x) = \epsilon^a \mathrm{m}_a^\mu(x), \qquad (15.141)$$

by using a generator function $\mathrm{m}_a^\mu(x)$, in complete analogy to symmetry transformations. By applying a partial integration in the first integral above, we obtain

$$\int_U (\delta \mathscr{L})\, \mathrm{d}^d x = \int_U \frac{\delta S}{\delta \phi_i}\, \delta \phi_i\, \mathrm{d}^d x + \int_U \partial_\mu\big(\Pi_i^\mu \delta \phi_i - \delta M^\mu \big)\, \mathrm{d}^d x. \qquad (15.142)$$

Assuming that the equations of motion are fulfilled, $\delta S/\delta \phi_i = 0$, the corresponding integral above vanishes. Summarizing the above partial results for $\delta S\,[\phi_i; x]$, we have

$$\delta S = \int_U \partial_\mu\big(\Pi_i^\mu \delta \phi_i + \mathscr{L}\, \delta x^\mu - \delta M^\mu \big)\, \mathrm{d}^d x, \qquad (15.143)$$

or, written with $\overline{\delta} \phi_i(x)$,

$$\delta S = \int_U \partial_\mu\Big\{ \Pi_i^\mu \big(\overline{\delta} \phi_i - (\partial_\nu \phi_i)\, \delta x^\nu \big) + \mathscr{L}\, \delta x^\mu - \delta M^\mu \Big\}\, \mathrm{d}^d x, \qquad (15.144)$$

and further, by using the symmetry generators,

$$\delta S = \int_U \partial_\mu \left\{ \left[\Pi_i^\mu \Phi_{i,a} - \left(\Pi_i^\mu (\partial_\nu \phi_i) - \delta_\nu^\mu \mathscr{L} \right) X_a^\nu - \mathrm{m}_a^\mu \right] \epsilon^a \right\} \mathrm{d}^d x. \tag{15.145}$$

At this point we introduce the *canonical Noether currents* $j_a^\mu(x)$, $a = 1, \ldots, r$, in the form

$$\boxed{ j_a^\mu \equiv \Theta^{\mu\nu} X_{\nu,a} - \Pi_i^\mu \Phi_{i,a} + \mathrm{m}_a^\mu } \tag{15.146}$$

where we use the notation $X_{\mu,a}(x) \equiv \eta_{\mu\nu} X_a^\nu(x)$. The new quantity $\Theta^{\mu\nu}(x)$ is the *canonical energy-momentum tensor*, abbreviated *CEMT* and defined by

$$\boxed{ \Theta^{\mu\nu} \equiv \Pi_i^\mu (\partial^\nu \phi_i) - \eta^{\mu\nu} \mathscr{L} } \tag{15.147}$$

Each Noether current is comprised of three terms. The first term is the canonical energy-momentum tensor multiplied by the change in the spacetime coordinates. The second term is the conjugate momentum multiplied by the change in the fields. The third term corresponds to the divergence freedom of the Lagrangian density. By using the Noether currents $j_a^\mu(x)$, the variation of the action becomes

$$\delta S = - \int_U \left[(\partial_\mu j_a^\mu) \epsilon^a + j_a^\mu (\partial_\mu \epsilon^a) \right] \mathrm{d}^d x. \tag{15.148}$$

For a *global* symmetry transformation, i.e. one in which the ϵ^a are independent of the coordinates x^μ, the second term on the above rhs vanishes. Note that for a *local* symmetry transformation this term is in general not zero. If we suppose that the action is invariant under such a global symmetry transformation, i.e. $\delta S [\phi_i; x] = 0$, we obtain the sought after local conservation equation. The *Noether conservation equation* in field theory reads

$$\boxed{ \partial_\mu j_a^\mu = 0, \quad a = 1, ..., r } \tag{15.149}$$

Noether's theorem in field theory can be phrased by saying:

> *For every global symmetry transformation of the action, there is a locally conserved Noether current.*

In the following two chapters we will apply Noether's theorem to spacetime transformations and to internal gauge transformations.

Comparison of Field Theory and Mechanics

It is illuminating to compare Noether's theorem for field theory with the one we derived for mechanics. The field-theoretic result as a continuity equation is explicitly written as

$$\partial_\mu \left\{ \left[\frac{\partial \mathscr{L}}{\partial (\partial_\mu \phi_i)} \partial_\nu \phi_i - \delta_\nu^\mu \mathscr{L} \right] X_a^\nu - \frac{\partial \mathscr{L}}{\partial (\partial_\mu \phi_i)} \Phi_{i,a} + \mathrm{m}_a^\mu \right\} = 0, \tag{15.150}$$

with $\Phi_{i,a}$ being the changes in the fields as our dynamical variables, and X_a^ν being the changes in the field label, which is the spacetime coordinate x^μ. The mechanical result for the Noether conservation equation is

$$\frac{\mathrm{d}}{\mathrm{d}t} \left\{ \left[\frac{\partial L}{\partial \dot{q}^j} \dot{q}^j - L \right] \mathrm{T}_a - \frac{\partial L}{\partial \dot{q}^j} \Phi_a^j + \mathrm{m}_a \right\} = 0, \tag{15.151}$$

with Φ_a^j being the changes in the dynamical variables and T_a being the changes in the time parameter t. Despite the fact that the involved quantities are purposefully constructed, the analogy between field theory and mechanics is still remarkable.

Canonical Noether Charges

The local conservation equation 15.149 can be integrated over a finite or an infinite spacetime volume. According to our analysis laid out in Section 6.2, we deduce that the *canonical Noether charge* $G(\Sigma)$, which is defined as the integral

$$G(\Sigma) \equiv \int_{\Sigma} \epsilon^a j_a^{\mu} \, d\Sigma_{\mu}, \tag{15.152}$$

is a constant global quantity of the field theory over any spacelike hypersurface Σ. The only requirement is that the Noether current $\epsilon^a j_a^{\mu}(x)$ vanishes on the hypersurface Σ at infinity. The canonical Noether charge

$$G(\Sigma) = \epsilon^a G_a(\Sigma), \tag{15.153}$$

as a global constant of the theory at hand, has a definite physical meaning, which we would like to understand. Let us consider the simplest case of a pure spacetime translation $\delta x_{\nu} = a^{\rho} X_{\nu,\rho} = a_{\nu}$, for which it is

$$\epsilon^a j_a^{\mu} = a^{\rho} j_{\rho}^{\mu} = a^{\rho} \Theta^{\mu\nu} X_{\nu,\rho} = \Theta^{\mu\nu} a_{\nu}. \tag{15.154}$$

For such a spacetime translation, we obtain the formula

$$G(\Sigma) = a_{\nu} P^{\nu}(\Sigma), \tag{15.155}$$

where we have introduced the *total canonical four-momentum of the field* $P^{\nu}(\Sigma)$ over the hypersurface Σ, defined as the integral quantity

$$P^{\nu}(\Sigma) \equiv \int_{\Sigma} \Theta^{\mu\nu} d\Sigma_{\mu}. \tag{15.156}$$

Since the translation parameters a_{ν} are arbitrary, we obtain that the total canonical field momentum $P_{\nu}(\Sigma)$ is a constant quantity for any spacelike hypersurface Σ. Translations are of course only one of many possible spacetime transformations. In the next chapter we will derive locally and globally conserved quantities of spacetime symmetric field theories.

Further Reading

In this chapter we treated only a certain subset of the questions, methods, and results of classical field theory. For a thorough discussion of classical electrodynamics, the reader should turn to the references of Jackson [46], Landau and Lifshitz [51], or Schwinger et al. [81]. The problem of self-interaction of charged particles and the conceptual limits of classical electrodynamics are discussed in Barut [6] and Kosyakov [49]. Classical field theory, treated from different points of view, can be found in Barut [6], Davis [20], and Felsager [27].

Spacetime Symmetries of Fields

In this chapter we study the consequences of Noether's theorem in field theory for the case of pure spacetime transformations. The symmetry groups considered are the Poincaré group and the conformal group. In this context, the energy-momentum tensor turns out to be the central quantity to focus on, since it allows the definition of all conserved quantities under the conformal group. After a detailed description of the Belinfante prescription for the construction of a symmetric energy-momentum tensor, numerous concrete examples from relevant field theories are given. In the last section, we discuss the meaning and the necessary conditions for achieving conformal symmetry in a field theory.

16.1 Spacetime Symmetries and Currents

Definition of Spacetime Transformations

In the following, we will specialize Noether's theorem to *pure spacetime transformations* and derive the conserved quantities under the various transformations of the conformal group. We consider transformations, for which

$$\delta x^\mu(x) \neq 0 \quad \text{and} \quad \delta\phi_i(x) = 0 \tag{16.1}$$

holds. One should bear in mind that even if we rule out explicit changes of the field functions, still the total variation $\bar{\delta}\phi_i(x)$, which contains the directional derivative, is not necessarily zero. Therefore, we must consider the Noether current

$$j_a^\mu = \Theta^{\mu\nu}X_{\nu,a} - \Pi_i^\mu \Phi_{i,a} + m_a^\mu \tag{16.2}$$

in its full generality. For simplicity, however, we will leave out the term $m_a^\mu(x)$ stemming from the freedom in the definition of a Lagrangian density. The generator functions $X_a^\mu(x)$ and $\Phi_{i,a}(x)$ for the various types of conformal transformations will be determined by the group-theoretical analysis that we have completed already in the Chapters 12, 13, and 14. Historically, it was *Erich P. W. Bessel-Hagen* who first derived the conserved currents under the transformations of the conformal group for the case of Maxwell electrodynamics. The following derivations apply to a field theory defined on Minkowski space $\mathbb{M}_4$, while in Section 16.4 we will consider the more general space $\mathbb{M}_d$.

DOI: 10.1201/9781003087748-16

Translation Invariance and the Canonical Energy-Momentum Tensor
The invariance of a field theory under spacetime translations, the latter in the form

$$\delta x^\rho = a^\rho, \tag{16.3}$$

$$\bar\delta \phi_i = 0, \tag{16.4}$$

or, for the generators,

$$\mathrm{X}^\rho_\sigma = \left.\frac{\partial}{\partial a^\sigma} \delta x^\rho\right|_{a^\sigma=0} = \delta^\rho_\sigma, \tag{16.5}$$

$$\Phi_{i,\sigma} = \left.\frac{\partial}{\partial a^\sigma} \bar\delta \phi_i\right|_{a^\sigma=0} = 0, \tag{16.6}$$

leads to the conservation equation

$$\boxed{\partial_\mu \Theta^{\mu\nu} = 0} \tag{16.7}$$

Thus, the CEMT $\Theta^{\mu\nu}(x)$ is locally conserved.

Lorentz Invariance and the Canonical Angular Momentum Tensor
The invariance of a theory under Lorentz rotations, which have the form

$$\delta x^\rho = \omega^\rho{}_\sigma x^\sigma, \tag{16.8}$$

$$\bar\delta \phi_i = -\frac{i}{2} \omega^{\alpha\beta} (S_{\alpha\beta})_{ij} \phi_j, \tag{16.9}$$

or, for the generators,

$$\mathrm{X}^\rho_{\alpha\beta} = \left.\frac{\partial}{\partial \omega^{\alpha\beta}} \delta x^\rho\right|_{\omega^{\alpha\beta}=0} = \frac{1}{2}(\delta^\rho_\alpha x_\beta - \delta^\rho_\beta x_\alpha), \tag{16.10}$$

$$\Phi_{i,\alpha\beta} = \left.\frac{\partial}{\partial \omega^{\alpha\beta}} \bar\delta \phi_i\right|_{\omega^{\alpha\beta}=0} = -\frac{i}{2}(S_{\alpha\beta})_{ij}\phi_j, \tag{16.11}$$

leads to the conservation equation

$$\boxed{\partial_\mu \Theta^{\mu\nu\rho} = 0} \tag{16.12}$$

Here we have introduced the *canonical angular momentum tensor* $\Theta^{\mu\nu\rho}(x)$ as

$$\Theta^{\mu\nu\rho} \equiv x^\nu \Theta^{\mu\rho} - x^\rho \Theta^{\mu\nu} + \Sigma^{\mu\nu\rho}. \tag{16.13}$$

It consists of the part $x^\nu \Theta^{\mu\rho}(x) - x^\rho \Theta^{\mu\nu}(x)$, which is the *canonical orbital angular momentum tensor*, and the part $\Sigma^{\mu\nu\rho}(x)$, which is the *canonical spin angular momentum tensor*, defined as

$$\Sigma^{\mu\nu\rho} \equiv -i\,\Pi^\mu_i\,(S^{\nu\rho})_{ij}\,\phi_j. \tag{16.14}$$

We see immediately that the canonical angular momentum tensor $\Theta^{\mu\nu\rho}(x)$ is antisymmetric in its last two indices. The equation 16.12 states that only the total canonical angular momentum tensor $\Theta^{\mu\nu\rho}(x)$ is conserved. In general, the orbital part and the spin part are not conserved individually, since there is an exchange of angular momentum between them during the evolution of a system. We observe also that the two conservation equations 16.7 and 16.12 lead to the result

$$\partial_\mu \Sigma^{\mu\nu\rho} = \Theta^{\rho\nu} - \Theta^{\nu\rho}. \tag{16.15}$$

It means that the indices symmetry of the canonical energy-momentum tensor is important: the CEMT is symmetric exactly if the spin tensor is conserved.

For real or complex scalar fields, the canonical spin tensor vanishes, which simplifies matters a lot. Therefore, for scalar fields, the CEMT is always symmetric. In addition, for scalar fields the local conservation of angular momentum is indeed valid for the orbital part.

In the case of the free Dirac spinor field, we start with the symmetric Dirac Lagrangian 15.64. The conjugate momentum $\Pi^\mu(x)$ of this theory is

$$\Pi^\mu = \Pi^\mu_\Psi + \Pi^\mu_{\overline\Psi} = \frac{i}{2}(\overline\Psi\gamma^\mu - \gamma^\mu\Psi), \tag{16.16}$$

containing the parts from both Dirac fields, $\Psi(x)$ and $\overline\Psi(x)$. For simplicity, we suppress Dirac indices here and in the following. The calculation of the Dirac spin tensor $\Sigma^{\mu\nu\rho}(x)$ requires a little bit of care concerning the order of factors. According to the definition it is

$$\Sigma^{\mu\nu\rho} = -i\,(\Pi^\mu_\Psi)(S^{\nu\rho}_\Psi)\Psi - i\,\overline\Psi(S^{\nu\rho}_{\overline\Psi})(\Pi^\mu_{\overline\Psi}). \tag{16.17}$$

By inserting the previous formula for the conjugate momentum and the representations of the Lorentz group generators from Section 12.5, we obtain the *canonical Dirac spin tensor*

$$\Sigma^{\mu\nu\rho} = \frac{i}{8}\,\overline\Psi(\gamma^\mu[\gamma^\nu,\gamma^\rho] + [\gamma^\nu,\gamma^\rho]\gamma^\mu)\Psi. \tag{16.18}$$

The expression $\{\gamma^\mu, [\gamma^\nu,\gamma^\rho]\}$ is not only antisymmetric in ν and ρ, but also antisymmetric under the exchange of any pair of indices. Indeed, only if the three indices μ, ν, ρ are distinct, the expression is equal $4\gamma^\mu\gamma^\nu\gamma^\rho$. In all other cases the expression vanishes.

For the pure Maxwell field described by the Lagrangian 15.104, the conjugate momentum $\Pi^\mu_\alpha(x)$ is

$$\Pi^\mu_\alpha = -\frac{1}{4\pi}F^\mu{}_\alpha, \tag{16.19}$$

which is gauge invariant. For the calculation of the Maxwell spin tensor $\Sigma^{\mu\nu\rho}(x)$, we have according to the definition

$$\Sigma^{\mu\nu\rho} = -i\,(\Pi^\mu_\alpha)(S^{\nu\rho})^\alpha{}_\beta A^\beta. \tag{16.20}$$

By using the Lorentz group generator $(S^{\nu\rho})^\alpha{}_\beta$ from Section 12.4, we obtain the *canonical Maxwell spin tensor*

$$\Sigma^{\mu\nu\rho} = \frac{1}{4\pi}(A^\nu F^{\mu\rho} - A^\rho F^{\mu\nu}). \tag{16.21}$$

Unfortunately, the last expression is not gauge invariant, a deficiency that should be remedied by a suitable definition of the set of basic quantities used in the theory. For the Dirac field and the Maxwell field, the canonical spin tensor is in general not locally conserved and consequently the CEMT for these fields is not symmetric. We will address the question about indices symmetry of the EMT in the next section.

Scale Invariance and the Canonical Dilatation Current

We are moving on to conformal transformations. Dilatations have the form

$$\delta x^\rho = \alpha\,x^\rho, \tag{16.22}$$

$$\overline\delta\phi_i = -\alpha\,\phi_i\,d_\phi, \tag{16.23}$$

or, for the generators,

$$\mathrm{X}^\rho = \left.\frac{\partial}{\partial\alpha}\delta x^\rho\right|_{\alpha=0} = x^\rho, \tag{16.24}$$

$$\Phi_i = \left.\frac{\partial}{\partial\alpha}\overline\delta\phi_i\right|_{\alpha=0} = -\phi_i\,d_\phi. \tag{16.25}$$

The corresponding conservation equation is therefore

$$\boxed{\partial_\mu \Theta^\mu = 0} \tag{16.26}$$

with the *canonical dilatation current* $\Theta^\mu(x)$ being defined by

$$\Theta^\mu \equiv \Theta^{\mu\nu} x_\nu + \Sigma^\mu, \tag{16.27}$$

and the *internal dilatation current* $\Sigma^\mu(x)$ being given by

$$\Sigma^\mu \equiv \Pi_i^\mu \, \phi_i \, d_\phi \, . \tag{16.28}$$

As discussed in Section 14.3, the scaling dimension d_ϕ is specific to each type of field present. The two conservation equations 16.7 and 16.26 lead to the relation

$$\partial_\mu \Sigma^\mu = -\Theta^\mu{}_\mu, \tag{16.29}$$

i.e. the dilatation current is conserved exactly if the trace of the CEMT vanishes. This indicates already that the tracelessness of the EMT is a crucial property.

Conformal Invariance and the Canonical Conformal Current
Finally, we consider special conformal transformations. SCTs have the form

$$
\begin{aligned}
\delta x^\rho &= 2(c \cdot x) x^\rho - c^\rho x^2, & (16.30) \\
\bar\delta \phi_i &= 2\mathrm{i}\, c^\sigma (\mathrm{i}\, x_\sigma d_\phi + x^\rho S_{\sigma\rho})_{ij} \phi_j \, , & (16.31)
\end{aligned}
$$

or, for the generators,

$$
\begin{aligned}
\mathrm{X}_\sigma^\rho &= \left. \frac{\partial}{\partial c^\sigma} \delta x^\rho \right|_{c^\sigma = 0} = 2x_\sigma x^\rho - \delta_\sigma^\rho x^2, & (16.32) \\
\Phi_{i,\sigma} &= \left. \frac{\partial}{\partial c^\sigma} \bar\delta \phi_i \right|_{c^\sigma = 0} = 2\mathrm{i}\, (\mathrm{i}\, x_\sigma d_\phi + x^\rho S_{\sigma\rho})_{ij} \phi_j \, . & (16.33)
\end{aligned}
$$

In this case, the corresponding conservation equation is

$$\boxed{\partial_\mu \Omega^{\mu\nu} = 0} \tag{16.34}$$

with the *canonical conformal current* $\Omega^{\mu\nu}(x)$ being defined by

$$\Omega^{\mu\nu} \equiv \Theta^{\mu\rho}(2x^\nu x_\rho - x^2 \eta^\nu{}_\rho) + 2(\Sigma^\mu x^\nu + \Sigma^{\mu\nu\rho} x_\rho). \tag{16.35}$$

Here we use the previous definitions for the internal dilatation current $\Sigma^\mu(x)$ and the canonical spin tensor $\Sigma^{\mu\nu\rho}(x)$. The canonical conformal current can be expressed also through the conserved currents $\Theta^{\mu\nu}(x)$, $\Theta^{\mu\nu\rho}(x)$ and $\Theta^\mu(x)$, and this formula reads (*exercise 16.1*)

$$\Omega^{\mu\nu} = -\Theta^{\mu\rho}(2x^\nu x_\rho - x^2 \eta^\nu{}_\rho) + 2(\Theta^\mu x^\nu + \Theta^{\mu\nu\rho} x_\rho). \tag{16.36}$$

Despite the fact that the current $\Omega^{\mu\nu}(x)$ can be expressed by the conserved currents $\Theta^{\mu\nu}(x)$, $\Theta^{\mu\nu\rho}(x)$, and $\Theta^\mu(x)$, the conservation law 16.34, arising from the invariance under SCTs, is independent of the conservation laws associated with translations, Lorentz rotations, and dilatations.

This concludes the derivation of the general formulae for the locally conserved field quantities in the case of conformal symmetry. We defer the discussion of the physical interpretation of these quantities to Section 16.3, where we will consider the corresponding integrated quantities over a finite or infinite volume.

16.2 Versions of the Energy-Momentum Tensor

Examples of the Canonical EMT

In the last section we have seen that we can use the CEMT as the basic quantity for constructing all conserved canonical currents under conformal symmetry. Let us now calculate some concrete examples of $\Theta^{\mu\nu}(x)$. For a free real scalar field, the CEMT is

$$\Theta^{\mu\nu} = \partial^\mu \phi \, \partial^\nu \phi - \eta^{\mu\nu} \mathscr{L}, \tag{16.37}$$

with the Lagrangian given in 15.36. Obviously, this CEMT is symmetric. The same applies for the CEMT of the free complex scalar field theory. For the complex scalar case we have

$$\Theta^{\mu\nu} = (\partial^\mu \phi)^* (\partial^\nu \phi) + (\partial^\nu \phi)^* (\partial^\mu \phi) - \eta^{\mu\nu} \mathscr{L}, \tag{16.38}$$

with the Lagrangian given in 15.47. The calculations become somewhat more complicated for fields carrying a spin. For the free Dirac field, the CEMT is calculated as

$$\Theta^{\mu\nu} = \frac{i}{2} \left(\overline{\Psi} \gamma^\mu (\partial^\nu \Psi) - (\partial^\nu \overline{\Psi}) \gamma^\mu \Psi \right) - \eta^{\mu\nu} \mathscr{L}, \tag{16.39}$$

with the Dirac Lagrangian given by 15.64. Obviously, this CEMT is not symmetric. For the free Maxwell field, the CEMT is denoted $\Theta_{\text{EM}}^{\mu\nu}(x)$ here and given by

$$\Theta_{\text{EM}}^{\mu\nu} = \frac{1}{4\pi} \left(-F^{\mu\rho} \partial^\nu A_\rho + \frac{1}{4} \eta^{\mu\nu} F^{\rho\sigma} F_{\rho\sigma} \right). \tag{16.40}$$

Unfortunately, this electromagnetic CEMT has two deficiencies. First, it is not symmetric, despite that symmetry represents a basic requirement for the components of an energy-momentum tensor to be consistent. Secondly, this CEMT is not gauge invariant. Gauge invariance is a crucial property that any physically observable quantity must have. After all, the energy-momentum tensor is supposed to capture the flow of energy and momentum through spacetime. So our basic definition 15.147 of the CEMT allows also results that may be unacceptable from a physical point of view. Is there a way to redefine the energy-momentum tensor and avoid these issues? We need to exploit a certain freedom, which we have to identify first, in order to define a physically acceptable EMT.

Belinfante Prescription for a Symmetric EMT

First, let us note that the CEMT $\Theta^{\mu\nu}(x)$ as defined in 15.147 is not uniquely determined, since the underlying Lagrangian possesses the divergence freedom 15.32. More specifically, adding a divergence term $\partial_\rho M^\rho(\phi_i)$ to the Lagrangian, leads to a new CEMT $\overline{\Theta}^{\mu\nu}(x)$, which is given as

$$\overline{\Theta}^{\mu\nu} = \Theta^{\mu\nu} - \eta^{\mu\nu} \partial_\rho M^\rho. \tag{16.41}$$

This an indication that we should add suitable terms to the CEMT that correct its deficiencies. The above divergence freedom, however, is not sufficient because the added term above is symmetric and thus cannot change the symmetry behavior of the original CEMT. Let us instead focus on the properties of an EMT and take the local conservation equation as the central requirement. The equation

$$\partial_\mu \Theta^{\mu\nu} = 0 \tag{16.42}$$

captures the physical content of the local conservation of energy and momentum. When we start from the CEMT $\Theta^{\mu\nu}(x)$, we can add to it a tensor field, call it $B^{\mu\nu}(x)$, with the property of being locally conserved,

$$\partial_\mu B^{\mu\nu} = 0. \tag{16.43}$$

Then, the redefinition

$$\Theta^{\mu\nu} \mapsto T^{\mu\nu} \equiv \Theta^{\mu\nu} + B^{\mu\nu} \tag{16.44}$$

yields a new energy-momentum tensor, denoted $T^{\mu\nu}(x)$, which is locally conserved too,

$$\partial_\mu T^{\mu\nu} = 0. \tag{16.45}$$

The tensor $B^{\mu\nu}(x)$ can be realized as

$$B^{\mu\nu} = \partial_\rho b^{\rho\mu\nu}, \tag{16.46}$$

with a rank-three tensor $b^{\rho\mu\nu}(x)$ that is antisymmetric in its first two indices,

$$b^{\rho\mu\nu} = -b^{\mu\rho\nu}. \tag{16.47}$$

By considering the *total field momentum* $P^\nu(\Sigma)$, defined as the integral

$$P^\nu(\Sigma) \equiv \int_\Sigma T^{\mu\nu} d\Sigma_\mu \tag{16.48}$$

of the new EMT $T^{\mu\nu}(x)$ over a spacelike hypersurface Σ, we see that it has the same value as the integral of the original $\Theta^{\mu\nu}(x)$, provided that the integral

$$\int_\Sigma B^{\mu\nu} d\Sigma_\mu \tag{16.49}$$

vanishes. For instance, in the case that the hypersurface Σ is actually 3-space, the last integral, due to the Gauss divergence formula, becomes

$$\int_\Sigma \partial_\rho b^{\rho\mu\nu} d\Sigma_\mu = \int_\Sigma \partial_k b^{k0\nu} d^3x = \oint_{\partial\Sigma} b^{k0\nu} d^2 S_k. \tag{16.50}$$

The last integral is evaluated over the 2-dimensional closed spatial surface $\partial\Sigma$ at infinite distance. Assuming that the tensor field $b^{\rho\mu\nu}(x)$ vanishes at spatial infinity, the total field momentum $P^\nu(\Sigma)$ is unaffected by the new terms in the EMT.

So the idea is to add to the CEMT certain terms that vanish when ones takes the divergence. This ensures that the local conservation equation holds. The symmetry of the newly defined EMT is achieved by a suitable choice of these additional terms. This method was developed in 1940 by *Frederik Jozef Belinfante* and leads to a *symmetric energy-momentum tensor* $T^{\mu\nu}(x)$, abbreviated *SEMT*. We start with the ansatz

$$T^{\mu\nu} = \Theta^{\mu\nu} + \partial_\rho b^{\rho\mu\nu}. \tag{16.51}$$

Instead of the tensor $b^{\rho\mu\nu}(x)$, it is helpful to use the equivalent ansatz

$$T^{\mu\nu} = \Theta^{\mu\nu} + \frac{1}{2}\partial_\rho(\sigma^{\rho\mu\nu} + \sigma^{\mu\nu\rho} - \sigma^{\nu\rho\mu}), \tag{16.52}$$

with the rank-three tensor $\sigma^{\rho\mu\nu}(x)$ being antisymmetric in its last two indices,

$$\sigma^{\rho\mu\nu} = -\sigma^{\rho\nu\mu}. \tag{16.53}$$

The connection between the two tensors $b^{\rho\mu\nu}(x)$ and $\sigma^{\rho\mu\nu}(x)$ is obviously

$$b^{\rho\mu\nu} = \frac{1}{2}(\sigma^{\rho\mu\nu} + \sigma^{\mu\nu\rho} - \sigma^{\nu\rho\mu}). \tag{16.54}$$

Now we want to determine how the tensor $\sigma^{\rho\mu\nu}(x)$ must look like explicitly in order to make the new energy-momentum tensor $T^{\mu\nu}(x)$ symmetric. Observe that our ansatz yields

$$T^{\mu\nu} - T^{\nu\mu} = \Theta^{\mu\nu} - \Theta^{\nu\mu} + \partial_\rho \sigma^{\rho\mu\nu}. \tag{16.55}$$

The local conservation of the CEMT and the definition of the canonical angular momentum tensor lead to

$$\partial_\rho \Theta^{\rho\mu\nu} = \Theta^{\mu\nu} - \Theta^{\nu\mu} + \partial_\rho \Sigma^{\rho\mu\nu}. \tag{16.56}$$

By combining the last two equations, we obtain the relation

$$T^{\mu\nu} - T^{\nu\mu} = \partial_\rho \Theta^{\rho\mu\nu} + \partial_\rho(\sigma^{\rho\mu\nu} - \Sigma^{\rho\mu\nu}). \tag{16.57}$$

In a Poincaré invariant theory, where the canonical angular momentum tensor $\Theta^{\rho\mu\nu}(x)$ is locally conserved, the symmetry of the newly formed EMT,

$$T^{\mu\nu} = T^{\nu\mu}, \tag{16.58}$$

is achieved if the *sufficient* condition

$$\sigma^{\rho\mu\nu} = \Sigma^{\rho\mu\nu} \tag{16.59}$$

is fulfilled. The above identification leads to the explicit formula we are looking for. The SEMT, according to Belinfante, is given by

$$\boxed{T^{\mu\nu} = \Theta^{\mu\nu} + \frac{1}{2}\partial_\rho\big(\Sigma^{\rho\mu\nu} + \Sigma^{\mu\nu\rho} - \Sigma^{\nu\rho\mu}\big)} \tag{16.60}$$

An energy-momentum tensor defined by the *Belinfante formula* above, employing the spin tensor of all fields of the theory, is locally conserved and symmetric. Interestingly, it is the spin tensor $\Sigma^{\mu\nu\rho}(x)$, a quantity originating from Lorentz invariance, that modifies the CEMT $\Theta^{\mu\nu}(x)$, which in contrast is rooted in translation invariance.

Now that we have forged the energy-momentum tensor symmetric, let us investigate what happens to the angular momentum tensor. In the transition from the CEMT to the SEMT, the angular momentum tensor changes like

$$\Theta^{\mu\nu\rho} \mapsto T^{\mu\nu\rho} \equiv x^\nu T^{\mu\rho} - x^\rho T^{\mu\nu} + \Sigma^{\mu\nu\rho}. \tag{16.61}$$

One can infer in a straightforward way that the tensor field $T^{\mu\nu\rho}(x)$ is locally conserved iff the spin tensor $\Sigma^{\mu\nu\rho}(x)$ is locally conserved. However, the spin tensor is not always locally conserved. On the other hand, the newly defined *total angular momentum tensor* $M^{\mu\nu\rho}(x)$, defined as

$$M^{\mu\nu\rho} \equiv x^\nu T^{\mu\rho} - x^\rho T^{\mu\nu}, \tag{16.62}$$

is indeed always locally conserved,

$$\partial_\mu M^{\mu\nu\rho} = 0, \tag{16.63}$$

as one can inspect. The two essential properties of the SEMT of being locally conserved and symmetric enter here. Therefore, the tensor field $M^{\mu\nu\rho}(x)$ is the more interesting angular momentum quantity to focus on. The new angular momentum tensor $M^{\mu\nu\rho}(x)$ can be expressed also through the original canonical angular momentum tensor $\Theta^{\mu\nu\rho}(x)$ by including spin tensor correction terms, (*exercise 16.2*)

$$M^{\mu\nu\rho} = \Theta^{\mu\nu\rho} + \frac{1}{2}\partial_\alpha\Big[x^\nu\big(\Sigma^{\alpha\mu\rho} + \Sigma^{\mu\rho\alpha} - \Sigma^{\rho\alpha\mu}\big) - x^\rho\big(\Sigma^{\alpha\mu\nu} + \Sigma^{\mu\nu\alpha} - \Sigma^{\nu\alpha\mu}\big)\Big]. \tag{16.64}$$

We will discuss the physical interpretation of $M^{\mu\nu\rho}(x)$ in Section 16.3.

Improved EMT

Let us summarize what we have achieved so far. By applying Noether's theorem, we obtain locally conserved canonical currents that use the CEMT $\Theta^{\mu\nu}(x)$ as the basic quantity. The conserved canonical currents are $\Theta^{\mu\nu}(x)$, $\Theta^{\mu\nu\rho}(x)$, $\Theta^{\mu}(x)$, and $\Omega^{\mu\nu}(x)$. However, as we have seen, the usage of the CEMT can result to tensors, which may be not acceptable on physical grounds. Gauge invariance is a property we certainly want to maintain in a theory like Maxwell electrodynamics. Symmetry of the spacetime indices of the energy-momentum tensor is another property, which we insist to have, if the EMT is to describe physically correct densities and flux densities of energy and momentum. Due to the freedom we have in defining the EMT, we can use a symmetric EMT instead of the CEMT for constructing conserved currents. It is always possible to obtain a SEMT by following the Belinfante procedure. The SEMT $T^{\mu\nu}(x)$ has the two crucial properties of being *locally conserved* and *symmetric*:

$$\boxed{\partial_\mu T^{\mu\nu} = 0 \quad \text{and} \quad T^{\mu\nu} = T^{\nu\mu}} \tag{16.65}$$

As explained above, by using the SEMT, we can define the *total angular momentum tensor* $M^{\mu\nu\rho}(x)$ by

$$M^{\mu\nu\rho} \equiv x^\nu T^{\mu\rho} - x^\rho T^{\mu\nu}, \tag{16.66}$$

without any explicit spin part. This total angular momentum tensor is locally conserved

$$\boxed{\partial_\mu M^{\mu\nu\rho} = 0} \tag{16.67}$$

due to the two defining properties of the SEMT.

In the context of dilatation invariance, we have seen that the tracelessness of the EMT is a desirable property. In fact, tracelessness of the EMT ensures the complete conformal invariance, as we will see in Section 16.4. The importance of symmetry and tracelessness of the EMT was first recognized by Bessel-Hagen. So, let us consider an energy-momentum tensor, denoted again as $T^{\mu\nu}(x)$, with the three properties of being *locally conserved, symmetric,* and *traceless*:

$$\boxed{\partial_\mu T^{\mu\nu} = 0, \quad T^{\mu\nu} = T^{\nu\mu} \quad \text{and} \quad T^\mu{}_\mu = 0} \tag{16.68}$$

This type of EMT is called the *improved energy-momentum tensor*, abbreviated *IEMT*. The total angular momentum tensor $M^{\mu\nu\rho}(x)$ is constructed as before. In addition, we define a *dilatation current*, denoted $D^\mu(x)$, by

$$D^\mu \equiv T^{\mu\nu} x_\nu, \tag{16.69}$$

without any internal dilatation part. By using the local conservation of the IEMT, the divergence of $D^\mu(x)$ is calculated as

$$\partial_\mu D^\mu = T^\mu{}_\mu. \tag{16.70}$$

Invoking additionally the property of tracelessness of the IEMT, we obtain the local conservation law

$$\boxed{\partial_\mu D^\mu = 0} \tag{16.71}$$

Furthermore, we define a *conformal current*, denoted $K^{\mu\nu}(x)$, without the internal dilatation part and the spin part as

$$K^{\mu\nu} \equiv T^{\mu\rho}(2x^\nu x_\rho - x^2 \eta^\nu{}_\rho). \tag{16.72}$$

By using the local conservation and symmetry of the IEMT, the divergence of the conformal current $K^{\mu\nu}(x)$ is calculated as

$$\partial_\mu K^{\mu\nu} = 2\, T^\mu{}_\mu x^\nu. \tag{16.73}$$

Finally, by using additionally the tracelessness of the IEMT, we arrive to the local conservation law

$$\boxed{\partial_\mu K^{\mu\nu} = 0} \tag{16.74}$$

This means that both currents, $D^\mu(x)$ and $K^{\mu\nu}(x)$, are locally conserved exactly when the trace of the EMT vanishes, the local conservation and symmetry of the IEMT always assumed.

In summary, we can state that a locally conserved EMT, which is gauge invariant, symmetric, and traceless, possesses the crucial properties for defining the currents of spacetime symmetries and significantly simplifies the derivation of the corresponding conservation equations. It is actually possible to systematically construct an IEMT, similar to the way we constructed the SEMT, by using the Belinfante procedure. However, we will not lay out the technical details of this construction here and the interested reader should turn to [13] or [61]. The symmetrization and improvement approaches discussed so far may seem rather ad hoc. So the question arises whether there is a more fundamental way to define an energy-momentum tensor. Later, in the part on general relativity, we will define another type of EMT, the so-called *metric energy-momentum tensor*. Since the EMT is the quantity that couples to the gravitational field, the EMT can be naturally defined through this gravitational interaction. But now let us move on to some concrete SEMT examples.

Symmetric EMT of Free Scalar Fields

As we have seen, the CEMT of free scalar fields is symmetric from the outset. The symmetrization procedure is trivial, since the spin tensor vanishes for scalar fields. Hence, we can write down the SEMT $T^{\mu\nu}(x)$ of the real scalar field immediately as

$$\boxed{T^{\mu\nu} = \partial^\mu\phi\,\partial^\nu\phi - \eta^{\mu\nu}\mathscr{L}} \tag{16.75}$$

The Lagrangian given by 15.36 is considered here in d spacetime dimensions. In the case the mass vanishes, $m = 0$, the trace of the SEMT is

$$T^\mu{}_\mu = \frac{1}{2}(2-d)\partial^\mu\phi\,\partial_\mu\phi, \tag{16.76}$$

where we have used that $\eta^\mu{}_\mu = d$. So the trace vanishes exactly in $d = 2$ spacetime dimensions. In the case of the complex scalar field, the SEMT is

$$\boxed{T^{\mu\nu} = (\partial^\mu\phi)^*(\partial^\nu\phi) + (\partial^\nu\phi)^*(\partial^\mu\phi) - \eta^{\mu\nu}\mathscr{L}} \tag{16.77}$$

with the Lagrangian 15.47 considered again in d spacetime dimensions. By assuming a vanishing mass, the trace of the corresponding SEMT is

$$T^\mu{}_\mu = (2-d)(\partial^\mu\phi)^*(\partial_\mu\phi). \tag{16.78}$$

We have again the situation that the trace vanishes exactly in $d = 2$ spacetime dimensions. However, the condition $(m = 0, d = 2)$ is not the only one to ensure tracelessness. In Section 16.4, we will more systematically investigate under which conditions a scalar field theory can have a traceless EMT, and will explain why tracelessness is crucial in the first place.

Symmetric EMT of the Free Dirac Field

Let us determine the SEMT $T^{\mu\nu}(x)$ of the free Dirac field by using the Belinfante formula. We base our theory on the symmetric Lagrangian 15.64. First of all, we use the property of the canonical Dirac spin tensor $\Sigma^{\mu\nu\rho}(x)$ of being totally antisymmetric. This simplifies the Belinfante formula to

$$T^{\mu\nu} = \Theta^{\mu\nu} + \frac{1}{2}\partial_\rho \Sigma^{\mu\nu\rho}. \tag{16.79}$$

In the calculation of the divergence $\partial_\rho \Sigma^{\mu\nu\rho}$, we use the Dirac equations 15.63 and the basic algebraic relation 12.162 of the gamma matrices. We leave the calculational details to the reader and write down the result here, (*exercise 16.3*)

$$\partial_\rho \Sigma^{\mu\nu\rho} = -\frac{i}{2}\left(\overline{\Psi}\gamma^\mu(\partial^\nu\Psi) - \overline{\Psi}\gamma^\nu(\partial^\mu\Psi) + (\partial^\mu\overline{\Psi})\gamma^\nu\Psi - (\partial^\nu\overline{\Psi})\gamma^\mu\Psi\right). \tag{16.80}$$

By applying this correction to the CEMT $\Theta^{\mu\nu}(x)$, as given in 16.39, we obtain the SEMT $T^{\mu\nu}(x)$ of the free Dirac field with the Lagrangian 15.64 as

$$T^{\mu\nu} = \frac{i}{4}\left(\overline{\Psi}\gamma^\mu(\partial^\nu\Psi) + \overline{\Psi}\gamma^\nu(\partial^\mu\Psi) - (\partial^\mu\overline{\Psi})\gamma^\nu\Psi - (\partial^\nu\overline{\Psi})\gamma^\mu\Psi\right) - \eta^{\mu\nu}\mathscr{L}. \tag{16.81}$$

We see that the SEMT of the Dirac theory is simply the algebraically symmetrized version of the corresponding CEMT. In fact, we can simplify the SEMT of the Dirac field by remembering that the corresponding Lagrangian 15.64 is identically zero on-shell. So the SEMT can be written as

$$\boxed{T^{\mu\nu} = \frac{i}{4}\left(\overline{\Psi}\gamma^\mu(\partial^\nu\Psi) + \overline{\Psi}\gamma^\nu(\partial^\mu\Psi) - (\overline{\partial^\mu\Psi})\gamma^\nu\Psi - (\overline{\partial^\nu\Psi})\gamma^\mu\Psi\right)} \tag{16.82}$$

In the last formula we have used $\partial^\mu\overline{\Psi} = \overline{\partial^\mu\Psi}$, which is the more suited expression for generalizing the SEMT in the case of interactions, as we will see in the next chapter. Instead of the Lagrangian 15.64, we could have initially used the less symmetric version 15.62 for the free Dirac field. In this case the SEMT is given by (*exercise 16.4*)

$$T^{\mu\nu} = \frac{i}{2}\left(\overline{\Psi}\gamma^\mu(\partial^\nu\Psi) + \overline{\Psi}\gamma^\nu(\partial^\mu\Psi)\right) \tag{16.83}$$

This SEMT is also symmetric and locally conserved due to the equations of motion. The trace $T^\mu{}_\mu(x)$ of the SEMT for both Lagrangians 15.64 and 15.62 is calculated as

$$T^\mu{}_\mu = m\overline{\Psi}\Psi. \tag{16.84}$$

This means that the trace of the SEMT vanishes exactly if the mass vanishes.

Symmetric EMT of the Free Maxwell Field

An important example of a SEMT is that of the free electromagnetic field. The Maxwell field carries energy and momentum and the SEMT is the physically correct expression for the densities and flux densities of energy and momentum of the field. Note that it is not the CEMT but the SEMT that gives the correct values of energy and momentum, as measured in physical experiments. For obtaining the SEMT, we can either use the correction term $\partial_\rho b^{\rho\mu\nu}$ with $b^{\rho\mu\nu} = (4\pi)^{-1}F^{\mu\rho}A^\nu$, or calculate explicitly the Belinfante formula 16.60 with the $\Sigma^{\rho\mu\nu}$-terms. Both approaches lead to the central result for the *SEMT of the free Maxwell field*, denoted $T_{\text{EM}}^{\mu\nu}(x)$ and given by the formula

$$\boxed{T_{\text{EM}}^{\mu\nu} = \frac{1}{4\pi}\left(F^{\mu\rho}F_\rho{}^\nu + \frac{1}{4}\eta^{\mu\nu}F^{\rho\sigma}F_{\rho\sigma}\right)} \tag{16.85}$$

This SEMT of the electromagnetic field is locally conserved, symmetric, and gauge invariant. In addition it is indeed traceless, (*exercise 16.5*)

$$(T_{\text{EM}})^\mu{}_\mu = 0. \tag{16.86}$$

In terms of the 3-vector electric and magnetic fields, $\boldsymbol{E}(x)$ and $\boldsymbol{B}(x)$, the components of the SEMT are given by:

$$T_{\text{EM}}^{00} = \frac{1}{8\pi}\left(\boldsymbol{E}^2 + \boldsymbol{B}^2\right), \tag{16.87}$$

$$T_{\text{EM}}^{0k} = T_{\text{EM}}^{k0} = \frac{1}{4\pi}\left(\boldsymbol{E} \times \boldsymbol{B}\right)^k, \tag{16.88}$$

and

$$T_{\text{EM}}^{kl} = \frac{1}{4\pi}\left[E^k E^l + B^k B^l + \frac{1}{2}\delta^{kl}\left(\boldsymbol{E}^2 + \boldsymbol{B}^2\right)\right]. \tag{16.89}$$

The quantity $T_{\text{EM}}^{00}(x)$ is the *energy density* of the electromagnetic field. The *energy flux density* $T_{\text{EM}}^{0k}(x)$ is called also the *Poynting vector* (*John Henry Poynting*). The *momentum flux density* $T_{\text{EM}}^{kl}(x)$ is also known as the *Maxwell stress tensor*. We like to note that the SEMT of the free Maxwell theory can be written in a very symmetrically looking way by using the dual Faraday tensor $\widetilde{F}^{\mu\nu}(x)$ in the form

$$T_{\text{EM}}^{\mu\nu} = \frac{1}{8\pi}\left(F^{\mu\rho}F_\rho{}^\nu + \widetilde{F}^{\mu\rho}\widetilde{F}_\rho{}^\nu\right). \tag{16.90}$$

The proof is left to the reader. (*exercise 16.6*)

EMT of Maxwell Field with an External Source

Let us examine the Belinfante EMT of the *Maxwell field driven by a predefined, external source*. The total system is described by the Lagrangian 15.107, with the *given, external current* $j^\mu(x)$. The equation of motion is the Maxwell equation 15.82, and the local current conservation 15.80 holds. However, this is not a closed physical system, since the external current permanently transfers energy and momentum to the electromagnetic field. We anticipate that the total EMT of this system will *not* be conserved. The total CEMT of the system, denoted $\Theta_{\text{Total}}^{\mu\nu}(x)$, is

$$\Theta_{\text{Total}}^{\mu\nu} = \Theta_{\text{EM}}^{\mu\nu} + \eta^{\mu\nu}j_\alpha A^\alpha, \tag{16.91}$$

where we use the expression for the CEMT $\Theta_{\text{EM}}^{\mu\nu}(x)$, as written in equation 16.40. The symmetrization procedure is similar to the free-field case. For the Belinfante correction terms, we note that it is $b^{\rho\mu\nu} = (4\pi)^{-1}F^{\mu\rho}A^\nu$ again, since the external current does not contribute anything. The divergence $\partial_\rho b^{\rho\mu\nu}$ in the present case yields

$$\partial_\rho b^{\rho\mu\nu} = \frac{1}{4\pi}F^{\mu\rho}\partial_\rho A^\nu - j^\mu A^\nu. \tag{16.92}$$

Hence, the Belinfante EMT of the total system, denoted $T_{\text{Total}}^{\mu\nu}(x)$, is given by

$$T_{\text{Total}}^{\mu\nu} = T_{\text{EM}}^{\mu\nu} + \eta^{\mu\nu}j_\alpha A^\alpha - j^\mu A^\nu. \tag{16.93}$$

The expression for the SEMT $T_{\text{EM}}^{\mu\nu}(x)$ is provided by the formula 16.85. Unfortunately, the above Belinfante EMT $T_{\text{Total}}^{\mu\nu}(x)$ is *not* symmetric, which can be traced back to the fact that the predefined source is not a dynamic field of the theory. We note also that the last term on the rhs, $-j^\mu A^\nu$, spoils gauge invariance, even if we integrate over a spacelike hypersurface. So, it only remains to check if the local conservation of the EMT $T^{\mu\nu}_{\text{Total}}(x)$ is still valid.

Before we investigate the divergence of the above $T^{\mu\nu}_{\text{Total}}(x)$, let us first carry out the calculation for the part $T^{\mu\nu}_{\text{EM}}(x)$ only. By using the Maxwell equations, we obtain

$$\partial_\mu T^{\mu\nu}_{\text{EM}} = F^{\alpha\nu} j_\alpha - \frac{1}{4\pi} \left(\partial^\beta F^{\alpha\nu} - \frac{1}{2} \partial^\nu F^{\alpha\beta} \right) F_{\alpha\beta}. \tag{16.94}$$

The last term on the rhs side vanishes, because the expression in the bracket is symmetric in the indices α and β, due to the integrability condition, while the tensor $F_{\alpha\beta}$ is antisymmetric. (*exercise 16.7*) Thus, we obtain that

$$\partial_\mu T^{\mu\nu}_{\text{EM}} = -F^{\nu\alpha} j_\alpha. \tag{16.95}$$

Whenever there are no external sources, $j^\mu = 0$, the pure Maxwell SEMT is locally conserved. In the case that external sources are present, $j^\mu \neq 0$, the Maxwell field is influenced by the sources and there is an exchange of energy and momentum. The pure Maxwell SEMT is not conserved anymore then. What is in fact conserved is an EMT consisting of the externally given mechanical system plus the Maxwell field. We write the corresponding conservation equation in the form

$$\partial_\mu (T^{\mu\nu}_{\text{EM}} + T^{\mu\nu}_{\text{mech}}) = 0, \tag{16.96}$$

where we have used the implicit definition

$$\partial_\mu T^{\mu\nu}_{\text{mech}} \equiv F^{\nu\alpha} j_\alpha \tag{16.97}$$

for the *SEMT of the external system*, denoted $T^{\mu\nu}_{\text{mech}}(x)$. What is expressed by the above formula? The lhs is the mechanical *force density* associated with the external current, whereas the rhs corresponds to the *Lorentz force law* in its form for a current density. In order to analyze this, we can use the delta distribution representation of a single electrically charged particle as the source of the electromagnetic field. Then, the mechanical SEMT $T^{\mu\nu}_{\text{mech}}(x)$ is given by

$$T^{\mu\nu}_{\text{mech}}(x) = \int_{-\infty}^{\infty} m\, u^\mu(s)\, u^\nu(s)\, \delta\left(x - z(s)\right) \mathrm{d}s, \tag{16.98}$$

while the electric current density $j_\alpha(x)$ is given by the formula

$$j_\alpha(x) = \int_{-\infty}^{\infty} q\, u_\alpha(s)\, \delta\left(x - z(s)\right) \mathrm{d}s. \tag{16.99}$$

By using the result

$$\partial_\mu T^{\mu\nu}_{\text{mech}}(x) = \int_{-\infty}^{\infty} m \frac{\mathrm{d}^2 z^\nu(s)}{\mathrm{d}s^2} \delta\left(x - z(s)\right) \mathrm{d}s \tag{16.100}$$

from Section 15.1, we recover the Lorentz force law,

$$m \frac{\mathrm{d}^2 z^\nu}{\mathrm{d}s^2} = q F^{\nu\alpha} u_\alpha, \tag{16.101}$$

as desired. In the present situation, the above Lorentz force equation expresses that the trajectory $z^\nu(s)$ of the electrically charged particle is prescribed and thus the resulting electromagnetic field $F^{\nu\alpha}(z(s))$ is determined.

Now let us come back to the initial question about the local behavior of the total EMT $T^{\mu\nu}_{\text{Total}}(x)$ of the system. The divergence is calculated in a straightforward way and yields

$$\partial_\mu T^{\mu\nu}_{\text{Total}} = A^\alpha \partial^\nu j_\alpha \neq 0. \tag{16.102}$$

As anticipated, the total EMT of the theory with an external source is *not* conserved. This is due to the fact that the Lagrangian 15.107 takes into account how the external source influences the Maxwell field, but not the other way around. So the missing piece is how the Maxwell field acts back on the source, which can only be captured if the source is not predetermined but a dynamical variable. This completed dynamics of a closed system is described by a set of coupled differential equations, each one containing both the Maxwell field and the electrically charged matter field. In the next chapter, we will derive this complete interaction by applying the gauge principle.

16.3   Conserved Integrals

Globally Conserved Charges

Our discussion in the previous two sections was general and somewhat abstract. Now we would like to get a better understanding of what the locally conserved currents mean physically. A possible approach is to integrate the currents over a spacelike hypersurface Σ, or a spatial volume and thus define corresponding *globally conserved charges*. Here the reader should look again at the integral definitions in Section 6.2. We consider field theories possessing a SEMT $T^{\mu\nu}(x)$. In a Poincaré invariant field theory, one obtains a conserved *total four-momentum*, and a conserved *total four-angular momentum* of the field. If the field theory is additionally invariant under dilatations and SCTs, one obtains in addition a globally conserved *dilatation charge* and *conformal charge*.

Conservation of Total Four-Momentum

The local conservation of the SEMT $T^{\mu\nu}(x)$, as expressed by the first equation in 16.65, is integrated over a spacelike hypersurface Σ, for instance a hyperplane defined by the condition $x^0 = \text{const}$. This integral is

$$\int_\Sigma \partial_\mu T^{\mu\nu} \mathrm{d}^3 x = \int_\Sigma \partial_0 T^{0\nu} \mathrm{d}^3 x + \int_\Sigma \partial_k T^{k\nu} \mathrm{d}^3 x. \tag{16.103}$$

The last term on the rhs vanishes because it is a 2-dimensional surface integral over the boundary $\partial\Sigma$, which is supposed to be sufficiently far away so that the SEMT vanishes there. Thus, the global quantity $P^\nu(x^0)$, defined as

$$P^\nu(x^0) \equiv \int_\Sigma T^{0\nu}(x)\, \mathrm{d}^3 x, \tag{16.104}$$

and interpreted as the *total four-momentum of the field* contained in the 3-dimensional spatial volume $\Sigma = \{x \in \mathbb{M}_4 \mid x^0 = \text{const}\}$ and evaluated at the time $t = x^0$, is conserved in time. We can write

$$\frac{\mathrm{d}P^\nu(t)}{\mathrm{d}t} = 0. \tag{16.105}$$

Let us recall that $T^{00}(x) \equiv \mathcal{E}(x)$ is the *energy density of the field* and $T^{0k}(x) \equiv \mathcal{P}^k(x)$ is the 3-*momentum density of the field*, which we can combine to the *four-momentum density of the field* $\mathcal{P}^\mu(x)$, written out as a column vector as

$$\mathcal{P}^\mu(x) = \begin{pmatrix} \mathcal{E}(x) \\ \mathcal{P}^k(x) \end{pmatrix}. \tag{16.106}$$

Thus, the above conservation equation expresses the constancy in time for the *total energy* $E(t)$ and the *total 3-momentum* $P^k(t)$ of the field,

$$E(t) = \int_\Sigma \mathcal{E}(x)\, \mathrm{d}^3x = \text{const} \quad \text{and} \quad P^k(t) = \int_\Sigma \mathcal{P}^k(x)\, \mathrm{d}^3x = \text{const}. \tag{16.107}$$

More generally, one can consider the integral over an arbitrary spacelike hypersurface Σ and define the quantity $P^\nu(\Sigma)$ as

$$\boxed{P^\nu(\Sigma) \equiv \int_\Sigma T^{\mu\nu}(x)\, \mathrm{d}\Sigma_\mu} \tag{16.108}$$

Then we can state: *the total four-momentum of the field $P^\nu(\Sigma)$ is constant for any spacelike hypersurface Σ.*

Conservation of Total Four-Angular Momentum

Let us turn to the consequences of Lorentz invariance and the resulting local conservation of the angular momentum tensor $M^{\mu\nu\rho}(x)$, as expressed in equation 16.67. We consider the integral over a spacelike hypersurface Σ and define the *total four-angular momentum of the field $M^{\nu\rho}(\Sigma)$* in the form

$$\boxed{M^{\nu\rho}(\Sigma) \equiv \int_\Sigma M^{\mu\nu\rho}(x)\, \mathrm{d}\Sigma_\mu} \tag{16.109}$$

We can carry out an integration exactly as before and conclude: *the total four-angular momentum of the field $M^{\nu\rho}(\Sigma)$ is constant for any spacelike hypersurface Σ.* Again we can choose a hyperplane $x^0 = \text{const}$ and derive that the total four-angular momentum of the field

$$M^{\nu\rho}(x^0) \equiv \int_\Sigma M^{0\nu\rho}(x)\, \mathrm{d}^3x, \tag{16.110}$$

as contained in the 3-dimensional spatial volume $\Sigma = \{x \in \mathbb{M}_4 \mid x^0 = \text{const}\}$ remains constant for all times t,

$$\frac{\mathrm{d}M^{\nu\rho}(t)}{\mathrm{d}t} = 0. \tag{16.111}$$

The $\nu\nu$-components, for all $\nu = 0, 1, 2, 3$, are trivial. The kl-components, for $k \neq l = 1, 2, 3$, lead to the constancy of the *total orbital 3-angular momentum* $L^{kl}(t) = M^{kl}(t)$ of the field,

$$L^{kl}(t) = \int_\Sigma \left\{ x^k \mathcal{P}^l(x) - x^l \mathcal{P}^k(x) \right\} \mathrm{d}^3x = \text{const}. \tag{16.112}$$

The $0k$-components, for $k = 1, 2, 3$, lead to the constancy of the *total Galilei momentum* $G^k(t) = M^{0k}(t)$ of the field,

$$G^k(t) = \int_\Sigma \left\{ t\mathcal{P}^k(x) - x^k \mathcal{E}(x) \right\} \mathrm{d}^3x = tP^k(t) - E(t)X^k(t) = \text{const}. \tag{16.113}$$

Here we have introduced the 3-*dimensional center of energy* $X^k(t)$ as

$$X^k(t) \equiv \frac{1}{E(t)} \int_\Sigma x^k \mathcal{E}(x)\, \mathrm{d}^3x. \tag{16.114}$$

In the present relativistic case, where we consider the total energy and not only the rest mass, the Galilei momentum $G^k(t)$ should be called the *center of energy momentum*. The

astancy of $G^k(t)$ during the course of time means that the center of energy $X^k(t)$ of the
d moves with the constant velocity $P^k(t)/E(t)$,

$$\frac{\mathrm{d}X^k(t)}{\mathrm{d}t} = \frac{P^k(t)}{E(t)}. \tag{16.115}$$

e above results obtained for the field-theoretic total four-angular momentum $M^{\mu\nu}(t)$
 consistent with the interpretation of the four-angular momentum $L^{\mu\nu}$ of relativistic
chanics, as discussed in Section 6.3.

nservation of Dilatation Charge and Conformal Charge
Poincaré invariant field theory, which is also invariant under scale transformations and
Ts, possesses two additional conserved global quantities. We assume the field theory to
ve an IEMT $T^{\mu\nu}(x)$, as described by the conditions 16.68. The two locally conserved
rents associated with the specific symmetries of the conformal group are the dilatation
rent in 16.69 and the conformal current in 16.72. The conserved *dilatation charge of the
'd* $D(\Sigma)$ associated with scale invariance is

$$\boxed{D(\Sigma) \equiv \int_\Sigma D^\mu(x)\,\mathrm{d}\Sigma_\mu} \tag{16.116}$$

en evaluated over an $x^0 = $ const hyperplane, the dilatation charge $D(t)$ satisfies the
servation equation

$$\frac{\mathrm{d}D(t)}{\mathrm{d}t} = 0. \tag{16.117}$$

 can easily see that we can write this constant dilatation charge $D(t)$ as

$$D(t) = \int_\Sigma \{t\mathcal{E}(x) - x^k \mathcal{P}^k(x)\}\,\mathrm{d}^3x = \int_\Sigma x_\mu \mathcal{P}^\mu(x)\,\mathrm{d}^3x = \text{const.} \tag{16.118}$$

the last expression we have used the four-momentum density of the field $\mathcal{P}^\mu(x)$ again.
 The conserved *conformal charge of the field* $K^\nu(\Sigma)$ associated with the invariance under
T is

$$\boxed{K^\nu(\Sigma) \equiv \int_\Sigma K^{\mu\nu}(x)\,\mathrm{d}\Sigma_\mu} \tag{16.119}$$

e conformal charge $K^\nu(t)$, when evaluated over an $x^0 = $ const hyperplane, satisfies the
servation equation

$$\frac{\mathrm{d}K^\nu(t)}{\mathrm{d}t} = 0. \tag{16.120}$$

straight calculation shows that we can express the constant conformal charge $K^\nu(t)$ as
ercise 16.8)

$$K^\nu(t) = \int_\Sigma \{2x^\nu x_\mu \mathcal{P}^\mu(x) - x^2 \mathcal{P}^\nu(x)\}\,\mathrm{d}^3x = \text{const.} \tag{16.121}$$

is completes our derivation of the basic conserved integral quantities.

calizing Energy and Momentum
w that we have discussed the differential form as well as the integral form of conservation
s, let us pause for a moment and look at these constructions. Classical field theory allows
 definition of continuous quantities like energy density, momentum density, etc. for each

spacetime point. However, although it is possible to formally assign to each spacetime point a definite energy density value, momentum density value, etc., it is physically less obvious what this means. This is because for the wave of a field to define physically an energy or a momentum, it has to extend over a full wavelength, if not multiple wavelengths. But when we consider one or more wavelengths, we are no longer considering points, but finite regions of spacetime. These local regions may be small compared to the total system, but they are supposed to be large enough to contain at least one wavelength of the field present. Hence, from a practical standpoint, the quantities of energy density, momentum density, etc. are physically valid only for certain small regions of spacetime. The size of a small finite region of spacetime to be considered depends on the actual physical situation at hand. We should also note here that the question of localization of energy in spacetime appears again within general relativity in a different form and is discussed in Section 23.2.

16.4   Conditions for Conformal Symmetry

Conformal Symmetry and Tracelessness of the EMT

In this section we consider Minkowski space $\mathbb{M}_d$ with dimensionality $d = 1 + D$. Let us start with a general field theory defined through an action. There are indeed different ways to define conformal symmetry of a field theory. One way is to require invariance of the physical laws on the level of the equations of motion. Here we will define the symmetry of the theory at the level of the action. By definition, a field theory possesses conformal symmetry if the action is invariant under arbitrary conformal transformations. Let us perform a general coordinate transformation,

$$x'^{\mu} = x^{\mu} + \delta x^{\mu} = x^{\mu} + \epsilon^a X_a^{\mu}, \qquad (16.122)$$

with constant parameters ϵ^a. The variation $\delta S\,[\phi_i; x]$ of the action, according to Section 15.6, is then

$$\delta S = -\int_U (\partial_\mu j_a^\mu)\epsilon^a \, \mathrm{d}^d x. \qquad (16.123)$$

For pure spacetime transformations, the Noether current $j_a^\mu(x)$ is given by

$$j_a^\mu = \Theta^{\mu\nu} X_{\nu,a}, \qquad (16.124)$$

and thus we obtain

$$\delta S = -\int_U \Theta^{\mu\nu}(\partial_\mu \delta x_\nu) \, \mathrm{d}^d x, \qquad (16.125)$$

where we have used the local conservation of the CEMT. In the next step, we express the CEMT $\Theta^{\mu\nu}(x)$ through the SEMT $T^{\mu\nu}(x)$, as introduced in Section 16.2,

$$\Theta^{\mu\nu} = T^{\mu\nu} - \partial_\rho b^{\rho\mu\nu}, \qquad (16.126)$$

by using the tensor field $b^{\rho\mu\nu}(x) = -b^{\mu\rho\nu}(x)$. With this change, the variation of the action becomes

$$\delta S = -\int_U T^{\mu\nu}(\partial_\mu \delta x_\nu) \, \mathrm{d}^d x, \qquad (16.127)$$

since the integral containing the factor $\partial_\rho b^{\rho\mu\nu}(x)$ vanishes after a partial integration. Due to the symmetry of the SEMT, we can write the variation of the action as

$$\delta S = -\frac{1}{2} \int_U T^{\mu\nu}(\partial_\mu \delta x_\nu + \partial_\nu \delta x_\mu) \, \mathrm{d}^d x. \qquad (16.128)$$

In the particular case of conformal transformations, we have a special condition in place, namely the conformal Killing equation 14.19,

$$\partial_\mu \delta x_\nu + \partial_\nu \delta x_\mu = \frac{2}{d}(\partial_\rho \delta x^\rho)\, \eta_{\mu\nu}. \tag{16.129}$$

Thus, the variation of the action becomes

$$\delta S = -\frac{1}{d}\int_U T^\mu{}_\mu (\partial_\rho \delta x^\rho)\, \mathrm{d}^d x. \tag{16.130}$$

Since the integration domain U is arbitrary, we obtain that the tracelessness of the SEMT,

$$\boxed{T^\mu{}_\mu = 0} \tag{16.131}$$

is a necessary and sufficient condition for the conformal invariance of the action, $\delta S = 0$. As a consequence, if we have a field theory for which we can find a traceless SEMT, then this theory is conformally invariant. This justifies why the improved EMT, as introduced in Section 16.2, is so interesting.

Conformal Symmetry of Free Maxwell Theory

In 1909 and 1910, *Ebenezer Cunningham* and *Harry Bateman* discovered that the Maxwell free-field equations are invariant under conformal transformations of spacetime. We are going to prove this here and investigate under which conditions this holds. First, we need to formulate electrodynamics in arbitrary $d = 1 + D$ spacetime dimensions, with D being the number of space dimensions. The underlying Minkowski metric is $\eta_{\mu\nu}$. We start with a few facts about the surface area of a D-dimensional *unit ball* B^D within the D-dimensional Euclidean space, see definitions in appendix B.7. Our 4-dimensional Maxwell Lagrangian in Gaussian units is given by

$$\mathscr{L} = -\frac{1}{16\pi}F^2, \tag{16.132}$$

where we have used the shorthand notation $F^2 \equiv F_{\mu\nu}F^{\mu\nu}$. The factor $(4\pi)^{-1}$ in the above rhs is of geometric origin and represents the surface area of a 3-dimensional unit ball B^3, living in three space dimensions. In order to generalize this geometric factor to D space dimensions, we need the surface content of a D-dimensional unit ball B^D. In appendix B.7 we derive this surface content $\mathrm{Area}(B^D)$ as

$$\mathrm{Area}(B^D) = \frac{2\pi^{\frac{D}{2}}}{\Gamma\left(\frac{D}{2}\right)}, \tag{16.133}$$

where $\Gamma(\,.\,)$ is the gamma function. For $D = 3$, for instance, one obtains $\mathrm{Area}(B^3) = 4\pi$. Now the generalization of the free Maxwell theory is straightforward. The Maxwell free-field Lagrangian in $d = 1 + D$ spacetime dimensions is given by

$$\boxed{\mathscr{L} = -\frac{F^2}{4\,\mathrm{Area}(B^D)}} \tag{16.134}$$

The associated SEMT is correspondingly

$$T^{\mu\nu} = \frac{1}{\mathrm{Area}(B^D)}\left(F^{\mu\rho}F_\rho{}^\nu + \frac{1}{4}\eta^{\mu\nu}F^2\right). \tag{16.135}$$

By taking the trace of this SEMT, we obtain

$$T^\mu{}_\mu = \frac{F^2}{\text{Area}(B^D)} \left(\frac{d}{4} - 1\right). \tag{16.136}$$

This means that tracelessness is reached if and only if $d = 4$. We obtain the fascinating result:

> The free Maxwell theory on Minkowski space is conformally invariant exactly in $d = 4$ spacetime dimensions.

We should point out again that this result is only valid for the free Maxwell theory and does not hold if electric currents or masses are included.

Conformal Symmetry of Scalar Field Theory

The real scalar field theory in d spacetime dimensions, although one of the simplest field theories, has a rich structure when it comes to the behavior under conformal transformations. First, we will investigate under which conditions the scalar field theory is dilatation invariant and then we proceed to the question about full conformal invariance.

Let us view the following scale transformation of coordinates x and the scalar field $\phi(x)$:

$$\left. \begin{array}{rcl} x & \mapsto & x' = \omega x \\ \phi(x) & \mapsto & \phi'(x') = \omega^{-d_\phi}\phi(x) \end{array} \right\}. \tag{16.137}$$

As usual, d_ϕ denotes the scaling dimension of the field. The corresponding action $S\,[\phi; x]$ transforms as

$$S\,[\phi'; x'] = \int_U \mathscr{L}\left(\phi'(x'), \partial'_\mu \phi'(x')\right) \mathrm{d}^d x'. \tag{16.138}$$

Note that the Lagrangian density, as a function, remains unchanged. By considering that the differential operator transforms as $\partial'_\mu = \omega^{-1}\partial_\mu$, we obtain

$$S\,[\phi'; x'] = \omega^d \int_U \mathscr{L}\left(\omega^{-d_\phi}\phi(x), \omega^{-d_\phi-1}\partial_\mu\phi(x)\right) \mathrm{d}^d x. \tag{16.139}$$

Now the typical Lagrangian density of a scalar field theory has the form

$$\mathscr{L} = \frac{1}{2}\partial_\mu\phi\,\partial^\mu\phi + \kappa\phi^n, \tag{16.140}$$

with the exponent n being an integer number and κ being a scale invariant constant. Let us make a distinction between different cases. In the case $n = 0$, scale symmetry is reached iff the scaling dimension d_ϕ fulfills

$$d_\phi = \frac{d-2}{2}. \tag{16.141}$$

For $d = 4$, for instance, the resulting value is $d_\phi = 1$. For $d = 2$ the resulting scaling dimension is $d_\phi = 0$, and we obtain again the result that the 2-dimensional scalar theory without interactions is scale invariant. Compare with Section 16.2, where we have derived the tracelessness of the corresponding SEMT. In the case $n \neq 0$, the additional term $\kappa\phi^n$ is present. Each term of the total Lagrangian density must be scale invariant, and this leads to the additional condition

$$n = \frac{2d}{d-2} \tag{16.142}$$

for attaining scale invariance. Solving with respect to d yields an equation of the same form, but with the parameters interchanged,

$$d = \frac{2n}{n-2}. \tag{16.143}$$

Let us summarize the most interesting values for the pair (d, n). These are $(d \to 2, n \to \infty)$, $(d = 3, n = 6)$, $(d = 4, n = 4)$, $(d = 6, n = 3)$, and $(d \to \infty, n \to 2)$. In the important case $d = 4$, we can add a ϕ^4 self-interaction term in the form

$$-\frac{\lambda}{4!} \phi^4 \tag{16.144}$$

to the Lagrangian density and still retain scale invariance, compare with the definition 15.44. However, adding a mass term like

$$-\frac{m}{2} \phi^2 \tag{16.145}$$

breaks the scale invariance. We have discussed this effect of symmetry breaking in the presence of masses already in Section 14.3. A field theory can be scale invariant only if all static masses vanish.

For the rest of this section, let us consider the scalar field theory described by the Lagrangian density

$$\boxed{\mathscr{L} = \frac{1}{2}(\partial \phi)^2 + \kappa \phi^{\frac{2d}{d-2}}} \tag{16.146}$$

in $d \geq 3$ dimensions. As we just have seen, this theory is scale invariant. The question is now if it is even fully conformally invariant. Actually, if we can find a traceless IEMT for this theory, then we know that it is conformally invariant, and this is going to be our approach. For the above Lagrangian density, the different definitions of the EMT, i.e. the canonical and the symmetric, coincide and yield

$$\Theta^{\mu\nu} = \partial^\mu \phi \, \partial^\nu \phi - \eta^{\mu\nu} \mathscr{L}, \tag{16.147}$$

which represents a symmetric, locally conserved but not traceless EMT. The canonical dilatation current $\Theta^\mu(x)$ is calculated as

$$\Theta^\mu = \Theta^{\mu\nu} x_\nu + \frac{d-2}{4} \partial^\mu(\phi^2). \tag{16.148}$$

Here we have used our previous result $d_\phi = (d-2)/2$ for the scaling dimension of the field. The procedure for finding a traceless IEMT is similar to the Belinfante method. We add to the existing EMT $\Theta^{\mu\nu}(x)$ a correction term, denoted $B^{\mu\nu}(x)$, such that the resulting new EMT $T^{\mu\nu}(x)$, given as

$$T^{\mu\nu} = \Theta^{\mu\nu} + B^{\mu\nu}$$

acquires all desired properties, including a vanishing trace. A concrete realization for the sought after correction term was discovered by *Curtis G. Callan*, *Sidney R. Coleman* and *Roman W. Jackiw* in 1970 and is described in their article [13]. This suited tensor field is

$$B^{\mu\nu} = -\frac{d-2}{4d-4}(\partial^\mu \partial^\nu - \eta^{\mu\nu} \partial^2)(\phi^2). \tag{16.149}$$

The tensor field $B^{\mu\nu}(x)$ has all the desired properties, as one can inspect. (*exercise 16.9*) It is symmetric, locally conserved, it does not contribute to the total energy and momentum of the field, and in particular it renders the new EMT $T^{\mu\nu}(x)$ traceless,

$$T^\mu{}_\mu = 0. \tag{16.150}$$

Hence, the desired *IEMT of the conformally symmetric scalar field* is

$$T^{\mu\nu} = \Theta^{\mu\nu} - \frac{d-2}{4d-4}(\partial^\mu\partial^\nu - \eta^{\mu\nu}\partial^2)(\phi^2). \tag{16.151}$$

The corresponding *improved dilatation current*, denoted $D^\mu(x)$, is given by the general formula

$$D^\mu = T^{\mu\nu}x_\nu, \tag{16.152}$$

as explained in Section 16.2. By calculating explicitly, one finds (*exercise 16.10*)

$$D^\mu = \Theta^\mu + \frac{d-2}{4d-4}\,\partial_\nu(x^\mu\partial^\nu - x^\nu\partial^\mu)(\phi^2). \tag{16.153}$$

The above construction of a traceless IEMT is the proof point that the classical massless scalar theory, as defined by the Lagrangian 16.146, is indeed fully conformally invariant. More generally, one can ask under which conditions a scale invariant theory is also fully conformally invariant. Actually, this is the case for a number of theories of practical use. However, there are also counterexamples where this is not the case. For a discussion of these dependencies the interested reader is referred to the overview article [45].

Further Reading

The original derivation of the conserved quantities under the conformal group goes back to Erich Bessel-Hagen [9]. Field theory texts with a particular focus on symmetry and conservation include Barut [6], Davis [20], and Felsager [27]. The spacetime symmetries we have discussed in this chapter actually do not encompass all conserved quantities. Daniel M. Lipkin discovered in 1964 additional symmetries and conserved tensor fields of the free Maxwell theory that cannot be expressed by Noether's theorem. For this topic, the interested reader should consult the monograph of Fushchich and Nikitin [32]. An accessible discussion of scale and conformal invariance of particular field theory cases is provided in the article by Jackiw and Pi [45].

17

Gauge Symmetry

The framework of field theory allows the definition of a class of transformations other than spacetime transformations. These are the internal transformations which, due to Noether's theorem, also lead to conserved currents and charges. We first discuss global phase transformations and subsequently the more general local phase transformations of fields, also called gauge transformations. Gauge symmetry provides a powerful guiding principle in order to define the interaction between different types of fields. The gauge symmetry leads to a coupling prescription that fully determines the interaction between fields. The Maxwell electromagnetic field interacting with a matter field represents the prime example. In this chapter, we discuss the resulting Lagrangian densities, the equations of motion, and the associated conserved currents and energy-momentum tensors.

17.1 Internal Symmetries and Charge Conservation

Definition of Internal Transformations

In this chapter we restrict Noether's theorem to *pure internal transformations*. This means that the following two conditions hold:

$$\delta x^\mu(x) = 0 \quad \text{and} \quad \delta \phi_i(x) \neq 0. \tag{17.1}$$

Consequently, the Noether current $j_a^\mu(x)$ to be considered is

$$j_a^\mu = -\Pi_i^\mu \Phi_{i,a}. \tag{17.2}$$

For simplicity, we again omit the contributions to the Noether current arising from pure divergence terms in the Lagrangian density. The generator functions $\Phi_{i,a}(x)$ are determined explicitly by the specific internal symmetry at hand.

Noether Theorem for Global Phase Transformations

We assume a Lagrangian field theory containing complex-valued fields. These fields can be complex scalar fields $\phi(x)$ or spinor fields $\Psi(x)$. For definiteness, let us work with a set of complex scalar fields, as described by the Lagrangian density 15.47. The results are readily transferable to spinor fields. We consider a *global phase transformation* of the field functions

$\phi(x)$ and $\phi^*(x)$ in the form

$$\left.\begin{array}{rcl}\phi(x) & \mapsto & \phi'(x) = \mathrm{e}^{\mathrm{i}q\theta}\,\phi(x) \\[4pt] \phi^*(x) & \mapsto & \phi'^*(x) = \mathrm{e}^{-\mathrm{i}q\theta}\,\phi^*(x)\end{array}\right\}. \tag{17.3}$$

The angle θ is a real number parametrizing the complex phase, whereas q is a real number acting as the generator of the symmetry transformation. Later, we will identify the generator q with the electric charge carried by the field. Note that there are no coordinate transformations included here. Obviously, this global phase transformation leaves the Lagrangian density 15.47 and the corresponding EL-Eqs 15.48 invariant and we can apply Noether's theorem. Infinitesimally, the symmetry transformation reads

$$\left.\begin{array}{l}\delta\phi(x) = \mathrm{i}q\theta\,\phi(x) \\[4pt] \delta\phi^*(x) = -\mathrm{i}q\theta\,\phi^*(x)\end{array}\right\}. \tag{17.4}$$

According to Noether's theorem, there exists a locally conserved current $j^\mu(x)$ given by

$$j^\mu = -\Pi_i^\mu\Phi_i, \tag{17.5}$$

with the field index i enumerating the two fields $\phi(x)$ and $\phi^*(x)$. The quantity $\Pi_i^\mu(x)$ is the conjugate field momentum and $\Phi_i(x)$ is the generator of the symmetry transformation. A direct calculation yields the *locally conserved Noether current* $j^\mu(x)$. (*exercise 17.1*) It is

$$j^\mu = \mathrm{i}q\left(\phi^*(\partial^\mu\phi) - (\partial^\mu\phi)^*\phi\right) = 2q\,\mathrm{Im}[(\partial^\mu\phi)^*\phi]. \tag{17.6}$$

The last equality shows that the current is a real-valued quantity. According to Noether's theorem, the continuity equation

$$\partial_\mu j^\mu = 0 \tag{17.7}$$

holds. This local conservation leads to the *globally conserved charge* $Q(t)$, which stays constant in time and is given by

$$Q(t) \equiv \int_\Sigma j^0(x)\,\mathrm{d}^3x = \mathrm{const.} \tag{17.8}$$

Once again is the hyperplane Σ defined by $x^0 = \mathrm{const.}$ The global quantity $Q(t)$ is interpreted as the *electric charge* carried by the complex scalar field $\phi(x)$. By choosing a normalization of the field functions $\phi(x)$ and $\phi^*(x)$ such that

$$\int_\Sigma \mathrm{i}\left(\phi^*(x)\partial^0\phi(x) - \phi(x)\partial^0\phi^*(x)\right)\mathrm{d}^3x = 1, \tag{17.9}$$

we obtain that the conserved electric charge Q is identical to the generator q of the phase transformations. The case of a Dirac spinor field is treated in complete analogy and again the time-conserved electric charge of the field is obtained.

17.2   Interactions and the Gauge Principle

Interactions in a Lagrangian Theory

In almost all of our discussions in field theory so far, we have dealt with free, non-interacting fields. Let us ask the general question how we can properly introduce an interaction between fields. The Lagrangian framework allows us to formulate interactions in a very principal way.

Assume we have two separate field theories described by the Lagrangians $\mathscr{L}_1\left(\phi_i, \partial_\mu \phi_i\right)$ and $\mathscr{L}_2\left(\psi_k, \partial_\mu \psi_k\right)$, with the respective field collections $\phi_i(x)$ and $\psi_k(x)$. The Lagrangian of the total system without any interaction between the two subsystems is the sum

$$\mathscr{L}_1\left(\phi_i, \partial_\mu \phi_i\right) + \mathscr{L}_2\left(\psi_k, \partial_\mu \psi_k\right). \tag{17.10}$$

The variation with respect to the $\phi_i(x)$ fields yields the EL-Eqs of system 1 only, while the variation with respect to the $\psi_k(x)$ fields yields the EL-Eqs of system 2 only. An interaction between the two systems can be achieved by adding an interaction term $\mathscr{L}_{\text{Int}}(\cdot)$ and consequently by using a total Lagrangian of the form

$$\mathscr{L}_1\left(\phi_i, \partial_\mu \phi_i\right) + \mathscr{L}_2\left(\psi_k, \partial_\mu \psi_k\right) + \mathscr{L}_{\text{Int}}\left(\phi_i, \psi_k, \partial_\mu \phi_i, \partial_\mu \psi_k\right). \tag{17.11}$$

Now the variation of the $\phi_i(x)$ fields provides the EL-Eqs for these fields including their coupling to the $\psi_k(x)$ fields, and vice versa. So the task is to find a suitable interaction Lagrangian $\mathscr{L}_{\text{Int}}(\cdot)$ capturing correctly the interaction of the involved subsystems.

Local Phase Transformations and the Gauge Principle

We introduce now the powerful concept of *local gauge invariance*, which leads to the definition of interactions between so-called *matter fields* and *gauge fields* in a fundamental way. We consider again the case of complex scalar fields described by the Lagrangian 15.47, which here represent the *matter fields*. As we have seen in the previous section, global phase transformations leave this Lagrangian invariant. We consider in contrast now a *local phase transformation* of the field functions $\phi(x)$ and $\phi^*(x)$ in the form

$$\left.\begin{array}{rcl} \phi(x) & \mapsto & \phi'(x) = e^{iq\theta(x)}\,\phi(x) \\ \phi^*(x) & \mapsto & \phi'^*(x) = e^{-iq\theta(x)}\,\phi^*(x) \end{array}\right\}. \tag{17.12}$$

Here the angle $\theta(x)$ is "gauged" and becomes a locally varying real scalar function. Under the above local phase transformation, the Lagrangian 15.47 is not invariant anymore. To recover invariance, one has to incorporate additional terms in the Lagrangian to balance the changes. This is accomplished in the most elegant way by introducing the *(gauge-)covariant derivative* $D_\mu(x)$, which is the differential operator

$$\boxed{D_\mu(x) \equiv \partial_\mu + iqA_\mu(x)} \tag{17.13}$$

that includes the *gauge field* $A_\mu(x)$ as a four-vector field. We require the gauge field $A_\mu(x)$ to transform as

$$A_\mu(x) \mapsto A'_\mu(x) = A_\mu(x) - \partial_\mu \theta(x) \tag{17.14}$$

under a local phase transformation. Note that there is no change in the coordinates. Correspondingly, the covariant derivative $D_\mu(x)$ transforms as

$$D_\mu(x) \mapsto D'_\mu(x) = \partial_\mu + iqA_\mu(x) - iq\partial_\mu \theta(x). \tag{17.15}$$

As a consequence, we have the following transformations for the covariant derivatives of the fields: (*exercise 17.2*)

$$\left.\begin{array}{rcl} D_\mu(x)\phi(x) & \mapsto & D'_\mu(x)\phi'(x) = e^{iq\theta(x)}\,D_\mu(x)\phi(x) \\ \left[D_\mu(x)\phi(x)\right]^* & \mapsto & \left[D'_\mu(x)\phi'(x)\right]^* = e^{-iq\theta(x)}\left[D_\mu(x)\phi(x)\right]^* \end{array}\right\}. \tag{17.16}$$

The covariant derivatives of the field functions are transforming exactly like the field functions itself. This is what we need, because now the recipe to achieve invariance under local phase transformations is to implement the basic exchange

$$\partial_\mu \;\to\; D_\mu(x)$$　　　　　　　　　　(17.17)

in the Lagrangian 15.47. The exchange 17.17 is called the *minimal coupling prescription*, and in the present case it leads to the new Lagrangian

$$\mathscr{L} = (D_\mu \phi)^* (D^\mu \phi) - m^2 \phi^* \phi.$$　　　　　　　　(17.18)

Now under a combined *local gauge transformation* of the matter fields $\phi(x)$, $\phi^*(x)$, and the gauge field $A_\mu(x)$, as specified above, the Lagrangian remains gauge invariant. The just introduced gauge field $A_\mu(x)$ is identified with the vector potential of Maxwell theory. Indeed, the local gauge transformation 17.14 is identical to the gauge transformation 15.86 of the vector potential as encountered previously within Maxwell theory. Thus, we have a recipe how to define the interaction between a given scalar (or a spinor) field and the electromagnetic gauge field. This principle is called the *gauge principle*, sometimes also the *Weyl principle*, due to Hermann Weyl who originally proposed it in a preliminary form in 1918 and then developed it in its final form in 1929, see [96]. Condensed in one sentence, the gauge principle says:

> *For the physics of matter fields to be insensitive to local phase transformations,*
> *a gauge field and the corresponding interaction must be introduced.*

Because the local phases $\exp(iq\theta(x))$ for each point x are elements of the abelian group $U(1)$, the resulting new theory with interactions is called the $U(1)$ *gauge theory*. One can extend to other symmetry groups, such as the non-abelian group $SU(n)$, which then leads to the gauge theories underlying the *Standard Model* of elementary particle interactions.

Properties of the Covariant Derivative

The most important property of the gauge-covariant derivative $D_\mu(x)$ is its transformation behavior under gauge transformations. which we can write as

$$D'_\mu(x) = e^{iq\theta(x)}\, D_\mu(x)\, e^{-iq\theta(x)}.$$　　　　　　　(17.19)

The covariant derivative is a linear first-order differential operator with some additional properties that we like to summarize here. It is for the commutator

$$[D_\mu, D_\nu] \equiv D_\mu D_\nu - D_\nu D_\mu = iq F_{\mu\nu},$$　　　　　（17.20)

where

$$F_{\mu\nu} \equiv \partial_\mu A_\nu - \partial_\nu A_\mu$$　　　　　　　　（17.21)

is the usual definition of the *field strength* $F_{\mu\nu}(x)$ of the $U(1)$ gauge field. One can deduce from each of the last two equations that the field strength is gauge invariant. Furthermore, the covariant derivative fulfills the *Jacobi identity*,

$$[[D_\mu, D_\nu], D_\rho] + [[D_\nu, D_\rho], D_\mu] + [[D_\rho, D_\mu], D_\nu] = 0,$$　　　（17.22)

as one can easily prove. (*exercise 17.3*) The Jacobi identity is equivalent to the *integrability condition*

$$\partial_\mu F_{\nu\rho} + \partial_\nu F_{\rho\mu} + \partial_\rho F_{\mu\nu} = 0,$$　　　　　　（17.23)

which we have encountered already in Section 15.5 when we discussed the Maxwell theory.

17.3   Scalar Electrodynamics

Scalar Electrodynamics Lagrangian

In the previous section, we have introduced the interaction between the electrically charged complex scalar field and the Maxwell gauge field. In order to write down the complete Lagrangian of the total system, we need to incorporate also the pure Maxwell part. The resulting *scalar electrodynamics Lagrangian* $\mathscr{L}\,(\phi, \phi^*, A^\mu)$ is

$$\mathscr{L} = -\frac{F^2}{16\pi} + (D_\mu \phi)^* (D^\mu \phi) - m^2 \phi^* \phi \tag{17.24}$$

We use again the notation $F^2 \equiv F_{\mu\nu} F^{\mu\nu}$. By writing out the covariant derivatives, the total Lagrangian becomes

$$\mathscr{L} = -\frac{F^2}{16\pi} + \partial_\mu \phi^* \partial^\mu \phi - m^2 \phi^* \phi - iq(\phi^* \partial_\mu \phi - \phi \, \partial_\mu \phi^*) A^\mu + q^2 \phi^* \phi A_\mu A^\mu. \tag{17.25}$$

The interaction term $\mathscr{L}_{\text{Int}}\,(\phi, \phi^*, A^\mu)$ is given by

$$\mathscr{L}_{\text{Int}} = -iq(\phi^* \partial_\mu \phi - \phi \, \partial_\mu \phi^*) A^\mu + q^2 \phi^* \phi A_\mu A^\mu. \tag{17.26}$$

The new term $q^2 |\phi|^2 A^2$ in the Lagrangian is imposed by the gauge symmetry and is the one responsible for the mechanism of *spontaneous symmetry breaking* in the Standard Model of particle physics. The basic workings of spontaneous symmetry breaking are indeed established already at the level of classical field theory. The reader interested in this topic should turn to quantum field theory texts, for example Huang [43] or Maggiore [56].

Field Equations

Now let us derive the classical equations of motion, separately for the Maxwell gauge field and for the scalar matter field. Variation with respect to the Maxwell field $A^\nu(x)$ yields the field equation

$$\partial_\mu F^{\mu\nu} = 4\pi iq \left(\phi^* (D^\nu \phi) - (D^\nu \phi)^* \phi \right) \tag{17.27}$$

The variations with respect to the scalar fields $\phi^*(x)$ and $\phi(x)$ deliver the two scalar field equations

$$(D^2 + m^2)\phi = 0 \quad \text{and} \quad (D^2 + m^2)^* \phi^* = 0 \tag{17.28}$$

We use the abbreviation $D^2 \equiv D_\mu D^\mu$. The above three equations 17.27 and 17.28 are a set of coupled PDEs that describe the complete closed system. The equation for the Maxwell field can be rewritten as

$$\partial_\mu F^{\mu\nu} = 4\pi(j^\nu - 2q^2 \phi^* \phi A^\nu), \tag{17.29}$$

with the current $j^\nu(x)$ denoting again the one of the free-field case, given by

$$j^\nu = iq \left(\phi^* (\partial^\nu \phi) - (\partial^\nu \phi)^* \phi \right). \tag{17.30}$$

Conservation of Electric Charge

The new, locally gauge invariant four-vector current $J^\mu(x)$ appearing in the EL-Eqs is

$$J^\mu = iq \left(\phi^* (D^\mu \phi) - (D^\mu \phi)^* \phi \right). \tag{17.31}$$

Interestingly, this new current $J^\mu(x)$ consists of the expression for $j^\mu(x)$, stemming from the free scalar field, plus the term $-2q^2 |\phi|^2 A^\mu$ containing also the gauge field. So the question arises if the current $J^\mu(x)$ can still be considered as representing the electric current flow of the scalar field. To answer the question, we should observe that the local conservation

$$\partial_\mu J^\mu = 0 \qquad (17.32)$$

holds, provided that the scalar field EL-Eqs 17.28 are fulfilled. (*exercise 17.4*) So the local conservation of the new current crucially depends on the dynamical properties of the scalar field. Thus, the interpretation remains that $J^\mu(x)$ represents the correct electric current density of the complex scalar field.

Conservation of Energy and Momentum

The complete, closed system consisting of the complex scalar field interacting with the Maxwell field has a total SEMT $T^{\mu\nu}_{\text{Total}}(x)$, which we like to derive now. The CEMT $\Theta^{\mu\nu}_{\text{Total}}(x)$ of the total coupled system is calculated as

$$\Theta^{\mu\nu}_{\text{Total}} = (D^\mu\phi)^*(\partial^\nu\phi) + (\partial^\nu\phi)^*(D^\mu\phi) - \frac{1}{4\pi}F^{\mu\rho}\partial^\nu A_\rho - \eta^{\mu\nu}\mathscr{L}. \qquad (17.33)$$

Unfortunately, this total CEMT is neither symmetric nor gauge invariant. Both deficiencies are remedied by introducing a SEMT. For the derivation of the SEMT, we can follow the *Belinfante prescription*. The three individual fields $\phi(x)$, $\phi^*(x)$ and $A^\mu(x)$ are all treated together and on equal footing in the calculation. We do not divide up the system into a scalar sector and an electromagnetic sector. Obviously, only the electromagnetic field contributes a non-vanishing canonical Maxwell spin tensor for the Belinfante corrections terms. In the calculation, also the Maxwell equation 17.27 has to be used. By applying this prescription, we obtain the desired *total SEMT of scalar electrodynamics*:

$$T^{\mu\nu}_{\text{Total}} = \frac{1}{4\pi}\left(F^{\mu\rho}F_\rho{}^\nu + \frac{1}{4}\eta^{\mu\nu}F^2\right)$$
$$+ (D^\mu\phi)^*(D^\nu\phi) + (D^\nu\phi)^*(D^\mu\phi) - \eta^{\mu\nu}\left[(D_\rho\phi)^*(D^\rho\phi) - m^2\phi^*\phi\right]. \quad (17.34)$$

The derivation is left to the reader. (*exercise 17.5*) Formally, we could have obtained this result by adding the free-field SEMTs of the scalar field and the Maxwell field and by replacing the partial derivatives in the scalar part by covariant derivatives. The total SEMT $T^{\mu\nu}_{\text{Total}}(x)$ is gauge invariant, symmetric, and locally conserved:

$$\boxed{\partial_\mu T^{\mu\nu}_{\text{Total}} = 0} \qquad (17.35)$$

The local conservation of energy and momentum, as expressed through the above continuity equation, is a confirmation of the physical validity of the specific interaction and the way it follows from the gauge principle.

17.4   Spinor Electrodynamics

Spinor Electrodynamics Lagrangian

In this section, we apply the gauge principle to the interacting system consisting of a Dirac spinor field and the Maxwell electromagnetic field. The complete *spinor electrodynamics Lagrangian* $\mathscr{L}\left(\Psi, \overline{\Psi}, A^\mu\right)$ according to the gauge principle is

$$\boxed{\mathscr{L} = -\frac{F^2}{16\pi} + \frac{i}{2}\left(\overline{\Psi}\gamma^\mu(D_\mu\Psi) - (\overline{D_\mu\Psi})\gamma^\mu\Psi\right) - m\overline{\Psi}\Psi} \qquad (17.36)$$

Here, we have used the symmetric version 15.64 of the Dirac Lagrangian. Written out, the total Lagrangian reads

$$\mathscr{L} = -\frac{F^2}{16\pi} + \frac{i}{2}\left(\overline{\Psi}\gamma^\mu(\partial_\mu\Psi) - (\partial_\mu\overline{\Psi})\gamma^\mu\Psi\right) - m\overline{\Psi}\Psi - q\overline{\Psi}\gamma^\mu\Psi A_\mu. \tag{17.37}$$

The interaction term $\mathscr{L}_{\text{Int}}\left(\Psi, \overline{\Psi}, A^\mu\right)$, which is given as

$$\mathscr{L}_{\text{Int}} = -q\overline{\Psi}\gamma^\mu\Psi A_\mu, \tag{17.38}$$

can be written in the form $-J^\mu A_\mu$, provided the electric current density $J^\mu(x)$ is given by

$$J^\mu = q\overline{\Psi}\gamma^\mu\Psi. \tag{17.39}$$

This is indeed the case and we will derive this in the following.

Field Equations

For the derivation of the classical equations of motion, separately for the Maxwell gauge field and for the Dirac spinor matter field, one proceeds along the known lines. The variation with respect to the Maxwell field $A^\nu(x)$ yields the field equation

$$\boxed{\partial_\mu F^{\mu\nu} = 4\pi q\overline{\Psi}\gamma^\nu\Psi} \tag{17.40}$$

The variations with respect to the Dirac fields $\overline{\Psi}(x)$ and $\Psi(x)$ deliver the two spinorial field equations

$$\boxed{i\gamma^\mu(D_\mu\Psi) - m\Psi = 0 \quad \text{and} \quad i(\overline{D_\mu\Psi})\gamma^\mu + m\overline{\Psi} = 0} \tag{17.41}$$

The above three equations 17.40 and 17.41 constitute a set of coupled PDEs that describe the complete closed system.

Conservation of Electric Charge

The locally gauge invariant four-vector field $J^\mu(x)$ given as

$$J^\mu = q\overline{\Psi}\gamma^\mu\Psi \tag{17.42}$$

represents the electric current density of the Dirac spinor field carrying the electric charge q. The local conservation

$$\partial_\mu J^\mu = 0 \tag{17.43}$$

holds, provided that the Dirac field EL-Eqs 17.41 are fulfilled. (*exercise 17.6*)

Conservation of Energy and Momentum

The total SEMT $T^{\mu\nu}_{\text{Total}}(x)$ of the spinor electrodynamics system, consisting of the Dirac spinor field interacting with the Maxwell electromagnetic field, can once again be calculated with the aid of the *Belinfante prescription*. For the derivation it is important that the system is not divided into a spinor sector and an electromagnetic sector, but that all fields are treated in a unified way. We outline the derivation below and leave some details of the calculation to the reader. (*exercise 17.7*) The CEMT $\Theta^{\mu\nu}_{\text{Total}}(x)$ of the total coupled Dirac-Maxwell system is calculated as

$$\Theta^{\mu\nu}_{\text{Total}} = \frac{i}{2}\left(\overline{\Psi}\gamma^\mu(\partial^\nu\Psi) - (\partial^\nu\overline{\Psi})\gamma^\mu\Psi\right) - \frac{1}{4\pi}F^{\mu\rho}\partial^\nu A_\rho - \eta^{\mu\nu}\mathscr{L}. \tag{17.44}$$

Unfortunately, this CEMT not only fails to be symmetric but it is also not gauge invariant. We will remedy both deficiencies by redefining the EMT according to the Belinfante formula 16.60. For determining the correction terms, we note that the conjugate momenta of the total system are simply the momenta of the single free fields. This is because the interaction term of the present Lagrangian does not contribute anything to the conjugate momenta. Thus, the total canonical spin tensor $\Sigma^{\mu\nu\rho}_{\text{Total}}(x)$ of the coupled system consists simply of a portion due to the electromagnetic field and a portion due to the spinor field,

$$\Sigma^{\mu\nu\rho}_{\text{Total}} = \Sigma^{\mu\nu\rho}_{\text{EM}} + \Sigma^{\mu\nu\rho}_{\text{Spinor}}. \tag{17.45}$$

In fact, we have calculated the conjugate momenta and the corresponding canonical spin tensors already in Section 16.1. The contribution to the correction terms stemming from the Maxwell spin tensor $\Sigma^{\mu\nu\rho}_{\text{EM}}(x)$ is equal

$$\frac{1}{4\pi}\partial_\rho(A^\nu F^{\mu\rho}) = \frac{1}{4\pi}(\partial_\rho A^\nu)F^{\mu\rho} - qA^\nu\overline{\Psi}\gamma^\mu\Psi. \tag{17.46}$$

The above rhs holds because of the Maxwell equation 17.40. We observe that the electromagnetic contribution to the correction terms contains also the spinor fields. The contribution to the correction terms stemming from the spinor fields needs a bit more work. We use the complete antisymmetry of the Dirac spin tensor $\Sigma^{\mu\nu\rho}_{\text{Spinor}}(x)$ and conclude that the corresponding correction term is equal

$$\frac{1}{2}\partial_\rho\Sigma^{\mu\nu\rho}_{\text{Spinor}} = \frac{i}{16}\partial_\rho\left(\overline{\Psi}(\gamma^\mu[\gamma^\nu,\gamma^\rho] + [\gamma^\nu,\gamma^\rho]\gamma^\mu)\Psi\right). \tag{17.47}$$

In the subsequent calculation, we use repeatedly the basic algebraic identity $\{\gamma^\mu,\gamma^\nu\} = 2\eta^{\mu\nu}$ and the Dirac equations 17.41 until all products of Dirac matrices are eliminated. The result of this longer calculation is

$$\frac{1}{2}\partial_\rho\Sigma^{\mu\nu\rho}_{\text{Spinor}} = \frac{i}{4}\left((\partial^\nu\overline{\Psi})\gamma^\mu\Psi - (\partial^\mu\overline{\Psi})\gamma^\nu\Psi + \overline{\Psi}\gamma^\nu(\partial^\mu\Psi) - \overline{\Psi}\gamma^\mu(\partial^\nu\Psi)\right)$$
$$+ \frac{q}{2}\left(A^\nu\overline{\Psi}\gamma^\mu\Psi - A^\mu\overline{\Psi}\gamma^\nu\Psi\right). \tag{17.48}$$

We observe that the spinorial contribution to the correction terms contains also the electromagnetic field. By applying these correction terms to the CEMT $\Theta^{\mu\nu}_{\text{Total}}(x)$, we obtain the corresponding SEMT $T^{\mu\nu}_{\text{Total}}(x)$ in the form

$$T^{\mu\nu}_{\text{Total}} = \frac{i}{4}\left(\overline{\Psi}\gamma^\mu(\partial^\nu\Psi) + \overline{\Psi}\gamma^\nu(\partial^\mu\Psi) - (\partial^\mu\overline{\Psi})\gamma^\nu\Psi - (\partial^\nu\overline{\Psi})\gamma^\mu\Psi\right)$$
$$- \frac{q}{2}\left(A^\nu\overline{\Psi}\gamma^\mu\Psi + A^\mu\overline{\Psi}\gamma^\nu\Psi\right) + \frac{1}{4\pi}F^{\mu\rho}F_\rho{}^\nu - \eta^{\mu\nu}\mathscr{L}. \tag{17.49}$$

In a final step, we take into account that due to the Dirac equations it is

$$\frac{i}{2}\left(\overline{\Psi}\gamma^\mu(D_\mu\Psi) - (\overline{D_\mu\Psi})\gamma^\mu\Psi\right) - m\overline{\Psi}\Psi = 0, \tag{17.50}$$

and we introduce the covariant derivative $D_\mu(x)$ into all expressions. In this way, we obtain the desired *total SEMT of spinor electrodynamics*:

$$T^{\mu\nu}_{\text{Total}} = \frac{1}{4\pi}\left(F^{\mu\rho}F_\rho{}^\nu + \frac{1}{4}\eta^{\mu\nu}F^2\right)$$
$$+ \frac{i}{4}\left(\overline{\Psi}\gamma^\mu(D^\nu\Psi) + \overline{\Psi}\gamma^\nu(D^\mu\Psi) - (\overline{D^\mu\Psi})\gamma^\nu\Psi - (\overline{D^\nu\Psi})\gamma^\mu\Psi\right). \tag{17.51}$$

The resulting total SEMT corresponds again to the sum of the single free-field SEMTs with the minimal coupling 17.17 being applied. The total SEMT $T^{\mu\nu}_{\text{Total}}(x)$ is gauge invariant, symmetric, and locally conserved:

$$\boxed{\partial_\mu T^{\mu\nu}_{\text{Total}} = 0} \tag{17.52}$$

The non-conservation we encountered at the end of Section 16.2 for the open Maxwell system interacting with a fixed external source no longer occurs. The spinor electrodynamics system considered here is closed and the associated SEMT is locally conserved.

Further Reading

The original proposal of the gauge principle goes back to Weyl [96]. The rich theory of classical gauge fields, including topological objects, is treated in Rubakov [76]. With the background provided so far, a possible path forward is to study quantum field theory. Nice introductions discussing both, classical and quantum fields, are provided by Maggiore [56] and Ramond [69]. An accessible exposition on spontaneous symmetry breaking, quantum fields, and their application to the Standard Model of particle physics is given by Huang [43].

V

Riemannian Geometry

Connection and Geodesics

In this chapter we turn back to differential geometry and introduce the notions of connection and covariant derivative on a general manifold. These additional structures allow us to formulate tensor field equations on a manifold in an invariant way. There is a natural connection on every geometric manifold, the Levi-Civita connection, which we develop in detail. In addition, the integral formulae of Gauss and Stokes are revisited for the case of geometric manifolds. In the last section we consider the notion of parallel transport of a vector along a curve. It is particularly fruitful to consider the parallel transport of the tangent vector of a curve along that curve. This leads to geodesics and a generalized concept of minimal distance between points in a geometric manifold.

18.1 Connection and the Covariant Derivative

Field Equations on a Manifold

In the last chapters on field theory, we have formulated various physical laws through PDEs considered on Minkowski space $\mathbb{M}_d$. Some examples were the scalar field equations 15.37, 15.45, 15.48, the spinorial equations 15.54, 15.60, 15.63, or the Maxwell equations 15.82, 15.83. Also, the local conservation laws we have encountered, like 15.149, and numerous equations in the Chapters 16 and 17, are all formulated as PDEs. If we subsumed all these field equations within a general scheme, we could write them as a PDE in the form

$$\partial_{\lambda_1} \cdots \partial_{\lambda_l} \partial_{\mu_1} \cdots \partial_{\mu_m} F^{\mu_1 \cdots \mu_m}{}_{\nu_1 \ldots \nu_n}(x) = 0, \tag{18.1}$$

with the free indices being possibly further contracted with copies of the metric tensor $\eta_{\mu\nu}$, or other tensorial objects, like gamma matrices. The (m, n)-tensor field $F^{\mu_1 \cdots \mu_m}{}_{\nu_1 \ldots \nu_n}(x)$ is essentially a function of the basic fields $\phi_i(x)$ of the theory at hand. As we know, the Poincaré transformations

$$x'^{\mu} = \Lambda^{\mu}{}_{\nu} x^{\nu} + a^{\mu} \tag{18.2}$$

are the symmetry transformations of Minkowski space $\mathbb{M}_d$. These Poincaré transformations are also the ones that leave the tensor field equation 18.1 invariant. Indeed, under a Poincaré transformation, the expression in the lhs of 18.1 transforms like a $(0, l + n)$-tensor field. This means that the combination of Minkowski space, Poincaré transformations, and field equations in the form of 18.1, is a framework for a consistent physical description from an spacetime symmetry point of view.

DOI: 10.1201/9781003087748-18

The situation changes substantially if we consider a tensor field equation as above on a general differentiable manifold $\mathcal{M}$. At the present stage, we do not require the existence of a metric. We want to allow the most general coordinate transformations,

$$x'^{\mu} = \xi^{\mu}(x),\tag{18.3}$$

where $\xi^{\mu}(x)$ is an arbitrary differentiable function of x, labeled by the indices $\mu = 0, 1, \ldots, D$. Under such general coordinate transformations, the lhs of equation 18.1 does *not* transform anymore as a tensor field. Even if an equation like 18.1 was valid in one coordinate system x^{μ}, it would not be valid in another, transformed coordinate system x'^{μ}. We can see this already in the simple example of the expression $\partial_{\rho} V^{\mu}$, with $V^{\mu}(x)$ being a four-vector field. Each object by itself, the partial derivative operator ∂_{ρ} and the vector field $V^{\mu}(x)$, transforms tensorially. But the application of the partial derivative operator on the vector field produces an object that is not a tensor. Indeed, under a general coordinate transformation, the expression $\partial_{\rho} V^{\mu}$ changes to

$$\partial'_{\rho} V'^{\mu} = \frac{\partial x^{\sigma}}{\partial x'^{\rho}} \frac{\partial}{\partial x^{\sigma}} \left(\frac{\partial x'^{\mu}}{\partial x^{\nu}} V^{\nu} \right) = \frac{\partial x^{\sigma}}{\partial x'^{\rho}} \frac{\partial x'^{\mu}}{\partial x^{\nu}} \partial_{\sigma} V^{\nu} + \frac{\partial x^{\sigma}}{\partial x'^{\rho}} \frac{\partial^2 x'^{\mu}}{\partial x^{\sigma} \partial x^{\nu}} V^{\nu}.\tag{18.4}$$

The first term on the above rhs represents exactly the tensorial transformation law but the presence of the second term spoils the desired behavior. So the question arises if we are able to re-formulate the physical laws in a way that they remain invariant, and thus valid, under *arbitrary changes of coordinates*. As we have just seen, the question boils down to whether we can find a derivative operator whose application on tensor fields produces again tensor fields. In addition, the newly defined derivative operator should reduce to the simple partial derivative as soon as one considers only Minkowski space and Poincaré transformations. Let us denote this new, still to be determined differential operator by the symbol ∇_{ρ}. The defining property of this new differential operator is to produce properly transforming tensor fields. In our example above, the expression $\nabla_{\rho} V^{\mu}$ must transform as

$$\nabla'_{\rho} V'^{\mu} = \frac{\partial x^{\sigma}}{\partial x'^{\rho}} \frac{\partial x'^{\mu}}{\partial x^{\nu}} \nabla_{\sigma} V^{\nu},\tag{18.5}$$

i.e. like a $(1,1)$-tensor field. In the following, we are going to construct this new differential operator step by step.

Introduction of the Covariant Derivative

Let us consider a general manifold $\mathcal{M}$, which does not need to possess a metric. The local bases for the tangent and cotangent vector fields are denoted $\{\partial_{\mu}\}$ and $\{dx^{\mu}\}$ respectively. We are going to define the desired derivative operator ∇_{ρ} in such a way that the expression $\nabla_{\rho} V^{\mu}$ transforms as a $(1,1)$-tensor field. We want to provide the definition for arbitrary (m, n)-tensor fields as defined in 1.93 from the outset. The *covariant derivative* ∇ maps an (m, n)-tensor field ϕ to an $(m, n + 1)$-tensor field $\nabla \phi$ given in component representation as

$$\boxed{\nabla \phi = \nabla_{\rho} \phi^{\mu_1 \cdots \mu_m}{}_{\nu_1 \ldots \nu_n}(x)\, \partial_{\mu_1} \otimes \cdots \otimes \partial_{\mu_m} \otimes dx^{\rho} \otimes dx^{\nu_1} \otimes \cdots \otimes dx^{\nu_n}}\tag{18.6}$$

Here we use the notation

$$\nabla_{\rho} \phi^{\mu_1 \cdots \mu_m}{}_{\nu_1 \ldots \nu_n} \equiv (\nabla \phi)^{\mu_1 \cdots \mu_m}{}_{\rho\, \nu_1 \ldots \nu_n}.\tag{18.7}$$

The covariant derivative is required to satisfy the following four basic properties. *Linearity*:

$$\nabla(a\phi + b\psi) = a\nabla\phi + b\nabla\psi,\tag{18.8}$$

id for any numbers a, b and any tensor fields ϕ and ψ of the same valence. *Leibniz rule*:

$$\nabla(\phi \otimes \psi) = (\nabla\phi) \otimes \psi + \phi \otimes (\nabla\psi), \tag{18.9}$$

tensor fields ϕ and ψ of arbitrary valence. *Commutativity with index contraction*:

$$\nabla_\rho(\phi^{\mu_1...\lambda...\mu_m}{}_{\nu_1...\lambda...\nu_n}) = (\nabla\phi)^{\mu_1...\lambda...\mu_m}{}_{\rho\,\nu_1...\lambda...\nu_n}. \tag{18.10}$$

duction to partial derivative for scalar fields $f(x)$:

$$\nabla_\rho f = \partial_\rho f. \tag{18.11}$$

e above basic properties will be used during the course of determining the explicit form the covariant derivative operator. We carry out this analysis first for the simplest case of vector field $V^\mu(x)$ and generalize then to (m,n)-tensor fields. Our initial ansatz for the variant derivative formula of a vector field $V^\mu(x)$ is

$$\nabla_\rho V^\mu \equiv \partial_\rho V^\mu + \Gamma^\mu_{\nu\rho} V^\nu. \tag{18.12}$$

e new expression $\Gamma^\mu_{\nu\rho} V^\nu$ in the above rhs must be non-tensorial to compensate the non-sorial transformation behavior of the expression $\partial_\rho V^\mu$. In the following, we will address question of how the new quantity $\Gamma^\mu_{\nu\rho}(x)$ must transform.

nnection on a Manifold

viously, for the definition of a covariant derivative the d^3 quantities $\Gamma^\mu_{\nu\rho}(x)$ are decisive. ese quantities are called the *connection coefficients* or the *Christoffel symbols* (*Elwin uno Christoffel*). We will derive their transformation properties now by requiring that covariant derivative expression $\nabla_\rho V^\mu$ transforms as a $(1,1)$-tensor field. We take the satz 18.12 and transform both sides of the equation. The transformed lhs yields

$$\nabla'_\rho V'^\mu = \frac{\partial x^\sigma}{\partial x'^\rho}\frac{\partial x'^\mu}{\partial x^\nu}(\partial_\sigma V^\nu + \Gamma^\nu_{\alpha\sigma}V^\alpha), \tag{18.13}$$

ereas the rhs becomes

$$\partial'_\rho V'^\mu + \Gamma'^\mu_{\nu\rho}V'^\nu = \frac{\partial x^\sigma}{\partial x'^\rho}\frac{\partial x'^\mu}{\partial x^\nu}\partial_\sigma V^\nu + \frac{\partial x^\sigma}{\partial x'^\rho}\frac{\partial^2 x'^\mu}{\partial x^\sigma \partial x^\nu}V^\nu + \Gamma'^\mu_{\nu\rho}\frac{\partial x'^\nu}{\partial x^\alpha}V^\alpha. \tag{18.14}$$

e lhs and rhs are to be equated and the first two terms on each side are identical. By nceling the common factor V^α and solving for the transformed quantity $\Gamma'^\mu_{\nu\rho}(x')$, we tain the result

$$\boxed{\Gamma'^\mu_{\nu\rho} = \frac{\partial x'^\mu}{\partial x^\alpha}\frac{\partial x^\beta}{\partial x'^\nu}\frac{\partial x^\gamma}{\partial x'^\rho}\Gamma^\alpha_{\beta\gamma} - \frac{\partial^2 x'^\mu}{\partial x^\alpha \partial x^\beta}\frac{\partial x^\alpha}{\partial x'^\nu}\frac{\partial x^\beta}{\partial x'^\rho}} \tag{18.15}$$

is is how the connection coefficients $\Gamma^\mu_{\nu\rho}(x)$ must transform under general transformati-s of the points of the manifold. Except in the case of affine-linear transformations, the ristoffel connection coefficients do *not* transform like a $(1,2)$-tensor due to the second m in the rhs of 18.15. This non-tensorial term* can be written differently. Starting from identity

$$\frac{\partial x'^\mu}{\partial x^\alpha}\frac{\partial x^\alpha}{\partial x'^\nu} = \delta^\mu_\nu, \tag{18.16}$$

---

*This is why the indices placement is like $\Gamma^\mu_{\nu\rho}$ and not like $\Gamma^\mu{}_{\nu\rho}$ as it would be in the usual tensor notation.

we apply the operator $\partial/\partial x^\beta$ to it and obtain

$$\frac{\partial^2 x'^\mu}{\partial x^\alpha \partial x^\beta}\frac{\partial x^\alpha}{\partial x'^\nu} + \frac{\partial x'^\mu}{\partial x^\alpha}\frac{\partial^2 x^\alpha}{\partial x'^\nu \partial x'^\gamma}\frac{\partial x'^\gamma}{\partial x^\beta} = 0. \tag{18.17}$$

By contracting with $\partial x^\beta/\partial x'^\rho$, we get

$$\frac{\partial^2 x'^\mu}{\partial x^\alpha \partial x^\beta}\frac{\partial x^\alpha}{\partial x'^\nu}\frac{\partial x^\beta}{\partial x'^\rho} = -\frac{\partial x'^\mu}{\partial x^\alpha}\frac{\partial^2 x^\alpha}{\partial x'^\nu \partial x'^\rho}. \tag{18.18}$$

Therefore, the transformation formula 18.15 can be written equivalently as

$$\boxed{\Gamma'^\mu_{\nu\rho} = \frac{\partial x'^\mu}{\partial x^\alpha}\frac{\partial x^\beta}{\partial x'^\nu}\frac{\partial x^\gamma}{\partial x'^\rho}\Gamma^\alpha_{\beta\gamma} + \frac{\partial x'^\mu}{\partial x^\alpha}\frac{\partial^2 x^\alpha}{\partial x'^\nu \partial x'^\rho}} \tag{18.19}$$

Note that with $\Gamma^\mu_{\nu\rho}(x)$, the index-flipped quantity $\Gamma^\mu_{\rho\nu}(x)$ is also a valid connection. The difference

$$\Gamma^\mu_{\nu\rho} - \Gamma^\mu_{\rho\nu} \tag{18.20}$$

is in fact a $(1,2)$-tensor field, called the *torsion tensor*. More generally, for any two connections $\Gamma^\mu_{\nu\rho}(x)$ and $\widetilde{\Gamma}^\mu_{\nu\rho}(x)$ the difference $\Gamma^\mu_{\nu\rho} - \widetilde{\Gamma}^\mu_{\nu\rho}$ constitutes a $(1,2)$-tensor field. We note also that we can add to any given connection $\Gamma^\mu_{\nu\rho}(x)$ an arbitrary $(1,2)$-tensor field $G^\mu{}_{\nu\rho}(x)$ and the sum

$$\Gamma^\mu_{\nu\rho} + G^\mu{}_{\nu\rho} \tag{18.21}$$

is also a connection. Note also that if we restrict ourselves to the set of rigid Poincaré transformations, a connection transforms like a tensor field, which is a manifestation of how the connection is present also in this simple case. In the particular case of Minkowski space, the vanishing of the connection in one coordinate system means that it vanishes in any other coordinate system obtained through Poincaré transformations.

The *connection* $\Gamma^\mu_{\nu\rho}(x)$ is an additional structure on a manifold, independent of other structures, such as a metric. A manifold with a connection, $(\mathcal{M}, \Gamma^\mu_{\nu\rho})$, is called a *connected manifold*. However, if we consider a geometric manifold $(\mathcal{M}, g_{\mu\nu})$ possessing a metric $g_{\mu\nu}(x)$, then there exists a natural connection which is specific for this geometric manifold. We will discuss this fact in the next section. Whenever a connection is present, a covariant derivative can be defined. Conversely, a covariant derivative defines a certain connection, as we will demonstrate in the following.

18.2   Formulae for the Covariant Derivative

Covariant Derivative of Tensors

We summarize here the basic formulae for the *covariant derivative*, which we employ in practical calculations. For a vector field $V^\mu(x)$, it is

$$\boxed{\nabla_\rho V^\mu = \partial_\rho V^\mu + \Gamma^\mu_{\nu\rho}V^\nu} \tag{18.22}$$

Considering the coordinate-independent result of the covariant differentiation operation, the expression $\nabla_\rho V^\mu$ represents the components of the $(1,1)$-tensor field ∇V, which we write in components as

$$\nabla V = (\nabla_\rho V^\mu)\, \partial_\mu \otimes \mathrm{d}x^\rho. \tag{18.23}$$

Here, we have used the usual $(1,1)$-tensor field basis $\{\partial_\mu \otimes \mathrm{d}x^\rho\}$. The covariant derivative of a scalar field $f(x)$ is

$$\boxed{\nabla_\rho f = \partial_\rho f} \qquad (18.24)$$

For the covariant derivative of a covector field $\omega_\nu(x)$ it is not immediately obvious what the correct formula is. Let us start with the ansatz

$$\nabla_\rho \omega_\nu = \partial_\rho \omega_\nu + \tilde{\Gamma}^\mu_{\nu\rho} \omega_\mu, \qquad (18.25)$$

in which we have to determine the unknown coefficients $\tilde{\Gamma}^\mu_{\nu\rho}(x)$. Consider the scalar contraction $\omega_\nu V^\nu$ of the covector field ω with a vector field V. Applying the covariant derivative to this scalar field yields

$$\nabla_\rho(\omega_\nu V^\nu) = \partial_\rho(\omega_\nu V^\nu) = (\partial_\rho \omega_\nu)V^\nu + \omega_\nu(\partial_\rho V^\nu). \qquad (18.26)$$

On the other hand, the Leibniz rule yields

$$\nabla_\rho(\omega_\nu V^\nu) = (\partial_\rho \omega_\nu + \tilde{\Gamma}^\mu_{\nu\rho} \omega_\mu)V^\nu + \omega_\nu(\partial_\rho V^\nu + \Gamma^\nu_{\lambda\rho}V^\lambda). \qquad (18.27)$$

By equating both sides and solving for the unknown coefficients, we obtain

$$\tilde{\Gamma}^\mu_{\nu\rho} = -\Gamma^\mu_{\nu\rho}. \qquad (18.28)$$

Therefore, the formula for the covariant derivative of a covector field $\omega_\nu(x)$ reads

$$\boxed{\nabla_\rho \omega_\nu = \partial_\rho \omega_\nu - \Gamma^\mu_{\nu\rho} \omega_\mu} \qquad (18.29)$$

The covariant derivative of a $(1,1)$-tensor field $\phi^\mu{}_\nu(x)$ is clear now and we can write down the corresponding formula. This can be derived also by considering a suitable scalar contraction. It is

$$\boxed{\nabla_\rho \phi^\mu{}_\nu = \partial_\rho \phi^\mu{}_\nu + \Gamma^\mu_{\alpha\rho} \phi^\alpha{}_\nu - \Gamma^\alpha_{\nu\rho} \phi^\mu{}_\alpha} \qquad (18.30)$$

The generalization of the above formula to higher order tensor fields is clear. Starting from an (m,n)-tensor field $\phi^{\mu_1\dots\mu_m}{}_{\nu_1\dots\nu_n}(x)$, its covariant derivative is the $(m, n+1)$-tensor field $\nabla_\rho\phi^{\mu_1\dots\mu_m}{}_{\nu_1\dots\nu_n}(x)$ given as:

$$
\begin{aligned}
\nabla_\rho \phi^{\mu_1\dots\mu_m}{}_{\nu_1\dots\nu_n} =\ & \partial_\rho \phi^{\mu_1\dots\mu_m}{}_{\nu_1\dots\nu_n} \\
& + \Gamma^{\mu_1}_{\alpha\rho} \phi^{\alpha\,\mu_2\dots\mu_m}{}_{\nu_1\dots\nu_n} + \dots + \Gamma^{\mu_m}_{\alpha\rho} \phi^{\mu_1\dots\mu_{m-1}\alpha}{}_{\nu_1\dots\nu_n} \\
& - \Gamma^\alpha_{\nu_1\rho} \phi^{\mu_1\dots\mu_m}{}_{\alpha\,\nu_2\dots\nu_n} - \dots - \Gamma^\alpha_{\nu_n\rho} \phi^{\mu_1\dots\mu_m}{}_{\nu_1\dots\nu_{n-1}\alpha}.
\end{aligned}
\qquad (18.31)
$$

This is the general formula for the covariant derivative of proper tensor fields.

Covariant Derivative of Tensor Densities

For tensor density fields the formula for the covariant derivative is slightly more complicated. We recall the Jacobian determinant $J(x)$ of a general coordinate transformation, defined as

$$J = \det\left(\frac{\partial x'}{\partial x}\right), \qquad (18.32)$$

appearing in the transformation law of tensor densities. Our analysis here follows the approach as laid out in [55]. In order to derive the formula for general tensor densities, we consider the simplest case of a scalar density $S(x)$ transforming as

$$S'(x') = J(x)^w S(x),$$
(18.33)

with the weight $w \in \mathbb{Z}$. We require that the covariant derivative of the scalar density $S(x)$ is a $(0,1)$-tensor density $\nabla_\rho S(x)$ with the same weight w. Starting from

$$S(x) = S'(x') J(x)^{-w},$$
(18.34)

we differentiate partially with respect to x^μ and obtain

$$\frac{\partial S}{\partial x^\mu} = \frac{\partial S'}{\partial x'^\rho} \frac{\partial x'^\rho}{\partial x^\mu} J^{-w} - w J^{-w-1} \frac{\partial J}{\partial x^\mu} S'.$$
(18.35)

If the gradient operation would lead to a $(0,1)$-tensor density, only the first term in the rhs would appear. The second term shows that we need to take care of the partial derivative $\partial J / \partial x^\mu$ of the Jacobian explicitly. Let us denote the cofactor of the matrix element $\partial x'^\sigma / \partial x^\nu$ by $C^\nu{}_\sigma$, see appendix B.1. It satisfies

$$\frac{\partial x'^\sigma}{\partial x^\nu} C^\nu{}_\tau = J \delta^\sigma_\tau.$$
(18.36)

Because of

$$\frac{\partial x'^\sigma}{\partial x^\nu} \frac{\partial x^\nu}{\partial x'^\tau} = \delta^\sigma_\tau,$$
(18.37)

it is

$$C^\nu{}_\tau = J \frac{\partial x^\nu}{\partial x'^\tau}.$$
(18.38)

According to the partial derivative formula for a determinant $a(t) = \det A(t)$, see appendix B.1,

$$\frac{\partial a}{\partial t} = C^\nu{}_\tau \frac{\partial A^\tau{}_\nu}{\partial t},$$
(18.39)

we have in our case

$$\frac{\partial J}{\partial x^\mu} = J \frac{\partial x^\nu}{\partial x'^\tau} \frac{\partial}{\partial x^\mu} \left(\frac{\partial x'^\tau}{\partial x^\nu} \right).$$
(18.40)

Thus, we have

$$\frac{\partial S}{\partial x^\mu} = J^{-w} \left(\frac{\partial S'}{\partial x'^\rho} \frac{\partial x'^\rho}{\partial x^\mu} - w \frac{\partial x^\nu}{\partial x'^\tau} \frac{\partial^2 x'^\tau}{\partial x^\mu \partial x^\nu} S' \right).$$
(18.41)

Now the factor

$$\frac{\partial x^\nu}{\partial x'^\tau} \frac{\partial^2 x'^\tau}{\partial x^\mu \partial x^\nu}$$
(18.42)

appears in the transformation law 18.19 of the connection coefficients, which in the present case reads

$$\Gamma^\nu_{\mu\nu} = \frac{\partial x^\nu}{\partial x'^\alpha} \frac{\partial x'^\beta}{\partial x^\mu} \frac{\partial x'^\gamma}{\partial x^\nu} \Gamma'^\alpha_{\beta\gamma} + \frac{\partial x^\nu}{\partial x'^\tau} \frac{\partial^2 x'^\tau}{\partial x^\mu \partial x^\nu},$$
(18.43)

leading to the equation

$$\frac{\partial x^\nu}{\partial x'^\tau} \frac{\partial^2 x'^\tau}{\partial x^\mu \partial x^\nu} = \Gamma^\nu_{\mu\nu} - \frac{\partial x'^\beta}{\partial x^\mu} \Gamma'^\alpha_{\beta\alpha}.$$
(18.44)

Note that the quantity $\Gamma^\nu_{\mu\nu}(x)$ is actually a sum over the repeated index,

$$\Gamma^\nu_{\mu\nu} \equiv \sum_\nu \Gamma^\nu_{\mu\nu}.$$
(18.45)

We have therefore

$$\frac{\partial S}{\partial x^\mu} = J^{-w} \left(\frac{\partial S'}{\partial x'^\rho} \frac{\partial x'^\rho}{\partial x^\mu} - w \Gamma^\nu_{\mu\nu} S' + w \frac{\partial x'^\beta}{\partial x^\mu} \Gamma'^\alpha_{\beta\alpha} S' \right). \tag{18.46}$$

By using $J^{-w} S' = S$ and by arranging the terms above in a suitable way, we arrive at the relation

$$\frac{\partial S}{\partial x^\mu} + w \Gamma^\nu_{\mu\nu} S = J^{-w} \frac{\partial x'^\rho}{\partial x^\mu} \left(\frac{\partial S'}{\partial x'^\rho} + w \Gamma'^\alpha_{\rho\alpha} S' \right). \tag{18.47}$$

This relation can be inverted and written as

$$\frac{\partial S'}{\partial x'^\rho} + w \Gamma'^\alpha_{\rho\alpha} S' = J^w \frac{\partial x^\mu}{\partial x'^\rho} \left(\frac{\partial S}{\partial x^\mu} + w \Gamma^\alpha_{\mu\alpha} S \right). \tag{18.48}$$

We realize that the form of the expression in the above lhs provides the recipe for the sought after covariant derivative of a scalar density. Thus, the covariant derivative $\nabla_\rho S(x)$ of a scalar density $S(x)$ with weight w is introduced as

$$\boxed{\nabla_\rho S = \partial_\rho S + w \Gamma^\alpha_{\rho\alpha} S} \tag{18.49}$$

The additional term $w \Gamma^\alpha_{\rho\alpha} S$ appearing for $w \neq 0$ in fact applies generally for any tensor density. For instance, for a $(1,1)$-tensor density $T^\mu{}_\nu(x)$ with weight w, the corresponding covariant derivative formula reads

$$\boxed{\nabla_\rho T^\mu{}_\nu = \partial_\rho T^\mu{}_\nu + \Gamma^\mu_{\alpha\rho} T^\alpha{}_\nu - \Gamma^\alpha_{\nu\rho} T^\mu{}_\alpha + w \Gamma^\alpha_{\rho\alpha} T^\mu{}_\nu} \tag{18.50}$$

The additional term for a general tensor density $T_i(x)$ of arbitrary valence is again equal $w \Gamma^\alpha_{\rho\alpha} T_i$ and for $w = 0$ we recover the formula for proper tensor fields.

Directional Covariant Derivative

One can introduce also the *directional covariant derivative* of a vector field $V = V^\mu \partial_\mu$ with respect to a vector field $X = X^\rho \partial_\rho$ as the vector field $\nabla_X V$ defined as

$$\nabla_X V \equiv X^\rho (\nabla_\rho V^\mu) \, \partial_\mu. \tag{18.51}$$

Geometrically, one projects the covariant derivative $\nabla_\rho V^\mu$ along the direction of the vector field X^ρ. Written out, this is

$$\nabla_X V = X^\rho (\partial_\rho V^\mu + \Gamma^\mu_{\nu\rho} V^\nu) \, \partial_\mu. \tag{18.52}$$

Here we have used the usual basis vectors $\{\partial_\mu\}$. The directional covariant derivative does not change the tensor valence. Generally, it maps an (m, n)-tensor field ϕ to a tensor field $\nabla_X \phi$ of the same valence. We list some of the basic properties of the directional covariant derivative, which can be directly deduced from the previous results. For any numbers a, b and any tensor fields ϕ and ψ of the same valence the *linearity*

$$\nabla_X (a\phi + b\psi) = a\nabla_X \phi + b\nabla_X \psi \tag{18.53}$$

holds. Moreover, the linearity property

$$\nabla_{aX+bY} \phi = a\nabla_X \phi + b\nabla_Y \phi \tag{18.54}$$

holds, where again a and b are numbers and ϕ any tensor field. Remarkably, the last linearity property holds even if one replaces the numbers a and b by scalar functions $f(x)$ and $h(x)$. It is then

$$\nabla_{fX+hY}\phi = f\nabla_X\phi + h\nabla_Y\phi. \tag{18.55}$$

A connection fulfilling the two properties 18.53 and 18.55 is called a *Koszul connection* (*Jean-Louis Koszul*). The *Leibniz rule* holds generally for any two tensor fields ϕ and ψ of arbitrary valence,

$$\nabla_X(\phi \otimes \psi) = (\nabla_X\phi) \otimes \psi + \phi \otimes (\nabla_X\psi). \tag{18.56}$$

For a scalar function $f(x)$ and a tensor field ϕ, the Leibniz rule becomes

$$\nabla_X(f\phi) = (Xf)\phi + f\nabla_X\phi. \tag{18.57}$$

Here we have used that covariant directional derivative $\nabla_X f$ of a scalar function is

$$\nabla_X f = X^\rho \partial_\rho f = Xf, \tag{18.58}$$

i.e. it coincides with the simple directional derivative of the function. We note also the product formula for the natural pairing $\langle \omega, V \rangle$ of a covector field ω and a vector field V,

$$\nabla_X \langle \omega, V \rangle = \langle \nabla_X\omega, V \rangle + \langle \omega, \nabla_X V \rangle. \tag{18.59}$$

The directional covariant derivative can be isolated and written as an operator as

$$\nabla_X = X^\rho \nabla_\rho. \tag{18.60}$$

For the special case that X is a basis vector ∂_μ, we obtain

$$\nabla_{\partial_\rho} = \nabla_\rho. \tag{18.61}$$

In other words, the covariant derivative operator ∇_ρ is essentially the directional covariant derivative ∇_{∂_ρ} with respect to the basis vector field ∂_ρ. A straightforward calculation, which is left to the reader (*exercise 18.1*), reveals that

$$\nabla_\rho \partial_\nu = \Gamma^\mu_{\nu\rho} \partial_\mu. \tag{18.62}$$

It means that the connection coefficients are the components of a basis expansion, when one calculates the covariant derivatives of these basis vectors. Therefore, in differential geometric notation, the relation

$$\Gamma^\mu_{\nu\rho} = \langle \mathrm{d}x^\mu, \nabla_\rho \partial_\nu \rangle \tag{18.63}$$

holds, where we use the natural pairing of the local basis elements $\mathrm{d}x^\mu$ and $\nabla_\rho \partial_\nu$. This actually provides an explicit formula for the connection coefficients $\Gamma^\mu_{\nu\rho}(x)$ in the case a covariant derivative is known.

Lie Derivative vs. Directional Covariant Derivative

Let us compare the notion of the Lie derivative with the directional covariant derivative, for simplicity first for the case of a vector field V. Both derivative operations produce again vector fields, $\mathcal{L}_X V$ and $\nabla_X V$ respectively, given by the formulae

$$\mathcal{L}_X V^\mu = X^\alpha \partial_\alpha V^\mu - V^\alpha \partial_\alpha X^\mu, \tag{18.64}$$

and

$$\nabla_X V^\mu = X^\alpha \partial_\alpha V^\mu + V^\alpha \Gamma^\mu_{\alpha\beta} X^\beta. \tag{18.65}$$

We observe that both formulae contain the basic directional derivative $X^\alpha \partial_\alpha V^\mu$. The additional term in the Lie derivative depends on the rate of change of the vector field X, measured by the factor $\partial_\alpha X^\mu$. In contrast, the additional term in the directional covariant derivative depends only on the local value of the vector field X, but not on its rate of change. This is the reason why the directional covariant derivative has the linearity property 18.55 with functions $f(x)$ and $h(x)$, which does not hold for the Lie derivative.

The Lie derivative, despite the fact that it does not depend on the existence of a metric or a connection on the manifold, can still be expressed in a manifestly covariant way. We assume the existence of a *symmetric* connection here. It can be easily seen that for any two vector fields X and V, the Lie derivative of V with respect to X is given by

$$\mathcal{L}_X V^\mu = X^\alpha \nabla_\alpha V^\mu - V^\alpha \nabla_\alpha X^\mu. \tag{18.66}$$

Written independently of coordinates, this reads

$$\mathcal{L}_X V = \nabla_X V - \nabla_V X. \tag{18.67}$$

A similar formula holds for the Lie derivative of a covector field ω with respect to a given vector field X. It is then

$$\mathcal{L}_X \omega = \nabla_X \omega + \langle \omega, \nabla X \rangle. \tag{18.68}$$

Both formulae can be obtained easily by using the previously discussed properties of the directional covariant derivative. It is even more remarkable that the Lie derivative can be expressed by formulae where ∂_ρ is exchanged by ∇_ρ even in the case of arbitrary tensor density fields. To be concrete, let us consider a $(1,1)$-tensor density field $T^\mu{}_\nu(x)$ with weight w. Then, the Lie derivative $\mathcal{L}_X T^\mu{}_\nu$ with respect to X can be written as (*exercise 18.2*)

$$\mathcal{L}_X T^\mu{}_\nu = X^\alpha \nabla_\alpha T^\mu{}_\nu - T^\alpha{}_\nu \nabla_\alpha X^\mu + T^\mu{}_\alpha \nabla_\nu X^\alpha - w \left(\nabla_\alpha X^\alpha \right) T^\mu{}_\nu. \tag{18.69}$$

In other words, we can always write the Lie derivative of an arbitrary tensor density field by employing the covariant derivative operator instead of the partial derivative.

18.3   The Levi-Civita Connection

Introducing a Metric

So far we have introduced the connection and the associated covariant derivative without any reference to a metric. However, in our physical applications, the manifolds $(\mathcal{M}, g_{\mu\nu})$ considered do indeed have a metric. The metric tensor field $g_{\mu\nu}(x)$, as a real symmetric matrix, at each point x can be brought into the diagonal form

$$g_{\mu\nu}(x) = \text{diag}(\underbrace{+1, \ldots, +1}_{p \text{ times}}, \underbrace{-1, \ldots, -1}_{n \text{ times}}) \tag{18.70}$$

by a suitable similarity transformation. This coordinate transformation is varying from point to point, of course. However, the count p of eigenvalues $+1$ and the count n of eigenvalues -1 on the diagonal remains invariant throughout all points x. The pair (p, n) is called the *signature* of the metric. It is $p + n = d$, the dimension of the manifold. If $n = 0$ one speaks of a *Riemannian manifold*. If $n > 0$ one speaks of a *pseudo-Riemannian*, or *semi-Riemannian* manifold. In the special case that the signature is $(1, D)$ or $(D, 1)$, i.e. only one diagonal entry has an opposite sign compared to all other entries, the manifold is called *Lorentzian*. In the following, if not stated differently, we will assume to have a Lorentzian manifold with the signature $(1, -1, \ldots, -1)$ at hand, the value -1 appearing D times. The dimensionality of the Lorentzian manifold is $d = 1 + D$.

Metric Compatibility

Given a geometric manifold $(\mathcal{M}, g)$, we claim that it possesses a naturally defined and unique *metric connection* $\Gamma^\mu_{\nu\rho}(x)$ such that

$$\boxed{\nabla_\rho g_{\mu\nu} = 0 \quad \text{and} \quad \Gamma^\mu_{\nu\rho} = \Gamma^\mu_{\rho\nu}} \tag{18.71}$$

for all index values of ρ, μ, ν. We have to prove the existence of the metric connection and the validity of the above. The first condition,

$$\nabla_\rho g_{\mu\nu} = 0, \tag{18.72}$$

is called the *metric compatibility* or *metric concordance*. The second condition,

$$\Gamma^\mu_{\nu\rho} = \Gamma^\mu_{\rho\nu}, \tag{18.73}$$

is equivalent to the *vanishing of the torsion*, expressed through the symmetry of the two lower indices. Writing out the metric compatibility, it reads

$$\partial_\rho g_{\mu\nu} - \Gamma^\alpha_{\mu\rho} g_{\alpha\nu} - \Gamma^\alpha_{\nu\rho} g_{\mu\alpha} = 0. \tag{18.74}$$

Permutation of the indices ρ, μ, ν yields the two equations

$$\partial_\mu g_{\nu\rho} - \Gamma^\alpha_{\nu\mu} g_{\alpha\rho} - \Gamma^\alpha_{\rho\mu} g_{\nu\alpha} = 0, \tag{18.75}$$

and

$$\partial_\nu g_{\rho\mu} - \Gamma^\alpha_{\rho\nu} g_{\alpha\mu} - \Gamma^\alpha_{\mu\nu} g_{\rho\alpha} = 0. \tag{18.76}$$

By subtracting the last two equations from the previous one and by using the indices symmetry of the connection, we obtain

$$\partial_\rho g_{\mu\nu} - \partial_\mu g_{\nu\rho} - \partial_\nu g_{\rho\mu} + 2\Gamma^\alpha_{\mu\nu} g_{\rho\alpha} = 0. \tag{18.77}$$

Contracting with $g^{\rho\sigma}$ and renaming the indices subsequently yields the result:

$$\boxed{\Gamma^\mu_{\nu\rho} = \frac{1}{2} g^{\mu\sigma} (\partial_\nu g_{\rho\sigma} + \partial_\rho g_{\nu\sigma} - \partial_\sigma g_{\nu\rho})} \tag{18.78}$$

This formula for the metric connection $\Gamma^\mu_{\nu\rho}(x)$ is a direct consequence of two conditions 18.71. One can inspect explicitly that the above expression for $\Gamma^\mu_{\nu\rho}(x)$ indeed transforms as a connection. This connection is called the *Levi-Civita connection* (*Tullio Levi-Civita*) or the *Riemannian connection*. The explicit formula ensures the claimed uniqueness. Conversely, starting from the formula 18.78, the two conditions 18.71 are derived. Therefore, 18.71 and 18.78 are equivalent. The Levi-Civita connection has $d^2(d+1)/2$ independent components. *From here onward in this book we will deal solely with Levi-Civita connections possessing a vanishing torsion.*

Useful Formulae

The metric compatibility has some additional consequences. For the inverse metric $g^{\mu\nu}(x)$, we have also

$$\nabla_\rho g^{\mu\nu} = 0, \tag{18.79}$$

which follows from the identity $g_{\mu\nu} g^{\mu\nu} = d$. Another consequence of the metric compatibility, being immensely useful in practical calculations, is that the metric and the covariant derivative commute, so that we can write, for instance,

$$g^{\rho\sigma} \nabla_\rho V_\sigma = \nabla_\rho (g^{\rho\sigma} V_\sigma) = \nabla_\rho V^\rho. \tag{18.80}$$

We can use also the usual recipe for raising the index of ∇_ρ in the form

$$\nabla^\sigma = g^{\rho\sigma}\nabla_\rho. \tag{18.81}$$

For the Kronecker delta and the epsilon tensor, we have

$$\nabla_\rho \delta^\mu_\nu = 0 \quad \text{and} \quad \nabla_\rho \epsilon^{\mu_1\cdots\mu_d} = 0. \tag{18.82}$$

Now let us look at the determinant $g = \det(g_{\mu\nu})$ of the metric. In appendix B.1, we list a general formula for the derivative of the determinant $a \equiv \det A$ of a matrix A. This result can be written also as a variational equation as

$$\delta a = a\,(A^{-1})^\nu{}_\mu \delta A^\mu{}_\nu. \tag{18.83}$$

In the case of the metric determinant g, this reads

$$\delta g = g\,g^{\mu\nu}\delta g_{\mu\nu}. \tag{18.84}$$

Translated to the gradient $\partial_\rho g$ of the determinant, the formula becomes

$$\partial_\rho g = g\,g^{\mu\nu}\partial_\rho g_{\mu\nu}. \tag{18.85}$$

Starting from the identity $g^{\mu\nu}g_{\nu\sigma} = \delta^\mu_\sigma$, we deduce that

$$\delta g^{\mu\nu} = -g^{\mu\rho}g^{\nu\sigma}\delta g_{\rho\sigma}. \tag{18.86}$$

This is an important formula to remember. Consequently, the two previous formulae lead also to

$$\delta g = -g\,g_{\mu\nu}\delta g^{\mu\nu} \quad \text{and} \quad \partial_\rho g = -g\,g_{\mu\nu}\partial_\rho g^{\mu\nu}. \tag{18.87}$$

Next, we examine the square root $\sqrt{|g|}$, where $|g| \geq 0$ is the absolute value of the determinant, i.e. $g = |g|\,\mathrm{sgn}(g)$. A direct calculation shows that

$$\delta\sqrt{|g|} = \frac{1}{2}\sqrt{|g|}\,g^{\mu\nu}\delta g_{\mu\nu}, \tag{18.88}$$

and similarly

$$\partial_\rho\sqrt{|g|} = \frac{1}{2}\sqrt{|g|}\,g^{\mu\nu}\partial_\rho g_{\mu\nu}. \tag{18.89}$$

For the contracted Levi-Civita connection $\Gamma^\alpha_{\rho\alpha}(x)$ we have

$$\Gamma^\alpha_{\rho\alpha} = \Gamma^\alpha_{\alpha\rho} = \frac{1}{2}g^{\mu\nu}\partial_\rho g_{\mu\nu} = \frac{\partial_\rho\sqrt{|g|}}{\sqrt{|g|}} = \partial_\rho(\ln\sqrt{|g|}). \tag{18.90}$$

The square root $\sqrt{|g|}$ is a scalar density with weight $w = -1$, so according to our general formula 18.49, we obtain immediately

$$\nabla_\rho\sqrt{|g|} = 0. \tag{18.91}$$

Furthermore, given any vector field $V^\mu(x)$, the *covariant divergence* $\nabla_\rho V^\rho(x)$ of it can be calculated as

$$\nabla_\rho V^\rho = \frac{1}{\sqrt{|g|}}\partial_\rho(\sqrt{|g|}V^\rho). \tag{18.92}$$

This formula will be used soon in the Gauss divergence theorem. We note also a similar formula valid for any antisymmetric tensor field, $A^{\mu\nu}(x) = -A^{\nu\mu}(x)$,

$$\nabla_\rho A^{\rho\nu} = \frac{1}{\sqrt{|g|}}\partial_\rho(\sqrt{|g|}A^{\rho\nu}). \tag{18.93}$$

In the special case the vector field $V^\mu(x)$ above is given as the gradient of a scalar field $\phi(x)$, i.e.

$$V^\rho = g^{\rho\sigma}\partial_\sigma\phi = \nabla^\rho\phi, \tag{18.94}$$

we obtain the formula

$$\nabla_\rho\nabla^\rho\phi = \frac{1}{\sqrt{|g|}}\partial_\rho(\sqrt{|g|}g^{\rho\sigma}\partial_\sigma\phi) \tag{18.95}$$

for the *covariant d'Alembert operator*,

$$g^{\rho\sigma}\nabla_\rho\nabla_\sigma = \nabla^\sigma\nabla_\sigma. \tag{18.96}$$

For all practical purposes, the above formula for the covariant d'Alembert operator can be used equally well for curved manifolds as for curvilinear coordinates on Euclidean space.

Connection of the 2-Sphere

We have collected a large set of general results, so let us look at an example of a Levi-Civita connection. The 2-sphere $S^2(a)$ of radius a is a geometric manifold with the metric

$$ds^2 = a^2(d\theta^2 + \sin^2\theta\,d\varphi^2), \tag{18.97}$$

as derived in Section 2.3, with the two spherical coordinates $0 \le \theta \le \pi$ and $0 \le \varphi < 2\pi$. A straightforward calculation (*exercise 18.3*) shows that the only non-vanishing connection coefficients are

$$\left.\begin{array}{l} \Gamma^\theta_{\varphi\varphi} = -\sin\theta\,\cos\theta \\ \Gamma^\varphi_{\theta\varphi} = \Gamma^\varphi_{\varphi\theta} = \cot\theta \end{array}\right\}. \tag{18.98}$$

Note that there is no summation over repeated indices above. Within the used set of coordinates, the connection coefficients depend only on the θ-coordinate.

Gauss Divergence Formula for Vector Fields on Geometric Manifolds

Let us recall the Gauss divergence theorem 2.159 for a vector density, as derived in Section 2.6, and examine how it can be re-expressed when a metric is present. In our d-dimensional pseudo-Riemannian manifold, we consider a d-dimensional region U with a closed boundary ∂U. The boundary ∂U is a $(d-1)$-dimensional hypersurface. Further, we consider a vector density field $a^\mu(x)$ with weight $w = -1$, which we can always write as

$$a^\mu = \sqrt{|g|}\,V^\mu, \tag{18.99}$$

with $V^\mu(x)$ being a proper vector field with weight $w = 0$. For any vector density $a^\mu(x)$ with weight $w = -1$, the equation $\partial_\mu a^\mu = \nabla_\mu a^\mu$ holds. By using additionally the property $\nabla_\mu\sqrt{|g|} = 0$, we obtain

$$\partial_\mu a^\mu = \sqrt{|g|}\,\nabla_\mu V^\mu. \tag{18.100}$$

So the lhs of the Gauss divergence formula 2.159 reads now

$$\int_U (\nabla_\mu V^\mu)\sqrt{|g|}\,dx^0 \wedge \cdots \wedge dx^{d-1}. \tag{18.101}$$

e rhs of the divergence formula 2.159 consists of the integral expression

$$\oint_{\partial U} V^\mu \sqrt{|g|}\, d\sigma_\mu. \tag{18.102}$$

employing the d-dimensional invariant volume form

$$\mathrm{dvol}_g \equiv \sqrt{|g|}\, \mathrm{d}^d x = \sqrt{|g|}\, \mathrm{d}x^0 \wedge \cdots \wedge \mathrm{d}x^{d-1} \tag{18.103}$$

d the $(d-1)$-dimensional hypersurface element

$$\mathrm{d}\Sigma_\mu \equiv \sqrt{|g|}\, \mathrm{d}\sigma_\mu, \tag{18.104}$$

can write down the *Gauss divergence formula for a vector field* $V^\mu(x)$ *on a geometric nifold* as

$$\boxed{\int_U \nabla_\mu V^\mu \sqrt{|g|}\, \mathrm{d}^d x = \oint_{\partial U} V^\mu\, \mathrm{d}\Sigma_\mu} \tag{18.105}$$

is formula can be written in an alternative way by expressing the hypersurface element $_\mu$ in coordinates that are specific to the hypersurface. As shown in appendix B.8, the tric-independent hypersurface element $\mathrm{d}\sigma_\mu$ can be brought to the form

$$\mathrm{d}\sigma_\mu = n_\mu\, \mathrm{d}y^1 \wedge \cdots \wedge \mathrm{d}y^{d-1}. \tag{18.106}$$

e covector field $n_\mu(x)$ is orthogonal to the hypersurface ∂U and outward directed. The 1 coordinates $(y^1, \ldots, y^{d-1})$ label the points of the hypersurface. Within the integral over hypersurface, actually the restriction $g_{\mu\nu}(x)|_{\partial U}$ of the metric $g_{\mu\nu}(x)$ on ∂U is in effect. us, one can alternatively use the induced metric $h_{\alpha\beta}(y)$ on ∂U. Furthermore, instead of orthogonal covector field $n_\mu(x)$, the orthogonal covector field of unit norm $N_\mu(x)$ can used, their relation being

$$\sqrt{|g|}\, n_\mu = \sqrt{|h|}\, N_\mu. \tag{18.107}$$

e quantity $|h(y)|$ is the absolute value of the determinant $h(y) \equiv \det(h_{\alpha\beta}(y))$ of the luced metric. Consequently, the Gauss divergence formula 18.105 can alternatively be ressed as

$$\int_U \nabla_\mu V^\mu \sqrt{|g|}\, \mathrm{d}x^0 \wedge \cdots \wedge \mathrm{d}x^{d-1} = \oint_{\partial U} V^\mu N_\mu \sqrt{|h|}\, \mathrm{d}y^1 \wedge \cdots \wedge \mathrm{d}y^{d-1}, \tag{18.108}$$

playing the dependence on the respective metrics of the region U and the hypersurface . The hypersurface element $\mathrm{d}\Sigma_\mu$ is now written as

$$\mathrm{d}\Sigma_\mu \equiv N_\mu \sqrt{|h|}\, \mathrm{d}y^1 \wedge \cdots \wedge \mathrm{d}y^{d-1}, \tag{18.109}$$

d we note again that it is a proper vector field. The Gauss divergence formula in the form 108 is particularly useful for practical applications.

kes Formula for Antisymmetric Tensor Fields on Geometric Manifolds

us revisit the Stokes integral formula 2.174 for antisymmetric tensor densities. The nains of integration are a $(d-1)$-dimensional hypersurface Σ and its $(d-2)$-dimensional sed boundary $\partial\Sigma$. For the antisymmetric $(2,0)$-tensor density $a^{\mu\nu}(x)$ with weight $w = -1$, note that we can always write it as

$$a^{\mu\nu} = \sqrt{|g|}\, A^{\mu\nu}, \tag{18.110}$$

with $A^{\mu\nu}(x)$ being an antisymmetric $(2,0)$-tensor field with weight $w = 0$. By noting that it is

$$\partial_\nu a^{\mu\nu} = \sqrt{|g|}\, \nabla_\nu A^{\mu\nu}, \qquad (18.111)$$

we obtain equivalently

$$(\partial_\nu a^{\mu\nu})\, \mathrm{d}\sigma_\mu = \nabla_\nu A^{\mu\nu}\, \mathrm{d}\Sigma_\mu \qquad (18.112)$$

for the lhs of the Stokes formula 2.174. For the corresponding rhs, we see that it is

$$a^{\mu\nu}\, \mathrm{d}\sigma_{\mu\nu} = A^{\mu\nu}\, \mathrm{d}\Sigma_{\mu\nu}. \qquad (18.113)$$

Therefore, we can write down the *Stokes formula for an antisymmetric tensor field* $A^{\mu\nu}(x)$ *on a geometric manifold* as

$$\boxed{\int_\Sigma \nabla_\nu A^{\mu\nu}\, \mathrm{d}\Sigma_\mu = \frac{1}{2}\oint_{\partial\Sigma} A^{\mu\nu}\, \mathrm{d}\Sigma_{\mu\nu}} \qquad (18.114)$$

This equation will be put into practice in Section 23.3 when we discuss conserved quantities in general relativity.

18.4   Parallel Transport and Geodesic Curves

Parallel Transport

The covariant derivative not only provides an invariant differentiation measure on a geometric manifold, but it can be employed also to "parallel transport" a vector from one point to another point without altering that vector. This needs a bit of explanation. As we know, vectors attached to a point of a manifold belong to the tangent space of the manifold at that point. If we want to compare, in coordinates, a vector $V^\mu(x_1)$ at the point $x_1 \in \mathcal{M}$ with another vector $W^\mu(x_2)$ at another point $x_2 \in \mathcal{M}$, we do not have a natural way, because the two vectors belong to two different tangent spaces. In order to be able to compare $V^\mu(x_1)$ with $W^\mu(x_2)$, we need to transport $V^\mu(x_1)$ from the initial point x_1 to the new point x_2 without altering it. This is achieved by parallel transporting the vector $V^\mu(x_1)$ along a curve that connects the two points x_1 and x_2. Obviously, the employed curve connecting the two points x_1 and x_2 enters into the result and indeed this has a profound impact, as we will show in the next chapter. For now, let us first define the recipe for a parallel transport. We consider a curve $z^\mu(\tau)$ parametrized by a real parameter τ taking the values $\tau_1 \leq \tau \leq \tau_2$ with $z(\tau_1) = x_1$ and $z(\tau_2) = x_2$. We consider also a vector field $V^\mu(x)$, which is defined along the curve and within a domain of $\mathcal{M}$ around the curve. In the basic case of Minkowski space, we would express the parallel transport condition for $V^\mu(x)$ as

$$\frac{\mathrm{d}V^\mu}{\mathrm{d}\tau}(z(\tau)) = \frac{\mathrm{d}z^\rho}{\mathrm{d}\tau}(\tau)\, \partial_\rho V^\mu(z(\tau)) = 0, \qquad (18.115)$$

to be fulfilled along the curve. In the case of a geometric manifold, we generalize the condition for the *parallel transport* of the vector $V^\mu(x)$ along the curve, expressed by the symbol $\mathrm{D}V^\mu(z(\tau))/\mathrm{d}\tau$, to be the invariant requirement

$$\frac{\mathrm{D}V^\mu}{\mathrm{d}\tau}(z(\tau)) \equiv \frac{\mathrm{d}z^\rho}{\mathrm{d}\tau}(\tau)\, \nabla_\rho V^\mu(z(\tau)) = 0, \qquad (18.116)$$

i.e. by replacing the partial derivative by the covariant derivative. If we use the symbol $T^\rho(\tau)$ for the tangent vector of the curve,

$$T^\rho(\tau) \equiv \dot{z}^\rho(\tau) \equiv \frac{\mathrm{d}z^\rho}{\mathrm{d}\tau}(\tau), \qquad (18.117)$$

condition for the parallel transport can be written more compactly as

$$\nabla_T V = 0. \tag{18.118}$$

e sketch 18.1 below depicts the effect of the parallel transport of a vector field $V^\mu(x)$
ag a given curve $z^\mu(\tau)$ from the initial point x_1^μ to the final point x_2^μ. The curve lies
a 2-dimensional surface representing the manifold $\mathcal{M}$ here. The condition for parallel

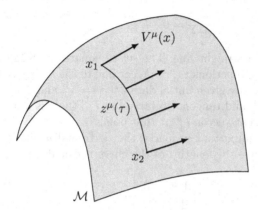

Figure 18.1: Parallel transport of a vector field

placement can be brought into a form more suited for practical applications. By inserting
formula 18.22 for the covariant derivative in 18.116, we obtain the condition for parallel
nsport in the form

$$\frac{dV^\mu}{d\tau}(\tau) + \Gamma^\mu_{\nu\rho}(z(\tau))V^\nu(z(\tau))\frac{dz^\rho}{d\tau}(\tau) = 0. \tag{18.119}$$

can parallel transport not only vector fields but arbitrary tensor fields, and the condition
parallel transport reads then

$$T^\rho(\tau)\,\nabla_\rho \phi^{\mu_1\cdots\mu_m}{}_{\nu_1\cdots\nu_n}(z(\tau)) = 0. \tag{18.120}$$

e notion of parallel transport allows us to define a metric-compatible connection in a diffe-
t way. Suppose a curve is given and we consider two vector fields $V^\mu(x)$ and $W^\mu(x)$, each
being parallel transported along the curve. Of course, their scalar product $V_\mu W^\mu$ can
o be considered along the curve. Then, the scalar product remains covariantly constant
ng the curve,

$$T^\rho\nabla_\rho(g_{\mu\nu}V^\mu W^\nu) = 0, \tag{18.121}$$

nd only if the metric compatibility condition $\nabla_\rho g_{\mu\nu} = 0$ holds, i.e. the covariant con-
ncy of the scalar product is equivalent to the metric compatibility.

odesic Curves

aat straight lines are in Minkowski (or Euclidean) space are *geodesic curves*, called also
ply *geodesics*, in a general geometric manifold. Straight lines parallel transport their
1 tangent vector along them and this is how we characterize geodesics. A geodesic is a
ve which continuously parallel transports its own tangent vector. This is why geodesics
also called *autoparallel curves*. Formally, a curve $z^\mu(\tau)$ with tangent vector $T^\mu(\tau)$ is a
desic iff

$$\nabla_T T = 0 \tag{18.122}$$

along the curve. In coordinates, this *geodesic equation* reads

$$\boxed{\frac{\mathrm{d}^2 z^\mu}{\mathrm{d}\tau^2} + \Gamma^\mu_{\nu\rho} \frac{\mathrm{d}z^\nu}{\mathrm{d}\tau} \frac{\mathrm{d}z^\rho}{\mathrm{d}\tau} = 0}$$

(18.123)

The above condition is in fact independent of the parametrization, here with τ, as long as one uses an *affine parameter* $\lambda = a\tau + b$, with real constants a and b. (*exercise 18.4*) In the basic case of Minkowski (or Euclidean) space, we can choose the connection coefficients to vanish and arrive at the geodesic equation

$$\frac{\mathrm{d}^2 z^\mu}{\mathrm{d}\tau^2} = 0,$$

(18.124)

which is exactly the equation of a straight line. The set of equations 18.123 is a set of d coupled second-order ODEs for the functions $z^\mu(\tau)$. According to the theory of ODEs, there exists always a unique solution for any given initial data $z^\mu(\tau_1) = x_1^\mu$ and $\dot{z}^\mu(\tau_1) = T_1^\mu$. This means that for any given point x_1^μ and tangent vector value T_1^μ, there is a unique geodesic curve through x_1^μ having the tangent vector T_1^μ at that point.

Geodesic curves have another important property, they extremalize the length value of the curve. Given an arbitrary curve, the length l of the curve is calculated as

$$l = \int_{\tau_1}^{\tau_2} \left| g_{\mu\nu}(z(\tau)) \frac{\mathrm{d}z^\mu}{\mathrm{d}\tau}(\tau) \frac{\mathrm{d}x^\nu}{\mathrm{d}\tau}(\tau) \right|^{\frac{1}{2}} \mathrm{d}\tau.$$

(18.125)

The result of the integral is independent of the parametrization, as one can inspect. Per definition, a curve is called *timelike / null / spacelike* iff its tangent vector is everywhere timelike / null / spacelike. Specifically for geodesic curves, we note that the norm $|g_{\mu\nu}T^\mu T^\nu|^{1/2}$ of the tangent vector, i.e. the scalar value of the velocity, remains constant along the curve. Hence, geodesics which are timelike / null / spacelike within a segment, keep this property everywhere. Let us now demonstrate that *geodesic curves make the curve length extremal*. For timelike geodesics, the proper time value is maximal. For spacelike geodesics, the path length value is minimal. For null geodesics, the curve length is trivially zero. We apply the variational method and consider the variation δl of the length functional l under variations of the curve, while the endpoints of the curve are held fixed. For definiteness, let us choose the case of a spacelike curve. The case of a timelike curve is similar. A straight calculation yields

$$\delta l = \int_{\tau_1}^{\tau_2} (-g_{\mu\nu}T^\mu T^\nu)^{-\frac{1}{2}} \left[-\frac{1}{2}(\partial_\alpha g_{\rho\sigma})\delta x^\alpha T^\rho T^\sigma - g_{\rho\sigma}T^\rho \frac{\mathrm{d}}{\mathrm{d}\tau}(\delta x^\sigma) \right] \mathrm{d}\tau,$$

(18.126)

where we have used the symmetry of the metric and the two identities

$$\delta g_{\rho\sigma} = (\partial_\alpha g_{\rho\sigma})\delta x^\alpha \quad \text{and} \quad \delta T^\sigma = \frac{\mathrm{d}}{\mathrm{d}\tau}(\delta x^\sigma).$$

(18.127)

To simplify the calculation, we can assume a parametrization for which $g_{\mu\nu}T^\mu T^\nu = -1$ holds throughout the curve. As we commented above, the value of the length is actually invariant under reparametrizations. Performing an integration by parts for the last term in the square brackets above gives

$$\delta l = \int_{\tau_1}^{\tau_2} \left[-\frac{1}{2}(\partial_\alpha g_{\rho\sigma})T^\rho T^\sigma + \frac{\mathrm{d}}{\mathrm{d}\tau}(g_{\rho\alpha}T^\rho) \right] \delta x^\alpha \, \mathrm{d}\tau.$$

(18.128)

There are no boundary terms, since the variation δx^α vanishes at the end points of the curve. The curve is extremal iff $\delta l = 0$ holds. This is equivalent to the vanishing of the terms in the square brackets,

$$-\frac{1}{2}(\partial_\alpha g_{\rho\sigma})T^\rho T^\sigma + (\partial_\sigma g_{\rho\alpha})T^\sigma T^\rho + g_{\rho\alpha}\frac{dT^\rho}{d\tau} = 0. \qquad (18.129)$$

Rearranging and inverting the equation with respect to the metric leads to the equation

$$\frac{dT^\beta}{d\tau} + \frac{1}{2}g^{\beta\alpha}(\partial_\rho g_{\sigma\alpha} + \partial_\sigma g_{\rho\alpha} - \partial_\alpha g_{\rho\sigma})T^\rho T^\sigma = 0. \qquad (18.130)$$

This is exactly the geodesic equation 18.123 with the Levi-Civita connection coefficients 18.78. So, in summary, in geometric manifolds the geodesic curves are the ones which possess an extremal curve length.

Isometries and Geodesics

In Euclidean geometry, one has the property that isometry transformations map straight lines to straight lines. In geometric manifolds, this is generalized to the property that *isometry transformations map geodesic curves to geodesic curves*. Let us consider an isometry transformation $F : \mathcal{M} \to \mathcal{M}$, implemented as an active coordinate transformation $x \mapsto x' \equiv F(x)$, with the defining property that the metric function stays invariant, i.e. $g'_{\mu\nu}(x) = g_{\mu\nu}(x)$, see Section 3.2. Then, the isometry F maps geodesics to geodesics. More specifically, if $z^\mu(\tau)$ is a geodesic with the tangent vector $\dot{z}^\mu(\tau)$, then the curve $\zeta^\mu(\tau) \equiv F^\mu(z(\tau))$ is also a geodesic with the tangent vector

$$\dot{\zeta}^\mu(\tau) = \frac{\partial F^\mu(z(\tau))}{\partial x^\nu}\dot{z}^\nu(\tau). \qquad (18.131)$$

We need to prove that the new curve $\zeta^\mu(\tau)$ satisfies the geodesic condition

$$\dot{\zeta}^\rho\nabla'_\rho\dot{\zeta}^\mu = 0. \qquad (18.132)$$

The primed covariant derivative is

$$\nabla'_\rho\xi^\mu = \partial'_\rho\xi^\mu + \Gamma'^\mu_{\nu\rho}\xi^\nu \qquad (18.133)$$

for any vector field $\xi^\mu(x)$, where the partial derivatives transform as usual and the primed connection coefficients are given by the formula 18.15. In the expression $\dot{\zeta}^\rho\nabla'_\rho\dot{\zeta}^\mu$, the contractions have to be carried out, which is a straightforward task (*exercise 18.5*). The resulting expression is

$$\frac{\partial x'^\lambda}{\partial x^\mu}\dot{z}^\rho(\partial_\rho\dot{z}^\mu + \Gamma^\mu_{\nu\rho}\dot{z}^\nu), \qquad (18.134)$$

which vanishes due to the geodesy property of the curve $z^\mu(\tau)$. The obtained result can be used in order to derive new geodesic curves from existing ones.

Geodesics of the 2-Sphere

As an example, we are going to derive the geodesic curves of the unit 2-sphere S^2 now, representing the "straightest path" between any two given points on the 2-sphere. Obviously, we should work in spherical coordinates (θ, φ) for which we already know the metric components and the connection coefficients. The two geodesic equations are easily obtained,

$$\ddot{\theta} - (\dot{\varphi})^2 \sin\theta\cos\theta = 0 \quad \text{and} \quad \ddot{\varphi} + 2\dot{\varphi}\dot{\theta}\cot\theta = 0, \qquad (18.135)$$

with $\theta(s)$ and $\varphi(s)$ being functions of the path length s. We use the shorthand notat
$\dot\theta = d\theta/ds$, $\dot\varphi = d\varphi/ds$, and so forth. Our approach for solving the two above coupled OD
is to eliminate one of the two unknown functions by expressing the function θ throug
Let us note the simple relations

$$\dot\theta = \dot\varphi \, \frac{d\theta}{d\varphi} \quad \text{and} \quad \ddot\theta = (\dot\varphi)^2 \frac{d^2\theta}{d\varphi^2} + \ddot\varphi \, \frac{d\theta}{d\varphi}. \tag{18.1}$$

The first equation in 18.135 becomes

$$(\dot\varphi)^2 \frac{d^2\theta}{d\varphi^2} + \ddot\varphi \, \frac{d\theta}{d\varphi} - (\dot\varphi)^2 \sin\theta \cos\theta = 0. \tag{18.1}$$

With the aid of the second equation in 18.135, we can eliminate $\ddot\varphi$ and obtain

$$\frac{d^2\theta}{d\varphi^2} - 2\cot\theta \left(\frac{d\theta}{d\varphi} \right)^2 - \sin\theta \cos\theta = 0. \tag{18.1}$$

We reached to a single ODE for the implicit function $\theta(\varphi)$. If, instead of θ, we use
function $\cot\theta$, the last differential equation reads

$$\frac{d^2}{d\varphi^2}(\cot\theta) + \cot\theta = 0. \tag{18.1}$$

This differential equation has the solution $\cot\theta = a\cos\varphi + b\sin\varphi$, with the two real
rameters a, b. By introducing Cartesian coordinates (x, y, z), this solution is writter
$ax + by - z = 0$. This is exactly the equation of a 2-dimensional plane passing through
coordinate origin and having the normal vector $(a, b, -1)$. Let us illustrate the result in
figure 18.2 below for a plane with the normal vector $(0, b, 0)$.

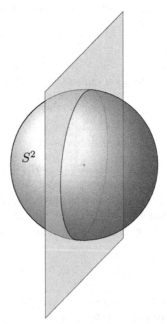

Figure 18.2: Great circles provide the geodesics of the 2-sphere

Hence, we have obtained: *the geodesic curves on the 2-sphere are segments of
at circles.* Exactly these are the curves obtained as intersections of the 2-sphere wit
dimensional planes which pass through the origin of the sphere.

Geodesics of the Upper Half-Plane Model

As another example, let us look at the upper half-plane model H^2 having the metric

$$ds^2 = \frac{1}{y^2} \left(dx^2 + dy^2 \right) \tag{18.140}$$

in Euclidean coordinates (x, y), with $y > 0$. The metric components are given by $g_{xx} = g_{yy} = 1/y^2$, $g_{xy} = g_{yx} = 0$. The inverse metric is obviously $g^{xx} = g^{yy} = y^2$, $g^{xy} = g^{yx} = 0$. The connection coefficients are quickly calculated as

$$\left. \begin{array}{l} \Gamma^x_{xx} = \Gamma^x_{yy} = \Gamma^y_{xy} = \Gamma^y_{yx} = 0 \\[2mm] \Gamma^x_{xy} = \Gamma^x_{yx} = \Gamma^y_{yy} = -\Gamma^y_{xx} = -1/y \end{array} \right\}. \tag{18.141}$$

This leads to the two coupled geodesic equations

$$\ddot{x} - \frac{2}{y}\dot{x}\dot{y} = 0 \quad \text{and} \quad \ddot{y} + \frac{1}{y}(\dot{x}^2 - \dot{y}^2) = 0. \tag{18.142}$$

Here the dots indicate again derivatives with respect to the path length s, used as the curve parameter, i.e. $\dot{x} = dx/ds$, etc. Looking at the first of these geodesic equations, we see that $\dot{x} = 0$ defines one class of solutions. These are *straight vertical half-lines* within the upper half range. For $\dot{x} \neq 0$, the first geodesic equation is equivalent to the condition $\dot{x} = cy^2$ with a real parameter c. This condition and the second geodesic equation can be used to determine the other class of geodesic curves. Instead of this route, it is easier to go back to the metric definition 18.140 and write it down as the differential equation $y^2 = \dot{x}^2 + \dot{y}^2$. This equation, together with the condition $\dot{x} = cy^2$, leads to

$$dx = \frac{cy}{\sqrt{1 - c^2 y^2}} \, dy. \tag{18.143}$$

This can be integrated immediately and yields the condition $(x - x_0)^2 + y^2 = a^2$. We see that the other type of geodesic curves are *semicircles* of radius $a = c^{-1}$, with their origin lying at the point $(x, y) = (x_0, 0)$. The origin x_0 on the x-axis and the radius a parametrize this set of solutions. An illustration of these geodesics has been given in Section 2.3. We need to be careful with our visualization on a 2-dimensional (Euclidean) sheet of paper. As we approach the line $y = 0$, distances become continuously larger and diverge in the limit when we reach points on $y = 0$. For instance, let us consider the situation as depicted in the figure 18.3 below.

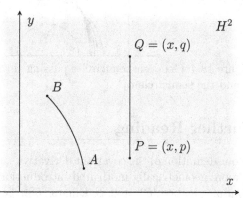

Figure 18.3: Two examples of geodesics in the UHP model

The diagram shows a pair of points, A and B, connected via their arc segment geode
The other points, P and Q, have the same x-coordinate and their geodesic is a piece
vertical straight line. The distance $s(P,Q)$ between the points P and Q is

$$s(P,Q) = \int_P^Q \sqrt{\frac{\dot{x}^2 + \dot{y}^2}{y^2}}\, ds = \int_p^q \frac{dy}{y} = \ln \frac{q}{p}. \qquad (18.1$$

In the limit $p \to 0$, the distance $s(P,Q)$ becomes infinite. Each point on the line $y = 0$
at infinity in this hyperbolic space model.

In order to generate new geodesic curves of the UHP model, we can employ isomet
of this geometry. Then, by starting from vertical half-lines or semicircles, we can gener
new geodesic curves by applying isometries to them. One type of isometries of H^2 are
translations $(x,y) \mapsto (x + x_0, y)$ along the x-axis, parametrized by the real number
Another type of isometries are the scale transformations $(x,y) \mapsto (\omega x, \omega y)$, parametri
by the real number ω. Under both transformations the metric ds^2 stays invariant. In
way, we can generate all geodesics of the UHP model. Actually, the isometries applied h
only recreate the two already known classes of geodesic solutions.

In Section 2.3 we mentioned Euclid's axiom about parallel lines, valid for Euclid
geometry in the plane. Within the hyperbolic geometry of the UHP model, we now see t
a completely different constellation is in place. For the UHP model, the following appli

> *Given a geodesic line and a point outside that line, there exist infinitely ma-*
> *ny geodesic lines that pass through that point and remain parallel to the given*
> *geodesic line.*

This remarkable fact of the UHP model is visualized in the sketch 18.4. For an arbitrary g
desic semicircle c_0 and a point P outside of c_0, there are infinitely many geodesic semicir
passing though the point P being parallel to c_0. The same is valid if the initial geodesic
vertical half-line l_0. In both cases there are uncountably many semicircles passing thro
the point P and never intersecting the given geodesic line.

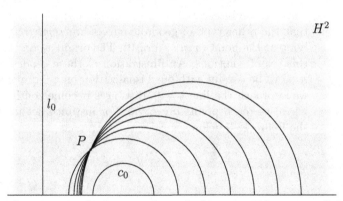

Figure 18.4: Geodesic semicircles passing through the point P being parallel to the half-
l_0 and the semicircle c_0

Further Reading

In our definition of the covariant derivative we have been guided by invariance requireme
A more geometrically motivated introduction of the covariant derivative can be found,
example, in Lee [52] and Nakahara [58]. For a discussion of the geodesics of hyperb
geometry models in dimensions greater than two, one can turn to Lee [52].

19

Riemannian Curvature

In this chapter we introduce the central notion of curvature according to Riemann, first by considering the effect of parallel transport of vectors, and then by pursuing the question of whether one can bring the metric tensor locally into a Euclidean form. Formally, we introduce the Riemann curvature tensor through the commutator of covariant derivatives applied to arbitrary tensor densities. After a discussion of the algebraic properties of the Riemann tensor, we study its contractions, namely the Ricci tensor and the curvature scalar. We derive the Bianchi identity as the central differential equation for curvature. Then we introduce the Einstein tensor, which is essential to the formulation of general relativity. In the last part, we discuss the Ricci decomposition of the Riemann tensor and introduce another important quantity, the Weyl tensor.

19.1   Manifestation of Curvature

Path-Dependence of Parallel Transport
In the last chapter we introduced the parallel transport of vectors and we noted that the result of this operation principally depends on the path along which the vector is transported. This path dependence of the parallel transport operation can now be employed to discriminate between manifolds with a non-vanishing curvature and those manifolds with a zero curvature. Let us consider the 2-sphere manifold as an example, as depicted in the sketch 19.1. The vector initially attached at the point A is parallel transported to the final point C. This can be done, for instance, by parallel transporting the vector directly from A to C along the shortest geodesic path. Another possible way to parallel transport is to shift the vector first to point B and then finally to point C along the geodesic segments. It is obvious that the result of the second transport route will be a different final vector. This path dependence is characteristic for a manifold with a non-vanishing curvature. The 2-sphere is indeed a manifold with curvature. An alternative view is to consider the vector attached at the point C and to parallel transport it along the closed path $CABC$. The end result of parallel transporting around the loop $CABC$ will be a vector which is rotated compared to the original vector. We realize that the order in which parallel transports or covariant differentiations are carried out is crucial. On the most basic level, this means that we need to consider the commutator of covariant derivatives. The associated quantity measuring the non-commutativity of covariant differentiation is provided by the Riemann

DOI: 10.1201/9781003087748-19

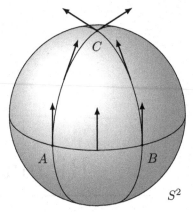

Figure 19.1: Parallel transport of a vector on a 2-sphere

curvature tensor. But before we formalize this notion, let us study how a geometric manif
looks like from a metrical point of view.

Local Flatness of Riemannian Manifolds

We consider a d-dimensional pseudo-Riemannian manifold $(\mathcal{M}, g_{\mu\nu})$ with a metric of ar
trary signature. Then the following theorem holds: for any point P of the manifold we
find a set of local coordinates $\{x'\}$ such that

$$\left.\begin{array}{l} g'_{\mu\nu}(x'_P) = \eta_{\mu\nu} \\[2mm] \partial'_\rho g'_{\mu\nu}(x'_P) = 0, \text{ for all } \rho, \mu, \nu \end{array}\right\}$$

(19

in these coordinates. In other words, the metric $g_{\mu\nu}(x)$ can be made *locally flat* and att
a Euclidean (or pseudo-Euclidean) form. In general, it is however

$$\partial'_\rho \partial'_\sigma g'_{\mu\nu}(x'_P) \neq 0,$$

(19

which is a manifestation of the curvature of the manifold. In order to derive this result,
consider the original coordinates $\{x\}$ and the new coordinates $\{x'\}$ and count the freed
we have in the coordinate transformation compared to the number of components we n
to fix for the metric tensor and its derivatives. A Taylor expansion of the old coordina
$\{x\}$ around the point x_P is given by

$$x^\mu(x') = x^\mu_P + \frac{\partial x^\mu}{\partial x'^\rho}(x'_P)(x'^\rho - x'^\rho_P)$$

$$+ \frac{1}{2}\frac{\partial^2 x^\mu}{\partial x'^\rho \partial x'^\sigma}(x'_P)(x'^\rho - x'^\rho_P)(x'^\sigma - x'^\sigma_P)$$

(19

$$+ \frac{1}{6}\frac{\partial^3 x^\mu}{\partial x'^\rho \partial x'^\sigma \partial x'^\tau}(x'_P)(x'^\rho - x'^\rho_P)(x'^\sigma - x'^\sigma_P)(x'^\tau - x'^\tau_P) + \mathcal{O}(4).$$

The coefficients $\partial x^\mu/\partial x'^\rho$ in the first-order term correspond to d^2 independent valu
The coefficients $\partial^2 x^\mu/\partial x'^\rho \partial x'^\sigma$ provide $d^2(d+1)/2$ independent values. The coefficie
$\partial^3 x^\mu/\partial x'^\rho \partial x'^\sigma \partial x'^\tau$ correspond to $d^2(d+1)(d+2)/6$ independent values. These degree
freedom in the coordinate transformation have to be compared with the number of val
we need to fix for the metric and its derivatives. Generally, the metric transforms as in 2.
i.e. the value of the metric is determined by the coefficients of the form $\partial x^\mu/\partial x'^\rho$. In orde
fix the value $g'_{\mu\nu}(x'_P)$, we need $d(d+1)/2$ real numbers. For $d \geq 2$, the number d^2 is grea

than $d(d+1)/2$ and thus it is sufficient to achieve $g'_{\mu\nu}(x'_P) = \eta_{\mu\nu}$. The remaining $d(d-1)/2$ degrees of freedom in the $\partial x^\mu/\partial x'^\rho$ coefficients correspond to the (Lorentz) rotations, which leave $\eta_{\mu\nu}$ invariant. Next is the question if we can always achieve $\partial'_\rho g'_{\mu\nu}(x'_P) = 0$, which corresponds to fixing $d^2(d+1)/2$ real numbers. Indeed, the coefficients $\partial^2 x^\mu/\partial x'^\rho \partial x'^\sigma$ provide exactly this count of degrees of freedom. Finally, we can ask if we can make the terms $\partial'_\rho \partial'_\sigma g'_{\mu\nu}(x'_P)$ to vanish, which corresponds to fixing $d^2(d+1)^2/4$ real numbers. Unfortunately, this is not possible, since $d^2(d+1)^2/4$ is greater than $d^2(d+1)(d+2)/6$ for $d \geq 2$. In general, $d^2(d^2-1)/12$ of these terms $\partial'_\rho \partial'_\sigma g'_{\mu\nu}(x'_P)$ cannot be made to vanish.

This means that for every point in a geometric manifold we can "trivialize" the value of the metric and its first-order derivatives. Consequently, the connection coefficients can also be made to vanish in each point. However, we cannot achieve this for the second-order derivatives of the metric. In this sense, these second-order derivatives are more descriptive of the geometry than the value of the metric and its first-order derivatives. The second-order derivatives of the metric will prove to be crucial for defining the notion of curvature.

19.2 The Riemann Curvature Tensor

Repeated Covariant Differentiation

In the last section we have seen that the order in which two different directional covariant derivatives are carried out is essential. Thus, let us study what the effect of a repeated covariant differentiation is. The main result of this analysis is the Riemann tensor, which measures the non-commutativity of two different paths in a manifold and thus the curvature of the manifold. For the simple partial differentiation of any smooth function $f(x)$ we have the basic rule $\partial_\rho \partial_\sigma f = \partial_\sigma \partial_\rho f$. However, this commutativity does not hold for the covariant derivative in general. Let us analyze what happens if we consider tensor densities of an increasing valence. First, for a proper scalar field $\phi(x)$ it is $\nabla_\rho \phi = \partial_\rho \phi$, and it is quickly found that it is always

$$[\nabla_\rho, \nabla_\sigma]\,\phi \equiv (\nabla_\rho \nabla_\sigma - \nabla_\sigma \nabla_\rho)\phi = 0. \tag{19.4}$$

Next, let us consider a scalar density $S(x)$ transforming as $S'(x') = J(x)^w S(x)$, with the weight $w \in \mathbb{Z}$. The covariant derivative $\nabla_\sigma S$, which is calculated as

$$\nabla_\sigma S = \partial_\sigma S + w\Gamma^\alpha_{\sigma\alpha} S, \tag{19.5}$$

is a covector density with the same weight w. Its covariant derivative with respect to ∇_ρ is

$$\nabla_\rho \nabla_\sigma S = \partial_\rho(\nabla_\sigma S) - \Gamma^\beta_{\sigma\rho}\nabla_\beta S + w\Gamma^\beta_{\rho\beta}\nabla_\sigma S. \tag{19.6}$$

By inserting the formula for $\nabla_\sigma S$ into the first and third term on the rhs, we obtain

$$\nabla_\rho \nabla_\sigma S = \partial_\rho \partial_\sigma S + w\partial_\rho(\Gamma^\alpha_{\sigma\alpha} S) - \Gamma^\beta_{\sigma\rho}\nabla_\beta S + w\Gamma^\beta_{\rho\beta}\partial_\sigma S + w^2\Gamma^\beta_{\rho\beta}\Gamma^\alpha_{\sigma\alpha} S. \tag{19.7}$$

Considering the index-flipped expression $\nabla_\sigma \nabla_\rho S$ and subsequently omitting terms canceling each other yields

$$[\nabla_\rho, \nabla_\sigma]\,S = w(\partial_\rho \Gamma^\alpha_{\sigma\alpha} - \partial_\sigma \Gamma^\alpha_{\rho\alpha})S. \tag{19.8}$$

Here we have used the symmetry of the connection. By using also the formula $\Gamma^\alpha_{\sigma\alpha} = \partial_\sigma(\ln\sqrt{|g|})$, valid for a Levi-Civita connection which we assume to have, we deduce that the commutator vanishes,

$$\boxed{[\nabla_\rho, \nabla_\sigma]\,S = 0} \tag{19.9}$$

So, for a scalar density $S(x)$, exactly as for a proper scalar field $\phi(x)$, the order of the covariant derivatives is immaterial. The situation changes as soon as we consider tensor densities with a non-vanishing valence. Let us view a vector density $B^\mu(x)$ with the weight w. The covariant derivative $\nabla_\sigma B^\mu$ is computed as

$$\nabla_\sigma B^\mu = \partial_\sigma B^\mu + \Gamma^\mu_{\alpha\sigma} B^\alpha + w\Gamma^\alpha_{\sigma\alpha} B^\mu, \tag{19.10}$$

and it represents a $(1,1)$-tensor density with the same weight w. Thus, its covariant derivative with respect to ∇_ρ is

$$\nabla_\rho \nabla_\sigma B^\mu = \partial_\rho(\nabla_\sigma B^\mu) + \Gamma^\mu_{\beta\rho} \nabla_\sigma B^\beta - \Gamma^\beta_{\sigma\rho} \nabla_\beta B^\mu + w\Gamma^\beta_{\rho\beta} \nabla_\sigma B^\mu. \tag{19.11}$$

We proceed similarly to the scalar density case and insert the formula for $\nabla_\sigma B^\mu$ into the rhs above. By writing out the index-flipped expression $\nabla_\sigma \nabla_\rho B^\mu$ and by building the commutator, we obtain the final result,

$$[\nabla_\rho, \nabla_\sigma]\, B^\mu = (\partial_\rho \Gamma^\mu_{\nu\sigma} - \partial_\sigma \Gamma^\mu_{\nu\rho} + \Gamma^\mu_{\alpha\rho} \Gamma^\alpha_{\nu\sigma} - \Gamma^\mu_{\alpha\sigma} \Gamma^\alpha_{\nu\rho}) B^\nu. \tag{19.12}$$

The details of this calculation are left to the reader. (*exercise 19.1*) Remarkably, there are no terms including the factor w anymore. The fact that $B^\mu(x)$ is a vector density and not a proper vector proves to be immaterial. Therefore, the same formula will be obtained for a proper vector field $V^\mu(x)$. The lhs of equation 19.12 is a $(1,2)$-tensor density with weight w. In the rhs, $B^\nu(x)$ is a vector density with weight w. Hence, the quantity in the round brackets in the rhs must be a proper $(1,3)$-tensor field. This tensor field is immensely important in differential geometry and it deserves a formal definition. We introduce the *Riemann curvature tensor* $R^\mu{}_{\nu\rho\sigma}(x)$ as the $(1,3)$-tensor field defined as

$$\boxed{R^\mu{}_{\nu\rho\sigma} \equiv \partial_\rho \Gamma^\mu_{\nu\sigma} - \partial_\sigma \Gamma^\mu_{\nu\rho} + \Gamma^\mu_{\alpha\rho} \Gamma^\alpha_{\nu\sigma} - \Gamma^\mu_{\alpha\sigma} \Gamma^\alpha_{\nu\rho}} \tag{19.13}$$

Written as a multilinear basis expansion, the Riemann tensor field is

$$R^\mu{}_{\nu\rho\sigma}(x)\, \partial_\mu \otimes \mathrm{d}x^\nu \otimes \mathrm{d}x^\rho \otimes \mathrm{d}x^\sigma. \tag{19.14}$$

The tensorial transformation behavior of the Riemann tensor can be inspected also directly from the definition 19.13 by inserting the transformation formula 18.19 of the connection coefficients. Having now introduced the Riemann tensor, the commutator relation 19.12 can be written as

$$\boxed{[\nabla_\rho, \nabla_\sigma]\, B^\mu = R^\mu{}_{\nu\rho\sigma} B^\nu} \tag{19.15}$$

In this form, it is known as the *Ricci identity* (*Gregorio Ricci-Curbastro*), written here for the case of a vector density. It is remarkable that the application of the commutator of covariant derivatives is equivalent to the simple algebraic contraction with the Riemann tensor. In this way, the Riemann tensor is the quantity that fully captures the curvature of a manifold.

Ricci Identity for Tensor Densities

Repeating the previous calculation for the commutator of covariant derivatives now in the case of a covector density $Q_\nu(x)$ with the weight w we obtain (*exercise 19.2*)

$$\boxed{[\nabla_\rho, \nabla_\sigma]\, Q_\nu = -R^\mu{}_{\nu\rho\sigma}\, Q_\mu} \tag{19.16}$$

This is the Ricci identity for a covector density. Hence, for a $(1,1)$-tensor density $T^\mu{}_\nu(x)$ with weight w, the Ricci identity reads

$$\boxed{[\nabla_\rho, \nabla_\sigma] T^\mu{}_\nu = R^\mu{}_{\alpha\rho\sigma} T^\alpha{}_\nu - R^\alpha{}_{\nu\rho\sigma} T^\mu{}_\alpha} \tag{19.17}$$

The reader should note the plus and minus signs in front of each term, depending on the contraction of co- and contravariant indices of the original tensor. The Ricci identity for a general (m,n)-tensor density field $T^{\mu_1\cdots\mu_m}{}_{\nu_1\ldots\nu_n}(x)$, with weight w is

$$[\nabla_\rho, \nabla_\sigma] T^{\mu_1\cdots\mu_m}{}_{\nu_1\ldots\nu_n} =$$

$$+ R^{\mu_1}{}_{\alpha\rho\sigma} T^{\alpha\,\mu_2\cdots\mu_m}{}_{\nu_1\ldots\nu_n} + \cdots + R^{\mu_m}{}_{\alpha\rho\sigma} T^{\mu_1\cdots\mu_{m-1}\,\alpha}{}_{\nu_1\ldots\nu_n} \tag{19.18}$$

$$- R^\alpha{}_{\nu_1\rho\sigma} T^{\mu_1\cdots\mu_m}{}_{\alpha\,\nu_2\ldots\nu_n} - \cdots - R^\alpha{}_{\nu_n\rho\sigma} T^{\mu_1\cdots\mu_m}{}_{\nu_1\ldots\nu_{n-1}\,\alpha}.$$

The Ricci identities will be important for all upcoming calculations.

Flatness of a Geometric Manifold

Formally, we define a geometric manifold of dimension $d \geq 1$ to be *flat*, iff its Riemann tensor vanishes everywhere,

$$R^\mu{}_{\nu\rho\sigma}(x) = 0. \tag{19.19}$$

A manifold which is not flat is called *curved*. These characterizations are coordinate-independent. Geometrically, in a flat manifold parallel transporting a vector along a closed loop brings back the original vector. In a curved manifold, this closed loop transport produces an altered vector. An alternative characterization is that for a flat manifold the parallel transport is path-independent, whereas for a curved manifold the result of the parallel transport depends on the path taken.

The flatness condition 19.19 essentially is equivalent to the metric tensor being of the form

$$g_{\mu\nu}(x) = \eta_{\mu\nu} \tag{19.20}$$

in the Lorentzian case (or $g_{\mu\nu}(x) = \delta_{\mu\nu}$ in the Riemannian case). Obviously, if the metric tensor is of the form $\eta_{\mu\nu}$ (or $\delta_{\mu\nu}$), the connection coefficients vanish everywhere and so does the Riemann tensor. Conversely, if the Riemann tensor vanishes, then one can always find local coordinates in which the connection coefficients vanish. Thus, due to the metric concordance, the metric is constant in these coordinates, $\partial_\rho g_{\mu\nu} = 0$. Carrying out a linear coordinate transformation, the metric tensor can be brought to its diagonal form. Let us outline the proof leading to the local vanishing of the connection coefficients. Choose an arbitrary point x_0 of the manifold and a set of d linearly independent vectors $e^\mu_{(\alpha)}$, $\alpha = 0, 1, \ldots, D$, attached at this point x_0. Now we transport these vectors from the point x_0 to another point x of the manifold by means of a parallel transport. This produces a set of d new vectors attached at the point x, which we denote $E^\mu_{(\alpha)}(x)$. The condition of parallel transport is expressed here as

$$\partial_\rho E^\mu_{(\alpha)} + \Gamma^\mu{}_{\nu\rho} E^\nu_{(\alpha)} = 0, \tag{19.21}$$

with $E^\mu_{(\alpha)}(x_0) = e^\mu_{(\alpha)}$ for the initial values. Due to the vanishing of the Riemann tensor, the result of the parallel transport from x_0 to x is independent of the path taken. Therefore, this procedure defines uniquely d vector fields $E^\mu_{(\alpha)}(x)$ within a local region $U \subset \mathcal{M}$ around the initial point x_0, as illustrated in 19.2 below. For the set $\{e^\mu_{(\alpha)}\}$ of linearly independent

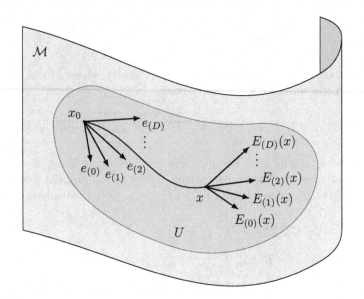

Figure 19.2: Defining new vector fields by starting from a set of basis vectors

vectors the determinant $\det(e^{\mu}_{(\alpha)})$ is non-vanishing. Furthermore, we consider the deter
nant $E(x) \equiv \det(E^{\mu}_{(\alpha)}(x))$, which is a scalar density of weight $w = 1$. Its covariant derivat
is

$$\nabla_{\rho} E = \partial_{\rho} E + \Gamma^{\beta}_{\rho\beta} E = \sum_{\alpha=0}^{D} C^{(\alpha)}_{\mu} \partial_{\rho} E^{\mu}_{(\alpha)} + \Gamma^{\beta}_{\rho\beta} E, \qquad (19.$$

where $C^{(\alpha)}_{\mu}(x)$ is the cofactor of $E^{\mu}_{(\alpha)}(x)$. By inserting the expression $\partial_{\rho} E^{\mu}_{(\alpha)} = -\Gamma^{\mu}_{\nu\rho} E^{\nu}_{(\alpha)}$,
obtain readily $\nabla_{\rho} E = 0$. Therefore, the determinant E is non-vanishing and consequer
the set $\{E^{\mu}_{(\alpha)}(x)\}$ represents a vector field basis within the region U. Now let us introd
new coordinates x' by the coordinate transformation $x \mapsto x'$ defined through

$$\frac{\partial x^{\mu}(x')}{\partial x'^{\alpha}} = E^{\mu}_{(\alpha)}(x(x')). \qquad (19.$$

Due to $\det(\partial x^{\mu}/\partial x'^{\alpha}) \neq 0$, the above prescription represents a perfectly admissible co
dinate transformation. It is in these new coordinates x' where the connection coefficie
vanish. By using the formula 18.19 for the transformed connection coefficients $\Gamma'^{\mu}_{\nu\rho}(x')$ a
by noting that

$$\frac{\partial^2 x^{\mu}}{\partial x'^{\alpha} \partial x'^{\beta}} = \frac{\partial E^{\mu}_{(\alpha)}}{\partial x^{\rho}} \frac{\partial x^{\rho}}{\partial x'^{\beta}} = -\Gamma^{\mu}_{\nu\rho} \frac{\partial x^{\nu}}{\partial x'^{\alpha}} \frac{\partial x^{\rho}}{\partial x'^{\beta}}, \qquad (19.$$

we obtain that the connection coefficients vanish, $\Gamma'^{\mu}_{\nu\rho}(x') = 0$, as desired. Hence, loca
one can bring the metric into the form $g'_{\mu\nu}(x') = \eta_{\mu\nu}$.

Curvature Tensor Examples
We should note that a flat manifold is not necessarily a Euclidean space. The cylin
$C^2(a)$ of radius a is a good example. Embedded in 3-dimensional Euclidean space, the
dimensional cylinder appears to be curved but intrinsically it has a zero curvature. T
cylinder can be bent back to a flat 2-dimensional surface. We can use directly the r
tric $ds^2 = a^2 d\varphi^2 + dz^2$ to derive the flatness. A coordinate transformation $l = a$

with $0 \leq l < 2\pi a$, leads to the equivalent Euclidean metric $ds^2 = dl^2 + dz^2$, which indeed has a zero curvature.

A manifold with a non-vanishing curvature is the 2-sphere $S^2(a)$ of radius a. Using the angle coordinates (θ, φ), with $0 \leq \theta \leq \pi$ and $0 \leq \varphi < 2\pi$, we can readily obtain the Riemann curvature tensor, which has the components

$$R_{\theta\varphi\theta\varphi} = R_{\varphi\theta\varphi\theta} = -R_{\theta\varphi\varphi\theta} = -R_{\varphi\theta\theta\varphi} = a^2 \sin^2\theta, \tag{19.25}$$

with all other components being zero. Within the angle coordinates (θ, φ), the curvature tensor of the 2-sphere depends only on the θ-coordinate.

Another basic but nontrivial example is the upper half-plane model H^2 of hyperbolic space. Its Riemann curvature tensor is calculated to be

$$R_{xyxy} = R_{yxyx} = -R_{xyyx} = -R_{yxxy} = -\frac{1}{y^4}, \tag{19.26}$$

with all other components being zero. When the coordinate y asymptotically approaches the value zero, the above curvature tensor components become infinite. The directional nature of the Riemann tensor components obscures here the fact that the sphere and the UHP model are very symmetrical manifolds. In Section 19.4, we will introduce another quantity, the scalar curvature, which provides a very useful measure for describing a geometry. The above results are obtained from straightforward calculations. (*exercise 19.3*)

19.3 Algebraic Symmetries

Antisymmetric Indices

Henceforth, we will also use the version of the Riemann tensor with all indices being covariant, which is sometimes called the *covariant Riemann tensor*,

$$R_{\mu\nu\rho\sigma} = g_{\mu\alpha}R^\alpha{}_{\nu\rho\sigma}. \tag{19.27}$$

The antisymmetry in the second pair of indices,

$$R_{\mu\nu\rho\sigma} = -R_{\mu\nu\sigma\rho}, \tag{19.28}$$

follows directly from the definition 19.13 of the Riemann tensor. The antisymmetry in the first pair of indices,

$$R_{\mu\nu\rho\sigma} = -R_{\nu\mu\rho\sigma}, \tag{19.29}$$

follows from the metric concordance 18.72. (*exercise 19.4*)

Algebraic Bianchi Identity

The *algebraic Bianchi identity* (*Luigi Bianchi*), called also the *cyclic identity*, states that

$$\boxed{R^\mu{}_{\nu\rho\sigma} + R^\mu{}_{\rho\sigma\nu} + R^\mu{}_{\sigma\nu\rho} = 0} \tag{19.30}$$

In order to derive this important formula, one has to use the definition 19.13 and write out all the terms. (*exercise 19.5*) Furthermore, by using the cyclic identity and the previous antisymmetry in the indices, one can deduce that

$$R_{\mu\nu\rho\sigma} = R_{\rho\sigma\mu\nu}. \tag{19.31}$$

The proof is left to the reader. (*exercise 19.6*) It means that we can exchange the order of the first pair and the second pair of indices in the Riemann tensor.

Independent Components of the Riemann Tensor

Principally, the Riemann tensor $R_{\mu\nu\rho\sigma}(x)$ has d^4 components in d dimensions. However, its algebraic symmetries described above greatly reduce the number of the independent components. We can derive that the number of *independent components* of the Riemann tensor is equal

$$\frac{d^2(d^2-1)}{12}. \tag{19.32}$$

To this end, let us view the Riemann tensor as a rank-two tensor $R_{(\mu\nu)(\rho\sigma)}(x)$ with the multi-indices $(\mu\nu)$ and $(\rho\sigma)$. Each of the multi-indices takes $d(d-1)/2$ independent values, due to the antisymmetry within. On the other hand, the rank-two tensor $R_{(\mu\nu)(\rho\sigma)}(x)$ is symmetric under the exchange of its multi-indices. This amounts to

$$N_f = \frac{1}{2}\left[\frac{d(d-1)}{2}\right]\left[\frac{d(d-1)}{2}+1\right] \tag{19.33}$$

independent components. We need to take also into account the constraints imposed by the cyclic identity 19.30. The cyclic identity is trivial whenever any two indices are equal. In other words, the cyclic identity provides a true constraint only when all indices are distinct. This amounts to choosing 4 distinct elements out of d, which is equal to

$$N_c = \binom{d}{4} = \frac{d!}{4!(d-4)!}. \tag{19.34}$$

Therefore, the number of independent components of the Riemann tensor is $N_f - N_c$, which is equal to $d^2(d^2-1)/12$, as claimed.

For $d=1$, we obtain that the Riemann tensor has zero components. Indeed, for a 1-dimensional manifold the curvature vanishes, as the curvature describes the inner bending of the manifold, which in this case can always be straightened out by a reversible transformation. For $d=2$, the Riemann tensor has only one independent component, e.g. $R_{1212}(x)$. In three dimensions, the Riemann tensor has six independent components. For $d=4$, there are 20 independent Riemann tensor components out of the 256 possible. In all cases the number $d^2(d^2-1)/12$ of independent Riemann tensor components coincides with the number of second-order derivatives of the metric which cannot be made to vanish, compare with Section 19.1. This confirms that the Riemann tensor fully captures all curvature degrees of freedom of a manifold.

19.4   Bianchi Identity and the Einstein Tensor

Jacobi Identity

The following *Jacobi identity for covariant derivatives* is generally valid:

$$\boxed{\left[\left[\nabla_\mu, \nabla_\nu\right], \nabla_\rho\right] + \left[\left[\nabla_\nu, \nabla_\rho\right], \nabla_\mu\right] + \left[\left[\nabla_\rho, \nabla_\mu\right], \nabla_\nu\right] = 0} \tag{19.35}$$

For the proof, one has to write out all the terms for these linear differential operators. Despite its simplicity, the importance of this Jacobi identity is immense. It is the basis for the soon to be discussed differential Bianchi identity and the vanishing of the divergence of the Einstein tensor. The latter provides a justification for the Einstein gravitational field equations. An equivalent version of the Jacobi identity is expressed by using the directional covariant derivatives. It is

$$\left[\left[\nabla_X, \nabla_Y\right], \nabla_Z\right] + \left[\left[\nabla_Y, \nabla_Z\right], \nabla_X\right] + \left[\left[\nabla_Z, \nabla_X\right], \nabla_Y\right] = 0, \tag{19.36}$$

valid for the directional covariant derivatives with respect to the vector fields X, Y, Z.

Differential Bianchi Identity

The *differential Bianchi identity* is a consequence of the Jacobi identity 19.35 and reads

$$\boxed{\nabla_\lambda R_{\mu\nu\rho\sigma} + \nabla_\mu R_{\nu\lambda\rho\sigma} + \nabla_\nu R_{\lambda\mu\rho\sigma} = 0} \tag{19.37}$$

For the proof, one applies the Jacobi identity 19.35 to a vector density field and writes out the resulting terms. We lave this for the reader (*exercise 19.7*) The differential Bianchi identity describes the dynamics of the Riemann tensor field.

Ricci Tensor, Curvature Scalar and Einstein Tensor

The *Ricci tensor* $R_{\mu\nu}(x)$ is a rank-two tensor, defined as the contraction

$$R_{\mu\nu} \equiv R^\alpha{}_{\mu\alpha\nu}. \tag{19.38}$$

Obviously, the result is the same whether one contracts the first and third index or the second and fourth index of the Riemann tensor. By contracting the cyclic identity 19.30, one deduces immediately that the Ricci tensor is symmetric,

$$R_{\mu\nu} = R_{\nu\mu}. \tag{19.39}$$

A different way to obtain this symmetry is by expressing the Ricci tensor through the connection. (*exercise 19.8*) Therefore, the Ricci tensor has $d(d+1)/2$ independent components. The *Ricci scalar curvature* (or *curvature scalar*), denoted $R(x)$, is defined as the contraction

$$R \equiv R^\alpha{}_\alpha. \tag{19.40}$$

The *Einstein tensor* $G_{\mu\nu}(x)$ is defined as the rank-two tensor field

$$G_{\mu\nu} \equiv R_{\mu\nu} - \frac{1}{2} R g_{\mu\nu}. \tag{19.41}$$

It is symmetric and has obviously $d(d+1)/2$ independent components. Because of the differential Bianchi identity 19.37, it is (*exercise 19.9*)

$$\boxed{\nabla_\mu G^{\mu\nu} = 0} \tag{19.42}$$

In other words, the *Einstein tensor is divergence-free* and this is a central result for establishing the general theory of relativity. In Section 21.2 we will explain in which sense the Einstein tensor is unique.

Einstein Metric

A metric $g_{\mu\nu}(x)$ is called an *Einstein metric* if the associated Ricci tensor $R_{\mu\nu}(x)$ satisfies the condition

$$R_{\mu\nu} = f g_{\mu\nu}, \tag{19.43}$$

with a scalar function $f(x)$ as proportionality factor. By taking the trace of both sides, we conclude that the condition is actually

$$R_{\mu\nu} = \frac{R}{d} g_{\mu\nu}. \tag{19.44}$$

If the geometric manifold possesses an Einstein metric and the dimension is $d \geq 3$, then the scalar curvature $R(x)$ is constant over the entire manifold. Indeed, by applying the covariant derivative ∇_ρ to the last equation we obtain

$$\nabla_\rho R_{\mu\nu} = \frac{1}{d} g_{\mu\nu} \nabla_\rho R. \tag{19.45}$$

Contracting the indices ρ and μ and using the zero divergence of the Einstein tensor yields the equality

$$\frac{1}{2}g_{\mu\nu}\nabla^{\mu}R = \frac{1}{d}g_{\mu\nu}\nabla^{\mu}R. \tag{19.46}$$

Hence, for any dimension $d \geq 3$ the scalar curvature is a constant number. A further consequence is that the associated Einstein tensor is given by

$$G_{\mu\nu} = \left(\frac{1}{d} - \frac{1}{2}\right)Rg_{\mu\nu}. \tag{19.47}$$

Therefore, if a manifold of dimension $d \geq 3$ possesses an Einstein metric, the proportionality relation $G_{\mu\nu} \propto R_{\mu\nu} \propto g_{\mu\nu}$ holds. Geometric manifolds possessing an Einstein metric are called *Einstein spaces*. We have left out the case $d = 2$ so far but is a straightforward task to show that every 2-dimensional geometric manifold is an Einstein space. (*exercise 19.10*)

19.5   Ricci Decomposition and the Weyl Tensor

Decomposing the Riemann Tensor
In the last section we saw that in the case of an Einstein space, we can express the Ricci tensor in a simple way through the curvature scalar and the metric. Let us pose the question more generally and ask whether we can express the Riemann tensor by the Ricci tensor, the Ricci scalar, and the metric. This type of decomposition is indeed possible and it provides valuable insights into the individual components of the Riemann tensor. However, the result depends on the dimensionality of the manifold. In addition, we will see that there exist certain decompositions which are *orthogonal*, such that each of the tensors into which the Riemann tensor is decomposed is always orthogonal to all the other tensors of the corresponding decomposition. We will first study the special cases $d = 2$ and $d = 3$ and then proceed to the general theory in $d \geq 3$ dimensions.

Riemann Tensor in $d = 2$
For a 2-dimensional Riemannian manifold, the Riemann tensor $R_{abcd}(x)$, whose indices a, b, c, d take the values $1, 2$, acquires a particularly simple form. First, we note that in two dimensions, the Riemann tensor has only one non-vanishing component, let us pick $R_{1212}(x)$, all other components being either equal $\pm R_{1212}(x)$ or zero. More explicitly,

$$\left.\begin{array}{l} R_{1212} = R_{2121} = -R_{1221} = -R_{2112} \\ R_{1111} = R_{1122} = R_{2211} = R_{2222} = 0 \end{array}\right\}. \tag{19.48}$$

By using the determinant $\det(g_{ab}) = g_{11}g_{22} - g_{12}g_{21}$ of the 2-dimensional metric $g_{ab}(x)$, we can write the Riemann tensor in a unified way as

$$R_{abcd} = \frac{R_{1212}}{\det(g_{ab})}(g_{ac}g_{bd} - g_{ad}g_{bc}). \tag{19.49}$$

Contraction of the first and third index yields the Ricci tensor $R_{bd}(x)$,

$$R_{bd} = \frac{R_{1212}}{\det(g_{ab})}g_{bd}. \tag{19.50}$$

Further contraction yields the 2-dimensional curvature scalar $R(x)$,

$$R = \frac{2R_{1212}}{\det(g_{ab})}. \tag{19.51}$$

Hence, the 2-dimensional Riemann tensor can be decomposed in the form

$$R_{abcd} = \frac{R}{2}\left(g_{ac}g_{bd} - g_{ad}g_{bc}\right). \tag{19.52}$$

In particular, we infer again that any 2-dimensional geometric manifold is also an Einstein space. Actually in any 2-dimensional manifold the Einstein tensor vanishes identically. The property of 2-dimensional manifolds that the Riemann tensor can be expressed by the curvature scalar is based on the fact that these two quantities have the same number of independent components, while the rank-four tensor $g_{ac}g_{bd} - g_{ad}g_{bc}$ displays the same algebraic symmetries as the Riemann tensor.

Scalar Curvature Examples

Let us take again the 2-sphere $S^2(a)$ and the UHP model H^2 as examples, both being 2-dimensional geometric manifolds. The Ricci tensor $R_{cd}(\theta, \varphi)$ of the 2-sphere according to the previous results is

$$R_{cd} = \frac{1}{a^2}g_{cd}, \tag{19.53}$$

with $g_{cd}(\theta, \varphi)$ being the metric of $S^2(a)$. The Ricci curvature scalar $R(\theta, \varphi)$ is

$$R = \frac{2}{a^2}, \tag{19.54}$$

representing a *positive constant value*. The larger the radius a of the sphere, the smaller is its scalar curvature. The Ricci tensor $R_{cd}(x, y)$ of the UHP model is

$$R_{cd} = -g_{cd}, \tag{19.55}$$

with $g_{cd}(x, y)$ being the metric of H^2. The Ricci curvature scalar $R(x, y)$ is

$$R = -2. \tag{19.56}$$

The scalar curvature has a *constant negative value* throughout the UHP manifold. Both examples considered here represent *manifolds of constant curvature*.

Riemann Tensor in $d = 3$

In a 3-dimensional Riemannian manifold, the Riemann tensor has six independent components, which is also the number of independent components of the Ricci tensor. So we can be optimistic that we can express $R_{jklm}(x)$ through $R_{jk}(x)$, $R(x)$, and $g_{jk}(x)$, with the indices taking the values $j, k, l, m = 1, 2, 3$ here. We start with the ansatz

$$R_{jklm} = g_{jl}U_{km} - g_{jm}U_{kl} + g_{km}U_{jl} - g_{kl}U_{jm}. \tag{19.57}$$

The currently unknown symmetric tensor field $U_{jk}(x)$ is a function of the Ricci tensor, the curvature scalar, and the metric. Contraction of indices leads to the Ricci tensor

$$R_{km} = U_{km} + Ug_{km}, \tag{19.58}$$

with $U \equiv U^j{}_j$ denoting the trace. Contracting further gives $R = 4U$, and so we obtain

$$U_{km} = R_{km} - \frac{1}{4}Rg_{km} \tag{19.59}$$

for the sought after tensor. Hence, the unique decomposition of the Riemann tensor in three dimensions reads

$$R_{jklm} = (g_{jl}R_{km} - g_{jm}R_{kl} + g_{km}R_{jl} - g_{kl}R_{jm}) - \frac{R}{2}\left(g_{jl}g_{km} - g_{jm}g_{kl}\right). \tag{19.60}$$

The first term is sometimes called the *Ricci part*, whereas the second tensor is called the *scalar* (or *trace*) *part*. There is another decomposition possible, very similar to 19.60, if one uses the traceless part of the Ricci tensor instead of the Ricci tensor itself. (*exercise 19.11*) The formula 19.60 is a so-called *Ricci decomposition* and it will be embedded into a more general result in the following.

Decomposition with the Ricci Tensor

Let us consider geometric manifolds with dimensionality equal or higher than three. For $d \geq 4$, the number of independent components of the Riemann tensor is strictly larger than the number of independent components of the Ricci tensor, the difference being

$$\frac{d^2(d^2 - 1)}{12} - \frac{d(d+1)}{2} = \frac{d(d+1)(d+2)(d-3)}{12}. \tag{19.61}$$

For $d = 4$, this is equal to 10 additional degrees of freedom that are contained in the Riemann tensor but cannot be described by the Ricci tensor. So an additional rank-four tensor is needed for a decomposition scheme of the Riemann tensor. We therefore introduce the *Weyl tensor* $W_{\mu\nu\rho\sigma}(x)$, also called the *conformal tensor*, through the general *Ricci decomposition* of the Riemann tensor in the form

$$R_{\mu\nu\rho\sigma} = W_{\mu\nu\rho\sigma} + \frac{1}{d-2}\left(g_{\mu\rho}R_{\nu\sigma} - g_{\mu\sigma}R_{\nu\rho} + g_{\nu\sigma}R_{\mu\rho} - g_{\nu\rho}R_{\mu\sigma}\right)$$

$$- \frac{R}{(d-1)(d-2)}\left(g_{\mu\rho}g_{\nu\sigma} - g_{\mu\sigma}g_{\nu\rho}\right). \tag{19.62}$$

This decomposition formula implicitly defines the Weyl tensor and once again uses prominently the Ricci tensor and the Ricci scalar. Comparing the general decomposition 19.62 with the result 19.60, we realize that for $d = 3$ the Weyl tensor vanishes identically.

The Weyl tensor possesses the same algebraic symmetries as the Riemann tensor, since this is true for the second and third tensor in the rhs of 19.62. Hence, the Weyl tensor displays the antisymmetry in the first and the second pair of indices respectively,

$$W_{\mu\nu\rho\sigma} = W_{[\mu\nu][\rho\sigma]}, \tag{19.63}$$

the cyclic identity,

$$W_{\mu[\nu\rho\sigma]} = 0, \tag{19.64}$$

and the indices pairs flipping symmetry,

$$W_{\mu\nu\rho\sigma} = W_{\rho\sigma\mu\nu}. \tag{19.65}$$

Additionally, the Weyl tensor has the property that all its traces vanish. It is easily seen that

$$W^{\mu}{}_{\nu\mu\sigma} = 0, \tag{19.66}$$

and by virtue of this, all other traces of the Weyl tensor vanish as well. (*exercise 19.12*) This tracelessness property corresponds to $d(d+1)/2$ constraints restricting the freedom of the Weyl tensor compared to the Riemann tensor. Thus, we obtain again that the Weyl tensor has $d(d+1)(d+2)(d-3)/12$ independent components. *The central property of the Weyl tensor is that it describes the curvature of the manifold even in the case that the Ricci tensor vanishes.* Another important property of the Weyl tensor is that it stays invariant under Weyl rescalings, which we will discuss in Section 20.2.

Decomposition with the Traceless Part of the Ricci Tensor

Assuming again geometric manifolds with dimensionality $d \geq 3$, there is another way to carry out the algebraic decomposition of the Riemann tensor. In the formula 19.62, we used the Weyl tensor plus the Ricci tensor and the Ricci scalar as the basic ingredients. Now we are aiming to find a decomposition with the Weyl tensor, the traceless part of the Ricci tensor and a trace term being proportional to the Ricci scalar. Thus, let us consider the *traceless part of the Ricci tensor*, denoted $S_{\mu\nu}(x)$ and given by

$$S_{\mu\nu} \equiv R_{\mu\nu} - \frac{1}{d}R\, g_{\mu\nu}, \tag{19.67}$$

which is a symmetric rank-two tensor field with a vanishing trace,

$$S^{\mu}{}_{\mu} = 0. \tag{19.68}$$

The tensor $S_{\mu\nu}(x)$ has $(d+2)(d-1)/2$ independent components. Furthermore, we can construct the rank-four tensor $S_{\mu\nu\rho\sigma}(x)$ defined as

$$S_{\mu\nu\rho\sigma} \equiv \frac{1}{d-2}\left(g_{\mu\rho}S_{\nu\sigma} - g_{\mu\sigma}S_{\nu\rho} + g_{\nu\sigma}S_{\mu\rho} - g_{\nu\rho}S_{\mu\sigma}\right). \tag{19.69}$$

The tensor $S_{\mu\nu\rho\sigma}(x)$ has the same algebraic symmetries as the Riemann tensor. The simple trace of $S_{\mu\nu\rho\sigma}(x)$ gives back the tensor $S_{\mu\nu}(x)$ again, $g^{\mu\rho}S_{\mu\nu\rho\sigma} = S_{\nu\sigma}$, and thus its complete trace vanishes,

$$g^{\mu\rho}g^{\nu\sigma}S_{\mu\nu\rho\sigma} = 0. \tag{19.70}$$

Finally, we construct the rank-four tensor $G_{\mu\nu\rho\sigma}(x)$ defined as

$$G_{\mu\nu\rho\sigma} \equiv \frac{R}{d(d-1)}\left(g_{\mu\rho}g_{\nu\sigma} - g_{\mu\sigma}g_{\nu\rho}\right). \tag{19.71}$$

The tensor $G_{\mu\nu\rho\sigma}(x)$ has also the same symmetries as the Riemann tensor and its complete trace is identical with the one of the Riemann tensor,

$$g^{\mu\rho}g^{\nu\sigma}G_{\mu\nu\rho\sigma} = R. \tag{19.72}$$

With these definitions, the following decomposition of the Riemann tensor holds (*exercise 19.13*)

$$\boxed{R_{\mu\nu\rho\sigma} = W_{\mu\nu\rho\sigma} + S_{\mu\nu\rho\sigma} + G_{\mu\nu\rho\sigma}} \tag{19.73}$$

This alternative decomposition implicitly defines again the Weyl tensor $W_{\mu\nu\rho\sigma}(x)$. Written out, the decomposition reads

$$R_{\mu\nu\rho\sigma} = W_{\mu\nu\rho\sigma} + \frac{1}{d-2}\left(g_{\mu\rho}S_{\nu\sigma} - g_{\mu\sigma}S_{\nu\rho} + g_{\nu\sigma}S_{\mu\rho} - g_{\nu\rho}S_{\mu\sigma}\right)$$
$$+ \frac{R}{d(d-1)}\left(g_{\mu\rho}g_{\nu\sigma} - g_{\mu\sigma}g_{\nu\rho}\right). \tag{19.74}$$

Note that the last term in 19.74 being proportional to the scalar curvature has a plus sign, compared to the analogous term in the decomposition formula 19.62, which has a minus sign. Due to the vanishing of the Weyl tensor in $d = 3$, one obtains again the previous result 19.60.

The specific Ricci decomposition 19.73 employing the traceless tensor $S_{\mu\nu}(x)$ has the additional property of being *orthogonal*. This means that the three tensors in the rhs of 19.73

are orthogonal to each other according to the definition of the tensorial inner product, see appendix B.1. I.e. it is $\langle W, S \rangle = 0$, $\langle W, G \rangle = 0$, and $\langle S, G \rangle = 0$. As a consequence, the Euclidean norm $|R|$ of the Riemann tensor decomposes always into three pieces, like

$$|R|^2 = |W|^2 + |S|^2 + |G|^2. \tag{19.75}$$

In general, it is $|S|^2 \neq 0$ and $|G|^2 \neq 0$. As it is also $|R|^2 \neq 0$ in general, we conclude that it is $|W|^2 \neq 0$ in general. An orthogonality property like 19.75 does not hold for the decomposition 19.62 though. The proof is left to the reader. (*exercise 19.14*)

Decomposition with the Schouten Tensor

We introduce the rank-two *Schouten tensor* $P_{\mu\nu}(x)$, sometimes also called the *Rho tensor*, for dimensions $d \geq 3$ as

$$P_{\mu\nu} \equiv \frac{1}{d-2} \left(R_{\mu\nu} - \frac{1}{2d-2} R g_{\mu\nu} \right). \tag{19.76}$$

The Schouten tensor is symmetric and possesses, like the Ricci tensor, $d(d+1)/2$ independent components. The trace $P(x) \equiv P^\alpha{}_\alpha(x)$ of the Schouten tensor is generally non-vanishing,

$$P = \frac{R}{2d-2}. \tag{19.77}$$

By employing the Schouten tensor, the Ricci decomposition takes the form (*exercise 19.15*)

$$\boxed{R_{\mu\nu\rho\sigma} = W_{\mu\nu\rho\sigma} + (g_{\mu\rho} P_{\nu\sigma} - g_{\mu\sigma} P_{\nu\rho} + g_{\nu\sigma} P_{\mu\rho} - g_{\nu\rho} P_{\mu\sigma})} \tag{19.78}$$

Here the Riemann tensor splits into the Weyl tensor and only one additional term containing the Schouten tensor. Nice properties of this decomposition are its independence of the dimension of the manifold and its orthogonality. Let us remark that this decomposition is also unique, there is no other way to obtain the Riemann tensor from the Weyl tensor and the Schouten tensor. With this decomposition, the Ricci tensor $R_{\mu\nu}(x)$ can be written as

$$R_{\mu\nu} = (d-2) P_{\mu\nu} + P g_{\mu\nu}. \tag{19.79}$$

By contraction, we obtain again the previous result where the curvature scalar is expressed through the trace of the Schouten tensor, $R = (2d-2)P$. The decomposition 19.78 allows one to prove the vanishing of the Weyl tensor in three dimensions in an elegant way. (*exercise 19.16*) Beyond its algebraic properties displayed here, the Schouten tensor becomes especially interesting when we consider the differential Bianchi identity.

Bianchi Identity for the Weyl Tensor

We consider geometric manifolds with dimensionality $d \geq 3$. By contracting the differential Bianchi identity 19.37, we obtain the formula

$$\nabla_\mu R^\mu{}_{\nu\rho\sigma} = \nabla_\rho R_{\sigma\nu} - \nabla_\sigma R_{\rho\nu}. \tag{19.80}$$

This should be understood as the dynamical equation for the Riemann tensor $R^\mu{}_{\nu\rho\sigma}(x)$ as expressed through the Ricci tensor. We can translate this formula into an analogous dynamical equation for the Weyl tensor $W^\mu{}_{\nu\rho\sigma}(x)$ by inserting the decomposition 19.62. This results into the *differential Bianchi identity for the Weyl tensor*,

$$\nabla_\mu W^\mu{}_{\nu\rho\sigma} = \frac{d-3}{d-2} \left[\nabla_\rho R_{\sigma\nu} - \nabla_\sigma R_{\rho\nu} - \frac{1}{2d-2} (g_{\sigma\nu} \nabla_\rho R - g_{\rho\nu} \nabla_\sigma R) \right]. \tag{19.81}$$

By using the Schouten tensor $P_{\mu\nu}(x)$, we can write this differential Bianchi identity for the Weyl tensor more concisely as

$$\boxed{\nabla_\mu W^\mu{}_{\nu\rho\sigma} = (d-3)\left(\nabla_\rho P_{\sigma\nu} - \nabla_\sigma P_{\rho\nu}\right)} \qquad (19.82)$$

All above formulae are obtained in a straightforward way. (*exercise 19.17*) The Bianchi identity 19.82 can be written even denser by employing the *Cotton tensor* $C_{\rho\sigma\nu}(x)$ (*Émile Clément Cotton*), which is defined as the rank-three tensor field

$$C_{\nu\rho\sigma} \equiv \nabla_\rho P_{\sigma\nu} - \nabla_\sigma P_{\rho\nu}. \qquad (19.83)$$

The Cotton tensor is antisymmetric in its last two indices, $C_{\nu\rho\sigma} = -C_{\nu\sigma\rho}$, and obeys the cyclic identity

$$C_{[\nu\rho\sigma]} = 0. \qquad (19.84)$$

By using the Cotton tensor, the Bianchi identity for the Weyl tensor becomes

$$\nabla_\mu W^\mu{}_{\nu\rho\sigma} = (d-3)\, C_{\nu\rho\sigma}. \qquad (19.85)$$

From this we can infer that all traces of the Cotton tensor vanish. Let us note here also the two divergence-like formulae

$$\nabla_\mu P^{\mu\nu} = \nabla^\nu P \qquad (19.86)$$

for the Schouten tensor and

$$\nabla_\nu C^\nu{}_{\rho\sigma} = 0 \qquad (19.87)$$

for the Cotton tensor. Both formulae essentially originate from the contracted Bianchi identity for the Ricci tensor. (*exercise 19.18*) In the case of three dimensions, the Bianchi identity 19.85 is trivial. In the case of four dimensions, the geometry is governed by the dynamical equation $\nabla^\mu W_{\mu\nu\rho\sigma} = C_{\nu\rho\sigma}$. In our upcoming discussion of general relativity, we will discuss in which sense the identity 19.85 is the gravitational analogue of the Maxwell equations. In the same spirit, the divergence formula 19.87 can be viewed as the analogue of electric current conservation. We will also consider the important case where the Ricci tensor vanishes, corresponding to a matter-free spacetime. In this special case, the Bianchi identity for the Weyl tensor becomes

$$\nabla_\mu W^\mu{}_{\nu\rho\sigma} = 0. \qquad (19.88)$$

This dynamical equation 19.88 for the Weyl tensor carves out the specific part of the Bianchi identity 19.37 that is relevant even when the Ricci tensor vanishes.

Further Reading

In this chapter, we have focused on the most fundamental notions of Riemannian curvature and have omitted certain topics, such as the construction of Riemann normal coordinates or the notion of geodesic deviation. A treatment of these topics can be found in many books on Riemannian geometry, e.g. in Eisenhart [26], Frankel [30], Lee [52], Lovelock, Rund [55], Nakahara [58], or Willmore [98]. Additional references for pseudo-Riemannian geometry are mathematically oriented texts on general relativity, like de Felice, Clarke [21], Hawking, Ellis [39], Straumann [84], and Wald [91]. The notion of the conformal tensor was originally introduced by Weyl [95] and was later developed much further by Schouten [80].

20

Symmetries of Riemannian Manifolds

In this chapter we discuss the notion of symmetric space, as a manifold with geometric symmetries expressed through the existence of Killing vector fields. We derive the maximum possible number of distinct Killing vector fields that can exist in a geometric manifold and provide examples. Then we move to a different question and consider Weyl rescalings of the metric and the corresponding Weyl group acting on a manifold. We investigate under which conditions a geometric manifold is conformally flat. The answer is given by the classical Weyl-Schouten theorem, whose complete proof is given. Finally, we discuss the algebraic structure of the set of general coordinate transformations of a given manifold forming the infinite-dimensional group of diffeomorphisms.

20.1 Symmetric Spaces

Isometry Condition in Covariant Form

In Chapter 3, we saw how isometries of a geometric manifold $(\mathcal{M}, g_{\mu\nu})$ can be expressed by using Killing vector fields (KVFs) $K^\mu(x)$. It is possible and useful to write the defining Killing condition $\mathcal{L}_K g_{\mu\nu} = 0$ by employing the covariant derivative. First, let us note that the Lie derivative of the metric $g_{\mu\nu}(x)$ with respect to an *arbitrary vector field* $V^\mu(x)$ is always given by (*exercise 20.1*)

$$\boxed{\mathcal{L}_V g_{\mu\nu} = \nabla_\mu V_\nu + \nabla_\nu V_\mu} \tag{20.1}$$

Consequently, the Killing equation $\mathcal{L}_K g_{\mu\nu} = 0$ can be written covariantly as

$$\boxed{\nabla_\mu K_\nu + \nabla_\nu K_\mu = 0} \tag{20.2}$$

This is the *covariant form of the Killing equation* and explicitly shows that the Killing condition is an invariant statement. On the other hand, the formulation that uses the Lie derivative is by no means less general, as the Lie derivative exists without any reference to a metric or connection on the manifold. We note also the following two basic properties of a KVF $K^\mu(x)$,

$$\nabla_\nu K^\nu = 0 \quad \text{and} \quad \nabla_\mu \nabla_\nu K^\nu = 0, \tag{20.3}$$

which will be useful for later purposes.

DOI: 10.1201/9781003087748-20

Maximal Symmetry of a Manifold

Given a geometric manifold, one can ask if there is a way to determine how many geometric symmetries (i.e. isometries) the manifold can have. Actually, there exists an upper bound of possible distinct isometries:

> *For a geometric manifold of dimension $d \geq 2$, the maximum possible number of linearly independent KVFs is equal $d(d+1)/2$.*

In other words, $d(d+1)/2$ is the maximum number of distinct symmetries that can exist on a geometric manifold, always based purely on the structure of the manifold. In order to prove this statement, we consider a KVF $K^\mu(x)$ satisfying the condition 20.2 and employ the Ricci identity 19.16 together with the cyclic identity 19.30. By writing the Ricci identity three times with permuted indices and by summing up, we obtain (*exercise 20.2*)

$$\boxed{\nabla_\rho \nabla_\sigma K_\mu = R^\alpha{}_{\rho\sigma\mu} K_\alpha} \tag{20.4}$$

This formula is valid for a KVF $K^\mu(x)$, but it is not generally true for an arbitrary vector field. We realize that any KVF of the algebra can be expressed as a linear combination of terms proportional to $K_\mu(x)$ and $\nabla_\rho K_\mu(x)$. All higher covariant derivatives in a Taylor series expansion reduce to zero-order or to first-order covariant derivatives. Hence, the maximum number of linearly independent KVFs is equal

$$d + \frac{d(d-1)}{2} = \frac{d(d+1)}{2}. \tag{20.5}$$

The actual number of linearly independent KVFs in a concrete case is not determined by the dimension d only. The number $d(d+1)/2$ represents only an upper bound. It is the actually present symmetry of the manifold that determines the number of independent KVFs. Geometric manifolds which possess the maximum number $d(d+1)/2$ of independent KVFs are called *maximally symmetric spaces*. A basic example of a maximally symmetric space is the four-dimensional Minkowski space with its ten-parameter group of Poincaré symmetry transformations.

By contracting the equation 20.4 in the indices ρ and μ, we see that for a KVF $K^\mu(x)$ the following identity holds:

$$\nabla_\rho \nabla_\sigma K^\rho = R_{\rho\sigma} K^\rho. \tag{20.6}$$

By using in addition the Einstein divergence formula 19.42 and the covariant Killing equation 20.2, we consequently obtain (*exercise 20.3*)

$$K^\mu \nabla_\mu R = 0. \tag{20.7}$$

This expresses the fact that the directional derivative of the scalar curvature $R(x)$ with respect to the Killing vector $K^\mu(x)$ vanishes. In other words, the scalar curvature does not change along the symmetry directions of the manifold and it is $\mathcal{L}_K R = 0$. Let us remark here that the Killing condition 20.2 and the formula 20.4 can be used to prove again that the KVFs of a manifold constitute a real Lie algebra. (*exercise 20.4*)

Curvature Structure of a Maximally Symmetric Space

The following theorem describes the curvature structure of a maximally symmetric space in detail. For a maximally symmetric space $(\mathcal{M}, g_{\mu\nu})$ of dimension $d \geq 2$, the scalar curvature, the Ricci tensor, and the Riemann tensor have the following form:

$$R = \text{const}, \tag{20.8}$$

$$R_{\mu\nu} = \frac{R}{d} g_{\mu\nu}, \tag{20.9}$$

$$R_{\mu\nu\rho\sigma} = \frac{R}{d(d-1)} (g_{\mu\rho}g_{\nu\sigma} - g_{\mu\sigma}g_{\nu\rho}). \tag{20.10}$$

In particular, a maximally symmetric space possesses an Einstein metric. In order to prove the above formulae, we employ the $(0,2)$-tensor field $\nabla_\mu K_\nu$, where $K_\nu(x)$ is a KVF. By using the identity 20.4, the commutator yields

$$[\nabla_\rho, \nabla_\sigma] \nabla_\mu K_\nu = \nabla_\rho(R^\alpha{}_{\sigma\mu\nu} K_\alpha) - \nabla_\sigma(R^\alpha{}_{\rho\mu\nu} K_\alpha) =$$
$$(\nabla_\rho R^\alpha{}_{\sigma\mu\nu} - \nabla_\sigma R^\alpha{}_{\rho\mu\nu})K_\alpha + R^\alpha{}_{\sigma\mu\nu}\nabla_\rho K_\alpha - R^\alpha{}_{\rho\mu\nu}\nabla_\sigma K_\alpha. \tag{20.11}$$

On the other hand, the Ricci identity and the Killing property 20.2 lead to

$$[\nabla_\rho, \nabla_\sigma] \nabla_\mu K_\nu = R^\alpha{}_{\mu\rho\sigma}\nabla_\nu K_\alpha - R^\alpha{}_{\nu\rho\sigma}\nabla_\mu K_\alpha. \tag{20.12}$$

By combining the two last equations, we obtain the condition

$$(\nabla_\rho R^\alpha{}_{\sigma\mu\nu} - \nabla_\sigma R^\alpha{}_{\rho\mu\nu})K_\alpha$$
$$+(R^\alpha{}_{\sigma\mu\nu}\delta^\beta_\rho - R^\alpha{}_{\rho\mu\nu}\delta^\beta_\sigma + R^\alpha{}_{\mu\sigma\rho}\delta^\beta_\nu - R^\alpha{}_{\nu\sigma\rho}\delta^\beta_\mu)\nabla_\beta K_\alpha = 0. \tag{20.13}$$

This condition holds for any KVF $K_\alpha(x)$ and its first-order derivative $\nabla_\beta K_\alpha(x)$. Thus, each of the factors of $K_\alpha(x)$ and $\nabla_\beta K_\alpha(x)$ must vanish individually. For the factor of $\nabla_\beta K_\alpha(x)$, due the Killing property 20.2, we obtain the condition

$$R^\alpha{}_{\sigma\mu\nu}\delta^\beta_\rho - R^\beta{}_{\sigma\mu\nu}\delta^\alpha_\rho - R^\alpha{}_{\rho\mu\nu}\delta^\beta_\sigma + R^\beta{}_{\rho\mu\nu}\delta^\alpha_\sigma$$
$$+R^\alpha{}_{\mu\sigma\rho}\delta^\beta_\nu - R^\beta{}_{\mu\sigma\rho}\delta^\alpha_\nu - R^\alpha{}_{\nu\sigma\rho}\delta^\beta_\mu + R^\beta{}_{\nu\sigma\rho}\delta^\alpha_\mu = 0. \tag{20.14}$$

By contracting the indices β and ρ, this yields

$$R^\alpha{}_{\sigma\mu\nu}d - R^\alpha{}_{\sigma\mu\nu} + R^\alpha{}_{\sigma\nu\mu}$$
$$+R^\alpha{}_{\mu\sigma\nu} + R_{\mu\sigma}\delta^\alpha_\nu + R^\alpha{}_{\nu\mu\sigma} - R_{\nu\sigma}\delta^\alpha_\mu = 0. \tag{20.15}$$

If we use the cyclic identity, this equation is simplified to

$$(d-1)R_{\alpha\sigma\mu\nu} = R_{\nu\sigma}g_{\alpha\mu} - R_{\mu\sigma}g_{\alpha\nu}. \tag{20.16}$$

Since the Riemann tensor is antisymmetric in the index pair $(\alpha\sigma)$, while the Ricci tensor is symmetric, we equivalently have

$$(d-1)R_{\alpha\sigma\mu\nu} = -R_{\nu\alpha}g_{\sigma\mu} + R_{\mu\alpha}g_{\sigma\nu}. \tag{20.17}$$

Contraction of the indices α and μ leads to the desired formula 20.9. By inserting this result into the equation 20.17, we obtain the second sought after formula 20.10. Finally, we need to employ also the fact that the factor of $K_\alpha(x)$ vanishes. This is the condition

$$\nabla_\rho R^\alpha{}_{\sigma\mu\nu} - \nabla_\sigma R^\alpha{}_{\rho\mu\nu} = 0. \tag{20.18}$$

By applying the form 20.10 of the Riemann tensor plus the metric concordance to the above condition, we obtain immediately

$$\nabla_\rho R = 0, \tag{20.19}$$

and therefore the claimed constancy property 20.8. We note that for a maximally symmetric manifold it is also $\nabla_\lambda R_{\mu\nu} = 0$ and $\nabla_\lambda R_{\mu\nu\rho\sigma} = 0$, which is easily obtained from the formulae 20.9 and 20.10. For a maximally symmetric space, the Weyl tensor vanishes. (*exercise 20.5*) We also note that all KVFs of a maximally symmetric space are eigenvectors of the covariant Laplace / d'Alembert operator. (*exercise 20.6*)

Conformality Condition in Covariant Form

In Section 3.3 we introduced conformal Killing vector fields (CKVFs), for which we use the same symbol $K^\mu(x)$ again. Let us cast the associated formulae into a covariant form. The condition $\mathcal{L}_K g_{\mu\nu} = 2\tau g_{\mu\nu}$ for conformal symmetry, now covariantly formulated, reads

$$\nabla_\mu K_\nu + \nabla_\nu K_\mu = 2\tau g_{\mu\nu}. \tag{20.20}$$

By taking the trace of both sides, we identify the scalar function $\tau(x)$ as

$$\tau = \frac{1}{d} \nabla_\alpha K^\alpha. \tag{20.21}$$

Thus, we can write down the desired *covariant form of the conformal Killing equation*,

$$\boxed{\nabla_\mu K_\nu + \nabla_\nu K_\mu = \frac{2}{d}(\nabla_\alpha K^\alpha) g_{\mu\nu}} \tag{20.22}$$

It is clear that CKVFs represent less stringent structures than KVFs. For example, the properties 20.3, which are valid for KVFs are, in general, not valid for CKVFs. By carrying out similar steps as for the derivation of relation 20.4, we can see that for a CKVF its second-order covariant derivative can be expressed as (*exercise 20.7*)

$$\nabla_\rho \nabla_\sigma K_\mu = R^\alpha{}_{\rho\sigma\mu} K_\alpha + \frac{1}{d}\Big[\nabla_\rho(\nabla_\alpha K^\alpha) g_{\sigma\mu} - \nabla_\mu(\nabla_\alpha K^\alpha) g_{\rho\sigma} + \nabla_\sigma(\nabla_\alpha K^\alpha) g_{\mu\rho}\Big]. \tag{20.23}$$

The contraction of the indices ρ and μ in the above equation leads to

$$\nabla_\rho \nabla_\sigma K^\rho = R_{\rho\sigma} K^\rho + \nabla_\sigma \nabla_\rho K^\rho, \tag{20.24}$$

which is nothing new, however, as it is only the contracted Ricci identity for the CKVF.

Moving on, one can ask what the maximum possible number of linearly independent CKVFs is on any given d-dimensional manifold. We would expect this number to be higher than the number $d(d+1)/2$ of proper KVFs, as conformal transformations represent a weaker symmetry requirement than isometry transformations. If we consider the maximally symmetric Minkowski space in $d \geq 3$ dimensions, we know already from Chapter 14 that the number of independent CKVFs is equal $(d+1)(d+2)/2$. This corresponds to the representation of the conformal group as $SO(2,d)$. Therefore, in the general case of a d-dimensional geometric manifold, we can expect that the number of independent CKVFs has $(d+1)(d+2)/2$ as an upper bound.

Examples of Maximally Symmetric Spaces

We have encountered already various maximally symmetric spaces. Maximally symmetric spaces of zero curvature, $R = 0$, are for instance the D-dimensional Euclidean space $\mathbb{E}^D$ and the d-dimensional Minkowski space $\mathbb{M}_d$. The corresponding KVFs generate the Euclidean group $ISO(D)$ and the Poincaré group $ISO(1,D)$, which are the respective isometry groups, see the Chapters 10 and 13.

An example of a maximally symmetric space of constant positive curvature, $R > 0$, is the 2-dimensional sphere $S^2(a)$. The 2-sphere possesses the maximum number of three independent Killing vector fields. Comparing with our results in Section 10.2 where we derived the generators of rotations, we see that the corresponding Killing vector fields generate the group $SO(3)$, which represents the isometry group of the 2-sphere. The KVFs of the 2-sphere can be derived explicitly in suitable (e.g. in spherical) coordinates. (*exercise 20.8*) More generally, we have that the number $D(D+1)/2$ of distinct KVFs of the hypersphere

$S^D(a)$ matches the dimension $n(n-1)/2$ of the rotation group $SO(n)$ for $n = D+1$, i.e. the group $SO(D+1)$ represents the isometry group of the D-sphere. The curvature quantities of the D-sphere are summarized in appendix B.7.

An example of a maximally symmetric space of constant negative curvature, $R < 0$, is provided by the 2-dimensional UHP model H^2. Its isometry group is $SO(1,2)$. More generally, the upper half-space model H^D in D dimensions (see Section 2.3) has the group $SO(1,D)$ as its isometry group.

20.2   Weyl Rescalings

Definition of Weyl Rescalings

If a geometric manifold $(\mathcal{M}, g_{\mu\nu})$ is given, a *Weyl rescaling* is the basic transformation

$$\boxed{g_{\mu\nu} \mapsto \widehat{g}_{\mu\nu} = \Omega^2 g_{\mu\nu}} \tag{20.25}$$

of the metric tensor, with the *conformal factor* $\Omega(x)$ being a differentiable, strictly positive scalar function defined on the manifold, see Section 2.3. Note that this is not a coordinate transformation, but the assignment of an entirely new metric tensor $\widehat{g}_{\mu\nu}(x)$ with the corresponding change of the geometry. Simply stated, the geometric manifold $(\mathcal{M}, \widehat{g}_{\mu\nu})$ possesses a new geometry. In the context of general relativity, Weyl rescalings are sometimes called also *conformal rescalings*, but we must again point out that Weyl rescalings do not represent any type of coordinate transformation. In order to keep this distinction clear, we will always use the terms "Weyl rescaling" and "Weyl symmetry" for the transformations 20.25, and reserve the terms "conformal transformation" and "conformal symmetry" for the coordinate transformations treated in Chapter 14.

Technically, a Weyl rescaling affects all tensorial quantities depending on the metric tensor. For instance, a "fundamental" vector field $V^\mu(x)$ can be defined without any reference to the metric and consequently it does not transform under a Weyl rescaling, it is simply $\widehat{V}^\mu = V^\mu$ then. In contrast, the "derived" covector field $V_\mu = g_{\mu\nu}V^\nu$ does depend on the metric and transforms as

$$V_\mu \mapsto \widehat{V}_\mu = \widehat{g}_{\mu\nu}\widehat{V}^\nu = \Omega^2 V_\mu. \tag{20.26}$$

Similar Weyl transformations apply for tensor fields of higher valence. When we consider a tensor field, say $\phi^{\mu\nu}(x)$, we must agree upfront if $\phi^{\mu\nu}(x)$, or $\phi_{\mu\nu}(x)$, or a different indices placement represents the "fundamental" field, while the remaining indices combinations belong to "derived" versions. For the inverse metric $\widehat{g}^{\mu\nu}(x)$, the defining property is

$$\widehat{g}^{\mu\nu}\widehat{g}_{\nu\rho} = \delta^\mu_\rho, \tag{20.27}$$

from which we infer the unique solution

$$\widehat{g}^{\mu\nu} = \Omega^{-2} g^{\mu\nu}. \tag{20.28}$$

The original metric $g_{\mu\nu}(x)$ is used to raise and lower indices of the original tensor fields, while the Weyl-transformed metric $\widehat{g}_{\mu\nu}(x)$ is used to raise and lower indices of the Weyl-transformed tensor fields. It is convenient to write the conformal factor $\Omega(x)$ as

$$\Omega(x) = \exp s(x), \tag{20.29}$$

with the real scalar function $s(x)$. It is also useful to introduce the symbol $\Upsilon_\mu(x)$ ("ypsilon") for the gradient of the logarithm of the conformal factor, i.e.

$$\Upsilon_\mu \equiv \nabla_\mu \ln \Omega = \frac{\nabla_\mu \Omega}{\Omega} = \nabla_\mu s. \tag{20.30}$$

If one prefers, all covariant derivatives in the above definition can be replaced by ordinary derivatives. Obviously, the quantity $\Upsilon_\mu(x)$ is an absolute covector field.* The contravariant version is $\Upsilon^\mu = g^{\mu\nu}\Upsilon_\nu$. We like to introduce yet another quantity, which will prove to be useful in expressing the effect of Weyl transformations. It is the $(0,2)$-tensor field $\mathcal{Z}_{\mu\nu}(x)$ ("zeta"), which we define here as

$$\mathcal{Z}_{\mu\nu} \equiv \Upsilon_\mu\Upsilon_\nu - \nabla_\mu\Upsilon_\nu - \frac{1}{2}g_{\mu\nu}\Upsilon_\alpha\Upsilon^\alpha. \tag{20.31}$$

Due to the fact that it is $\nabla_\mu\Upsilon_\nu = \nabla_\nu\Upsilon_\mu$, the tensor field $\mathcal{Z}_{\mu\nu}(x)$ is symmetric, $\mathcal{Z}_{\mu\nu} = \mathcal{Z}_{\nu\mu}$. We denote the trace by $\mathcal{Z} \equiv g^{\mu\nu}\mathcal{Z}_{\mu\nu}$. We emphasize that the quantities $s(x)$, $\Upsilon_\mu(x)$, and $\mathcal{Z}_{\mu\nu}(x)$ are fixed as constituents of a Weyl rescaling and do not transform like other tensors.

The Effect of Weyl Rescalings

Let us now see what the effect of Weyl rescalings is for various relevant tensorial quantities. Starting from the metric, we gradually move up to the curvature tensor and its related quantities. For the Levi-Civita connection $\Gamma^\mu_{\nu\rho}(x)$, one obtains the following transformation behavior under a Weyl rescaling (*exercise 20.9*)

$$\boxed{\widehat{\Gamma}^\mu_{\nu\rho} = \Gamma^\mu_{\nu\rho} + \delta^\mu_\nu\Upsilon_\rho + \delta^\mu_\rho\Upsilon_\nu - g_{\nu\rho}\Upsilon^\mu} \tag{20.32}$$

Consequently, the Weyl-transformed covariant derivative $\widehat{\nabla}_\rho\widehat{V}^\mu$ of a fundamental vector field $V^\mu = \widehat{V}^\mu$ is

$$\widehat{\nabla}_\rho V^\mu = \nabla_\rho V^\mu + \Upsilon_\rho V^\mu - \Upsilon^\mu V_\rho + \delta^\mu_\rho\Upsilon_\alpha V^\alpha. \tag{20.33}$$

In a similar way, the Weyl-transformed covariant derivative $\widehat{\nabla}_\rho\widehat{\omega}_\nu$ of a fundamental covector field $\omega_\mu = \widehat{\omega}_\mu$ is

$$\widehat{\nabla}_\rho\omega_\nu = \nabla_\rho\omega_\nu - \Upsilon_\rho\omega_\nu - \Upsilon_\nu\omega_\rho + g_{\rho\nu}\Upsilon^\alpha\omega_\alpha. \tag{20.34}$$

The antisymmetrized combination $\nabla_\rho\omega_\nu - \nabla_\nu\omega_\rho$ is *Weyl rescaling invariant*. We can infer this also in an alternative way, since the antisymmetric combination is simply the exterior derivative $d\omega$ of the covector $\omega_\mu(x)$ and as such it is defined without any reference to a geometry. For a fundamental $(0,2)$-tensor field $\phi_{\mu\nu}(x)$, we have the formula

$$\widehat{\nabla}_\rho\phi_{\mu\nu} = \nabla_\rho\phi_{\mu\nu} - 2\Upsilon_\rho\phi_{\mu\nu} - \Upsilon_\mu\phi_{\rho\nu} - \Upsilon_\nu\phi_{\mu\rho} + g_{\rho\mu}\Upsilon^\alpha\phi_{\alpha\nu} + g_{\rho\nu}\Upsilon^\alpha\phi_{\mu\alpha}. \tag{20.35}$$

Once again is the fully antisymmetrized combination $\nabla_{[\rho}\phi_{\mu\nu]}$ Weyl rescaling invariant,

$$\widehat{\nabla}_{[\rho}\phi_{\mu\nu]} = \nabla_{[\rho}\phi_{\mu\nu]}. \tag{20.36}$$

Considering the divergence $\nabla^\mu\phi_{\mu\nu} = g^{\rho\mu}\nabla_\rho\phi_{\mu\nu}$ of the fundamental tensor field $\phi_{\mu\nu}(x)$, one obtains the formula

$$\widehat{\nabla}^\mu\phi_{\mu\nu} = \Omega^{-2}\big[\nabla^\mu\phi_{\mu\nu} + (d-3)\,\Upsilon^\mu\phi_{\mu\nu} + \Upsilon^\mu\phi_{\nu\mu} - \Upsilon_\nu\phi^\mu{}_\mu\big]. \tag{20.37}$$

Specifically for an *antisymmetric* $(0,2)$-tensor field $A_{\mu\nu} = -A_{\nu\mu}$, this yields

$$\widehat{\nabla}^\mu A_{\mu\nu} = \Omega^{-2}\big[\nabla^\mu A_{\mu\nu} + (d-4)\,\Upsilon^\mu A_{\mu\nu}\big]. \tag{20.38}$$

---

*Although the symbol $\Upsilon_\mu(x)$ is used in the literature, the quantity lacks a name. For our purposes we can call it the covector field *ypsilon*.

We see that, if the dimensionality is $d = 4$, it is identically

$$\widehat{\nabla}^\mu A_{\mu\nu} = \Omega^{-2}\nabla^\mu A_{\mu\nu}, \tag{20.39}$$

which corresponds to *Weyl rescaling covariance*. I.e. for an antisymmetric fundamental tensor $A_{\mu\nu}(x)$, its divergence $\nabla^\mu A_{\mu\nu}$ transforms along with the metric and picks up a factor that is a certain power of $\Omega(x)$. Later, in Section 23.4, we will make use of this fact when we consider the behavior of the free Maxwell equations under Weyl rescalings.

A Weyl rescaling has the following effect on the Riemann curvature tensor $R_{\mu\nu\rho\sigma}(x)$:

$$\widehat{R}_{\mu\nu\rho\sigma} = \Omega^2 \Big[R_{\mu\nu\rho\sigma} + (g_{\mu\rho}Z_{\nu\sigma} - g_{\mu\sigma}Z_{\nu\rho} + g_{\nu\sigma}Z_{\mu\rho} - g_{\nu\rho}Z_{\mu\sigma}) \Big] \tag{20.40}$$

The derivation, despite long, is elementary. By applying suitable contractions, we obtain the Weyl rescaling formulae for the Ricci tensor and the Ricci scalar. (*exercise 20.10*) The Weyl rescaling of the Schouten tensor is very simple: (*exercise 20.11*)

$$\widehat{P}_{\mu\nu} = P_{\mu\nu} + Z_{\mu\nu} \tag{20.41}$$

Under Weyl rescalings, the Weyl tensor with purely covariant indices $W_{\mu\nu\rho\sigma}(x)$ is *Weyl covariant*:

$$\widehat{W}_{\mu\nu\rho\sigma} = \Omega^2 W_{\mu\nu\rho\sigma} \tag{20.42}$$

The Weyl tensor in the version $W^\mu{}_{\nu\rho\sigma}(x)$, with the first index up, is actually *Weyl invariant*:

$$\widehat{W}^\mu{}_{\nu\rho\sigma} = W^\mu{}_{\nu\rho\sigma} \tag{20.43}$$

The derivation of these results is straightforward. (*exercise 20.12*) More Weyl rescaling transformation formulae are summarized in appendix C.

Group of Weyl Rescalings

Whenever a geometric manifold $(\mathcal{M}, g_{\mu\nu})$ is given, the Weyl rescalings of the metric constitute a group, called the *Weyl group* Weyl$(\mathcal{M}, g_{\mu\nu})$. The Weyl group is an infinite-dimensional abelian group. It should not be confused with the finite-dimensional non-abelian group of conformal coordinate transformations that we discussed in Chapter 14. The generators and (trivial) commutators of the Lie algebra of the Weyl group can easily be derived. (*exercise 20.13*)

20.3 The Weyl-Schouten Theorem

Conformal Flatness

In Section 19.2 we introduced the notion of flatness of a metric of a geometric manifold. We can generalize this notion by considering flat metrics equipped with a conformal factor. More precisely, we call a geometric manifold $(\mathcal{M}, g_{\mu\nu})$ with the metric $g_{\mu\nu}(x)$ *conformally flat* if there exists a Weyl rescaling transformation 20.25 such that the transformed metric $\widehat{g}_{\mu\nu}(x)$ has a vanishing Riemann tensor,

$$\widehat{R}^\mu{}_{\nu\rho\sigma} = 0. \tag{20.44}$$

In other words, the metric $g_{\mu\nu}(x)$ is conformally flat iff it can be expressed as the product of a conformal factor times a flat metric. We should remark that the flat metric here is not necessarily the Minkowski metric or the Euclidean metric. The interest for conformally flat metrics arises from the fact that they describe geometries which are simpler than the general case, but are not as trivial as plain flat metrics. In the following, we will derive the precise conditions under which conformal flatness is fulfilled. Interestingly, this is different for the dimensions 1, 2, 3, and $d \geq 4$. For $d = 1$ conformal flatness is trivially fulfilled. Before we proceed to the classic Weyl-Schouten theorem, expressing the conditions for attaining conformal flatness, we need to state another theorem concerning the existence of solutions of certain first-order PDEs.

Theorem of Frobenius

Consider an open set $U \times V \subset \mathbb{R}^d \times \mathbb{R}^N$, with U containing the point $0 \in \mathbb{R}^d$. Assume the dimensionalities $d \geq 2$ and $N \geq 1$. Furthermore, consider d smooth functions $F_\mu : U \times V \to \mathbb{R}^N$, $(x, v) \mapsto F_\mu(x, v)$, where each $F_\mu(x, v)$ is an N-tuple with its components denoted $F_\mu^k(x, v)$, $k = 1, \ldots, N$, $\mu = 1, \ldots, d$. Then we can view the set of first-order PDEs

$$\frac{\partial u^k}{\partial x^\mu}(x) = F_\mu^k(x, u(x)), \tag{20.45}$$

with some initial conditions $u^k(0) = w^k$. This is a set of Nd single PDEs for N unknown functions $u^k(x)$. As we have $Nd > N$ here, we deal with an overdetermined set of PDEs. The *theorem of Frobenius (Ferdinand Georg Frobenius)*, as presented here in its simplest form, gives a necessary and sufficient criterion under which a unique set of solutions $u^k(x)$ exists for the above set of PDEs. For every point $w \in V$ (specifying the initial conditions), there exists a unique set of smooth solutions $u : U \to V$, $x \mapsto u(x) = (u^k(x))$ if and only if the *Frobenius integrability condition*

$$\frac{\partial F_\mu}{\partial x^\nu} - \frac{\partial F_\nu}{\partial x^\mu} + \frac{\partial F_\mu}{\partial v^l} F_\nu^l - \frac{\partial F_\nu}{\partial v^l} F_\mu^l = 0 \tag{20.46}$$

holds, for all $\mu, \nu = 1, \ldots, d$. In order to prove that 20.46 is a necessary condition, we simply differentiate the initial PDEs and consider $\partial_\nu \partial_\mu u$. We have

$$\frac{\partial^2 u}{\partial x^\nu \partial x^\mu}(x) = \frac{\partial F_\mu}{\partial x^\nu}(x, u(x)) + \frac{\partial F_\mu}{\partial v^l}(x, u(x)) \frac{\partial u^l}{\partial x^\nu}(x). \tag{20.47}$$

By employing the PDEs 20.45 and by equating with the index-flipped expression $\partial_\mu \partial_\nu u$, we arrive at the integrability condition. The demonstration that 20.46 represents also a sufficient condition for the existence of unique solutions is a technical proof and is provided for example in [55].

Weyl-Schouten Theorem in $d \geq 4$

We state the main result here:

> *A geometric manifold $(\mathcal{M}, g_{\mu\nu})$ of dimension $d \geq 4$ is conformally flat if and only if its Weyl tensor vanishes, $W_{\mu\nu\rho\sigma} = 0$.*

The proof consists of two directions. If the manifold $(\mathcal{M}, g_{\mu\nu})$ is conformally flat, then according to the definition there exists a Weyl rescaling such that $\widehat{R}_{\mu\nu\rho\sigma} = 0$. From this it follows that it is also $\widehat{P}_{\mu\nu} = 0$, and therefore $\widehat{W}_{\mu\nu\rho\sigma} = 0$. Since the Weyl tensor transforms as in 20.42, we obtain that it is $W_{\mu\nu\rho\sigma} = 0$. This proves the one logical direction. Now assume that it is $W_{\mu\nu\rho\sigma} = 0$ for the manifold at hand. Then, according to 20.42, also the transformed

Weyl tensor vanishes for any conceivable Weyl rescaling, $\widehat{W}_{\mu\nu\rho\sigma} = 0$. Therefore, due to the Ricci decomposition 19.78, for any Weyl rescaling the transformed Riemann tensor has the form

$$\widehat{R}_{\mu\nu\rho\sigma} = \widehat{g}_{\mu\rho}\widehat{P}_{\nu\sigma} - \widehat{g}_{\mu\sigma}\widehat{P}_{\nu\rho} + \widehat{g}_{\nu\sigma}\widehat{P}_{\mu\rho} - \widehat{g}_{\nu\rho}\widehat{P}_{\mu\sigma}. \tag{20.48}$$

Our goal is to show that it is $\widehat{R}_{\mu\nu\rho\sigma} = 0$. To this end, it is sufficient to find a Weyl rescaling for which it is $\widehat{P}_{\mu\nu} = 0$. According to 20.41, the requirement $\widehat{P}_{\mu\nu} = 0$ is equivalent to the fulfillment of the condition

$$P_{\mu\nu} = -\mathcal{Z}_{\mu\nu}. \tag{20.49}$$

Written out, this condition reads

$$\nabla_\mu u_\nu = P_{\mu\nu} + u_\mu u_\nu - \frac{1}{2}g_{\mu\nu}u_\alpha u^\alpha, \tag{20.50}$$

being valid for a covector field $u_\mu(x) = \nabla_\mu s(x)$ defining the sought after Weyl rescaling. The above equation is a set of $d(d+1)/2$ first-order PDEs for the d functions $u_\mu(x)$. By writing out the covariant derivative as a partial derivative plus a contraction with the connection, we realize that we can apply Frobenius' theorem here. We can consider the Frobenius integrability condition $(\partial_\rho\partial_\mu - \partial_\mu\partial_\rho)u_\nu = 0$ and inspect if it is fulfilled. As it is more pleasant to work with covariant derivatives, we consider the equivalent integrability condition

$$[\nabla_\rho, \nabla_\mu]\, u_\nu = -R_{\alpha\nu\rho\mu}u^\alpha \tag{20.51}$$

and inspect its validity. For the lhs of this condition, we calculate the second-order derivative term

$$\nabla_\rho \nabla_\mu u_\nu = \nabla_\rho P_{\mu\nu} + (\nabla_\rho u_\mu)u_\nu + u_\mu(\nabla_\rho u_\nu) - g_{\mu\nu}(\nabla_\rho u_\alpha)u^\alpha \tag{20.52}$$

and carry out the same for the index flipped term $\nabla_\mu \nabla_\rho u_\nu$. By combining the two terms of the lhs, we obtain

$$\text{lhs} = C_{\nu\rho\mu} + u_\mu(\nabla_\rho u_\nu) - u_\rho(\nabla_\mu u_\nu) - g_{\mu\nu}(\nabla_\rho u_\alpha)u^\alpha + g_{\rho\nu}(\nabla_\mu u_\alpha)u^\alpha. \tag{20.53}$$

Here we have used that it is $\nabla_\rho u_\mu = \nabla_\mu u_\rho$ and the definition of the Cotton tensor. In order to eliminate the remaining covariant derivatives, we insert the PDE 20.50 into the above equation. A straightforward algebraic calculation leads to

$$\text{lhs} = C_{\nu\rho\mu} + u_\mu P_{\rho\nu} - u_\rho P_{\mu\nu} - g_{\mu\nu}P_{\rho\alpha}u^\alpha + g_{\rho\nu}P_{\mu\alpha}u^\alpha. \tag{20.54}$$

Next, we focus on the rhs of the integrability condition 20.51. The Riemann tensor $R_{\alpha\nu\rho\mu}(x)$ is expressed solely through the metric and the Schouten tensor, because the Weyl tensor vanishes. By carrying out the contractions in $-R_{\alpha\nu\rho\mu}u^\alpha$, one obtains for the rhs

$$\text{rhs} = u_\mu P_{\nu\rho} - u_\rho P_{\nu\mu} - g_{\nu\mu}P_{\alpha\rho}u^\alpha + g_{\nu\rho}P_{\alpha\mu}u^\alpha. \tag{20.55}$$

By equating the lhs and rhs, we conclude that the integrability condition reads

$$C_{\nu\rho\mu} = 0, \tag{20.56}$$

i.e. the Cotton tensor has to vanish. But this is already fulfilled, as the assumed vanishing of the Weyl tensor leads to the vanishing of the Cotton tensor due to 19.85 in $d \geq 4$ dimensions. Therefore, the integrability condition 20.51 is fulfilled, the PDE 20.50 has a unique solution $u_\mu(x)$ and the Schouten tensor vanishes, $\widehat{P}_{\mu\nu} = 0$. Thus, it is $\widehat{R}_{\mu\nu\rho\sigma} = 0$, which is what we wanted to prove. The manifold is conformally flat.

Weyl-Schouten Theorem in $d = 3$

In three dimensions, the Weyl tensor vanishes identically, so we cannot use the above proof directly. In this case, the role of the Weyl tensor is actually taken over by the Cotton tensor and the Weyl-Schouten theorem states then:

A geometric manifold $(\mathcal{M}, g_{jk})$ of dimension $d = 3$ is conformally flat if and only if its Cotton tensor vanishes, $C_{jkl} = 0$.

For the proof, let us first assume that there is a Weyl rescaling such that $\widehat{R}_{jklm} = 0$. Consequently, all the quantities $\widehat{R}_{jk}$, $\widehat{R}$, and $\widehat{P}_{jk}$ vanish. Thus, it is also $\widehat{C}_{jkl} = 0$. Due to $\widehat{C}_{jkl} = C_{jkl}$ in three dimensions, see appendix C, we obtain also $C_{jkl} = 0$. This proves the one direction. For the other logical direction, let us assume that it is $C_{jkl} = 0$. In three dimensions, the Weyl tensor vanishes identically, $W_{jklm} = 0$ and $\widehat{W}_{jklm} = 0$. From here onward the proof proceeds in complete analogy to the higher dimensional case. The integrability condition is once again the vanishing of the Cotton tensor, which is indeed fulfilled. Hence, the manifold is conformally flat.

If we pause for a moment, we see that the proof we have just given for the case of three dimensions also works perfectly for $d \geq 4$. Thus, we can use the criterion of the vanishing of the Cotton tensor not only for $d = 3$, but actually for all dimensions $d \geq 3$.

Conformal Flatness in $d = 2$

The only case left to consider are manifolds with $d = 2$. In fact, all 2-dimensional geometries are conformally related to a flat geometry. As shown in the previous chapter, the Riemann tensor of a 2-dimensional manifold has always the form 19.52. So we need to show that we can always find a Weyl rescaling such that the resulting Riemann tensor vanishes, $\widehat{R}_{abcd} = 0$. Because in two dimensions the Riemann tensor has only one independent component, it suffices to consider one scalar equation to determine if a suitable Weyl rescaling exists. We can consider the scalar requirement

$$\widehat{R} = 0 \tag{20.57}$$

for the Ricci scalar, which in two dimensions is translated to

$$R = -2\mathcal{Z} \quad \text{with} \quad \mathcal{Z} = -\nabla^{\alpha}\nabla_{\alpha}s. \tag{20.58}$$

Thus, the requirement reads

$$\frac{R}{2} = \nabla^{\alpha}\nabla_{\alpha}s. \tag{20.59}$$

This is a Poisson equation, or an inhomogeneous wave equation (depending on the metric signature) for the scalar field $s(x)$ and has always a solution. Hence, it is fulfilled that $\widehat{R} = 0$ and $\widehat{R}_{abcd} = 0$. *Every 2-dimensional geometric manifold is conformally flat.*

In table 20.1 below we summarize the necessary and sufficient criteria for attaining conformal flatness for different dimensions of a geometric manifold.

Table **20.1**: Conditions for conformal flatness in different dimensions

Dimensionality of manifold	Conditions for conformal flatness
$d = 1$ and $d = 2$	Always fulfilled
$d = 3$	$C_{jkl} = 0$
$d \geq 4$	$C_{\mu\nu\rho} = 0$ or $W_{\mu\nu\rho\sigma} = 0$

Let us ask the question, what is the relation between maximally symmetric spaces and conformally flat manifolds? In the previous section we noted that for every maximally symmetric space the Weyl tensor vanishes. Therefore, *every maximally symmetric space is also a conformally flat manifold.* The practical usefulness of the notion of conformal flatness will be discussed in the general relativity part in the context of the question of how source-free spacetimes behave.

Examples of Conformally Flat Manifolds

All the symmetric spaces we have mentioned in Section 20.1 are also conformally flat manifolds. For the hyperbolic geometries, as expressed through the upper half-space model 2.88 or the Poincaré disk model 2.90, the conformal flatness is immediately apparent in the metric. Generally, for every maximally symmetric space one can find local coordinates such that the metric takes the form

$$g_{\mu\nu}(x) = \left(1 + \frac{R}{8}\eta_{\alpha\beta}x^\alpha x^\beta\right)^{-2} \eta_{\mu\nu} \tag{20.60}$$

in the pseudo-Riemannian case, or with $\eta_{\mu\nu}$ being replaced by $\delta_{\mu\nu}$ in the Riemannian case. The metric in this form is manifestly conformally flat. This formula is only valid for local coordinates that satisfy $1 + Rx^2/8 > 0$. The proof of the existence of these so-called *stereographic coordinates* is sketched in [26].

Let us note that a metric $g_{\mu\nu}(x)$ obtained through a Weyl rescaling from a flat metric (e.g. from $\eta_{\mu\nu}$) in general has a non-vanishing curvature, i.e. the resulting metric is in general non-flat. However, there are some special cases of the scale function $\Omega(x)$ where the resulting metric $\Omega^2(x)\,\eta_{\mu\nu}(x)$ is not only conformally flat, but also flat in the strict sense. For example, the special cases where the rescaled metric is defined with $\Omega(x) = $ const, corresponding to a *dilatation*, and with $\Omega(x) = x^{-2}$, for points with $x^2 \neq 0$, corresponding to an *inversion of norm*, represent two cases where the resulting geometries are flat in the Riemannian sense. (*exercise 20.14*)

20.4   Group of Diffeomorphisms

Group and Algebra of Diffeomorphisms

In this section we treat a special topic and complete the group-theoretical considerations we have started with the rotation group and have developed up the to conformal group in Chapters 10 to 14. Let us consider a d-dimensional differentiable manifold $\mathcal{M}$ and the set of all diffeomorphisms $F : \mathcal{M} \to \mathcal{M}$ on it. With the composition of maps as product, this set constitutes the *group of diffeomorphisms* $\mathrm{Diff}(\mathcal{M})$ on the manifold. Our plan is to elevate diffeomorphisms to the status of symmetry transformations. Let us introduce coordinates x^μ, $\mu = 0, 1, \ldots, D$, on the manifold. The diffeomorphisms can be viewed as depending on the coordinates x^μ, which are a collection of d continuous "indices". For this reason, the group of diffeomorphisms is a continuously infinite-dimensional group. An infinitesimal transformation of the coordinates is

$$x'^\mu = x^\mu + Y^\mu(x), \tag{20.61}$$

where the quantity $Y^\mu(x)$ is simply the variation of the coordinates, which we have denoted $\delta x^\mu(x)$ previously. We expect the $Y^\mu(x)$ to be small, $|Y(x)| \ll 1$. Considering an arbitrary vector field $V^\mu(x)$, the induced infinitesimal transformation is

$$V'^\mu(x) = V^\mu(x) + V^\nu \partial_\nu Y^\mu(x). \tag{20.62}$$

The above is a symmetry transformation of the vector field $V^\mu(x)$ iff it is $\mathcal{L}_Y V^\mu = 0$. Then, the infinitesimal symmetry transformation can be written equivalently as

$$V'^\mu(x) = V^\mu(x) + Y^\nu \partial_\nu V^\mu(x). \tag{20.63}$$

This variation formula will help us to derive the Lie algebra of the diffeomorphism group. Similarly to the group, the *algebra of diffeomorphisms*, denoted $\mathrm{diff}(\mathcal{M})$, is also infinite-dimensional, as we will see below. It is instructive to compare side by side discrete and continuous transformations, as in table 20.2.

Table **20.2**: Finite-dimensional vs. infinite-dimensional transformation

	Discrete r-dimensional	Continuous ∞-dimensional
Parameters	$\epsilon^a \in \mathbb{R}^r$	$Y^\mu(y) \in \mathcal{X}\mathcal{M}$
Indices values	$a = 1, \ldots, r$	$\mu = 0, 1, \ldots, D$ and y^μ
Generators	$\mathcal{G}_a(x)$	$\mathcal{G}_\mu(x; y)$
Contractions	$\mathrm{i}\,\epsilon^a \mathcal{G}_a(x)$	$\mathrm{i} \int Y^\mu(y)\,\mathcal{G}_\mu(x; y)\,\mathrm{d}^d y$

In the notation $\mathcal{G}_\mu(x; y)$, the two indices μ and y specify the generator, while the index x is free. In the continuous case, the contraction with the group parameters does not only require a summation over the discrete index μ, but also an integration over the continuous index y. In analogy to the discrete case, the variation of a tensor field $\phi_i(x)$ is given by the general expression

$$\delta\phi_i(x) = -\mathrm{i} \int_{\mathcal{M}} Y^\mu(y)\,\mathcal{G}_\mu(x; y)_{ij}\,\phi_j(y)\mathrm{d}^d y. \tag{20.64}$$

In order to determine the Lie algebra, we now specialize to the orbital representation $\mathcal{G}_\mu^{\mathrm{orbital}}(x; y)$. On one hand, the variation $\delta V^\rho(x)$ of a vector field is given by 20.63. On the other hand, the same variation within the orbital representation is also given by

$$\delta V^\rho(x) = -\mathrm{i} \int_{\mathcal{M}} Y^\mu(y)\,\mathcal{G}_\mu^{\mathrm{orbital}}(x; y)\,V^\rho(y)\mathrm{d}^d y. \tag{20.65}$$

In this way we can deduce the orbital generators $\mathcal{G}_\mu^{\mathrm{orbital}}(x; y)$ of diffeomorphisms,

$$\mathcal{G}_\mu^{\mathrm{orbital}}(x; y) = \mathrm{i}\,\delta(x - y)\frac{\partial}{\partial y^\mu}, \tag{20.66}$$

where $\delta(x - y)$ is the d-dimensional Dirac delta distribution. A basis of the Lie algebra of diffeomorphisms is the set $\{\mathrm{i}\delta(x - y)\partial/\partial y^\mu\}$, where we have suppressed the tensor indices i, j. This set of basis elements reveals that the algebra of diffeomorphisms is continuously infinite-dimensional. In the orbital representation, a tensor field $\phi_i(x)$ transforms as

$$\delta\phi_i(x) = Y^\mu(x)\partial_\mu\phi_i(x). \tag{20.67}$$

Certain contractions $Y^\mu(x)\partial_\mu\phi_i(x)$ have a particular origin. We Taylor-expand the variations $Y^\mu(x)$ with respect to x around the value zero and obtain

$$\begin{aligned} Y^\mu(x) &= Y^\mu(0) + x^\nu \partial_\nu Y^\mu(0) + x^\rho x^\nu \partial_\rho \partial_\nu Y^\mu(0) + \mathcal{O}(3) \\ &= v_0^\mu + x^\nu v_\nu^\mu + x^\rho x^\nu v_{\rho\nu}^\mu + \mathcal{O}(3), \end{aligned} \tag{20.68}$$

with the constants v_0^μ, v_ν^μ, $v_{\rho\nu}^\mu$, and so on. By considering the basis combinations

$$Y^\mu(x)\partial_\mu = v_0^\mu \partial_\mu + x^\nu v_\nu^\mu \partial_\mu + x^\rho x^\nu v_{\rho\nu}^\mu \partial_\mu + \mathcal{O}(3), \tag{20.69}$$

we can recognize terms with a specific role. The term proportional to ∂_μ corresponds to translations, the combination $x_\mu \partial_\nu - x_\nu \partial_\mu$ corresponds to rotations, and the term proportional to $x^\mu \partial_\mu$ corresponds to dilatations. Diffeomorphisms contain all these basic transformations as their first terms in a Taylor expansion.

Let us finally proceed to the discussion of the commutator relations of the algebra of diffeomorphisms. The continuous analogue to the discrete commutator relation

$$[\mathcal{G}_a, \mathcal{G}_b] = \mathrm{i} f_{abc} \mathcal{G}_c \tag{20.70}$$

is a commutator relation of the form

$$[\mathcal{G}_\mu(x; y), \mathcal{G}_\nu(x; z)] = \mathrm{i} \int_\mathcal{M} f_{\mu\nu}{}^\rho(y, z; w)\, \mathcal{G}_\rho(x; w)\, \mathrm{d}^d w, \tag{20.71}$$

where the functions $f_{\mu\nu}{}^\rho(y, z; w)$ represent the structure "constants". The index x in 20.71 is free for all involved generators. We can immediately infer the antisymmetry relation

$$f_{\mu\nu}{}^\rho(y, z; w) = -f_{\nu\mu}{}^\rho(z, y; w). \tag{20.72}$$

In fact, we can derive an explicit expression for the structure constants $f_{\mu\nu}{}^\rho(y, z; w)$. By inserting the formula 20.66 for the generators in the commutator, we obtain the result

$$[\mathcal{G}_\mu(x; y), \mathcal{G}_\nu(x; z)] = \delta(x - z) \frac{\partial \delta(x - y)}{\partial x^\nu} \frac{\partial}{\partial x^\mu} - \delta(x - y) \frac{\partial \delta(x - z)}{\partial x^\mu} \frac{\partial}{\partial x^\nu}. \tag{20.73}$$

The same result is indeed obtained if the formula

$$f_{\mu\nu}{}^\rho(y, z; w) = \delta_\nu^\rho\, \delta(w - y) \frac{\partial \delta(w - z)}{\partial w^\mu} - \delta_\mu^\rho\, \delta(w - z) \frac{\partial \delta(w - y)}{\partial w^\nu} \tag{20.74}$$

for the structure constants is inserted in the commutator relation 20.71. The above results express the basic Lie-algebraic structure of diffeomorphisms. In the following chapter we identify diffeomorphisms as the symmetry transformations of general relativistic spacetime.

Further Reading

Closely related to the notion of a maximally symmetric space is the notion of a homogeneous and isotropic space. The corresponding geometries have relevance for cosmological applications and are treated accessibly in Weinberg [94]. Weyl rescalings and the notion of conformal geometry constitute an active research field in mathematics and physics. An introduction to the basic elements of this topic with applications in relativity is given by Curry and Gover [17]. The group of diffeomorphisms is discussed in the lectures of DeWitt and Christensen [23].

VI

General Relativity and Symmetry

21

Einstein's Gravitation

We begin this chapter with a discussion of the equivalence principle, which historically motivated Einstein to his conception of the general theory of relativity. Within general relativity, the metric tensor is elevated to a central physical quantity describing a dynamical spacetime. This new situation requires one to formulate all fundamental physical laws in a geometric, diffeomorphism-invariant way, which is achieved by a minimal coupling prescription. The central theme of this chapter are the Einstein equations, which determine the spacetime metric for a given matter and energy distribution. We identify the degrees of freedom in Einstein's equations and discuss possible generalizations. Then we proceed to the Schwarzschild metric, which describes the spacetime outside a static, spherically symmetric mass distribution. Finally, we introduce the notion of an asymptotically flat spacetime, which is important for the formulation of conservation laws.

21.1 Physics in Curved Spacetimes

The Equivalence Principle

Within the special theory of relativity, Einstein had already established the fact that space and time are not absolute but dependent of the reference frame of the observer. The subsequent open question was how to incorporate gravitation into a relativistic framework. Einstein was intrigued by the picture of an observer in an elevator, in which both the elevator and the observer are "freely falling" in the direction of a constant gravitational field. As a consequence, the observer does not feel any gravitational force. The elevator is seen here as an accelerated reference frame, while the freely falling observer is subject to the external gravitational force. In this thought experiment, these two accelerations have the same value, so that the observer cannot measure any gravitational effects. This is also due to fact that the *inertial mass* m_I of the observer is numerically the same as the *gravitational mass* m_G. If g is the constant gravitational field and $a = g$ is the acceleration of the elevator, then the inertial force that the observer feels cancels exactly the gravitational force,

$$-m_I a + m_G g = 0. \tag{21.1}$$

If there was a difference between inertial mass and gravitational mass, the observer could measure a net acceleration. The equality of the two respective values of inertial and gravitational mass is called the *equivalence principle*.* These two masses can be measured physically, and experiments to date have measured that the masses have the same value with an accuracy of the order of 10^{-13}. In Newton's gravitation, the equivalence principle implies that all matter is gravitationally accelerated in the same way regardless of the mass. In the simple example above, we have an additional finding, however, namely that accelerated frames are fundamental ingredients that should be included into a universal description of gravitation. Hence, the type of reference frames employed in our theory is expanded from (constantly moving) inertial frames to arbitrary (accelerated) reference frames. From the point of view of spacetime transformations, this means that we allow all diffeomorphic coordinate transformations between reference frames. This of course calls for Riemannian geometry and the metric tensor acquires a central role as a physical quantity.

Metric Tensor = Gravitational Field

We are prepared to make the crucial step from a static, predefined spacetime toward a dynamical spacetime. Instead of the Minkowski space $\mathbb{M}_4$, we will consider a general four-dimensional geometric manifold $(\mathcal{M}, g_{\mu\nu})$ with the Lorentzian signature $(+ - - -)$. *The metric tensor $g_{\mu\nu}(x)$ represents Einstein's gravitational field.* This gravitational field is not like the other physical fields we have encountered so far, which are all defined on top of a fixed spacetime. Here, spacetime itself *is* the gravitational field. This is the point of departure in order to develop *Einstein's general theory of relativity (GR)*. The metric tensor represents the *gravitational potential* out of which all other relevant quantities can be derived. This role as a "potential" will be clear soon when we make contact with the Newtonian gravitational potential. Very much like in the Newtonian theory, the gravitational potential is determined by the masses being present. The corresponding field equations are the Einstein field equations, and introduced in the next section.

We need to make an important remark here. In the thought experiment discussed before, we saw that accelerated frames can cancel out the effects of gravity. Similarly, accelerated frames can also mimic the force of gravity even in the absence of masses. For example, starting from flat Minkowski space, the transition to an accelerated frame moving with constant acceleration (with respect to the Minkowski background) would mimic a globally constant gravitational field. However, the gravitational field generated by an accelerated frame is different, in principle, from the gravitational field generated by a mass distribution. In the accelerated case, the gravitational field can be transformed away everywhere, for the entire manifold. In the case masses are present, the gravitational field can at best be transformed away only locally, at certain points. The field generated from a mass distribution corresponds to a non-vanishing curvature tensor $R_{\mu\nu\rho\sigma} \neq 0$ in the curved manifold, and this curvature cannot be made to vanish by diffeomorphisms, except at individual points.

Diffeomorphism Invariance of a Physical Theory

Allowing general coordinate transformations in GR means that the spacetime symmetry group is no longer the Poincaré group $P = ISO(1,3)$, but the more general diffeomorphism group $\text{Diff}(\mathcal{M})$. In particular, we have to ensure that our fundamental equations transform covariantly under diffeomorphisms and not only under Poincaré transformations. *Diffeomorphism invariance is the main symmetry of general relativity.* Considering more broadly physical theories, per definition, *diffeomorphism invariance of a physical theory* is the property of invariance of the theory under arbitrary differential coordinate transformations (i.e. active transformations of the coordinates, the metric, and all involved tensor fields).

---

*In more technical terms, this is sometimes called the weak equivalence principle.

In mathematical terms, the diffeomorphism invariance of a physical theory can be very efficiently expressed through the invariance of the corresponding action of the theory. In the next chapter, diffeomorphism invariance of the total action is the principle based on which the Einstein equations are derived. Another way to express diffeomorphism invariance is to generalize the previous *principle of relativity* from Sections 4.1 and 6.1 and demand:

All physical laws have the same form in all reference systems.

This is also called the *general covariance* of physical laws. In practice, this means that under a diffeomorphism, the physical equations transform in such a way that they retain their form and validity with the new transformed quantities. It should be clear that the concept of general covariance is not really new, as we imposed the analogue requirement in Newtonian physics with Galilei invariance, as well as in special relativistic physics (mechanics and field theory) with Poincaré invariance. Diffeomorphism invariance is the natural generalization which goes beyond the previous notions of spacetime symmetries of a physical theory.

Expressing Physical Laws in Curved Spacetime

Let us examine how we can formulate known physical theories on a curved spacetime in such a way that they meet the requirement of general covariance. The first example we consider is a free relativistic particle, whose trajectory $z^\mu(s)$ is parametrized by the proper time s of the particle. In the case of Minkowski space, the free motion of a particle is described by

$$\frac{\mathrm{d}^2 z^\mu}{\mathrm{d}s^2} = 0. \tag{21.2}$$

The derivative in the lhs can always be written as

$$\frac{\mathrm{d}^2 z^\mu}{\mathrm{d}s^2} = \frac{\mathrm{d}u^\mu}{\mathrm{d}s} = \frac{\mathrm{d}z^\rho}{\mathrm{d}s} \partial_\rho u^\mu, \tag{21.3}$$

by using the velocity $u^\mu(s) \equiv \mathrm{d}z^\mu(s)/\mathrm{d}s$. In the case of a curved manifold, the above expression becomes

$$\frac{\mathrm{d}z^\rho}{\mathrm{d}s} \nabla_\rho u^\mu = \frac{\mathrm{d}z^\rho}{\mathrm{d}s}(\partial_\rho u^\mu + \Gamma^\mu_{\nu\rho} u^\nu) = \frac{\mathrm{d}u^\mu}{\mathrm{d}s} + \Gamma^\mu_{\nu\rho} u^\nu u^\rho. \tag{21.4}$$

Hence, the equation describing the free particle in a curved spacetime is

$$\boxed{\frac{\mathrm{d}^2 z^\mu}{\mathrm{d}s^2} + \Gamma^\mu_{\nu\rho} \frac{\mathrm{d}z^\nu}{\mathrm{d}s} \frac{\mathrm{d}z^\rho}{\mathrm{d}s} = 0} \tag{21.5}$$

which is nothing else than the geodesic equation 18.123. Mathematically equivalently we can say that we have replaced the total derivative $\mathrm{d}/\mathrm{d}s$ by the covariant derivative $\mathrm{D}/\mathrm{d}s$. It is appropriate to introduce here the notion of *acceleration* $b^\mu(s)$ of a particle in GR, which is defined by

$$b^\mu \equiv \frac{\mathrm{D}u^\mu}{\mathrm{d}s} = \frac{\mathrm{d}^2 z^\mu}{\mathrm{d}s^2} + \Gamma^\mu_{\nu\rho} \frac{\mathrm{d}z^\nu}{\mathrm{d}s} \frac{\mathrm{d}z^\rho}{\mathrm{d}s}. \tag{21.6}$$

In GR, the particle does not experience any acceleration, $b^\mu(s) = 0$, exactly if it moves along a geodesic trajectory.

As a second major example, let us consider Maxwell's theory of electromagnetism, whose equations 15.81 and 15.82 in Minkowski space have the form

$$\partial_\mu F_{\nu\rho} + \partial_\nu F_{\rho\mu} + \partial_\rho F_{\mu\nu} = 0 \quad \text{and} \quad \partial_\mu F^{\mu\nu} = 4\pi j^\nu. \tag{21.7}$$

How can these physical laws be generalized to the case of a curved spacetime? The above equations are valid in flat Minkowski space but equally well they are valid in locally flat coordinates around a point in a curved spacetime. In these locally flat coordinates we can exchange the partial derivatives by covariant derivatives, since they are identical there. However, the spacetime point considered is arbitrary. Thus, the equations employing the covariant derivative are valid for any point in the spacetime manifold. Hence, we can write down the *Maxwell equations in curved spacetime* in the form

$$\nabla_\mu F_{\nu\rho} + \nabla_\nu F_{\rho\mu} + \nabla_\rho F_{\mu\nu} = 0 \quad \text{and} \quad \nabla_\mu F^{\mu\nu} = 4\pi j^\nu \tag{21.8}$$

We should note that the first equation above, which is the integrability condition, is equivalent to the defining relation[*]

$$F_{\mu\nu} = \nabla_\mu A_\nu - \nabla_\nu A_\mu \tag{21.9}$$

between the field strength $F_{\mu\nu}(x)$ and the vector potential $A_\mu(x)$. This is can be seen in two different ways. One way to obtain the above relation is through the described principal change from partial to covariant derivatives. Another way is based on the circumstance that for any covector field $a_\nu(x)$ it is

$$\partial_\mu a_\nu - \partial_\nu a_\mu = \nabla_\mu a_\nu - \nabla_\nu a_\mu, \tag{21.10}$$

for a symmetric Levi-Civita connection. In addition, let us note here the following formula, being valid for any antisymmetric $(0,2)$-tensor field $A_{\mu\nu}(x)$:

$$\partial_\mu A_{\nu\rho} + \partial_\nu A_{\rho\mu} + \partial_\rho A_{\mu\nu} = \nabla_\mu A_{\nu\rho} + \nabla_\nu A_{\rho\mu} + \nabla_\rho A_{\mu\nu}. \tag{21.11}$$

This identity can be invoked to derive the integrability condition in 21.8 in an alternative way. However, in order to derive the dynamical Maxwell equation $\nabla_\mu F^{\mu\nu} = 4\pi j^\nu$, we must rely on the application of the rule $\partial_\mu \to \nabla_\mu$.

It should be noted that the Maxwell equation for the vector potential,

$$\partial_\mu \partial^\mu A_\nu - \partial_\mu \partial_\nu A^\mu = 4\pi j_\nu, \tag{21.12}$$

when the replacement $\partial_\mu \to \nabla_\mu$ is applied, becomes

$$\nabla_\mu \nabla^\mu A_\nu - \nabla_\nu \nabla_\mu A^\mu - R_{\nu\rho} A^\rho = 4\pi j_\nu, \tag{21.13}$$

due to the non-commutativity of covariant derivatives. We have still the gauge freedom to let the vector potential transform like $A'^\mu = A^\mu - \nabla^\mu \theta$, with a scalar gauge function $\theta(x)$. By choosing the Lorenz gauge $\nabla_\mu A^\mu = 0$, we obtain the simplified Maxwell equation

$$\nabla_\mu \nabla^\mu A_\nu - R_{\nu\rho} A^\rho = 4\pi j_\nu. \tag{21.14}$$

The term containing the Ricci tensor is unexpected. The last equation still reduces to the Maxwell equation 15.89 in the flat spacetime case. Conversely, however, if we had applied the recipe $\partial_\mu \to \nabla_\mu$ directly to the equation 15.89, we would have missed the term $-R_{\nu\rho} A^\rho$. So the application of the basic change $\partial_\mu \to \nabla_\mu$ requires some care and is not free of ambiguities. In fact, the term containing the Ricci tensor ensures that gauge invariance is fulfilled. (*exercise 21.1*)

---

[*]Regarding the application of Weyl rescalings, we consider $F_{\mu\nu}$ as the "fundamental" field and $F^{\mu\nu} = g^{\mu\rho} g^{\nu\sigma} F_{\rho\sigma}$ as the "derived" field.

Starting from 15.80, the *local conservation of electric current in curved spacetime* can be expressed as

$$\nabla_\mu j^\mu = 0 \tag{21.15}$$

In order to generalize the Lorentz force law 15.84, as the equation of motion of an electrically charged particle, we carry out again the transition from partial to covariant derivatives. Consequently, the *Lorentz force law in curved spacetime* takes the form

$$\frac{\mathrm{d}u^\mu}{\mathrm{d}s} + \Gamma^\mu_{\nu\rho} u^\nu u^\rho = \frac{q}{m} F^{\mu\nu} u_\nu \tag{21.16}$$

The electric charge of the particle is q and its rest mass is m. We see that in the absence of an electromagnetic field, or in the case of an electrically neutral particle, the above equation reduces to the geodesic equation 21.5.

The steps that have just been applied can be elevated to a general prescription, which then specifies how one starts from physical laws that are valid in flat Minkowski space and then generalizes them for the case of a curved spacetime. Therefore, in order to generalize an equation being valid in Minkowski space, one has to carry out the following combined formal replacements:

$$\begin{aligned} \eta_{\mu\nu} &\to g_{\mu\nu}(x) \\ \partial_\rho &\to \nabla_\rho \\ \sqrt{|\eta|}\,\mathrm{d}^4x &\to \sqrt{|g|}\,\mathrm{d}^4x \end{aligned} \tag{21.17}$$

This leads to the so-called *minimal coupling to the gravitational field*. How can we be sure that the resulting formulae are actually physically correct? The answer is the following: in order to apply the resulting covariant equations to a particular physical situation, one will employ an inertial reference frame for carrying out the measurements. But within an inertial frame, the covariant equations always reduce to the equations with partial derivatives, which are indeed physically valid. It should be noted, however, that in each concrete case the equations resulting from the prescription 21.17 must be inspected for their validity, because we have already seen in equation 21.14 that ambiguities can occur in certain situations.

Description of Matter and Energy in Curved Spacetime

In Newtonian theory, the mass density $\rho(x)$ is the source of the gravitational potential $\Phi(x)$, see Section 4.4. As much as we want to maintain the simplicity of a scalar quantity, in four-dimensional relativistic physics the energy-momentum tensor (EMT) takes over the role of the quantity that completely describes the distribution of matter. So we have to deal with the *energy-momentum tensor of matter* $T^{\mu\nu}(x)$, as introduced initially in Section 15.1. For theories compatible with a curved spacetime, we have to formulate everything in a generally covariant form. The matter EMT encompasses all physical fields carrying any sort of mass, energy and momentum. However, *this energy-momentum tensor does not include the energy and momentum of the gravitational field itself*. The EMT of the gravitational field in fact represents an open physical question and we discuss this more in Chapter 23. We saw in Section 15.1, that the matter EMT is symmetric, $T^{\mu\nu} = T^{\nu\mu}$, and that it fulfills the local conservation equation $\partial_\mu T^{\mu\nu} = 0$. We will continue to demand these properties from the matter EMT within GR. The *divergence equation of the matter EMT in curved spacetime* takes the form

$$\nabla_\mu T^{\mu\nu} = 0 \tag{21.18}$$

and represents a very stringent requirement. We must note here that the equation 21.18 does not represent, in general, a conservation equation, see also Section 23.1.

EMT of Discrete Particles

The EMT $T^{\mu\nu}(x)$ of a set of discrete relativistic point particles with masses m_n and trajectories $z_n^\mu(s_n)$, $n = 1, \ldots, N$, has been introduced already with the equation 15.8. In the present case of a curved manifold, one has to use the scalar delta distribution and obtains

$$T^{\mu\nu}(x) = \sum_n \frac{m_n}{\sqrt{|g(x)|}} \int_{-\infty}^{\infty} \frac{dz_n^\mu(s_n)}{ds_n} \frac{dz_n^\nu(s_n)}{ds_n} \delta\left(x - z_n(s_n)\right) ds_n. \tag{21.19}$$

The factor $|g|^{-1/2}$ ensures that $|g|^{-1/2}\delta(\,\cdot\,)$ is a scalar field, see appendix B.4. The symmetry of the above EMT is obvious. If the particle set consists of non-interacting point particles, it is called *dust*. Then, the local conservation 21.18 is immediately derived, see Section 15.1.

EMT of a Perfect Fluid

An important continuous matter model, especially for cosmology, is the so-called *perfect fluid*. From a physical standpoint, a perfect fluid has no viscosity and there is no conduction of heat energy. The transport of energy within the fluid happens only through the flow of mass. Moreover, the fluid is supposed to be electrically neutral throughout. For the description of a perfect fluid, one can use the *mass-energy density* $\rho(x)$, the *pressure* (= force per unit area) $p(x)$, and the normed timelike *velocity field* of the fluid $u^\mu(x)$, fulfilling $u_\mu u^\mu = 1$. If we pick any point in spacetime and use locally inertial coordinates, i.e. coordinates in which the fluid is momentarily at rest, then the pressure at this point is the same in all directions. In fact, $p(x)$ is a scalar field. Thus, the purely 3-dimensional part of the EMT is rotationally symmetric and given by $T^{jk}(x) = p(x)\delta^{jk}$. The time-time component is obviously $T^{00}(x) = \rho(x)$. All other components of the EMT of the perfect fluid vanish. Hence, in locally inertial coordinates, the EMT of a perfect fluid is written in matrix form as

$$(T^{\mu\nu}) = \begin{pmatrix} \rho & 0 & 0 & 0 \\ 0 & p & 0 & 0 \\ 0 & 0 & p & 0 \\ 0 & 0 & 0 & p \end{pmatrix}. \tag{21.20}$$

Taking into account that the normed velocity vector in inertial coordinates is $u^\mu = (1, \mathbf{0})^T$, we obtain readily the EMT of the perfect fluid for arbitrary coordinates,

$$T^{\mu\nu} = (\rho + p)u^\mu u^\nu - pg^{\mu\nu}. \tag{21.21}$$

In the non-relativistic limit, the motion of the constituent matter of the perfect fluid is such that the pressure is much smaller than the mass density, i.e. $p \ll \rho$. This leads to an EMT of the form $T^{\mu\nu} = \rho u^\mu u^\nu$. In locally inertial coordinates, this EMT has only one non-vanishing component $T^{00} = \rho$.

EMT of a Real Scalar Field

For a real scalar field in Minkowski space, we know already the EMT, given in 16.75. The corresponding generalized EMT $T^{\mu\nu}(x)$ in a curved spacetime is given by

$$T^{\mu\nu} = \nabla^\mu \phi \, \nabla^\nu \phi - \frac{1}{2} g^{\mu\nu} (\nabla^\alpha \phi \, \nabla_\alpha \phi - m^2 \phi^2). \tag{21.22}$$

The EMT of a more complicated scalar theory, such as the ϕ^4 theory, is obtained in a similar way.

EMT of the Maxwell Field

By employing the Belinfante formula, we have found the symmetric EMT of the electromagnetic field in Minkowski space, see 16.85. The corresponding generalized EMT $T^{\mu\nu}(x)$ in a curved spacetime has the expression

$$T^{\mu\nu} = \frac{1}{4\pi}\left(F^{\mu\alpha}F_\alpha{}^\nu + \frac{1}{4}g^{\mu\nu}F^{\alpha\beta}F_{\alpha\beta}\right). \tag{21.23}$$

The above energy-momentum tensors are derived again later by using a different method.

21.2 The Einstein Equations

Newtonian Limit

In the Newtonian theory of gravitation, the static field equation determining the gravitational potential $\Phi(x)$ from the mass density $\rho(x)$ is

$$\nabla^2\Phi(x) = 4\pi G_\mathrm{N}\rho(x), \tag{21.24}$$

as already discussed in Section 4.4. G_N is Newton's gravitational constant. Now we seek for a generalization of this field equation within the framework of GR, which demands a generally covariant theory with the EMT $T_{\mu\nu}(x)$ representing the local source of gravity. To this end, let us consider the equation of motion for a non-relativistic particle moving in a weak gravitational field. In the Newtonian theory, the equation of motion for the 3-dimensional trajectory $z(t)$ reads

$$\frac{\mathrm{d}^2 z}{\mathrm{d}t^2}(t) = -\nabla\Phi(z(t)), \tag{21.25}$$

with t being the Newtonian time. In GR, in the absence of any other interactions, the equation of motion for the 4-dimensional trajectory $z^\mu(s)$ within the gravitational field is the geodesic equation 21.5. Let us investigate how this equation is modified in the case of a slowly moving particle in a weak gravitational field. Non-relativistic motion means that

$$\left|\frac{\mathrm{d}z}{\mathrm{d}s}\right| \ll \left|\frac{\mathrm{d}t}{\mathrm{d}s}\right|, \tag{21.26}$$

so that the geodesic equation reduces to

$$\frac{\mathrm{d}^2 z^\mu}{\mathrm{d}s^2} + \Gamma^\mu_{00}\frac{\mathrm{d}z^0}{\mathrm{d}s}\frac{\mathrm{d}z^0}{\mathrm{d}s} = 0. \tag{21.27}$$

Concerning the gravitational field, we consider it to be static, so that all time derivatives vanish. Then, the connection coefficients are computed as

$$\Gamma^\mu_{00} = -\frac{1}{2}g^{\mu\sigma}\partial_\sigma g_{00}. \tag{21.28}$$

The gravitational field is assumed to be weak, so we can write for the metric tensor $g_{\mu\nu}(x)$ in this case

$$g_{\mu\nu}(x) \equiv \eta_{\mu\nu} + h_{\mu\nu}(x), \tag{21.29}$$

with $|h_{\mu\nu}(x)| \ll 1$ representing the deviation from the flat Minkowski metric. Therefore, to first order in $h_{\mu\nu}(x)$, the connection coefficients are

$$\Gamma^\mu_{00} = -\frac{1}{2}\eta^{\mu\sigma}\partial_\sigma h_{00}. \tag{21.30}$$

By inserting these connection coefficients into the simplified geodesic equation above and splitting the time and space parts, we obtain

$$\frac{\mathrm{d}^2 z^0}{\mathrm{d}s^2} = 0 \quad \text{and} \quad \frac{\mathrm{d}^2 z^k}{\mathrm{d}s^2} = -\frac{1}{2}\left(\frac{\mathrm{d}t}{\mathrm{d}s}\right)^2 \delta^{kl}\partial_l h_{00}. \tag{21.31}$$

The first equation means that $\mathrm{d}t/\mathrm{d}s = \text{const}$. By inserting this into the second equation, we obtain the equation of motion

$$\frac{\mathrm{d}^2 \boldsymbol{z}}{\mathrm{d}t^2}(t) = -\frac{1}{2}\boldsymbol{\nabla} h_{00}(\boldsymbol{z}(t)). \tag{21.32}$$

By comparing this equation of motion with the one employing the Newtonian potential, we obtain

$$h_{00}(\boldsymbol{x}) = 2\Phi(\boldsymbol{x}) + C, \tag{21.33}$$

where C is a real constant. We demand that the curved spacetime becomes similar to Minkowski space at large distances, i.e. $h_{00}(\boldsymbol{x}) \to 0$ and $\Phi(\boldsymbol{x}) \to 0$, for $|\boldsymbol{x}| \to \infty$. Therefore, it must be $C = 0$. We conclude to the relation

$$g_{00}(\boldsymbol{x}) = 1 + 2\Phi(\boldsymbol{x}) \tag{21.34}$$

between the metric component $g_{00}(\boldsymbol{x})$ and the gravitational potential $\Phi(\boldsymbol{x})$, being valid in the non-relativistic, weak-field limit. Now assuming that the mass distribution is a non-relativistic perfect fluid, we can employ the equation $T_{00}(\boldsymbol{x}) = \rho(\boldsymbol{x})$ for the EMT. The Newtonian gravitational potential equation 21.24 yields then

$$\nabla^2 g_{00}(\boldsymbol{x}) = 8\pi G_{\mathrm{N}} T_{00}(\boldsymbol{x}). \tag{21.35}$$

This equation gives us the necessary hints on how to obtain the general field equation of GR. In the lhs we have a geometric quantity, while in the rhs the proportionality factor $8\pi G_{\mathrm{N}}$ multiplies the EMT as the source of gravity.

Einstein Equations

Let us seek for a four-dimensional relativistic field equation of gravity. According to the previous findings, we should anticipate an equation of the form

$$\mathcal{G}_{\mu\nu} = 8\pi G_{\mathrm{N}} T_{\mu\nu}, \tag{21.36}$$

where the still unknown tensor field $\mathcal{G}_{\mu\nu}(x)$ is a measure of curved geometry and has the following properties: a) Like the EMT, the tensor $\mathcal{G}_{\mu\nu}(x)$ should be a symmetric tensor. b) Because the divergence formula 21.18 holds, we have to demand that $\nabla_\mu \mathcal{G}^{\mu\nu} = 0$ holds as well. c) The tensor field $\mathcal{G}_{\mu\nu}(x)$ should contain terms at most linear in the second-order derivatives of the metric and/or terms at most quadratic in the first-order derivatives of the metric. Candidate tensors which fulfill all three requirements are the metric $g_{\mu\nu}(x)$ itself and the Einstein tensor $G_{\mu\nu}(x)$. If we further restrict requirement c) and demand that the candidate tensor is exactly linear in the second-order derivatives, as in 21.35, then the metric tensor has to be given up and only the Einstein tensor remains. So it is indeed the Einstein tensor, which enters into the sought equations of gravity.* Thus, we conclude to

---

*Let us remark that there is no derivation of the Einstein equations in the strict sense. All paths to arrive to these field equations involve certain assumptions and require conscious choices. The somewhat more stringent way of derivation may seem to be through the action principle, but also this approach is not uniquely defined in many ways and offers actually more freedom of choices and more ambiguity.

the *Einstein field equations*:

$$R_{\mu\nu} - \frac{1}{2}Rg_{\mu\nu} = 8\pi G_{\mathrm{N}}T_{\mu\nu}$$ (21.37)

The lhs is an expression of the spacetime geometry, while the rhs describes the energy and matter distribution of the non-gravitational constituents. For the factor $8\pi G_{\mathrm{N}}$ it is customary to introduce the abbreviation with the *Einstein constant* $\kappa \equiv 8\pi G_{\mathrm{N}}$. The Einstein equations describe the effect of the gravitational field as a manifestation of spacetime curvature. Actually, the general theory of relativity states that there is no gravitational force, it is only the bending of spacetime, which accounts for the gravitational effects. A particle free from the influence of non-gravitational forces moves along a trajectory which is a geodesic curve. The deviation from geodesic motion is actually the effect of additional, non-gravitational forces.

Equations of Motion of Matter

Obviously, the Einstein equations 21.37 contain in particular the equation

$$\nabla_\mu T^{\mu\nu} = 0$$ (21.38)

for the EMT, and this equation in turn describes the motion of the matter particles and fields. It is straightforward to demonstrate that the EMT 21.19 for the simple case of a single relativistic particle, if it is inserted in the above divergence equation, leads to the geodesic equation 21.5. (*exercise 21.2*) On one hand, the Einstein field equations provide a solution for the spacetime metric $g_{\mu\nu}(x)$ for every given EMT of matter $T^{\mu\nu}(x)$. On the other hand, when the spacetime is fixed, the Einstein equations lead to the equations of motion for matter, i.e. to the geodesic equation for particles and matter distributions. In other words, the Einstein field equations 21.37 provide a closed and complete description of the coupled dynamics of gravity and matter.

Gravitational Degrees of Freedom

From a mathematical point of view, the Einstein field equations 21.37 are a set of 16 coupled nonlinear second-order PDEs determining the metric $g_{\mu\nu}(x)$ for a given matter source $T_{\mu\nu}(x)$. Because of the symmetry of the tensor fields involved, there are only 10 distinct equations, which matches the number of independent components of the metric tensor. Of these 10 degrees of freedom, 4 are dependent on a choice of coordinates, which corresponds to the freedom due to diffeomorphic coordinate transformations $x^\mu \mapsto x'^\mu(x)$. The remaining 6 degrees of freedom are the *dynamical degrees of freedom* of the gravitational field.

The Einstein field equations can be brought into a form displaying a time dynamics, similar to other equations of classical physics. In this so-called *initial value formulation*, a set of values $\{g_{kl}(x), \partial_0 g_{kl}(x)\}$, with $k, l = 1, 2, 3$, is initially given for a spacelike hypersurface Σ. The 4 equations $G_{0\nu} = 8\pi G_{\mathrm{N}}T_{0\nu}$, with $\nu = 0, 1, 2, 3$, are considered to be constraints. In this way, the 6 independent equations in $G_{kl} = 8\pi G_{\mathrm{N}}T_{kl}$, with $k, l = 1, 2, 3$, determine the time dynamics. A discussion of the initial value formulation, which is also known as the *Cauchy problem*, is beyond our scope and the interested reader can turn to Wald [91]. Finding exact solutions for the Einstein field equations is a highly nontrivial task. Very often only approximation schemes (e.g. the post-Newtonian approximation) or numerical methods remain as the only practical methods in order to find solutions for certain physical situations. In Section 21.3 we discuss the historically first and still very important exact solution of the Einstein equations, the Schwarzschild metric.

Lovelock Theorem

In our quest for gravitational field equations, we have selected the Einstein tensor as a well-suited quantity. We can ask again how unique this choice is. Remarkably, it turns out that this choice is almost the only one possible. Let us recall the requirements we imposed on the candidate tensor, denoted

$$\mathcal{G}_{\mu\nu}\,[g_{\rho\sigma}, \partial_\alpha g_{\rho\sigma}, \partial_\alpha\partial_\beta g_{\rho\sigma}]. \tag{21.39}$$

The square brackets indicate the functional dependence of $\mathcal{G}_{\mu\nu}(x)$ on the metric tensor and the associated first-order and second-order derivatives. Previously, we have required indices symmetry, a vanishing covariant divergence and the strict linearity in the second-order derivatives of the metric. For a moment, let us abandon the requirement on linearity entirely, i.e. we allow any power of the metric and its first-order and second-order derivatives. *David Lovelock* proved a theorem in 1972 which states that under these conditions and for the special case of *four spacetime dimensions*, the tensor $\mathcal{G}_{\mu\nu}(x)$ is *uniquely* fixed at

$$\mathcal{G}_{\mu\nu} = aG_{\mu\nu} + bg_{\mu\nu}, \tag{21.40}$$

where a, b are real numbers and $G_{\mu\nu}(x)$ is the Einstein tensor. The rather technical proof is provided in the book by Lovelock and Rund [55]. It is even more remarkable that the one can drop also the requirement of indices symmetry and the same result 21.40 continues to hold, always for the case of *four spacetime dimensions*, see [54]. This means that the mere functional dependence 21.39 of the candidate tensor in combination with a vanishing divergence are sufficient conditions to determine the form of the Einstein field equations. Besides the Einstein tensor, also the metric tensor appears in the result 21.40, so this calls for a separate discussion.

Cosmological Constant, More Dimensions

According to Lovelock's theorem, there is the possibility to add a term proportional to the metric tensor in the lhs of the Einstein field equations 21.37. The correspondingly generalized field equations read

$$\boxed{R_{\mu\nu} - \frac{1}{2}Rg_{\mu\nu} + \Lambda g_{\mu\nu} = 8\pi G_{\mathrm{N}}T_{\mu\nu}} \tag{21.41}$$

The constant factor Λ is called the *cosmological constant* and has the physical dimension $[L]^{-2}$ of an inverse length squared. The question if Λ is finite or zero can be answered only by observation. Based on today's experience, the term $\Lambda g_{\mu\nu}$ becomes relevant only for cosmological considerations at very large distance and time scales. Cosmological observations show that the value of Λ is extremely small. The Planck Collaboration group has published in their 2015 report a value of $|\Lambda| \lesssim 10^{-52}\,\mathrm{m}^{-2}$, see also [57]. For all applications and phenomena outside of cosmology, the cosmological constant Λ can be set to zero, and we will do this from the next paragraph onward.

Another type of generalization of Einstein's field equations is to consider them on a d-dimensional spacetime, with D space dimensions. The Einstein field equations 21.41 have to be written then in the form

$$R_{\mu\nu} - \frac{1}{2}Rg_{\mu\nu} + \Lambda g_{\mu\nu} = \kappa T_{\mu\nu}, \tag{21.42}$$

because the Einstein constant κ is not equal $8\pi G_{\mathrm{N}}$ in $d \neq 4$. The Einstein constant in four spacetime dimensions contains the geometric factor 4π which stems from the Poisson equation in three space dimensions. The factor 4π is the surface area of the usual 2-dimensional unit sphere S^2. In d spacetime dimensions (or equivalently in D space dimensions) the geometric factor is determined by the surface area of the $(D-1)$-dimensional unit hypersphere S^{D-1}, which is equal $2\pi^{D/2}/\Gamma(D/2)$, see the overview in appendix B.7.

One can solve for the Ricci tensor $R_{\mu\nu}(x)$ and rewrite the Einstein field equations 21.42 in the alternative form

$$R_{\mu\nu} = \kappa \left(T_{\mu\nu} - \frac{1}{d-2} T g_{\mu\nu} \right) + \frac{2\Lambda}{d-2} \, g_{\mu\nu}, \tag{21.43}$$

for dimensions $d \geq 3$. (*exercise 21.3*) We use the symbol $T(x) \equiv T^\mu{}_\mu(x)$ for the trace of the matter EMT. For the physically important case $d = 4$, the Einstein equations take the form

$$\boxed{R_{\mu\nu} = 8\pi G_{\mathrm{N}} \left(T_{\mu\nu} - \frac{1}{2} T g_{\mu\nu} \right) + \Lambda g_{\mu\nu}} \tag{21.44}$$

The gravitational field equations in this form are considerably simplified if one considers vacuum spacetimes.

Source-Free Spacetime

Per definition, a *source-free spacetime* is present if the matter EMT vanishes throughout the considered spacetime, i.e. $T_{\mu\nu} = 0$. Synonymous terms are *vacuum spacetime* or *empty spacetime*. Remembering that we set $\Lambda = 0$ from now on, the Einstein equations for a source-free spacetime are

$$\boxed{R_{\mu\nu} = 0} \tag{21.45}$$

and are called the *Einstein vacuum field equations*. In this particular case, one has an Einstein metric at hand and the results of Section 19.4 apply. In four spacetime dimensions, the Einstein vacuum field equations determine the 10 independent components of the Ricci tensor. However, the Riemann tensor possesses another 10 independent components that have to be determined. This matches the number of independent components of the Weyl tensor in four dimensions, compare with Section 19.5. Hence, we can conclude:

> *In the absence of matter sources, the gravitational degrees of freedom are described by the Weyl tensor.*

On the other hand, in dimensions two and three, the Weyl tensor vanishes and the Riemann curvature tensor is entirely determined by the Ricci tensor. This means that only in spacetime dimensions equal to four, or greater, the gravitational field possesses a dynamics that can exist without sources. From a physical point of view, one can say that gravitational waves can only exist in four or higher dimensions.

The dynamics of the gravitational degrees of freedom associated with the Weyl tensor is readily obtained. First, let us look at the differential Bianchi identity 19.81 for the Weyl tensor in four dimensions. This is a constraint equation, which we can combine with the Einstein field equations for the general case of non-vanishing sources. To do this, we just need to use the Ricci tensor and the Ricci scalar as expressed by a given EMT, i.e.

$$R_{\mu\nu} = 8\pi G_{\mathrm{N}} \left(T_{\mu\nu} - \frac{1}{2} T g_{\mu\nu} \right) \quad \text{and} \quad R = -8\pi G_{\mathrm{N}} T. \tag{21.46}$$

A quick algebraic calculation leads to the desired dynamical equation for the Weyl tensor for the case that sources are present,

$$\nabla_\mu W^\mu{}_{\nu\rho\sigma} = 4\pi G_{\mathrm{N}} \left[\nabla_\rho T_{\sigma\nu} - \nabla_\sigma T_{\rho\nu} - \frac{1}{3} \left(g_{\sigma\nu} \nabla_\rho T - g_{\rho\nu} \nabla_\sigma T \right) \right]. \tag{21.47}$$

We note, that in contrast to the Einstein equations, where the (trace-inverted) Ricci tensor is determined point-wise by the EMT, in the above field equation the Weyl tensor is determined non-locally by the derivatives of the EMT. This means that the Weyl tensor at a point is determined by the matter distribution at another point. The field equation 21.47 can be understood as a wave equation, in close analogy to the wave equation $\nabla_\mu F^{\mu\nu} = 4\pi j^\nu$ of the Maxwell theory. Using our results from Section 19.5, we immediately conclude that the rhs of 21.47 can be equated with the Cotton tensor $C_{\nu\rho\sigma}(x)$ in four dimensions. The Cotton tensor satisfies the divergence formula

$$\nabla_\nu C^\nu{}_{\rho\sigma} = 0, \tag{21.48}$$

which can now be understood as a conservation equation for energy and matter. In the case that the spacetime has no sources, the field equation for the Weyl tensor 21.47 is reduced to

$$\boxed{\nabla_\mu W^\mu{}_{\nu\rho\sigma} = 0} \tag{21.49}$$

which we have mentioned already in Section 19.5. This is the *wave propagation equation for a free gravitational field*.

21.3 Schwarzschild Metric

Spatially Isotropic Metric

In order to find exact solutions to Einstein's field equations, the existence of symmetries can be of great help. Here we assume that *spatial isotropy*, or in other words *spatial spherical symmetry*, is present in the metric. The spatially spherical gravitational field is meant to be generated by a spatially spherical mass distribution. The spatial rotational symmetry can be expressed by the existence of Killing vector fields $K^\mu(x)$ that fulfill the Killing equation $\mathcal{L}_K g_{\mu\nu} = 0$, or written in coordinates (see Section 3.3),

$$K^\rho \partial_\rho g_{\mu\nu} + g_{\rho\nu} \partial_\mu K^\rho + g_{\mu\rho} \partial_\nu K^\rho = 0. \tag{21.50}$$

In terms of coordinates, it is fitting to use spatially spherical coordinates. The metric components become functions of the time coordinate $-\infty < t < \infty$, the spacelike radial distance coordinate $0 \le r < \infty$ and the two angle coordinates $0 \le \theta \le \pi$ and $0 \le \varphi < 2\pi$. A hypersurface which is defined by the combined conditions $t = $ const and $r = $ const is the usual 2-sphere in three-dimensional space and has the surface area $4\pi r^2$. We are aiming to determine the form of the metric components $g_{\mu\nu}(t, r, \theta, \varphi)$ in the case of spherical symmetry. To this end, we need to use the specific KVFs for the 3-dimensional rotational symmetry. In exercise 20.8, we have derived the three KVFs of the 2-sphere, showing that they constitute the Lie algebra $so(3)$ of 3-dimensional rotations. Written as four-vectors, these KVFs $K_1^\mu(t, r, \theta, \varphi)$, $K_2^\mu(t, r, \theta, \varphi)$ and $K_3^\mu(t, r, \theta, \varphi)$ are

$$K_1 = \begin{pmatrix} 0 \\ 0 \\ \sin\varphi \\ \cot\theta\cos\varphi \end{pmatrix}, \quad K_2 = \begin{pmatrix} 0 \\ 0 \\ \cos\varphi \\ -\cot\theta\sin\varphi \end{pmatrix}, \quad K_3 = \begin{pmatrix} 0 \\ 0 \\ 0 \\ 1 \end{pmatrix}. \tag{21.51}$$

There is no dependence on the coordinates t and r. Each of the KVFs is inserted into the Killing equation above. It is beneficial to distinguish the various cases regarding the two free indices in the Killing equation beforehand. The case $(\mu, \nu) = (t, t)$ leads to the condition

$$K^j \partial_j g_{tt} = 0, \tag{21.52}$$

with the index j taking the spatial values r, θ, φ. The case $(\mu, \nu) = (t, n)$ leads to the condition

$$K^j \partial_j g_{tn} + g_{tj} \partial_n K^j = 0, \tag{21.53}$$

with j and n taking the values r, θ, φ. Finally, the case case $(\mu, \nu) = (m, n)$ yields the condition

$$K^j \partial_j g_{mn} + g_{jn} \partial_m K^j + g_{mj} \partial_n K^j = 0, \tag{21.54}$$

where j, m, and n take the values r, θ, φ again. The three above KVFs are used in each of the three above conditions with all possible index values j, m, n being realized. The obtained equations can be greatly simplified by elementary algebraic operations. The resulting conditions for the metric components are

$$g_{t\theta} = g_{t\varphi} = g_{r\theta} = g_{r\varphi} = g_{\theta\varphi} = 0 \quad \text{and} \quad g_{\varphi\varphi} = g_{\theta\theta} \sin^2 \theta, \tag{21.55}$$

and in addition

$$\partial_\theta g_{tt} = \partial_\theta g_{tr} = \partial_\theta g_{rr} = \partial_\theta g_{\theta\theta} = 0 \quad \text{and} \quad \partial_\varphi g_{\mu\nu} = 0. \tag{21.56}$$

One obtains also the condition $\partial_\theta g_{\varphi\varphi} = 2 g_{\varphi\varphi} \cot \theta$, but this is included already in the equation $g_{\varphi\varphi} = g_{\theta\theta} \sin^2 \theta$. So we can deduce that only the metric components g_{tt}, g_{rr}, $g_{\theta\theta}$, $g_{\varphi\varphi}$ and g_{tr} are non-zero. The components g_{tt}, g_{rr}, $g_{\theta\theta}$ and g_{tr} depend only on the coordinates t and r. The metric component $g_{\varphi\varphi}$ depends on t, r and θ, where the angle dependence is determined by $g_{\varphi\varphi} = g_{\theta\theta} \sin^2 \theta$. Thus, we can write down the spatially isotropic metric as

$$ds^2 = g_{tt} \, dt^2 + g_{tr} \, dt \, dr + g_{rr} \, dr^2 + g_{\theta\theta}(d\theta^2 + \sin^2 \theta \, d\varphi^2). \tag{21.57}$$

In order to simplify this metric, we define a new spacelike radial coordinate $\bar{r}$ by $\bar{r} \equiv \sqrt{g_{\theta\theta}(t, r)}$ and note that $g_{\theta\theta}(t, r)$ is always positive. Thus, without loss of generality, we can write the metric as

$$ds^2 = A(t, \bar{r}) \, dt^2 - B(t, \bar{r}) \, dt \, d\bar{r} - C(t, \bar{r}) \, d\bar{r}^2 - \bar{r}^2(d\theta^2 + \sin^2 \theta \, d\varphi^2), \tag{21.58}$$

with the new component functions A, B, and C depending on t and $\bar{r}$. Next, we define a new timelike coordinate $\bar{t}$ by the equation

$$d\bar{t} = F(t, \bar{r}) \left(A(t, \bar{r}) \, dt - \frac{1}{2} B(t, \bar{r}) \, d\bar{r} \right), \tag{21.59}$$

with F being a function of t and $\bar{r}$. Thus, the squared differential $d\bar{t}^2$ is

$$d\bar{t}^2 = F^2 \left(A^2 dt^2 + \frac{1}{4} B^2 d\bar{r}^2 - AB \, dt \, d\bar{r} \right). \tag{21.60}$$

The expression we like to simplify is

$$A \, dt^2 - B \, dt \, d\bar{r} = \frac{1}{AF^2} \, d\bar{t}^2 - \frac{B^2}{4A} \, d\bar{r}^2. \tag{21.61}$$

By introducing the new component functions $\overline{A}(\bar{t}, \bar{r})$ and $\overline{B}(\bar{t}, \bar{r})$ as

$$\overline{A} \equiv \frac{1}{AF^2} \quad \text{and} \quad \overline{B} \equiv \frac{B^2}{4A} + C, \tag{21.62}$$

we obtain a metric where the off-diagonal term proportional to $\mathrm{d}t\,\mathrm{d}\bar{r}$ is absorbed, and we get

$$\mathrm{d}s^2 = \overline{A}(\bar{t},\bar{r})\,\mathrm{d}\bar{t}^2 - \overline{B}(\bar{t},\bar{r})\,\mathrm{d}\bar{r}^2 - \bar{r}^2(\mathrm{d}\theta^2 + \sin^2\theta\,\mathrm{d}\varphi^2). \tag{21.63}$$

Now, we can omit the bars on the symbols and conclude to the final form of the *metric with spatial isotropy*:

$$\boxed{\mathrm{d}s^2 = A(t,r)\,\mathrm{d}t^2 - B(t,r)\,\mathrm{d}r^2 - r^2\mathrm{d}\Phi^2} \tag{21.64}$$

Here we have employed the squared solid angle element $\mathrm{d}\Phi^2 = \mathrm{d}\theta^2 + \sin^2\theta\,\mathrm{d}\varphi^2$. The two functions $A(t,r)$ and $B(t,r)$ depend only on the (timelike) coordinate t and the (spacelike) coordinate r. The range and the role of the these two coordinates has to be examined in each concrete physical situation.

Derivation of the Schwarzschild Solution

The *Schwarzschild solution* describes a *spherically symmetric* and *static* spacetime, which in particular means that the metric $g_{\mu\nu}(x)$ is independent of the timelike coordinate t. Consequently, the spatially isotropic and static metric exhibits the general form

$$\mathrm{d}s^2 = A(r)\,\mathrm{d}t^2 - B(r)\,\mathrm{d}r^2 - r^2\mathrm{d}\Phi^2, \tag{21.65}$$

where the functions $A(r)$ and $B(r)$ depend only on the radial coordinate r. Now we are aiming to determine the metric *outside* a spherically symmetric, static mass distribution. Thus, we have to solve the vacuum Einstein equation

$$R_{\mu\nu} = 0, \tag{21.66}$$

with the Ricci tensor being expressed by the connection coefficients

$$R_{\mu\nu} = \partial_\beta\Gamma^\beta_{\mu\nu} - \partial_\nu\Gamma^\beta_{\mu\beta} + \Gamma^\beta_{\alpha\beta}\Gamma^\alpha_{\mu\nu} - \Gamma^\beta_{\alpha\nu}\Gamma^\alpha_{\mu\beta}, \tag{21.67}$$

which in turn are expressed by the metric, as prescribed in 18.78. The resulting 10 coupled PDEs determine the sought metric components. In order to calculate the connection coefficients, we first read the metric components from the formula 21.65. They are

$$g_{tt} = A(r), \quad g_{rr} = -B(r), \quad g_{\theta\theta} = -r^2, \quad g_{\varphi\varphi} = -r^2\sin^2\theta. \tag{21.68}$$

For the inverse metric components it is simply $g^{\mu\nu}(r,\theta) = 1/g_{\mu\nu}(r,\theta)$. The calculation of the connection coefficients according to the formula 18.78 is a straightforward task. Of the forty possible connection coefficients, only nine are non-zero:

$$\left.\begin{array}{lll} \Gamma^t_{tr} = (2A)^{-1}A', & \Gamma^r_{tt} = (2B)^{-1}A', & \Gamma^r_{rr} = (2B)^{-1}B' \\ \Gamma^r_{\theta\theta} = -rB^{-1}, & \Gamma^r_{\varphi\varphi} = -rB^{-1}\sin^2\theta, & \Gamma^\theta_{r\theta} = r^{-1} \\ \Gamma^\theta_{\varphi\varphi} = -\sin\theta\cos\theta, & \Gamma^\varphi_{r\varphi} = r^{-1}, & \Gamma^\varphi_{\theta\varphi} = \cot\theta \end{array}\right\}. \tag{21.69}$$

Here we have used the notation $A'(r) \equiv \mathrm{d}A(r)/\mathrm{d}r$ and $B'(r) \equiv \mathrm{d}B(r)/\mathrm{d}r$. The above connection coefficients lead to the following four non-zero components of the Ricci tensor:

$$R_{tt} = \frac{A''}{2B} + \frac{A'}{rB} - \frac{A'}{4B}\left(\frac{A'}{A} + \frac{B'}{B}\right), \quad R_{rr} = -\frac{A''}{2A} + \frac{B'}{rB} + \frac{A'}{4A}\left(\frac{A'}{A} + \frac{B'}{B}\right),$$

$$R_{\theta\theta} = 1 - \frac{1}{B} - \frac{r}{2B}\left(\frac{A'}{A} - \frac{B'}{B}\right), \quad R_{\varphi\varphi} = R_{\theta\theta}\sin^2\theta. \tag{21.70}$$

All other components of the Ricci tensor vanish. Thus, we are led to the three independent ODEs

$$R_{tt} = 0, \qquad R_{rr} = 0 \quad \text{and} \quad R_{\theta\theta} = 0 \tag{21.71}$$

for determining the two functions $A(r)$ and $B(r)$. Multiplying the first equation above by B/A and adding it to the second equation leads to the equation $AB = k$, where k is a real constant to be determined. Inserting now $B = k/A$ in the third equation above leads to the ODE

$$\frac{\mathrm{d}}{\mathrm{d}r}(rA) = k, \tag{21.72}$$

which has the solution $rA = k(r + C)$, with a second real constant C. Hence, the sought line element takes the form

$$\mathrm{d}s^2 = k\left(1 + \frac{C}{r}\right)\mathrm{d}t^2 - \left(1 + \frac{C}{r}\right)^{-1}\mathrm{d}r^2 - r^2\mathrm{d}\Phi^2. \tag{21.73}$$

In order to derive the value of the two integration constants k and C, we consider the Newtonian weak field limit. In Section 21.2, we saw that in the static weak field limit, the relation $g_{tt}(r) = 1 + 2\Phi(r)$ holds. We can use the coordinate r, instead of x, due to the spherical symmetry being present. In the Newtonian limit, the spacelike coordinate r becomes the radial distance from the coordinate origin. At the same time, in this limit, the potential function becomes $\Phi(r) = -G_N M/r$, where M represents the *total mass* of the spherical mass distribution generating the field and G_N is Newton's gravitational constant. Based on this limit, the constants are determined and take the values $k = 1$ and $C = -2G_N M$. Therefore, we conclude to the *Schwarzschild metric*:

$$\boxed{\mathrm{d}s^2 = \left(1 - \frac{2G_N M}{r}\right)\mathrm{d}t^2 - \left(1 - \frac{2G_N M}{r}\right)^{-1}\mathrm{d}r^2 - r^2\mathrm{d}\Phi^2} \tag{21.74}$$

This metric was derived by *Karl Schwarzschild* a few months after Einstein had published his general theory of relativity. The Schwarzschild metric represents the most important exact solution of Einstein's equations and has a wide range of astrophysical applications.

Basic Properties of the Schwarzschild Metric

Let us look more closely at the adapted coordinates (t, r, θ, φ) for the Schwarzschild metric. Their range is $-\infty < t < \infty$, $2G_N M < r < \infty$, and as usually, $0 \leq \theta \leq \pi$ and $0 \leq \varphi < 2\pi$. The *proper time s*, as actually measured by a stationary observer within the Schwarzschild geometry, is related to Schwarzschild time coordinate t by the relation

$$\mathrm{d}s = \left(1 - \frac{2G_N M}{r}\right)^{1/2}\mathrm{d}t. \tag{21.75}$$

Since it is $r > 2G_N M$, the proper time interval $|\mathrm{d}s|$ is always shorter than the coordinate time interval $|\mathrm{d}t|$. This means that for an observer at the vicinity of a spherical mass generating the Schwarzschild geometry, the time runs slower than for an observer residing at a larger distance. An observer at infinite distance, $r \to \infty$, will measure $|\mathrm{d}s| = |\mathrm{d}t|$. Similarly, the physically measured *radial distance l* is related to the Schwarzschild radial coordinate r by the relation

$$\mathrm{d}l = \left(1 - \frac{2G_N M}{r}\right)^{-1/2}\mathrm{d}r. \tag{21.76}$$

Consequently, the measured radial distance dl is always greater than the radial coordinate dr. The closer the observer approaches the radial position $r = 2G_N M$, the more pronounced the effect becomes. The volume element $dvol_{(3)}$ of 3-dimensional space, as submanifold of the Schwarzschild manifold, is computed as

$$dvol_{(3)} = \left(1 - \frac{2G_N M}{r}\right)^{-1/2} r^2 \sin\theta \, dr \, d\theta \, d\varphi, \qquad (21.77)$$

and is obviously greater than the volume element of flat 3-space. In the limit $M \to 0$ of a vanishing mass, the Schwarzschild metric becomes the Minkowski metric. The same happens in the asymptotic region $r \to \infty$. The limit value $r \to 2G_N M$ represents a coordinate singularity, which can be transformed away in suitable coordinates. The special value $r_S \equiv 2G_N M$ defines the so-called *Schwarzschild radius* of a spherical object of mass M. The Schwarzschild radius of a "normal" astrophysical object, such as a planet or a star, is much smaller than the actual radius of the object in space. For instance, the Schwarzschild radius of the Sun has an approximate value of $2,95\,km$ and lies well below its outer surface. This ensures that the Schwarzschild metric can be applied for any points in empty space up to the surface of the massive object. Technically, the Schwarzschild metric is valid for radial values $r > r_S$ but practically, in the vicinity of a massive spherical object of radius R, we can apply the metric only for radial values $r > R$. For space points residing within the massive object, $r \leq R$, we would need to solve the Einstein equations with the non-vanishing matter EMT. For the limit, $r \to 0$, leading also to a singularity in the Schwarzschild metric, we have to note that this represents a true, physically relevant singularity, which points to the existence of black holes. This topic is far from complete and is still the subject of current research.

In terms of spacetime symmetries, the Schwarzschild metric is invariant under time translations and space rotations, exactly as designed from the original metric 21.64. We note here also the fact that the Weyl tensor of the Schwarzschild metric in non-vanishing and therefore this geometry is not conformally flat.

Birkhoff's Theorem
In the derivation of the Schwarzschild metric, we started with the isotropic and static metric 21.65. If we had started instead with the more general metric 21.64, being only isotropic, we would again have concluded that the Schwarzschild metric is the solution of the free Einstein equations. So, remarkably, the time-independent Schwarzschild metric 21.74 is also the solution for the spherically symmetric and non-static case. This is called *Birkhoff's theorem*, or the *Birkhoff-Jebsen theorem*, according to the discoverers *George David Birkhoff* and *Joerg Tofte Jebsen*. The derivation proceeds along the same lines as our previous derivation of the Schwarzschild metric and we will not repeat it here, see for instance [63]. Physically, a mass distribution which is not static but pulsating in time in a spherically symmetric way generates an external gravitational field, which is described by the time-independent Schwarzschild geometry. It can be concluded that a pulsating spherically symmetric mass distribution does not generate any gravitational radiation.

21.4   Asymptotically Flat Spacetimes

Isolated Objects
As in the example of Schwarzschild spacetime discussed above, we are very often interested in describing a physical object as if it were isolated from all other objects in the universe. This means that we simply neglect all gravitational and non-gravitational interactions of other objects far away and treat the massive object of interest, along with its spacetime,

as the only constituent parts of the universe. Such type of idealizations are of course very common in all areas of theoretical physics. In addition to the goal of developing analytic treatments for such simplified systems, there is another reason to consider isolated massive objects and their spacetime. When we tackle the question of how global (and conserved) quantities such as momentum, angular momentum, etc. can be defined in general relativity, then two major cases are known. One case is when there are symmetries within the spacetime at hand. This is explained in Chapter 23. The other case is when we are able to treat the total spacetime as a single isolated entity.

Concept of Asymptotic Flatness

So let us consider a massive object having a limited spatial extend. We can assume that the strength of the gravitational field of the object decreases as we move further away from it. With increasingly greater spatial distances from the object, the gravitational field becomes weaker and weaker and spacetime resembles more and more Minkowski space. The Schwarzschild spacetime is such a case. This behavior at great distances can be used for a definition. A spacetime $(\mathcal{M}, g_{\mu\nu})$ is called *asymptotically flat* if there exists a coordinate system $x^\mu = (t, x, y, z)$, such that the metric components display the asymptotic behavior

$$g_{\mu\nu}(x) = \eta_{\mu\nu} + \mathcal{O}(r^{-1}), \quad \text{for } r \to \infty, \tag{21.78}$$

for all lightlike and spacelike directions. We have to exclude timelike directions in order to allow the existence of a massive object at very early and very late times. In the above equation we have used the radial distance coordinate $r = \sqrt{x^2 + y^2 + z^2}$. The term $\mathcal{O}(r^{-1})$ behaves like $|\mathcal{O}(r^{-1})| = r^{-1}|f(t, \theta, \varphi)|$ for large values of r and the function $f(t, \theta, \varphi)$ does not depend on the coordinate r. In addition to the above condition, also the following fall-off conditions are expected to hold for large values of r:

$$\partial_\sigma g_{\mu\nu}(x) = \mathcal{O}(r^{-2}), \quad \partial_\rho \partial_\sigma g_{\mu\nu}(x) = \mathcal{O}(r^{-3}), \quad \text{for } r \to \infty. \tag{21.79}$$

Finally, also the condition characterizing a matter-free spacetime region,

$$R_{\mu\nu}(x) = 0, \quad \text{for } r \to \infty, \tag{21.80}$$

is expected to hold. In order to inspect the fulfillment of the above conditions in a concrete situation, one has to carry out the limit $r \to \infty$ explicitly.

Since we have used coordinates to define asymptotic flatness of a spacetime, it is natural to ask to which extend our definition depends on a suitable choice of coordinates. In the next paragraph we will give a coordinate-free definition of asymptotic flatness, but it is interesting to ask under which coordinate transformations the above asymptotic flatness property stays invariant. Since we consider spacetime regions far away from sources, we would expect the coordinate freedom to be captured exactly by the Poincaré transformations 13.2, as they represent exactly the allowed coordinate transformations of Minkowski space. It is surprising that the transformations leaving the asymptotic flatness property invariant are actually more general and given by

$$x'^\mu = \Lambda^\mu{}_\nu x^\nu + a^\mu(\theta, \varphi) + \mathcal{O}(r^{-1}), \tag{21.81}$$

with $a^\mu(\theta, \varphi)$ being angle-dependent four-translations. The above transformations constitute the *BMS group* as introduced 1962 by *Hermann Bondi, M. G. J. van der Burg, A. W. K. Metzner* in [11] and *Rainer K. Sachs* in [78] and [79]. Obviously, the BMS group generalizes the Poincaré group. The angle-dependent translations constitute the infinite-dimensional abelian *group of supertranslations*. Consequently, the BMS group is the semidirect product of the Lorentz group with the group of supertranslations. The existence of the BMS group

shows that asymptotically flat gravitational fields in the far-field region do not reduce unambiguously to simple Minkowski space and contain certain residual degrees of freedom.
The role and uses of the BMS group are still the subject of current research.

Conformal Infinity

The concept of asymptotic flatness as it is formulated in coordinates is easy to understand,
but it has the disadvantage that it is based on the limit $r \to \infty$, which is mathematically very delicate. There is an alternative (and essentially equivalent) notion of asymptotic flatness,
which is geometric and coordinate-free in its nature, introduced by Penrose in 1963 [65].
The basic idea is to attach the points at infinity in a suitable way to the spacetime manifold
of interest and treat infinity as a boundary of the manifold, the so-called *conformal infinity*
(sometimes also called *asymptotic infinity*). Technically, this is achieved by applying a Weyl
rescaling on the initial spacetime manifold with a suited conformal factor, which drags the
points at infinity to a finite distance within the Weyl-transformed manifold. More specifically, the original spacetime manifold $(\mathcal{M}, g_{\mu\nu})$ is transformed to the manifold denoted
$(\widehat{\mathcal{M}}, \widehat{g}_{\mu\nu})$, where the new spacetime manifold is the union $\widehat{\mathcal{M}} = \mathcal{M} \cup \partial\mathcal{M}$ and the boundary
$\partial\mathcal{M}$ represents conformal infinity. The basic idea of this so-called *conformal compactifica-
tion* (also called *conformal completion*) is illustrated in the figure 21.1. We should recall

Figure 21.1: Conformal compactification of a spacetime

that there are several types of infinity in a relativistic spacetime: past and future timelike
infinity, past and future null infinity and spacelike infinity. Ideally, all these points at infinity
should constitute the boundary $\partial\mathcal{M}$. The metric of the manifold shall be transformed as
we already know from Weyl rescalings, namely $\widehat{g}_{\mu\nu} = \Omega^2 g_{\mu\nu}$, where the conformal factor Ω
is a suited scalar function defined on $\widehat{\mathcal{M}}$. In mathematical terms, the initial manifold $\mathcal{M}$
is said to be *conformally compactified* to $\widehat{\mathcal{M}}$. The conformal factor Ω has to fall-off quickly
enough for points far away (along timelike / null / spacelike directions), so that the points
at infinity are rescaled by the special value $\Omega = 0$. The conformal factor has to possess some
additional properties in order to be applicable for the asymptotic construction and we will
discuss these properties soon. Before we tackle conformal infinity from a general standpoint,
however, we should first examine flat Minkowski space itself.

Compactified Minkowski Space

Let us start with the metric ds^2 of standard Minkowski space $\mathbb{M}_4$ in spherical coordinates,

$$ds^2 = dt^2 - dr^2 - r^2\, d\Phi^2, \tag{21.82}$$

where $d\Phi^2$ denotes the usual solid angle element. The ranges for the time and radial coordinate are $-\infty < t < \infty$ and $0 \le r < \infty$. We introduce the so-called *retarded* and *advanced* *null coordinates*, u and v respectively, defined as

$$u \equiv t - r \quad \text{and} \quad v \equiv t + r. \tag{21.83}$$

They have the ranges $-\infty < u < \infty$ and $-\infty < v < \infty$ and fulfill the condition $u \le v$. The Minkowski metric becomes

$$ds^2 = du\, dv - \frac{1}{4}(v - u)^2 \, d\Phi^2. \tag{21.84}$$

The new coordinates (u, v, θ, φ) describe the null cone structure of Minkowski space. More specifically, with the angles θ and φ held fixed, the surfaces defined by $u = \text{const}$ represent all future-directed null cones, while the surfaces with $v = \text{const}$ represent all past-directed null cones. We can visualize this null cone structure by considering the "surfaces" $u = \text{const}$ and $v = \text{const}$ in the (t, r)-plane by suppressing the angle coordinates θ and φ, see the diagram 21.2. Within this two-dimensional (t, r)-diagram, each point represents a 2-sphere

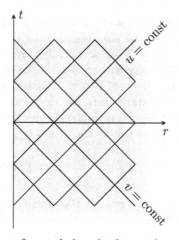

Figure 21.2: Surfaces of retarded and advanced constant null directions

with radius $r = (v - u)/2$. The asymptotic limits we are interested in are $u \to \infty$ and $v \to \infty$, corresponding respectively to *future null infinity* and *past null infinity*. Obviously, the metric diverges for these asymptotic values, so we have to make a clever choice of new coordinates under which the asymptotic values can be handled in a better way. We introduce the new coordinates U and V defined as

$$U \equiv \arctan u \quad \text{and} \quad V \equiv \arctan v, \tag{21.85}$$

taking the values $-\pi/2 < U, V < \pi/2$ and satisfying $U \le V$. A calculation gives

$$du = \frac{dU}{\cos^2 U}, \quad dv = \frac{dV}{\cos^2 V} \quad \text{and} \quad v - u = \frac{\sin(V - U)}{\cos U \cos V}, \tag{21.86}$$

so that the Minkowski metric becomes

$$ds^2 = (2\cos U \cos V)^{-2} \left[4\, dU\, dV - \sin^2(V - U)\, d\Phi^2 \right]. \tag{21.87}$$

The metric is not defined for the limit values $U \to \pm\pi/2$ and $V \to \pm\pi/2$, with each one representing null infinity. The crucial idea is now that we can overcome this limitation and tame the points of null infinity by including these points into a Weyl rescaled metric,

$$d\hat{s}^2 = \Omega^2 ds^2. \tag{21.88}$$

We use the conformal factor $\Omega = 2\cos U \cos V$, which becomes zero at the points of null infinity. The Weyl rescaled metric $d\widehat{s}^2$ is simply

$$d\widehat{s}^2 = 4\,dU\,dV - \sin^2(V - U)\,d\Phi^2 \tag{21.89}$$

and behaves perfectly regular at the null infinity points $U = \pm\pi/2$ and $V = \pm\pi/2$. Thus, we can achieve a finite description of the null infinity points through an extended Minkowski space $\widehat{\mathbb{M}}_4$ which contains null infinity as its boundary. To better understand the structure of the conformally extended Minkowski space, let us reintroduce timelike and spacelike coordinates T and R defined as

$$T \equiv V + U \quad \text{and} \quad R \equiv V - U, \tag{21.90}$$

possessing the compact ranges $-\pi \le T \le \pi$ and $0 \le R \le \pi$ and satisfying the constraint $-\pi \le T - R \le \pi$. The extended Minkowski space metric becomes

$$d\widehat{s}^2 = dT^2 - dR^2 - \sin^2 R\,d\Phi^2. \tag{21.91}$$

The *conformally extended Minkowski space* $\widehat{\mathbb{M}}_4$ can be considered to be a subspace of the *Einstein static universe*, the later being also described by the spacetime metric 21.91, using the wider coordinate ranges $-\infty < T < \infty$ and $0 \le R \le \pi$. In contrast to the static universe manifold, for the conformally extended Minkowski space the timelike coordinate is restricted to the compact interval $-\pi \le T \le \pi$. Einstein introduced the metric 21.91 in 1917 in order to describe a closed, static universe as an exact solution of the gravitational field equations. We remark that the static universe metric can be written differently by introducing the coordinate $L \equiv \sin R$ taking the values $0 \le L \le 1$, so that we have

$$d\widehat{s}^2 = dT^2 - \frac{1}{1 - L^2}\,dL^2 - L^2\,d\Phi^2. \tag{21.92}$$

The spatial part of the above metric is exactly the metric of the unit 3-sphere S^3, compare with Section 2.3. In other words, the closed static universe has essentially the structure of the product manifold $\mathbb{R} \times S^3$.

We can visualize the conformally extended Minkowski space $\widehat{\mathbb{M}}_4$ by looking at a two-dimensional diagram in the (T, R)-plane, with the angle coordinates being held fixed. The coordinate relations $T = \arctan(t+r) + \arctan(t-r)$ and $R = \arctan(t+r) - \arctan(t-r)$ for the original coordinate ranges $-\infty < t < \infty$ and $0 \le r < \infty$ and with the constraint $-\pi/2 < \arctan(t - r) < \pi/2$ lead to the spacetime diagram 21.3. Diagrams of this kind are called *Penrose-Carter* (or simply *Penrose*) *diagrams*, according to *Roger Penrose* and *Brandon Carter*. We can observe several things. First of all, the conformally extended Minkowski space $\widehat{\mathbb{M}}_4$ can now be visualized on a finite piece of paper including the points at infinity. Each point in the triangle, for any value of the pair (T, R), represents a 2-sphere. The maximum value that this radius can take is one, while the minimum value is zero, where the sphere shrinks to a point. The original Minkowski space is the interior of the triangle. The set of boundary points $i^- \cup i^+ \cup i^0 \cup \mathcal{I}^- \cup \mathcal{I}^+$ is called *conformal infinity*. We explain each of these point sets:

- The single point i^- describes *past timelike infinity* and corresponds to the values $t = -\infty$, r finite, or, $T = -\pi$, $R = 0$.
- The single point i^+ describes *future timelike infinity* and corresponds to the values $t = +\infty$, r finite, or, $T = +\pi$, $R = 0$.
- The single point i^0 describes *spacelike infinity* and corresponds to the values t finite, $r = +\infty$, or, $T = 0$, $R = +\pi$.

- The straight line segment $\mathcal{I}^-$ (called *scri minus*) describes *past null infinity* and corresponds to the values $t + r$ finite and $t = -\infty$, $r = +\infty$, or, $T - R = -\pi$ and $0 < R < \pi$.

- The straight line segment $\mathcal{I}^+$ (called *scri plus*) describes *future null infinity* and corresponds to the values $t - r$ finite and $t = +\infty$, $r = +\infty$, or, $T + R = +\pi$ and $0 < R < \pi$.

The points i^-, i^+, and i^0 are truly points, as the sphere shrinks down to a point there. The past and future null infinity regions $\mathcal{I}^-$ and $\mathcal{I}^+$, however, are topologically like $\mathbb{R} \times S^2$. All timelike geodesics start at i^- and end at i^+. All spacelike geodesics start and end at i^0. The diagram shows (horizontally oriented) lines with $t = $ const and (vertically oriented) lines with $r = $ const. For all the points at conformal infinity the conformal factor Ω vanishes. With this conformal compactification of Minkowski space, we have achieved a description of infinity in a unified geometric and coordinate-independent way.

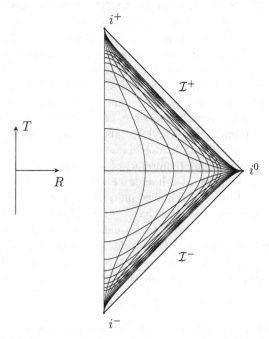

Figure 21.3: Penrose diagram for compactified Minkowski space

Asymptotic Flatness

The previous construction of a conformally compactified Minkowski space provides the clues as to how we can define asymptotic flatness for a general spacetime. We provide the definition: a spacetime manifold $(\mathcal{M}, g_{\mu\nu})$ is called *asymptotically flat* if there exists another spacetime manifold $(\widehat{\mathcal{M}}, \widehat{g}_{\mu\nu})$, with the following properties:

- The manifold $\mathcal{M}$ is an open submanifold of $\widehat{\mathcal{M}}$ and the boundary $\partial\mathcal{M}$ is such that $\widehat{\mathcal{M}} = \mathcal{M} \cup \partial\mathcal{M}$.

- There exists a smooth real scalar field Ω defined on $\widehat{\mathcal{M}}$ such that $\Omega > 0$, $\widehat{g}_{\mu\nu} = \Omega^2 g_{\mu\nu}$ on $\mathcal{M}$ and in addition $\Omega = 0$, $d\Omega \neq 0$ on $\partial\mathcal{M}$.

- Every null geodesic curve in $\widehat{\mathcal{M}}$ has a start point and an end point on the boundary $\partial\mathcal{M}$.

- In an open neighborhood of $\partial\mathcal{M}$ the Ricci tensor vanishes, $R_{\mu\nu} = 0$.

A few remarks are on order. First of all, the above definition of asymptotic flatness is independent of the Einstein field equations.* The usefulness of this definition is demonstrated by considering its consequences in concrete models of spacetime. In fact, there exist some variations of this definition in order to allow certain classes of spacetimes to be treated. The first and the second requirement in the above definition combined determine the point set of conformal infinity. The employed Weyl rescaling leads to the relation

$$\mathrm{d}\widehat{s}^2 = \Omega^2 \mathrm{d}s^2 \tag{21.93}$$

in $\mathcal{M}$, so that $\mathrm{d}\widehat{s}^2 = 0$ holds if and only if $\mathrm{d}s^2 = 0$ holds. In other words, the Weyl rescaling does not alter the causality of lightlike events and distances. Null geodesics in the original spacetime are mapped to null geodesics in the rescaled spacetime. Concentrating on the light cone structure of spacetime actually has proved to be very fruitful for the analysis of asymptotic flatness. The value $\Omega = 0$ on the conformal boundary $\partial\mathcal{M}$ ensures that for these initially infinitely far away points an infinite stretching of the metric is applied, in order to pull them to a finite distance in $\widehat{\mathcal{M}}$. The third requirement in the definition is a completeness requirement, so that all points of null infinity are included. The fourth requirement demands the vanishing of the Ricci tensor at infinity, which models the isolated spacetime. The condition $R_{\mu\nu} = 0$ still allows the existence of matter fields within spacetime, however these matter fields should not extend to infinity.

The manifold $(\mathcal{M}, g_{\mu\nu})$ is called the *physical spacetime*, whereas $(\widehat{\mathcal{M}}, \widehat{g}_{\mu\nu})$ is called the *unphysical spacetime*. This terminology is a bit misleading, since the unphysical spacetime is as physically relevant as the initial spacetime. On one hand, it is true that the manifold $\widehat{\mathcal{M}} \setminus \partial\mathcal{M}$ as image of $\mathcal{M}$ under the Weyl rescaling is not uniquely determined through the asymptotic flatness requirements. If one chooses a conformal factor $\Xi\,\Omega$, with a scalar function $\Xi > 0$ in $\mathcal{M}$, then the four requirements for asymptotic flatness are still fulfilled. On the other hand, it can be shown (although we will not do this here) that the boundary $\mathcal{I} \equiv \partial\mathcal{M}$, termed *conformal infinity*, is indeed uniquely defined. Conformal infinity has some more notable properties. First, we show that *conformal infinity $\mathcal{I}$ is a null hypersurface*. This can be proven by using the Weyl transformation formulae developed in the previous chapter. Since the conformal factor Ω is a scalar function, it is $\widehat{\nabla}_\rho\Omega = \nabla_\rho\Omega = \partial_\rho\Omega$. In contrast, it is $\widehat{\nabla}^\rho\Omega = \Omega^{-2}\nabla^\rho\Omega$. We employ the Weyl rescaling formula for the Ricci scalar, which in four dimensions reads, see appendix C,

$$\Omega^2 \widehat{R} = R + 6\mathcal{Z}. \tag{21.94}$$

The trace $\mathcal{Z} \equiv \mathcal{Z}^\alpha{}_\alpha$ of the zeta tensor in four dimensions is calculated as

$$\mathcal{Z} = -\Omega^{-1} g^{\alpha\beta}\,\nabla_\alpha\nabla_\beta\Omega. \tag{21.95}$$

In the next step, we express the trace $\mathcal{Z}$ through the Weyl-transformed quantities $\widehat{g}_{\mu\nu}$ and $\widehat{\nabla}_\rho\Omega$. This can be easily accomplished by using the formulae from appendix C. (*exercise 21.4*) We obtain the result

$$\Omega^2 \widehat{R} = R - 6\,\Omega\,\widehat{g}^{\alpha\beta}\,\widehat{\nabla}_\alpha\widehat{\nabla}_\beta\Omega + 12\,\widehat{g}^{\alpha\beta}\,\widehat{\nabla}_\alpha\Omega\,\widehat{\nabla}_\beta\Omega. \tag{21.96}$$

Since Ω and R are smooth functions, this also applies to $\widehat{R}$. On the conformal boundary it is $\Omega = 0$ and $\widehat{\nabla}_\rho\Omega \neq 0$. Thus, we obtain for all points on $\mathcal{I}$ the relation

$$\widehat{g}^{\alpha\beta}\,\widehat{\nabla}_\alpha\Omega\,\widehat{\nabla}_\beta\Omega = 0. \tag{21.97}$$

---

*The requirement on the vanishing of the Ricci tensor is of course motivated by the vanishing of the matter EMT in asymptotic regions.

Therefore, the normal vector field $\widehat{g}^{\alpha\beta}\widehat{\nabla}_\alpha\Omega$ on the boundary $\mathcal{I}$ is a null vector field.

Another property of the conformal boundary $\mathcal{I}$ is that the Weyl tensor vanishes there, which ensures that the entire Riemann tensor vanishes at conformal infinity. This fact justifies the definition of asymptotic flatness provided above. The complete proof is technically demanding, so we are only providing a plausibility argument here. In a neighborhood of the boundary $\mathcal{I}$, the spacetime $(\mathcal{M}, g_{\mu\nu})$ is conformally flat. Therefore, according to the Weyl-Schouten theorem, for this region of $\mathcal{M}$ it is $W_{\mu\nu\rho\sigma} = 0$. Equivalently, we can state that in a neighborhood of the boundary $\mathcal{I}$, the spacetime $(\widehat{\mathcal{M}}, \widehat{g}_{\mu\nu})$ is flat. By considering the rescaling formula $\widehat{W}_{\mu\nu\rho\sigma} = \Omega^2 W_{\mu\nu\rho\sigma}$, we can conclude that it must be also $\widehat{W}_{\mu\nu\rho\sigma} = 0$ near the boundary within $\widehat{\mathcal{M}}$. If we demand that the Weyl tensor is smooth (or at least differentiable up to some order) near and on the boundary, it must vanish for all points with $\Omega = 0$, which is exactly the conformal boundary.

Finally, we mention a truly remarkable application of conformal infinity that translates the geometric properties of the conformal boundary $\mathcal{I}$ into asymptotic fall-off properties of the gravitational field. This is known as the *peeling theorem*, which gives a deep insight into the structure of gravitational radiation. The peeling theorem was originally presented in [77] by Sachs and is a consequence of the Bianchi identity for the Weyl tensor. The theorem says that for a *vacuum spacetime*, when considering null geodesic directions, with the radial coordinate r being used as affine parameter of the geodesics considered, the Weyl tensor $W_{\mu\nu\rho\sigma}(x)$ in the asymptotic limit $r \to \infty$ behaves like

$$W_{\mu\nu\rho\sigma}(x) = \frac{W^{(4)}_{\mu\nu\rho\sigma}}{r} + \frac{W^{(3)}_{\mu\nu\rho\sigma}}{r^2} + \frac{W^{(2)}_{\mu\nu\rho\sigma}}{r^3} + \frac{W^{(1)}_{\mu\nu\rho\sigma}}{r^4} + \mathcal{O}(r^{-5}). \qquad (21.98)$$

In the above equation, each of the four Weyl components $W^{(i)}_{\mu\nu\rho\sigma}$ is finite and independent of the radial coordinate r. In the far-field region, for large values of the radial affine parameter r, the term $W^{(4)}_{\mu\nu\rho\sigma}/r$ dominates the gravitational field. As one approaches more and more the source of the gravitational field, the terms with the powers r^{-2}, r^{-3} and r^{-4} successively come into play and contribute to the Weyl curvature. The notion of conformal infinity, the basics of which we have outlined above, has provided satisfactory definitions for isolated objects and spacetimes from a conceptual and mathematical point of view and has contributed greatly to the understanding of gravitational radiation.

Further Reading

In this chapter, we have provided a basic overview of the central topics of Einstein's general theory of relativity. GR today has numerous applications in astrophysics, cosmology, and even in practical engineering applications such as satellite communication. Recommended textbooks on GR are the classics of Misner, Thorne, Wheeler [57] and Weinberg [94]. More modern and streamlined with different emphasis are Carroll [14] and Hobson, Efstathiou, Lasenby [40]. Mathematically more sophisticated treatments are provided by de Felice, Clarke [21], Straumann [84], and Wald [91]. Global techniques and results for general spacetimes are treated in the classic book by Hawking and Ellis [39]. A systematic discussion of cosmological solutions within the framework of GR can be found in Plebański, Krasiński [68]. The review article of Frauendiener [31] offers a gentle introduction to the use of conformal infinity. The current state of the development of conformal methods for the analysis of Einstein's equations is presented by Valiente Kroon [89].

22

Lagrangian Formulation

Here we develop the Lagrangian formalism for generally covariant physical theories who-
action remains invariant under arbitrary diffeomorphisms and we introduce the minimal
pling between matter fields and the metric field. We then consider the matter sector
arately and derive the Euler-Lagrange field equations. During this process, we are led to
ew notion for the energy-momentum tensor of matter fields, the so-called metric EMT.
sequently, we consider the pure metric theory and derive the Einstein field equations for
cetime manifolds which may have a boundary. In the last section, we turn to the question
how we can exploit diffeomorphism invariance to obtain conserved currents. We derive
identically conserved Noether currents and the associated Komar superpotentials.

.1   Action Principle in Curved Spacetimes

grangians and the Action

consider a general four-dimensional Lorentzian spacetime $(\mathcal{M}, g_{\mu\nu})$ possessing the space-
e metric $g_{\mu\nu}(x)$ and in addition we consider matter fields $\phi_i(x)$ defined on this spacetime
nifold as suitable tensor fields. In contrast to flat space field theory, which we discussed
Chapter 15, in the present case the metric is a dynamical field of the theory. In order
define field theories on curved spacetimes, it is very useful to develop the Lagrangian
malism. In adopting this approach, we are not only able to build a formal framework for
eral relativistic field theories, but furthermore we can gain central results regarding the
eraction of the various types of fields and regarding conserved quantities. So let us start
h a general relativistic *Lagrangian density* $\mathscr{L}$ displaying the functional dependence

$$\mathscr{L} = \mathscr{L}\left(\phi_i, \partial_\rho \phi_i, \dots, g_{\mu\nu}, \partial_\rho g_{\mu\nu}, \dots\right). \tag{22.1}$$

e matter fields enter the Lagrangian density along with their partial derivatives up to a
tain order. This Lagrangian density is supposed to depend explicitly on the spacetime
tric and its partial derivatives $\partial_\rho g_{\mu\nu}$. Remember that the covariant derivatives of the
tric vanish identically for Levi-Civita type metrics. For the moment, we do not want
restrict the order of derivatives contained within the Lagrangian. Very often we will
reviate the Lagrangian density as $\mathscr{L}\left(\phi_i, g_{\mu\nu}\right)$. The *total action* of the theory is the
ctional $S\left[\phi_i, g_{\mu\nu}\right]$, depending on the metric and the matter fields, and is defined as the

volume integral

$$S\left[\phi_i, g_{\mu\nu}\right] \equiv \int_U \mathscr{L}\left(\phi_i, \partial_\rho\phi_i, \ldots, g_{\mu\nu}, \partial_\rho g_{\mu\nu}, \ldots\right) \mathrm{d}^4 x \qquad (22.2)$$

The Lagrangian density $\mathscr{L}\left(\,\cdot\,\right)$ is a scalar density of weight -1, whereas the volume inte-gration measure $\mathrm{d}^4 x$ is a scalar density of weight $+1$. Consequently, the action $S\left[\,\cdot\,\right]$ is an absolute scalar, as desired. The four-dimensional domain of integration U can be the total manifold $\mathcal{M}$ or a suitable subset of it. We allow also the possibility that the domain of inte-gration U is a manifold with a boundary ∂U. One can express the total action alternatively as the integral

$$S\left[\phi_i, g_{\mu\nu}\right] \equiv \int_U L\left(\phi_i, \nabla_\rho\phi_i, \ldots, g_{\mu\nu}, \partial_\rho g_{\mu\nu}, \ldots\right) \sqrt{|g|}\, \mathrm{d}^4 x \qquad (22.3)$$

in which the *scalar Lagrangian* $L\left(\phi_i, g_{\mu\nu}\right)$ and the invariant volume integration measu-re $\sqrt{|g|}\, \mathrm{d}^4 x$ are employed. Within the scalar Lagrangian $L\left(\phi_i, g_{\mu\nu}\right)$, we treat each of the fields $g_{\mu\nu}$, $\partial_\rho g_{\mu\nu}$, ϕ_i, $\nabla_\rho\phi_i$, and higher order derivatives as independent fields. For instance, the matter fields ϕ_i are "fundamental" and independent of the metric. The field $\nabla_\rho\phi_i$ is considered to be independent of ϕ_i and independent of the metric. Obviously, it is

$$\mathscr{L}\left(\phi_i, \partial_\rho\phi_i, \ldots, g_{\mu\nu}, \partial_\rho g_{\mu\nu}, \ldots\right) = L\left(\phi_i, \nabla_\rho\phi_i, \ldots, g_{\mu\nu}, \partial_\rho g_{\mu\nu}, \ldots\right) \sqrt{|g|}, \qquad (22.4)$$

and depending on the situation, each one of the Lagrangian functions, $\mathscr{L}\left(\,\cdot\,\right)$ and $L\left(\,\cdot\,\right)$, has its merits. For a given theory, the Lagrangian is not uniquely defined, as it is always possible to add a divergence term to the Lagrangian without altering the value of the action. Indeed, the Lagrangian

$$\overline{L}\left(\phi_i, g_{\mu\nu}\right) = L\left(\phi_i, g_{\mu\nu}\right) + \nabla_\rho M^\rho(\phi_i) \qquad (22.5)$$

leads to the same value for the action, provided that the vector field $M^\rho(\phi_i)$ vanishes on the boundary ∂U. The major question is how we can construct *any* reasonable Lagrangian for a curved spacetime theory.

Minimal Coupling
Our curved spacetime theory contains the metric field and matter fields. We have discussed in Section 17.2 how an interaction between various fields can be introduced within the Lagrangian formalism in a general fashion. In our case, we have to add the Lagrangian of the free metric field with the Lagrangian of the free matter fields, and then add a third Lagrangian which describes the interaction between these two subsystems, see equation 17.11. As discussed in Section 17.2, the most natural way of introducing an interaction is by means of the *minimal coupling* principle, which amounts to replacing all partial derivatives ∂_ρ by (gauge) covariant derivatives D_ρ containing a suitably transforming "gauge field". The procedure is analogous for the definition of a coupling of the matter fields to the metric field. The usual derivatives ∂_ρ have to be replaced by covariant derivatives ∇_ρ and the Levi-Civita connection takes over the role of the "gauge field". In addition, the flat metric $\eta_{\mu\nu}$ (responsible for any Lorentz indices contractions in the Lagrangian) has to be replaced by the general curved metric $g_{\mu\nu}$. This is in fact exactly the prescription on how to attain generally covariant equations, as expressed in 21.17. A priori, there is no guarantee that this approach will lead to a correct or even a reasonable theory. The minimal coupling principle is justified only by its results, and fortunately it can indeed lead to useful theories. If the

reader now has enough confidence in the minimal coupling procedure, then we can proceed and write down the *total Lagrangian with minimal coupling*:

$$L\left(\phi_i, g_{\mu\nu}\right) = L_{\mathrm{M}}\left(\phi_i, g_{\mu\nu}\right) + L_{\mathrm{G}}\left(g_{\mu\nu}\right). \tag{22.6}$$

In the above equation, $L_{\mathrm{M}}\left(\phi_i, g_{\mu\nu}\right)$ includes both, the *Lagrangian of the free matter fields plus the Lagrangian describing the matter fields coupled to the gravitational field*, while $L_{\mathrm{G}}\left(g_{\mu\nu}\right)$ represents the *Lagrangian of the free gravitational field*. We have to emphasize that this is by no means the only conceivable form of interaction, and research is still being conducted to discover new forms of interaction between matter and geometry. Having determined the general form of the total Lagrangian, let us now turn to the question of suitable variations of the fields.

Variations of Fields and the Action
In order to apply the action principle for deriving the dynamical equations of the theory, we have to consider variations of the fields. The following infinitesimal variations of the fields are appropriate for this purpose:

$$\left. \begin{aligned} \phi_i' &= \phi_i + \delta\phi_i, & \text{with } \delta\phi_i = 0 \text{ on } \partial U \\ g_{\mu\nu}' &= g_{\mu\nu} + \delta g_{\mu\nu}, & \text{with } \delta g_{\mu\nu} = 0 \text{ on } \partial U \end{aligned} \right\}. \tag{22.7}$$

As usually, we require all variations of the fields to vanish on the boundary ∂U of the domain of integration. Note that there is *no transformation of the spacetime coordinates* x involved here. For the first-order derivatives we have the variations

$$\left. \begin{aligned} (\nabla_\rho \phi_i)' &= \nabla_\rho \phi_i' = \nabla_\rho \phi_i + \nabla_\rho(\delta\phi_i) \\ (\partial_\rho g_{\mu\nu})' &= \partial_\rho g_{\mu\nu}' = \partial_\rho g_{\mu\nu} + \partial_\rho(\delta g_{\mu\nu}) \end{aligned} \right\}, \tag{22.8}$$

where we have used the equations $\delta(\nabla_\rho \phi_i) = \nabla_\rho(\delta\phi_i)$ and $\delta(\partial_\rho g_{\mu\nu}) = \partial_\rho(\delta g_{\mu\nu})$. We do not impose any conditions on the behavior of the derivatives on the boundary ∂U. For variations of higher order derivatives one proceeds an a similar way. The combined field variations 22.7 and 22.8 induce a variation of the total action functional $S\left[\phi_i, g_{\mu\nu}\right]$. The variation $\delta S\left[\phi_i, g_{\mu\nu}\right]$ in turn is defined by

$$\delta S\left[\phi_i, g_{\mu\nu}\right] = S\left[\phi_i + \delta\phi_i,\ g_{\mu\nu} + \delta g_{\mu\nu}\right] - S\left[\phi_i, g_{\mu\nu}\right]. \tag{22.9}$$

The principle of stationary action states that the relevant field configurations are those for which

$$\boxed{\delta S\left[\phi_i, g_{\mu\nu}\right] = 0} \tag{22.10}$$

is satisfied. We point out that the above variation δ includes both, the variations of the matter fields and the variation of the metric. Assuming that our total Lagrangian is defined through a minimal coupling, as in 22.6, the total action $S\left[\phi_i, g_{\mu\nu}\right]$ is split in the form

$$S\left[\phi_i, g_{\mu\nu}\right] = S_{\mathrm{M}}\left[\phi_i, g_{\mu\nu}\right] + S_{\mathrm{G}}\left[g_{\mu\nu}\right], \tag{22.11}$$

where $S_{\mathrm{M}}\left[\phi_i, g_{\mu\nu}\right]$ is the *matter action minimally coupled to the gravitational field* and $S_{\mathrm{G}}\left[g_{\mu\nu}\right]$ is the *pure gravity action*. The variation of the total action becomes then

$$\delta S\left[\phi_i, g_{\mu\nu}\right] = \delta S_{\mathrm{M}}\left[\phi_i, g_{\mu\nu}\right] + \delta S_{\mathrm{G}}\left[g_{\mu\nu}\right]. \tag{22.12}$$

According to the definition of the functional derivative, see appendix B.6, the single variations are

$$\delta S_{\mathrm{M}}\left[\phi_i, g_{\mu\nu}\right] = \int_U \left(\frac{\delta S_{\mathrm{M}}}{\delta \phi_i}\right) \delta \phi_i \, \mathrm{d}^4 x + \int_U \left(\frac{\delta S_{\mathrm{M}}}{\delta g_{\mu\nu}}\right) \delta g_{\mu\nu} \, \mathrm{d}^4 x \qquad (22.13)$$

for the minimally coupled matter part, and

$$\delta S_{\mathrm{G}}\left[g_{\mu\nu}\right] = \int_U \left(\frac{\delta S_{\mathrm{G}}}{\delta g_{\mu\nu}}\right) \delta g_{\mu\nu} \, \mathrm{d}^4 x \qquad (22.14)$$

for the pure gravity part. The single functional derivatives lead to specific equations, and
these are examined in the next two sections. More specifically, we will see that $\delta S_{\mathrm{M}}/\delta \phi_i$
leads to the field equations of the free matter fields, that $\delta S_{\mathrm{M}}/\delta g_{\mu\nu}$ leads to a novel notion of
energy-momentum tensor of the matter fields, and that $\delta S_{\mathrm{G}}/\delta g_{\mu\nu}$ leads to the free Einstein
field equations.

22.2   The Action for Matter Fields

Euler-Lagrange Equations of Matter Fields

In this section, we focus on the *matter action* $S_{\mathrm{M}}\left[\phi_i, g_{\mu\nu}\right]$ which contains the matter fields
minimally coupled to the externally specified metric field. Here and in the following, we
assume that the matter action contains only up to first-order derivatives of the matter
fields, which is valid for the physically interesting cases we are going to consider. Let us first
examine the variations only with respect to the matter fields and then the variation only
with respect to the metric. If we vary the matter fields ϕ_i and use the Lagrangian density
$\mathscr{L}_{\mathrm{M}}\left(\phi_i, g_{\mu\nu}\right)$ and the action in the form 22.2, the action principle leads to the *Euler-Lagrange
equations* of the matter fields

$$\boxed{\frac{\delta S_{\mathrm{M}}}{\delta \phi_i} = \frac{\partial \mathscr{L}_{\mathrm{M}}}{\partial \phi_i} - \partial_\mu \frac{\partial \mathscr{L}_{\mathrm{M}}}{\partial(\partial_\mu \phi_i)} = 0} \qquad (22.15)$$

exactly as derived in 15.31 within the framework of flat space field theory. By using al-
ternatively the scalar Lagrangian $L_{\mathrm{M}}\left(\phi_i, g_{\mu\nu}\right)$ and the action in the form 22.3, the action
principle yields (*exercise 22.1*)

$$\boxed{\frac{1}{\sqrt{|g|}} \frac{\delta S_{\mathrm{M}}}{\delta \phi_i} = \frac{\partial L_{\mathrm{M}}}{\partial \phi_i} - \nabla_\mu \frac{\partial L_{\mathrm{M}}}{\partial(\nabla_\mu \phi_i)} = 0} \qquad (22.16)$$

The Euler-Lagrange equations in the form 22.16 are especially useful if we aim to express
our theory in a covariant way. As an example, consider the *scalar ϕ^4 field minimally coupled
to the metric field*, as described by the scalar Lagrangian[*]

$$L_{\mathrm{M}} = \frac{1}{2} g^{\mu\nu} \nabla_\mu \phi \, \nabla_\nu \phi - \frac{m^2}{2}\phi^2 - \frac{\lambda}{4!}\phi^4. \qquad (22.17)$$

The corresponding scalar field equations are easily obtained as (*exercise 22.2*)

$$g^{\mu\nu} \nabla_\mu \nabla_\nu \phi + m^2 \phi + \frac{\lambda}{6}\phi^3 = 0. \qquad (22.18)$$

---

[*]We could use partial derivatives but it is useful to keep a manifestly covariant notation throughout.

From the standpoint of general relativity, every field which acts as a source of gravity is considered to be a matter field. Therefore, the real and complex scalar fields $\phi(x)$ and $\phi^*(x)$, the Dirac spinor field $\Psi_a(x)$, and the Maxwell covector field $A_\mu(x)$ are all *matter fields* within general relativity. If the Euler-Lagrange field equations are satisfied, we say that the physical system is *on-shell*.

As a second example of a covariant Lagrangian theory, consider the *Maxwell field minimally coupled to the metric field*. The corresponding scalar Lagrangian is

$$L_{\mathrm{M}} = -\frac{1}{16\pi} g^{\mu\rho} g^{\nu\sigma} F_{\mu\nu} F_{\rho\sigma}, \tag{22.19}$$

where the Faraday tensor $F_{\mu\nu}(x)$ is considered to be "fundamental" (which means it is defined independently of the metric) and is given as

$$F_{\mu\nu} = \nabla_\mu A_\nu - \nabla_\nu A_\mu. \tag{22.20}$$

The electromagnetic vector potential $A_\mu(x)$ likewise is considered to be "fundamental" and represents the basic field of the theory. The variation with respect to $A_\nu(x)$ leads to the *free Maxwell equations in curved spacetime*, (*exercise 22.3*)

$$\nabla^\mu F_{\mu\nu} = 0. \tag{22.21}$$

In Section 23.4 we will examine how the real scalar field and the Maxwell field can each be coupled to the metric field and at the same time exhibit Weyl rescaling invariance.

Another interesting example is provided by *scalar electrodynamics minimally coupled to the metric field*. The corresponding scalar Lagrangian can be written down readily,

$$L_{\mathrm{M}} = -\frac{F^2}{16\pi} + g^{\mu\nu}(D_\mu\phi)^*(D_\nu\phi) - m^2\phi^*\phi. \tag{22.22}$$

Here we have used the notation $F^2 \equiv g^{\mu\rho}g^{\nu\sigma}F_{\mu\nu}F_{\rho\sigma}$ and we have introduced the *gauge-covariant derivative* $D_\mu(x)$ *in curved spacetime*,

$$D_\mu \equiv \nabla_\mu + \mathrm{i}qA_\mu, \tag{22.23}$$

as demanded by the covariantization prescription, in this way generalizing the initial definition 17.13. Application of the formula 22.16 for the EL-Eqs of the three involved fields $\phi(x)$, $\phi^*(x)$ and $A_\mu(x)$ leads to the coupled field equations

$$(D^2 + m^2)\phi = 0 \quad \text{and} \quad (D^2 + m^2)^*\phi^* = 0 \tag{22.24}$$

for the scalar fields, and

$$\nabla^\mu F_{\mu\nu} = 4\pi\mathrm{i}q\left(\phi^*(D_\nu\phi) - (D_\nu\phi)^*\phi\right) \tag{22.25}$$

for the Maxwell field, as anticipated from the results of Section 17.3. (*exercise 22.4*) We should note here that in contrast to the flat space case, all three examples above do not describe closed systems. In order to obtain a closed system, the interaction of the matter fields with the gravitational field must also be taken into account. This happens through the Einstein equations containing the respective energy-momentum tensor of matter.

Metric Energy-Momentum Tensor
We now consider the variation $\delta S_{\mathrm{M}}[\phi_i, g_{\mu\nu}]$ of the matter action only with respect to the metric field $g_{\mu\nu}$. We must emphasize that within the matter action, the metric field represents an externally specified field and therefore the variation of the metric does not lead, in

general, to a vanishing variation of the matter action. The functional derivative $\delta S_M / \delta g_{\mu\nu}$ appearing in the rhs of 22.13 takes on a special role. Let us write the variation of the matter action as

$$\delta S_M = -\frac{1}{2} \int_U \left(\frac{-2}{\sqrt{|g|}} \frac{\delta S_M}{\delta g_{\mu\nu}} \right) \delta g_{\mu\nu} \sqrt{|g|} \, d^4 x. \tag{22.26}$$

The expression within the above brackets is known as the *metric energy-momentum tensor* (*MEMT*) $T^{\mu\nu}(x)$ of the matter fields,

$$\boxed{T^{\mu\nu} \equiv \frac{-2}{\sqrt{|g|}} \frac{\delta S_M}{\delta g_{\mu\nu}}} \tag{22.27}$$

and has been introduced in 1940 by *Leon Rosenfeld* [74]. The variation of the matter action consequently is written as

$$\delta S_M = -\frac{1}{2} \int_U T^{\mu\nu} \delta g_{\mu\nu} \sqrt{|g|} \, d^4 x. \tag{22.28}$$

In other words, *the MEMT defines the linear part of the change of the matter action when the metric is varied.* The factor -2 in the definition 22.27 can be justified either by calculating specific examples, or by using the MEMT as the de facto source of the Einstein gravitational field. We will use both approaches below.* The definition 22.27 makes it apparent that the MEMT is a symmetric tensor. In addition, we want it to fulfill the divergence relation $\nabla_\mu T^{\mu\nu} = 0$ and in Section 22.4 we will see that this is a consequence of diffeomorphism invariance. Furthermore, we should expect that the formula for the MEMT leads to the same results for an energy-momentum tensor as the Belinfante method of Section 16.2.

It is useful to derive a formula for the MEMT that will simplify its practical calculation. Assuming that the matter Lagrangian $L_M(\phi_i, g_{\mu\nu})$ has only a dependence on the metric $g_{\mu\nu}$ but no dependence on any of its derivatives $\partial_\rho g_{\mu\nu}$, and so forth, the MEMT definition 22.27 yields

$$T^{\mu\nu} = \frac{-2}{\sqrt{|g|}} \frac{\partial(L_M \sqrt{|g|})}{\partial g_{\mu\nu}}. \tag{22.29}$$

By using the basic formula

$$\frac{\partial}{\partial g_{\mu\nu}} \sqrt{|g|} = \frac{1}{2} \sqrt{|g|} \, g^{\mu\nu}, \tag{22.30}$$

see Section 18.3, one obtains the expression

$$T^{\mu\nu} = -2 \frac{\partial L_M}{\partial g_{\mu\nu}} - g^{\mu\nu} L_M \tag{22.31}$$

for the MEMT. For the version with lowered indices, $T_{\mu\nu}(x)$, we use the identity

$$\frac{\partial f}{\partial g^{\mu\nu}} = -g_{\mu\rho} g_{\nu\sigma} \frac{\partial f}{\partial g_{\rho\sigma}}, \tag{22.32}$$

---

*We remark that the MEMT is very often defined with an opposite overall sign in the GR literature. The opposite sign is used when a spacetime metric of signature $(-+++)$ is employed, so that the CEMT and the MEMT each need a minus sign to achieve a positive energy density of the fields. Our definitions for the metric signature and the sign of the MEMT are the same as in [1], [20], [40], [51], and [68].

being valid for any scalar function $f(g_{\mu\nu})$, and obtain the alternative formula

$$T_{\mu\nu} = 2\frac{\partial L_M}{\partial g^{\mu\nu}} - g_{\mu\nu}L_M. \tag{22.33}$$

Both formulae 22.31 and 22.33 reveal that the MEMT is, as expected, an absolute tensor of rank two. This can also be inferred from the initial definition 22.27, since the functional derivative $\delta S_M/\delta g_{\mu\nu}$ is a $(2,0)$-tensor density with weight -1 and the factor $\sqrt{|g|}$ in the denominator of 22.27 is a scalar density also with weight -1, with the consequence that these weights neutralize each other. The usefulness of the MEMT conception and the associated formulae 22.31 and 22.33 is demonstrated in the following.

Examples for the Metric EMT

In Chapters 15 and 16, we have introduced the energy-momentum tensor by means of the canonical and the Belinfante definitions. In Chapter 17, we applied the Belinfante prescription to obtain the total energy-momentum tensor of interacting fields. Now we will employ the metric energy-momentum tensor to derive some of the previous results in a much simpler way. The first example is the real scalar field described by the Lagrangian 22.17. By using the formula 22.33, we obtain immediately the corresponding MEMT (*exercise 22.5*)

$$T_{\mu\nu} = \nabla_\mu\phi\,\nabla_\nu\phi - g_{\mu\nu}L_M. \tag{22.34}$$

This is the same result as in 21.22 obtained through the covariantization prescription 21.17, and it reduces to the Minkowski result 16.75 when the geometry is flat.

Another example is provided by the free Maxwell field described by the Lagrangian 22.19. Application of the MEMT formula 22.33 leads to (*exercise 22.6*)

$$T_{\mu\nu} = \frac{1}{4\pi}\left(F_{\mu\alpha}F^\alpha{}_\nu + \frac{1}{4}g_{\mu\nu}F^{\alpha\beta}F_{\alpha\beta}\right), \tag{22.35}$$

which is the same energy-momentum tensor as the one derived in 21.23 through the covariantization prescription and is consistent with the flat spacetime result 16.85.

A slightly more complicated example is provided by the coupled scalar-Maxwell gauge theory in curved spacetime as described by the Lagrangian 22.22. Application of the formula 22.33 leads to (*exercise 22.7*)

$$T_{\mu\nu} = \frac{1}{4\pi}\left(F_{\mu\alpha}F^\alpha{}_\nu + \frac{1}{4}g_{\mu\nu}F^2\right)$$
$$+ (D_\mu\phi)^*(D_\nu\phi) + (D_\nu\phi)^*(D_\mu\phi) - g_{\mu\nu}\left[(D_\alpha\phi)^*(D^\alpha\phi) - m^2\phi^*\phi\right] \tag{22.36}$$

for the MEMT of the total scalar electrodynamics system. Comparing with the flat spacetime result for the energy-momentum tensor 17.34, we see again that we can obtain the same MEMT by applying the general relativistic covariantization prescription 21.17 to the previous result 17.34. In any case, the application of the MEMT formula is very useful even in the case of flat spacetime theories, since one does not have to rely on the complicated Belinfante procedure to obtain a symmetric, divergence-free and gauge invariant energy-momentum tensor. In this case, the metric is first treated like a dynamic variable and the MEMT is obtained from the formulae 22.31 or 22.33. At the end of the calculation, the special case $g_{\mu\nu}(x) = \eta_{\mu\nu}$ of flat Minkowski space is implemented. The EMTs of the theories we have considered here are identical when derived as Belinfante SEMTs or as MEMTs. We should note, however, that there are field theories that contain derivatives of the fields higher than first order, where these two EMT prescriptions lead to different results.

22.3　The Action for the Gravitational Field

Einstein-Hilbert Action

We consider the *pure gravity action* $S_G[g_{\mu\nu}]$, which depends on the metric tensor $g_{\mu\nu}(x)$ as the only dynamical field of the theory. Our aim is, of course, to derive the free Einstein field equations from the variational principle. To this end we need to find a suitable Lagrangian. Starting from the metric and its partial derivatives of increasing order, one can build various scalars, but the most immediate nontrivial scalar is the Ricci scalar $R = R_{\mu\nu}g^{\mu\nu}$. The Ricci scalar contains first-order and second-order derivatives of the metric, with the latter entering only linearly into the expression for $R(x)$.[*] So our ansatz for the Lagrangian density is $\mathscr{L}_G(g_{\mu\nu}) = kR\sqrt{|g|}$, where the constant of proportionality k can be chosen as needed, which we will do later. The associated scalar Lagrangian is $L_G(g_{\mu\nu}) = kR$. The scalar Lagrangian $L_{EH}(g_{\mu\nu})$ without the proportionality factor

$$\boxed{L_{EH} \equiv R} \tag{22.37}$$

is called the *Einstein-Hilbert Lagrangian*. We could now proceed with the *Euler-Lagrange equations* that use the Lagrangian density $\mathscr{L}_G(g_{\mu\nu})$, which read in this case

$$\frac{\delta S_G}{\delta g_{\mu\nu}} = \frac{\partial \mathscr{L}_G}{\partial g_{\mu\nu}} - \partial_\rho \frac{\partial \mathscr{L}_G}{\partial(\partial_\rho g_{\mu\nu})} + \partial_\rho \partial_\sigma \frac{\partial \mathscr{L}_G}{\partial(\partial_\rho \partial_\sigma g_{\mu\nu})} = 0, \tag{22.38}$$

and then carry out the required derivations. However, it is much more instructive to apply the variational principle directly to the action functional. Hence, let us consider the *Einstein-Hilbert action* $S_{EH}[g_{\mu\nu}]$, which is defined as

$$\boxed{S_{EH} \equiv \int_U R\sqrt{|g|}\,\mathrm{d}^4x} \tag{22.39}$$

The domain of integration is the entire manifold $U = \mathcal{M}$ and we allow the possibility that the manifold has a boundary ∂U. The variation of the metric, $g'_{\mu\nu} = g_{\mu\nu} + \delta g_{\mu\nu}$, with the constraint $\delta g_{\mu\nu} = 0$ at the boundary ∂U, induces a variation of the action. From a technical point of view, it is more advantageous to consider the variation of the inverse metric, $g'^{\mu\nu} = g^{\mu\nu} + \delta g^{\mu\nu}$. So let us examine the variation $\delta S_{EH}[g_{\mu\nu}]$ directly,

$$\delta S_{EH} = \int_U \left[(\delta R_{\mu\nu})g^{\mu\nu}\sqrt{|g|} + R_{\mu\nu}(\delta g^{\mu\nu})\sqrt{|g|} + R(\delta\sqrt{|g|}) \right] \mathrm{d}^4x. \tag{22.40}$$

The second integral in the above rhs has already a form where the variation $\delta g^{\mu\nu}$ appears as a factor. In the third integral above we can apply the formula $\delta\sqrt{|g|} = -(1/2)\sqrt{|g|}\,g_{\mu\nu}\delta g^{\mu\nu}$ to obtain an explicit factor $\delta g^{\mu\nu}$. Only the first integral above needs a little more work.

We have to find an expression for the variation $\delta R_{\mu\nu}(x)$, which is triggered by the variation of the metric. Based on first principles, the variation of the Ricci tensor is defined as $\delta R_{\mu\nu}[g] = R_{\mu\nu}[g + \delta g] - R_{\mu\nu}[g]$ and can be readily calculated as

$$\delta R_{\mu\nu} = \delta R^\alpha{}_{\mu\alpha\nu} =$$
$$\delta(\partial_\alpha\Gamma^\alpha_{\mu\nu} - \partial_\nu\Gamma^\alpha_{\mu\alpha} + \Gamma^\alpha_{\beta\alpha}\Gamma^\beta_{\mu\nu} - \Gamma^\alpha_{\beta\nu}\Gamma^\beta_{\mu\alpha}) = \tag{22.41}$$
$$\partial_\alpha(\delta\Gamma^\alpha_{\mu\nu}) - \partial_\nu(\delta\Gamma^\alpha_{\mu\alpha}) + (\delta\Gamma^\alpha_{\beta\alpha})\Gamma^\beta_{\mu\nu} + \Gamma^\alpha_{\beta\alpha}(\delta\Gamma^\beta_{\mu\nu}) - (\delta\Gamma^\alpha_{\beta\nu})\Gamma^\beta_{\mu\alpha} - \Gamma^\alpha_{\beta\nu}(\delta\Gamma^\beta_{\mu\alpha}).$$

---

[*]It can be shown that the Ricci scalar is actually the only scalar that can be constructed from the metric and its first-order and second-order derivatives.

At this point we need to examine the variation $\delta\Gamma^\alpha_{\mu\nu}(x)$ of the connection coefficients. The variation $\delta\Gamma^\alpha_{\mu\nu}[g] = \Gamma^\alpha_{\mu\nu}[g + \delta g] - \Gamma^\alpha_{\mu\nu}[g]$ is a difference between two connections whose terms of second-order are identical and independent of the metric, see equation 18.15, and which consequently cancel each other. Therefore, the variation $\delta\Gamma^\alpha_{\mu\nu}(x)$ represents a proper $(1, 2)$-tensor field. We will see this also in a different way in the following paragraph. The covariant derivative of the variation can be computed as

$$\nabla_\sigma(\delta\Gamma^\alpha_{\mu\nu}) = \partial_\sigma(\delta\Gamma^\alpha_{\mu\nu}) + \Gamma^\alpha_{\beta\sigma}\delta\Gamma^\beta_{\mu\nu} - \Gamma^\beta_{\mu\sigma}\delta\Gamma^\alpha_{\beta\nu} - \Gamma^\beta_{\nu\sigma}\delta\Gamma^\alpha_{\mu\beta}. \tag{22.42}$$

By considering the difference $\nabla_\sigma(\delta\Gamma^\alpha_{\mu\nu}) - \nabla_\nu(\delta\Gamma^\alpha_{\mu\sigma})$, we obtain immediately that the variation $\delta R_{\mu\nu}(x)$ can be expressed in a covariant way as

$$\delta R_{\mu\nu} = \nabla_\alpha(\delta\Gamma^\alpha_{\mu\nu}) - \nabla_\nu(\delta\Gamma^\alpha_{\mu\alpha}). \tag{22.43}$$

This formula is known as the *Palatini identity* (*Attilio Palatini*). (*exercise 22.8*) Consequently, by using also the metric compatibility, we can write the first integral in the rhs of 22.40 as

$$\int_U \nabla_\sigma(g^{\mu\nu}\delta\Gamma^\sigma_{\mu\nu} - g^{\mu\sigma}\delta\Gamma^\nu_{\mu\nu})\sqrt{|g|}\,\mathrm{d}^4x. \tag{22.44}$$

This represents a volume integral of a divergence and can thus be converted into a surface integral over the closed boundary ∂U, so that it takes the form

$$\oint_{\partial U} (g^{\mu\nu}\delta\Gamma^\sigma_{\mu\nu} - g^{\mu\sigma}\delta\Gamma^\nu_{\mu\nu})\,\mathrm{d}\Sigma_\sigma. \tag{22.45}$$

Assuming that the variation of the connection coefficients is zero at the boundary, i.e. $\delta\Gamma^\alpha_{\mu\nu} = 0$ *on* ∂U, this surface integral vanishes. In the following paragraph we will see that the vanishing of the variation of the first-order derivatives of the metric at the boundary, i.e. $\delta(\partial_\rho g_{\mu\nu}) = 0$ on ∂U, represents a sufficient condition for this to happen. Despite the fact that each of these two just stated boundary conditions simplify the subsequent analysis, they actually represent additional assumptions that are not strictly necessary. Later in this section, we will examine how we can cope with the above surface integral without making additional assumptions about the behavior at the boundary.

In summary, our analysis of the variation of the Einstein-Hilbert action has led us to

$$\delta S_{\mathrm{EH}} = \int_U \left(R_{\mu\nu} - \frac{1}{2}Rg_{\mu\nu} \right) \delta g^{\mu\nu}\sqrt{|g|}\,\mathrm{d}^4x. \tag{22.46}$$

The variational principle demands that $\delta S_{\mathrm{EH}}[g_{\mu\nu}] = 0$ and in this way we can deduce that the Euler-Lagrange equations for the Einstein-Hilbert action are

$$\boxed{\frac{1}{\sqrt{|g|}}\frac{\delta S_{\mathrm{EH}}}{\delta g^{\mu\nu}} = G_{\mu\nu} = 0} \tag{22.47}$$

These equations are, as desired, the familiar gravitational vacuum field equations.

Variation of the Connection

We have remarked previously that the variation of the connection coefficients $\delta\Gamma^\alpha_{\mu\nu}(x)$ constitutes an absolute $(1, 2)$-tensor field. This can be seen directly from the manifestly covariant formula

$$\delta\Gamma^\alpha_{\mu\nu} = \frac{1}{2}g^{\alpha\beta}(\nabla_\mu\delta g_{\nu\beta} + \nabla_\nu\delta g_{\mu\beta} - \nabla_\beta\delta g_{\mu\nu}), \tag{22.48}$$

which, by the way, resembles very much the formula 18.78 for the Levi-Civita connection. In order to derive the identity 22.48, one has to write out explicitly the basic definition $\delta\Gamma^\alpha_{\mu\nu}[g] = \Gamma^\alpha_{\mu\nu}[g + \delta g] - \Gamma^\alpha_{\mu\nu}[g]$ for the variation of the metric. (*exercise 22.9*) Another formula for the variation of the connection, which is not manifestly covariant but also useful is

$$\delta\Gamma^\alpha_{\mu\nu} = \frac{1}{2}g^{\alpha\beta}(\partial_\mu\delta g_{\nu\beta} + \partial_\nu\delta g_{\mu\beta} - \partial_\beta\delta g_{\mu\nu} - 2\Gamma^\sigma_{\mu\nu}\delta g_{\sigma\beta}). \tag{22.49}$$

From the expression in the above rhs, we can see that if the two conditions $\delta g_{\mu\nu} = 0$ and $\delta(\partial_\rho g_{\mu\nu}) = 0$ hold for some region of spacetime (like a boundary ∂U), then the variation of the connection coefficients also vanishes in this region, $\delta\Gamma^\alpha_{\mu\nu} = 0$. We have used these special conditions above in the variation of the Einstein-Hilbert action. However, in the next paragraph we will see that we actually do not need to use the condition $\delta(\partial_\rho g_{\mu\nu}) = 0$ in order to successfully carry out the variation of the Einstein-Hilbert action.

Gibbons-Hawking-York Boundary Term

We have seen that the variation of the Einstein-Hilbert action leads to the result

$$\delta S_{\text{EH}} = \int_U G_{\mu\nu}\delta g^{\mu\nu}\sqrt{|g|}\,\mathrm{d}^4x + \oint_{\partial U} W^\sigma \mathrm{d}\Sigma_\sigma, \tag{22.50}$$

where the boundary integral over ∂U contains the vector field $W^\sigma(x)$, which we identified as

$$W^\sigma = g^{\mu\nu}\delta\Gamma^\sigma_{\mu\nu} - g^{\mu\sigma}\delta\Gamma^\nu_{\mu\nu}. \tag{22.51}$$

We have discarded this boundary integral on the grounds that all variations of the derivatives of the metric vanish on ∂U. This represents, however, an artificial requirement which we want to omit in the following. Therefore, let us examine the boundary integral

$$B \equiv \oint_{\partial U} W^\sigma \mathrm{d}\Sigma_\sigma \tag{22.52}$$

in more detail. We have written this as a surface integral over a *closed boundary* ∂U and this is the only special assumption we will make in the following. In other words, the manifold $U = \mathcal{M}$ we are considering should either have no boundary, in which case the boundary integral vanishes, or it should be a manifold with a closed boundary. By using the formula 22.48 for the variation of the connection, the vector field $W^\sigma(x)$ can be expressed as (*exercise 22.10*)

$$W^\sigma = g^{\rho\sigma}g^{\mu\nu}(\nabla_\mu\delta g_{\rho\nu} - \nabla_\rho\delta g_{\mu\nu}). \tag{22.53}$$

We insert this expression into the boundary integral 22.52 and obtain

$$B = \oint_{\partial U} N^\rho g^{\mu\nu}(\nabla_\mu\delta g_{\rho\nu} - \nabla_\rho\delta g_{\mu\nu})\sqrt{|h|}\,\mathrm{d}^3y, \tag{22.54}$$

where we have used the hypersurface element $\mathrm{d}\Sigma_\sigma$ in its explicit form, see appendix B.8,

$$\mathrm{d}\Sigma_\sigma = N_\sigma\sqrt{|h|}\,\mathrm{d}y^1 \wedge \mathrm{d}y^2 \wedge \mathrm{d}y^3. \tag{22.55}$$

In the above, $N_\sigma(y)$ is the unit normal field on the hypersurface, $h(y)$ is the determinant of the induced metric $h_{jk}(y)$, and (y^1, y^2, y^3) is a set of coordinates of the hypersurface. Now we employ the definition 2.48 of the projection operator $h_{\mu\nu}(x)$ on the hypersurface and write the metric as

$$g^{\mu\nu} = h^{\mu\nu} + N^2 N^\mu N^\nu. \tag{22.56}$$

When carrying out the contractions in the integrand of the boundary integral, one sees that the terms proportional to $N^\mu N^\nu N^\rho$ cancel each other. Thus, the boundary integral becomes

$$B = \oint_{\partial U} N^\rho h^{\mu\nu} \nabla_\mu \delta g_{\rho\nu} \sqrt{|h|}\, \mathrm{d}^3 y - \oint_{\partial U} N^\rho h^{\mu\nu} \nabla_\rho \delta g_{\mu\nu} \sqrt{|h|}\, \mathrm{d}^3 y. \qquad (22.57)$$

Within the first integral above it is $h^{\mu\nu} \nabla_\mu \delta g_{\rho\nu} = 0$, as the variation of the metric fulfills $\delta g_{\rho\nu} = 0 = \mathrm{const}$ on ∂U and thus the projection $h^{\mu\nu} \nabla_\mu \delta g_{\rho\nu}$ of the derivatives along ∂U vanishes. Hence, only the second integral contributes to the boundary term,

$$B = -\oint_{\partial U} N^\rho h^{\mu\nu} \nabla_\rho \delta g_{\mu\nu} \sqrt{|h|}\, \mathrm{d}^3 y. \qquad (22.58)$$

We aim to express this boundary integral as the variation of a suitable action term. In fact, we should look for an action term that is the volume integral of a total divergence. To this end, we are going to compute the variation of the integral expression

$$A \equiv \oint_{\partial U} h^{\mu\nu} \nabla_\mu N_\nu \sqrt{|h|}\, \mathrm{d}^3 y \qquad (22.59)$$

under variations of the metric, as specified above. The variation of the metric $g_{\mu\nu}(x)$ with the condition $\delta g_{\mu\nu} = 0$ on ∂U has the following effect on the quantities in the integral A:

$$\delta(\partial U) = 0, \quad \delta(\mathrm{d}^3 y) = 0, \quad \delta\sqrt{|h|} = 0 \quad \text{and} \quad \delta h^{\mu\nu} = 0. \qquad (22.60)$$

The first two variations are clear. For the variation $\delta\sqrt{|h|}$ we note how the induced metric $h_{jk}(y)$ of the boundary is expressed through the parent metric $g_{\mu\nu}(x)$, equation 2.42,

$$h_{jk}(y) = \frac{\partial x^\mu}{\partial y^j} \frac{\partial x^\nu}{\partial y^k} g_{\mu\nu}(x). \qquad (22.61)$$

The coordinates are not varied and therefore it is $\delta h_{jk} = 0$ on ∂U, which means that the induced metric remains unaffected. The same applies for its determinant then. For the variation $\delta h^{\mu\nu}(x)$ we use the formula for the projection operator with upper indices,

$$h^{\mu\nu}(x) = \frac{\partial x^\mu}{\partial y^j} \frac{\partial x^\nu}{\partial y^k} h^{jk}(y), \qquad (22.62)$$

where $h^{jk}(y)$ is the inverse metric of the boundary. Since $\delta h^{jk} = 0$ holds on the boundary, it is also $\delta h^{\mu\nu} = 0$. Therefore, the effect of the variation of the metric on the integral A is

$$\delta A = \oint_{\partial U} h^{\mu\nu} \delta(\nabla_\mu N_\nu) \sqrt{|h|}\, \mathrm{d}^3 y. \qquad (22.63)$$

We note that it is

$$\delta(\nabla_\mu N_\nu) = -\delta\Gamma^\alpha_{\nu\mu} N_\alpha. \qquad (22.64)$$

By contracting with $h^{\mu\nu}$ and by inserting the formula 22.48 for the variation of the connection, we obtain

$$2h^{\mu\nu} \delta(\nabla_\mu N_\nu) = N^\rho h^{\mu\nu} \nabla_\rho \delta g_{\mu\nu}. \qquad (22.65)$$

Here we have used again that the hypersurface projections of the derivatives of the metric variations vanish throughout. Thus, we have achieved the goal of expressing the boundary integral B as a variation. It is

$$B = -2\,\delta A, \qquad (22.66)$$

or, written out,

$$B = -2\,\delta \oint_{\partial U} h^{\mu\nu} \nabla_\mu N_\nu \sqrt{|h|}\, \mathrm{d}^3 y. \tag{22.67}$$

We note that it is $\nabla_\rho(N^2) = 0$ and therefore $(\nabla_\rho N_\mu)N^\mu = 0$ for the unit normal field. From this property, we obtain that it is always

$$h^{\mu\nu}\nabla_\mu N_\nu = g^{\mu\nu}\nabla_\mu N_\nu = \nabla_\mu N^\mu. \tag{22.68}$$

Hence, we can write the boundary integral more compactly as

$$B = -2\,\delta \oint_{\partial U} \nabla_\mu N^\mu \sqrt{|h|}\, \mathrm{d}^3 y. \tag{22.69}$$

We can summarize that in order to cancel the contribution of the boundary term B, which appears within the variation of the Einstein-Hilbert action, we have to expand the EH action by a counter-term that has has the property that it cancels itself in combination with the boundary term during the variation. Based on our analysis above, we are thus led to the pure gravity action (up to constant factors) in the form

$$S_{\text{EH}} + S_{\text{GHY}} = \int_U R\sqrt{|g|}\, \mathrm{d}^4 x + 2\oint_{\partial U} \nabla_\mu N^\mu \sqrt{|h|}\, \mathrm{d}^3 y. \tag{22.70}$$

The additional action term $S_{\text{GHY}}\,[g_{\mu\nu}]$, which is given as

$$\boxed{S_{\text{GHY}} \equiv 2\oint_{\partial U} \nabla_\mu N^\mu \sqrt{|h|}\, \mathrm{d}^3 y} \tag{22.71}$$

is called the *Gibbons-Hawking-York boundary term*. This integral was first discussed in 1972 in the article [100] by *James W. York* and later further developed by *Gary W. Gibbons* and *Stephen W. Hawking*. With the aid of the GHY boundary term and without any special assumptions regarding the derivatives of the metric, the variation of the metric provides the result

$$\delta(S_{\text{EH}} + S_{\text{GHY}}) = \int_U G_{\mu\nu}\delta g^{\mu\nu}\sqrt{|g|}\, \mathrm{d}^4 x, \tag{22.72}$$

which in turn leads to the free Einstein equations, as desired.

Let us also demonstrate that adding the GHY boundary term to the Einstein-Hilbert action corresponds to adding a pure divergence term to the Einstein-Hilbert Lagrangian. By introducing the notation $s \equiv N^2 = \pm 1$ and noting that $s^2 = 1$, we can write the GHY boundary term as

$$S_{\text{GHY}} = 2s \oint_{\partial U} N^\rho \nabla_\mu N^\mu\, \mathrm{d}\Sigma_\rho. \tag{22.73}$$

According to the Gauss divergence formula, this surface integral can be converted to a volume integral

$$S_{\text{GHY}} = 2s \int_U \nabla_\rho(N^\rho \nabla_\mu N^\mu)\sqrt{|g|}\, \mathrm{d}^4 x, \tag{22.74}$$

from which we can directly see that the GHY term indeed corresponds to a pure divergence in the Lagrangian.

We have introduced the GHY action term in order to avoid artificial boundary conditions for the metric. However, the GHY term provides some more advantages that come into play in modified theories of general relativity or in path integral calculations in Euclidean quantum gravity, although both are outside our scope here. Nevertheless, henceforth we will use the *action for the pure gravitational field* in the form

$$S_{\text{EH}} + S_{\text{GHY}} = \int_U \left\{ R + 2s\nabla_\rho (N^\rho \nabla_\mu N^\mu) \right\} \sqrt{|g|} \, \mathrm{d}^4 x \tag{22.75}$$

We remark here that within general relativity, the absolute value of the above action integral is less relevant than its functional dependency on the basic fields of the theory. For instance, the EH Lagrangian for the free gravitational field is always zero, $R = 0$. Still its functional form yields the free Einstein equations. The only remaining piece is now the determination of the proportionality factor k within the gravity action $S_{\text{G}}[g_{\mu\nu}]$, which we will address in the following.

Variation of the Complete Action for Matter and Gravity

We are now in a position to combine the results we have obtained so far from the variational approach in general relativity. The action principle, as expressed in 22.10, when applied to the total action $S[\phi_i, g_{\mu\nu}]$, consisting of the matter action $S_{\text{M}}[\phi_i, g_{\mu\nu}]$ with its coupling to gravity plus the pure gravity action $S_{\text{G}}[g_{\mu\nu}]$, leads to the complete set of dynamical equations of the matter-gravity system. These are the matter equations minimally coupled to gravity plus the Einstein field equations with a non-vanishing matter EMT. Let us lay down the details here. The variations with respect to the matter fields ϕ_i and the metric field $g_{\mu\nu}$, as specified before in 22.7 and 22.8, are applied to the total action $S[\phi_i, g_{\mu\nu}]$, which is constructed as the sum of the two single action parts,

$$S[\phi_i, g_{\mu\nu}] = S_{\text{M}}[\phi_i, g_{\mu\nu}] + S_{\text{G}}[g_{\mu\nu}]. \tag{22.76}$$

We do not specify the matter action in more detail here, while for the pure gravity action we employ the definition

$$S_{\text{G}}[g_{\mu\nu}] \equiv kS_{\text{EH}}[g_{\mu\nu}] + kS_{\text{GHY}}[g_{\mu\nu}], \tag{22.77}$$

which contains the Einstein-Hilbert action, the Gibbons-Hawking-York action and the soon to be determined constant factor k. If we put everything together, we can write the total action $S[\phi_i, g_{\mu\nu}]$ of the physical system as

$$S[\phi_i, g_{\mu\nu}] = S_{\text{M}}[\phi_i, g_{\mu\nu}] + kS_{\text{EH}}[g_{\mu\nu}] + kS_{\text{GHY}}[g_{\mu\nu}]. \tag{22.78}$$

The corresponding complete set of matter-gravity field equations consists of the equations

$$\frac{1}{\sqrt{|g|}} \frac{\delta S}{\delta \phi_i} = 0 \tag{22.79}$$

for the matter fields being minimally coupled to the metric field, compare with 22.16, and in addition the equations

$$\frac{1}{\sqrt{|g|}} \frac{\delta S}{\delta g^{\mu\nu}} = \frac{1}{\sqrt{|g|}} \frac{\delta S_{\text{M}}}{\delta g^{\mu\nu}} + \frac{1}{\sqrt{|g|}} \frac{\delta S_{\text{G}}}{\delta g^{\mu\nu}} = 0 \tag{22.80}$$

for the metric field coupled to the MEMT of matter. With the definition 22.27 of the MEMT and the result 22.72 for the gravity action, the gravitational field equations are written as

$$G_{\mu\nu} = -\frac{1}{2k} T_{\mu\nu}. \tag{22.81}$$

In order to recover the Einstein field equations 21.37, we have to set the constant k to the value

$$k = -\frac{1}{16\pi G_{\mathrm{N}}}. \tag{22.82}$$

Thus, the action principle has provided us the Einstein field equations with matter sources, which we like to write down again here,

$$G_{\mu\nu} = 8\pi G_{\mathrm{N}} T_{\mu\nu}. \tag{22.83}$$

This concludes our discussion about the derivation of the equations of motion from the action principle. The method of using variations of the (metric and matter) fields has proven to be very fruitful. In the next section we are going to employ the technique of variations in order to analyze diffeomorphism invariance, and this will lead to some more far-reaching consequences.

22.4   Diffeomorphisms and Noether Currents

Diffeomorphisms as Internal Transformations

We want to investigate the consequences of applying active diffeomorphisms to the action $S[\phi_i, g_{\mu\nu}]$ of a generally covariant theory. We have expressed active diffeomorphisms already in 9.39 and in 15.128. Within such diffeomorphisms, the coordinates transform as $x'^{\mu}(x) = x^{\mu} + \delta x^{\mu}(x)$, where the variation $\delta x^{\mu}(x) \equiv \epsilon V^{\mu}(x)$ is defined by means of an *arbitrary* vector field $V^{\mu}(x)$ on the spacetime manifold. Note that $V^{\mu}(x)$ is generally *not* a Killing vector field. Such diffeomorphisms induce, in a fundamental way, variations in all tensor fields of the theory, and in particular in the metric and the matter fields. However, one can decouple the effect of diffeomorphisms on coordinates from their effect on the metric and the matter fields. This is achieved by keeping the coordinates unchanged, i.e. by demanding $\delta x^{\mu}(x) = 0$, and by only considering diffeomorphic transformations of the metric and matter fields, as they are induced by arbitrary vector fields. Such diffeomorphism transformations can be readily written down by using the Lie derivative. They are, infinitesimally,

$$\delta_V \phi_i = -\mathcal{L}_{\epsilon V} \phi_i \tag{22.84}$$

for the matter fields, and

$$\delta_V g_{\mu\nu} = -\mathcal{L}_{\epsilon V} g_{\mu\nu} = -\epsilon(\nabla_\mu V_\nu + \nabla_\nu V_\mu) \tag{22.85}$$

for the metric field, compare with the formulae 3.25 and 20.1. The real parameter ϵ is assumed to be small, $|\epsilon| \ll 1$. We use the symbol δ_V to denote its dependence on the vector field $V^{\mu}(x)$ and to distinguish this type of variation from the other types of variations we have discussed so far. All fields are evaluated at the same spacetime point x. The just defined diffeomorphism transformations of the matter fields and the metric each represent internal transformations of these fields, which we can consequently call *internal diffeomorphisms* of ϕ_i and $g_{\mu\nu}$. In the following, we examine the effect of this type of diffeomorphism transformations in some detail.

Transformation of the Action under Internal Diffeomorphisms

Consider first an arbitrary total action $S[\phi_i, g_{\mu\nu}]$ of a theory given by the integral

$$S = \int_U L\sqrt{|g|}\, \mathrm{d}^4 x, \tag{22.86}$$

with a scalar Lagrangian function $L(\phi_i, g_{\mu\nu})$. The transformation of the action under an internal diffeomorphism δ_V is expressed by the variational change

$$\delta_V S = \int_U \left[(\delta_V L)\sqrt{|g|} + (\delta_V \sqrt{|g|})L \right] d^4x, \tag{22.87}$$

where the variations within the integrand are calculated as

$$\delta_V L = -\mathcal{L}_{\epsilon V} L = -\epsilon V^\rho \nabla_\rho L \tag{22.88}$$

for the first term, and

$$\delta_V \sqrt{|g|} = \frac{1}{2}\sqrt{|g|}\, g^{\mu\nu}\delta_V g_{\mu\nu} = -\epsilon\sqrt{|g|}\,\nabla_\rho V^\rho \tag{22.89}$$

for the second term. Hence, we obtain the generally valid result

$$\delta_V S = -\epsilon \int_U \nabla_\rho(V^\rho L)\sqrt{|g|}\, d^4x. \tag{22.90}$$

We observe that the variation $\delta_V S[\phi_i, g_{\mu\nu}]$ of the action corresponds to a pure divergence term for the Lagrangian and therefore the field equations are not affected. This internal diffeomorphism represents a symmetry transformation. Furthermore, in the case that the vector field $V^\mu(x)$ vanishes on the boundary ∂U of the integration domain, the associated hypersurface integral vanishes and it is $\delta_V S = 0$. Hence, there are constellations where the variation $\delta_V S[\phi_i, g_{\mu\nu}]$ vanishes, but in general this variation is not zero.

Diffeomorphism Invariance of the Matter Action

Let us consider the matter field sector only and the case where the matter action $S_M[\phi_i, g_{\mu\nu}]$ is invariant under the combined internal transformations 22.84 and 22.85. This can be achieved if we consider an internal diffeomorphism being defined by a vector field $V^\mu(x)$ that vanishes on the boundary ∂U, so that we have the situation that

$$\delta_V S_M = 0 \tag{22.91}$$

is fulfilled. Based on first principles, the variation $\delta_V S_M[\phi_i, g_{\mu\nu}]$ with respect to the involved fields ϕ_i and $g_{\mu\nu}$ is calculated as

$$\delta_V S_M = \int_U \left(\frac{\delta S_M}{\delta \phi_i}\delta_V \phi_i + \frac{\delta S_M}{\delta g_{\mu\nu}}\delta_V g_{\mu\nu} \right) d^4x. \tag{22.92}$$

By inserting factors of $\sqrt{|g|}$ and the definition of the metric EMT 22.27, we obtain readily

$$\delta_V S_M = \int_U \left(\frac{1}{\sqrt{|g|}}\frac{\delta S_M}{\delta \phi_i}\delta_V \phi_i - \frac{1}{2}T^{\mu\nu}\delta_V g_{\mu\nu} \right)\sqrt{|g|}\, d^4x. \tag{22.93}$$

Now if the matter system is on-shell and the matter EL-Eqs $\delta S_M/\delta \phi_i = 0$ hold, we obtain for the variation

$$\delta_V S_M = -\frac{1}{2}\int_U T^{\mu\nu}\delta_V g_{\mu\nu}\sqrt{|g|}\, d^4x. \tag{22.94}$$

From the formula 22.85 for the variation of the metric and the symmetry of the MEMT, we get

$$\delta_V S_M = \epsilon \int_U T^{\mu\nu}\nabla_\mu V_\nu \sqrt{|g|}\, d^4x. \tag{22.95}$$

We can carry out an integration by parts and write the above equation as

$$\delta_V S_\mathrm{M} = \epsilon \int_U \nabla_\mu (T^{\mu\nu} V_\nu) \sqrt{|g|}\, \mathrm{d}^4 x - \epsilon \int_U (\nabla_\mu T^{\mu\nu}) V_\nu \sqrt{|g|}\, \mathrm{d}^4 x. \qquad (22.96)$$

The first integral in the above rhs can be converted into a hypersurface integral over the boundary ∂U. This surface integral vanishes, since the vector field $V^\mu(x)$ is assumed to be zero on the boundary. Therefore, only the second integral in the above rhs remains. Since the integration domain U and the vector field $V^\mu(x)$ are arbitrary, the vanishing of the variation of the matter action, condition 22.91, leads to the equation

$$\boxed{\nabla_\mu T^{\mu\nu} = 0} \qquad (22.97)$$

In other words, *the diffeomorphism invariance of the matter action is equivalent to the divergence equation for the metric energy-momentum tensor*. This is a beautiful result, which confirms the meaningfulness of the MEMT, as introduced in Section 22.2.

Diffeomorphism Invariance of the Gravitational Action

Let us now consider the gravitational sector with the action $S_\mathrm{G}[g_{\mu\nu}]$ for the metric field staying invariant under internal transformations of the type 22.85. I.e. we consider internal diffeomorphisms defined by a vector field $V^\mu(x)$ vanishing on the boundary such that the condition

$$\delta_V S_\mathrm{G} = 0 \qquad (22.98)$$

is fulfilled. The variation $\delta_V S_\mathrm{G}[g_{\mu\nu}]$ with respect to the metric $g_{\mu\nu}$ can be calculated for the general gravitational action as given in 22.77, and we have derived the result already in the last section,

$$\delta_V S_\mathrm{G} = -k \int_U G^{\mu\nu} \delta_V g_{\mu\nu} \sqrt{|g|}\, \mathrm{d}^4 x. \qquad (22.99)$$

Compare here with 22.72 and 22.82 and note that we have used $G_{\mu\nu}\delta g^{\mu\nu} = -G^{\mu\nu}\delta g_{\mu\nu}$. By inserting the formula 22.85 for the variation of the metric, we obtain

$$\delta_V S_\mathrm{G} = 2\epsilon k \int_U G^{\mu\nu} \nabla_\mu V_\nu \sqrt{|g|}\, \mathrm{d}^4 x. \qquad (22.100)$$

Once again we can carry out an integration by parts and discard the hypersurface integral, since the vector field of the diffeomorphism vanishes on the boundary. We conclude that the vanishing of the variation of the gravitational action, condition 22.98, leads to the equation

$$\boxed{\nabla_\mu G^{\mu\nu} = 0} \qquad (22.101)$$

This means that *the diffeomorphism invariance of the gravitational action implies the contracted Bianchi identity for the Einstein tensor*. This is another nice result, and despite the fact that the vanishing of the divergence of the Einstein tensor is already known, it is intriguing that this result can also be obtained from the Lagrangian formalism.

Internal Diffeomorphisms and Associated Currents

In our analysis of internal diffeomorphisms so far, we have looked at cases in which the action stays invariant, since we have used vector fields that vanish at the boundary of the manifold. We will drop this last assumption in the following and consider completely general internal diffeomorphisms that are generated by vector fields $V^\mu(x)$, which can be non-zero at the boundary and, as said before, do not necessarily express any particular symmetry of the manifold. It is a remarkable fact that even in this general case the gravitational action leads to locally conserved currents without any additional assumptions. In particular, these currents exist independently of the equations of motion and are even valid off-shell. The derivation of these currents will be carried out in the following. For the sake of clarity, we shall first consider the simpler case of a manifold without boundary and then treat the general case of a manifold that may have a boundary.

Noether Currents for a Manifold without Boundary

Let us consider a spacetime manifold without boundary and an internal diffeomorphism generated by an *arbitrary vector field* $V^\mu(x)$. The gravitational action $S_G\,[g_{\mu\nu}]$ of such a system is simply

$$S_G = k \int_U R\sqrt{|g|}\, d^4x, \qquad (22.102)$$

as the GHY boundary term does not contribute. The constant k has the value $-1/(16\pi G_N)$, as derived in 22.82. According to the formula 22.90, the variation $\delta_V S_G\,[g_{\mu\nu}]$ of the action can be written formally as

$$\delta_V S_G = -\epsilon k \int_U \nabla_\rho (V^\rho R)\sqrt{|g|}\, d^4x. \qquad (22.103)$$

Note that this variation is in general not zero. On the other hand, we have seen in the last section in equation 22.50 that the variation of the gravitational action is calculated as

$$\delta_V S_G = -k \int_U G^{\rho\sigma}\delta_V g_{\rho\sigma}\sqrt{|g|}\, d^4x + k \int_U \nabla_\sigma W^\sigma \sqrt{|g|}\, d^4x, \qquad (22.104)$$

with the vector field $W^\sigma(x)$ being given as in 22.53. In the first integral above, we can insert the formula 22.85 for the variation $\delta_V g_{\rho\sigma}(x)$ and carry out an integration by parts to obtain the expression

$$2\epsilon k \int_U \left[\nabla_\rho (G^{\rho\sigma}V_\sigma) - (\nabla_\rho G^{\rho\sigma})V_\sigma \right]\sqrt{|g|}\, d^4x. \qquad (22.105)$$

We use the fact that the divergence of the Einstein tensor vanishes identically, so that only the first term remains and we can write the resulting integral as

$$\epsilon k \int_U \nabla_\rho (2R^{\rho\sigma}V_\sigma - V^\rho R)\sqrt{|g|}\, d^4x. \qquad (22.106)$$

For the second integral containing the divergence $\nabla_\sigma W^\sigma$ we can relabel indices and write

$$\nabla_\sigma W^\sigma = \nabla_\rho\left[(g^{\rho\mu}g^{\sigma\nu} - g^{\rho\sigma}g^{\mu\nu})\nabla_\sigma \delta_V g_{\mu\nu}\right]. \qquad (22.107)$$

By employing once again the formula 22.85 for the variation of the metric, we obtain

$$\nabla_\sigma W^\sigma = -\epsilon\nabla_\rho\left[(g^{\rho\mu}g^{\sigma\nu} - g^{\rho\sigma}g^{\mu\nu})\nabla_\sigma (\nabla_\mu V_\nu + \nabla_\nu V_\mu)\right]. \qquad (22.108)$$

We collect the two partial results for the integrals and can write down the variation $\delta_V S_G\,[g_{\mu\nu}]$ of the action as

$$\delta_V S_G = \epsilon k \int_U \nabla_\rho (2R^{\rho\sigma}V_\sigma - V^\rho R)\sqrt{|g|}\, d^4x$$
$$- \epsilon k \int_U \nabla_\rho\left[(g^{\rho\mu}g^{\sigma\nu} - g^{\rho\sigma}g^{\mu\nu})\nabla_\sigma (\nabla_\mu V_\nu + \nabla_\nu V_\mu)\right]\sqrt{|g|}\, d^4x. \qquad (22.109)$$

Now if we compare the two results 22.103 and 22.109 for the variation, we infer that it must be identically

$$\int_U \nabla_\rho [2R^{\rho\sigma}V_\sigma - (g^{\rho\mu}g^{\sigma\nu} - g^{\rho\sigma}g^{\mu\nu})\nabla_\sigma(\nabla_\mu V_\nu + \nabla_\nu V_\mu)]\sqrt{|g|}\,\mathrm{d}^4 x = 0. \qquad (22.110)$$

Since the integration domain is arbitrary, we can deduce that the following divergence relation must hold identically:

$$\boxed{\nabla_\rho J^\rho = 0} \qquad (22.111)$$

Here we have used the definition*

$$J^\rho \equiv R^{\rho\sigma}V_\sigma - \frac{1}{2}(g^{\rho\mu}g^{\sigma\nu} - g^{\rho\sigma}g^{\mu\nu})\nabla_\sigma(\nabla_\mu V_\nu + \nabla_\nu V_\mu) \qquad (22.112)$$

for the *Noether current* $J^\rho(x)$. The divergence relation 22.111 is called the *Noether identity*. We want to emphasize here that this Noether identity is valid for every vector field $V^\mu(x)$ that defines a diffeomorphism and it does not depend on the validity of the equations of motion. The Noether identity is a local divergence law that applies to any spacetime manifold. The Noether current $J^\rho(x)$ can be written in a nice and compact way as

$$\boxed{J^\rho = \frac{1}{2}\nabla_\sigma(\nabla^\rho V^\sigma - \nabla^\sigma V^\rho)} \qquad (22.113)$$

Indeed, it is easy to see that this formula is equivalent to

$$J^\rho = [\nabla_\sigma, \nabla^\rho]\,V^\sigma + \nabla^\rho \nabla_\sigma V^\sigma - \frac{1}{2}\nabla_\sigma(\nabla^\sigma V^\rho + \nabla^\rho V^\sigma). \qquad (22.114)$$

The first term in the above rhs can be written as

$$[\nabla_\sigma, \nabla^\rho]\,V^\sigma = R^{\rho\sigma}V_\sigma. \qquad (22.115)$$

The second term and third term combined can be expressed as

$$\nabla^\rho \nabla_\sigma V^\sigma - \frac{1}{2}\nabla_\sigma(\nabla^\sigma V^\rho + \nabla^\rho V^\sigma) = -\frac{1}{2}(g^{\rho\mu}g^{\sigma\nu} - g^{\rho\sigma}g^{\mu\nu})\nabla_\sigma(\nabla_\mu V_\nu + \nabla_\nu V_\mu), \quad (22.116)$$

and thus the formula 22.113 is proven. Of course, we could have obtained the Noether identity if we had started with the expression 22.113 and had carried out the divergence operation. (*exercise 22.11*) However, our above derivation using the principles of the Lagrange formalism provides a deeper insight into the structure of general relativity.

The Noether current $J^\rho(x)$ can itself be written as a divergence-like quantity as

$$J^\rho = \nabla_\sigma J_K^{\rho\sigma} \qquad (22.117)$$

if one introduces the so-called *Komar superpotential* $J_K^{\rho\sigma}(x)$, which is the antisymmetric second-rank tensor field defined as

$$J_K^{\rho\sigma} \equiv \frac{1}{2}(\nabla^\rho V^\sigma - \nabla^\sigma V^\rho). \qquad (22.118)$$

---

*The dimensionful constant k is omitted in this definition, but must be taken into account in order to achieve physically correct dimensions.

The notion "superpotential" was introduced very early in the development of GR, but the specific superpotential 22.118 was only discovered in 1959 by *Arthur B. Komar* [48]. If a Noether current $J^\rho(x)$ is given, one can add a divergenceless vector field $Q^\rho(x)$ to it without spoiling the Noether identity. A possible way to construct such a divergenceless vector field $Q^\rho(x)$ is through the formula

$$Q^\rho = \nabla_\sigma A^{\rho\sigma}, \tag{22.119}$$

in which an antisymmetric second-rank tensor field $A^{\rho\sigma}(x)$ is employed. Then it is

$$\nabla_\rho Q^\rho = \nabla_\rho \nabla_\sigma A^{\rho\sigma} = 0, \tag{22.120}$$

due to the antisymmetry of the tensor field $A^{\rho\sigma}(x)$. Therefore, the vector field

$$J^\rho + Q^\rho = \nabla_\sigma (J_K^{\rho\sigma} + A^{\rho\sigma}) \tag{22.121}$$

also represents an admissible Noether current fulfilling the Noether identity 22.111.

Noether Currents for Manifolds with Symmetry

Let us examine what happens if the spacetime manifold has geometric symmetries, as expressed by the existence of isometries or conformal maps. In the case the spacetime manifold has an *isometry* with the *Killing vector field* $K^\mu(x)$, we can employ this vector field to define an internal diffeomorphism. Due to the basic Killing equation

$$\nabla_\mu K_\nu + \nabla_\nu K_\mu = 0, \tag{22.122}$$

compare with Section 20.1, the associated Noether current 22.113 is simplified to

$$\boxed{J^\rho = \nabla^\sigma \nabla^\rho K_\sigma} \tag{22.123}$$

In addition, by using the identity

$$\nabla^\sigma \nabla^\rho K_\sigma = R^{\rho\sigma} K_\sigma, \tag{22.124}$$

which holds for any KVF, the associated Noether current can be equivalently written as

$$\boxed{J^\rho = R^{\rho\sigma} K_\sigma} \tag{22.125}$$

If we have the special case of a d-dimensional *maximally symmetric spacetime*, the Noether current takes the simple form $J^\rho = d^{-1} R K^\rho$. If the spacetime is d-dimensional and admits a *conformal map* with the *conformal Killing vector field* $K^\mu(x)$, then the condition

$$\nabla_\mu K_\nu + \nabla_\nu K_\mu = \frac{2}{d} (\nabla_\alpha K^\alpha) g_{\mu\nu} \tag{22.126}$$

is satisfied, see Section 20.1. Then the associated Noether current 22.113 takes the form

$$J^\rho = \nabla^\sigma \nabla^\rho K_\sigma - \frac{1}{d} \nabla^\rho \nabla^\sigma K_\sigma. \tag{22.127}$$

By using in addition the contracted Ricci identity

$$\nabla^\sigma \nabla^\rho K_\sigma = R^{\rho\sigma} K_\sigma + \nabla^\rho \nabla^\sigma K_\sigma, \tag{22.128}$$

which holds for any vector field $K^\mu(x)$, the Noether current can be written as

$$J^\rho = R^{\rho\sigma} K_\sigma + \left(1 - \frac{1}{d}\right) \nabla^\rho \nabla^\sigma K_\sigma. \tag{22.129}$$

For $d = 1$, the above Noether current is the same as for isometries. For $d \geq 2$, Noether currents for conformal maps differ from Noether currents for proper isometries because the factor $\nabla^\rho \nabla^\sigma K_\sigma$ is zero for isometries but is in general non-zero in the conformal case.

Noether Currents for a Manifold with Boundary

Let us now proceed to the general case of a spacetime manifold, which may have a boundary, and consider an internal diffeomorphism generated by an arbitrary vector field $V^\mu(x)$. The complete gravitational action $S_G [g_{\mu\nu}]$ of this system is, see equation 22.75,

$$S_G = k \int_U \{ R + 2s \nabla_\sigma M^\sigma \} \sqrt{|g|}\, d^4 x, \tag{22.130}$$

where the constant $s = \pm 1$ and the vector field $M^\sigma(x)$, defined as

$$M^\sigma \equiv N^\sigma \nabla_\alpha N^\alpha, \tag{22.131}$$

are determined by the manifold boundary. Since we have already dealt with the Einstein-Hilbert term $S_{EH} [g_{\mu\nu}]$, we now only need to concentrate on the Gibbons-Hawking-York boundary term $S_{GHY} [g_{\mu\nu}]$. By using the general formula 22.90, the variation $\delta_V S_{GHY} [g_{\mu\nu}]$ can be written formally as

$$\delta_V S_{GHY} = -2\epsilon s \int_U \nabla_\rho (V^\rho \nabla_\sigma M^\sigma) \sqrt{|g|}\, d^4 x. \tag{22.132}$$

On the other hand, we note that we have for the integrand

$$-\nabla_\sigma (V^\sigma \nabla_\rho M^\rho) = \nabla_\rho \left[M^\sigma \nabla_\sigma V^\rho - \nabla_\sigma (M^\rho V^\sigma) \right]. \tag{22.133}$$

The proof is straightforward and left to the reader. (*exercise 22.12*) Consequently, the variation $\delta_V S_{GHY} [g_{\mu\nu}]$ can be expressed alternatively as

$$\delta_V S_{GHY} = 2\epsilon s \int_U \nabla_\rho \left[M^\sigma \nabla_\sigma V^\rho - \nabla_\sigma (M^\rho V^\sigma) \right] \sqrt{|g|}\, d^4 x. \tag{22.134}$$

By combining the two results 22.132 and 22.134, we infer that it must be identically

$$2s \int_U \nabla_\rho \left[\nabla_\sigma (M^\sigma V^\rho - M^\rho V^\sigma) \right] \sqrt{|g|}\, d^4 x = 0. \tag{22.135}$$

Since the domain of integration is arbitrary, it must be identically

$$\nabla_\rho J^\rho_{GHY} = 0, \tag{22.136}$$

where the *Noether current $J^\rho_{GHY}(x)$ for the GHY boundary term* is defined as

$$J^\rho_{GHY} \equiv 2s \nabla_\sigma (M^\sigma V^\rho - M^\rho V^\sigma). \tag{22.137}$$

We realize that the correct inclusion of the boundary of a manifold contributes a nontrivial part to the Noether current.

We are finally ready to combine our previous single results for the EH-action and the GHY-action for a general spacetime manifold. The *Noether current $J^\rho(x)$ for a general spacetime manifold* can be defined as

$$J^\rho \equiv \frac{1}{2} \nabla_\sigma (\nabla^\rho V^\sigma - \nabla^\sigma V^\rho) - 2s \nabla_\sigma (M^\rho V^\sigma - M^\sigma V^\rho) \tag{22.138}$$

and has a vanishing divergence, as expressed in 22.111. By inserting the vector field $M^\mu(x)$ explicitly as a function of the unit normal field $N^\mu(x)$, the Noether current is given as

$$J^\rho = \frac{1}{2} \nabla_\sigma (\nabla^\rho V^\sigma - \nabla^\sigma V^\rho) - 2s \nabla_\sigma \left[(\nabla_\alpha N^\alpha)(N^\rho V^\sigma - N^\sigma V^\rho) \right]. \tag{22.139}$$

We can again write the Noether current $J^\rho(x)$ as a covariant divergence, as in 22.117, where the *Komar superpotential* $J_K^{\rho\sigma}(x)$ is now given by

$$J_K^{\rho\sigma} \equiv \frac{1}{2}(\nabla^\rho V^\sigma - \nabla^\sigma V^\rho) - 2s(\nabla_\alpha N^\alpha)(N^\rho V^\sigma - N^\sigma V^\rho). \qquad (22.140)$$

We would like to point out again that the existence of a Noether current is ensured for every diffeomorphism of the spacetime manifold. This is a nice result in itself, but how can we interpret these Noether currents? We will show soon that in the case of a symmetric spacetime, the Noether currents arising from isometries allow a physical interpretation. For example, for a timelike symmetry, a suitably defined integral of the Noether current can be identified with the total mass-energy of the spacetime considered. In the next chapter, we will examine more broadly how we can construct such locally and globally conserved quantities in general relativity.

Further Reading

In this chapter we have presented the core results of the Lagrangian formalism for general relativity. However, we have left out the first-order Palatini formalism, which treats the metric and the connection within the gravitational action as independent fields. Readers interested in this approach can consult the books on general relativity mentioned in the previous chapter. Also, we have not covered the Hamiltonian formalism, which is rich and leads to nontrivial results in general relativity. For a brief introduction to this subject, one can consult Padmanabhan [62] or Wald [91]. In addition to identically conserved Noether currents, we have introduced the notion of superpotential. For a discussion of the quantities that have historically led to the Komar superpotential, see Davis [20].

23

Conservation Laws and Further Symmetries

In this chapter we construct locally and globally conserved quantities based on spacetime symmetries of general relativity. The existence of Killing vector fields ensures that one can define conserved quantities from geodesics and from the energy-momentum tensor of matter fields. We discuss the notion of energy of the gravitational field and the challenges this conception poses. By employing the Noether currents of diffeomorphisms generated by suitable Killing vectors, we are able to define the so-called Komar integrals, which can measure the total energy and angular momentum of a spacetime. In the final part, we examine how Weyl rescalings affect diverse physical theories and under which conditions Weyl rescaling invariance can be achieved.

23.1 Locally and Globally Conserved Quantities

Conserved Currents from Killing Vectors
The existence of Killing vector fields not only allows us to identify the geometric directions of symmetry in a given spacetime, but also enables us to construct quantities which are locally conserved. More specifically, if the spacetime possesses a *Killing vector field* $K^\mu(x)$ and if there exists a geodesic curve $z^\mu(\tau)$, with tangent vector $u^\mu(\tau)$, then the scalar product $u^\mu K_\mu$ is a *constant scalar along the geodesic*,

$$u^\alpha \nabla_\alpha (u^\mu K_\mu) = 0. \tag{23.1}$$

Similarly, if there is a *conformal Killing vector field* $K^\mu(x)$ and a null geodesic curve $z^\mu(\tau)$, then the scalar product constructed as above is a *constant scalar along the null geodesic*. (*exercise 23.1*)

Conserved Currents from Killing Tensors
It is possible that there are symmetries that are not manifest in the spacetime metric but nevertheless lead to conserved quantities. In other words, these symmetries do not arise from (conformal) Killing vectors. These symmetries are called *hidden*. Let us start with a simple example. Suppose a rank-two tensor field $K_{\mu\nu}(x)$ which is symmetric, $K_{\mu\nu} = K_{\nu\mu}$,

DOI: 10.1201/9781003087748-23

fulfills the equation

$$\nabla_\alpha K_{\mu\nu} + \nabla_\mu K_{\nu\alpha} + \nabla_\nu K_{\alpha\mu} = 0. \tag{23.2}$$

This condition is obviously a generalization of the Killing condition $\nabla_\alpha K_\mu + \nabla_\mu K_\alpha = 0$ to the case where one index more is present. Also, the above condition, due to the symmetry of the tensor field $K_{\mu\nu}(x)$, is equivalent to

$$\nabla_{(\alpha} K_{\mu\nu)} = 0. \tag{23.3}$$

If we have a geodesic curve $z^\mu(\tau)$ with the tangent vector $u^\mu(\tau)$, then the scalar quantity $u^\mu u^\nu K_{\mu\nu}$ will be a constant along the geodesic,

$$u^\alpha \nabla_\alpha (u^\mu u^\nu K_{\mu\nu}) = 0. \tag{23.4}$$

A tensor field $K_{\mu\nu}(x)$ with this property is called a *Killing tensor field*. This Killing tensor field concept can be generalized to more indices. Suppose there is a totally symmetric tensor field $K_{\nu_1\ldots\nu_n}(x)$ with n indices, $K_{\nu_1\ldots\nu_n} = K_{(\nu_1\ldots\nu_n)}$, satisfying the condition

$$\nabla_{(\alpha} K_{\nu_1\ldots\nu_n)} = 0. \tag{23.5}$$

If, additionally, there exists a geodesic curve $z^\mu(\tau)$ with tangent vector $u^\mu(\tau)$, then the scalar contraction $u^{\nu_1}\cdots u^{\nu_n} K_{\nu_1\ldots\nu_n}$ is a constant quantity along the geodesic,

$$u^\alpha \nabla_\alpha (u^{\nu_1}\cdots u^{\nu_n} K_{\nu_1\ldots\nu_n}) = 0. \tag{23.6}$$

The proof is a straightforward application of the assumed properties. (*exercise 23.2*) We would like to note that, unlike Killing vectors, Killing tensors in general do not form Lie algebras and are not related to Noether symmetries.

Energy-Momentum Tensor and Conservation Laws

We have repeatedly used the EMT of the matter fields $T^{\mu\nu}(x)$ in all its versions in order to define locally and globally conserved quantities. The two basic defining relations of the EMT are

$$\nabla_\mu T^{\mu\nu} = 0 \quad \text{and} \quad T^{\mu\nu} = T^{\nu\mu}. \tag{23.7}$$

The above covariant divergence equation generalizes the continuity equation $\partial_\mu T^{\mu\nu} = 0$ of flat space. The divergence equation in flat space can be interpreted as the local conservation of energy and momentum and it leads also to globally conserved energy and momentum of the total matter system, compare with Sections 16.2 and 16.3. Unfortunately, this useful introduction of global quantities can not be maintained, in general, for the covariant divergence equation $\nabla_\mu T^{\mu\nu} = 0$. In order to see why this is the case, let us calculate the divergence term $\nabla_\mu T^\mu{}_\nu$ explicitly. We start with the general rule for the covariant derivative of a $(1,1)$-tensor and write

$$\nabla_\mu T^\mu{}_\nu = \partial_\mu T^\mu{}_\nu + \Gamma^\mu_{\alpha\mu} T^\alpha{}_\nu - \Gamma^\alpha_{\nu\mu} T^\mu{}_\alpha. \tag{23.8}$$

By using the formula for the contracted connection coefficients $\Gamma^\mu_{\alpha\mu}(x)$ and the basic formula 18.78, we obtain

$$\nabla_\mu T^\mu{}_\nu = \frac{1}{\sqrt{|g|}} \partial_\mu(\sqrt{|g|}\, T^\mu{}_\nu) - \frac{1}{2}(\partial_\nu g_{\mu\sigma}) T^{\mu\sigma}. \tag{23.9}$$

Consequently, the covariant divergence equation $\nabla_\mu T^{\mu\nu} = 0$ is equivalent to

$$\frac{1}{\sqrt{|g|}} \partial_\mu(\sqrt{|g|}\, T^\mu{}_\nu) = \frac{1}{2}(\partial_\nu g_{\mu\sigma}) T^{\mu\sigma}. \tag{23.10}$$

This equation can be integrated over a 4-dimensional region U of spacetime possessing a 3-dimensional closed boundary $\Sigma = \partial U$. We choose the hypersurface Σ to be like $\Sigma_a \cup \Sigma_b \cup \Sigma_\infty$ comprised of two spacelike hypersurface parts, Σ_a and Σ_b, and a timelike hypersurface part Σ_∞ lying at infinity, see sketch 23.1. A possible (although not the only) option for the spacelike hypersurfaces is to require $x^0 = \mathrm{const}$. We assume that the EMT vanishes

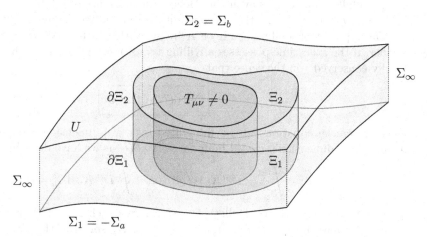

Figure 23.1: Four-dimensional integration domain U bounded by two spacelike hypersurfaces Σ_1 and Σ_2 and a timelike hypersurface Σ_∞

at spatial infinity. A more restrictive requirement would be that the matter sources are confined to a finite spatial volume for which $T_{\mu\nu} \neq 0$ holds, as depicted in the sketch. For the lhs of 23.10 we obtain, according to the theorem of Gauss,

$$\int_U \partial_\mu(\sqrt{|g|}\, T^\mu{}_\nu)\, \mathrm{d}^4 x = \oint_\Sigma \sqrt{|g|}\, T^\mu{}_\nu\, \mathrm{d}\sigma_\mu = H_\nu(\Sigma_2) - H_\nu(\Sigma_1), \qquad (23.11)$$

where the integral quantity $H_\nu(\Sigma_i)$ is defined as

$$H_\nu(\Sigma_i) \equiv \int_{\Sigma_i} T^\mu{}_\nu\, \mathrm{d}\Sigma_\mu \qquad (23.12)$$

and Σ_i, $i = 1, 2$, represents the two spacelike hypersurfaces. Here we have used the equation

$$\mathrm{d}\Sigma_\mu = \sqrt{|g|}\, \mathrm{d}\sigma_\mu = \sqrt{|g|}\, n_\mu\, \mathrm{d}y^1 \wedge \mathrm{d}y^2 \wedge \mathrm{d}y^3, \qquad (23.13)$$

with (y^1, y^2, y^3) denoting coordinates on the hypersurfaces, see appendix B.8. As depicted in the sketch, the hypersurface Σ_1 is the same as Σ_a but with flipped orientation, while the hypersurface Σ_2 is exactly the same as Σ_b. For the rhs of 23.10, the resulting volume integral is in general non-zero. This is the reason why the quantity $H_\nu(\Sigma_i)$ is in general not a constant for different hypersurfaces. However, if it is $\partial_\rho g_{\mu\nu} = 0$, or equivalently, if the connection coefficients vanish, then the integral of the rhs of 23.10 vanishes and the quantity $H_\nu(\Sigma_i)$ is constant over any spacelike hypersurface Σ_i that encloses all matter sources. If for instance, $g_{\mu\nu} = \eta_{\mu\nu} = \mathrm{const}$ holds, then the condition $\partial_\rho g_{\mu\nu} = 0$ is fulfilled. This is exactly what is implemented in the highly symmetric Minkowski space with its ten Poincaré symmetries. Now the important observation is that the condition $\partial_\rho g_{\mu\nu} = 0$ represents in a more general fashion a symmetry requirement for the spacetime. Indeed, suppose there is a Killing vector field $K^\mu(x)$ of the spacetime, with $\mathcal{L}_K g_{\mu\nu} = 0$. Then we can align the local coordinates in such a way that the ρth coordinate direction is along the KVF, and thus the identification $\mathcal{L}_K g_{\mu\nu} = K^\rho \partial_\rho g_{\mu\nu}$ holds. Note that there is no summation over ρ here, see also the comments at the end of Section 3.3. This means that the requirement $\partial_\rho g_{\mu\nu} = 0$ can be fulfilled if a KVF is available. Hence, the existence of a KVF ensures that the EMT leads to a quantity $H_\nu(\Sigma_i)$ that is constant over any spacelike hypersurface Σ_i.

The approach of combining a KVF $K^\mu(x)$ with the matter EMT $T^{\mu\nu}(x)$ can be developed further. If the spacetime possesses a Killing vector field $K^\mu(x)$, then the quantity $T^{\mu\nu} K_\nu$ is locally conserved*, in the sense that

$$\boxed{\nabla_\mu (T^{\mu\nu} K_\nu) = 0} \tag{23.14}$$

This divergence equation can be integrated over a 4-dimensional region U of spacetime having the 3-dimensional closed boundary $\Sigma = \partial U$, as we have done before. We obtain

$$\int_U \nabla_\mu (T^{\mu\nu} K_\nu) \sqrt{|g|}\, \mathrm{d}^4 x = \oint_\Sigma T^{\mu\nu} K_\nu\, \mathrm{d}\Sigma_\mu = 0. \tag{23.15}$$

We can again choose the constellation as shown in sketch 23.1. The spacelike hypersurfaces Σ_1 and Σ_2 have future-oriented normal vectors. By splitting the integral over the single hypersurface parts of the boundary, we obtain

$$\int_{\Sigma_1} T^{\mu\nu} K_\nu\, \mathrm{d}\Sigma_\mu = \int_{\Sigma_2} T^{\mu\nu} K_\nu\, \mathrm{d}\Sigma_\mu. \tag{23.16}$$

This equation expresses that the scalar integral quantity $G(\Sigma_i)$ which is defined as

$$\boxed{G(\Sigma_i) \equiv \int_{\Sigma_i} T^{\mu\nu} K_\nu\, \mathrm{d}\Sigma_\mu} \tag{23.17}$$

attains the same value independently of the specific spacelike hypersurface Σ_i on which it is calculated. If the Killing vector $K^\mu(x)$ is *timelike* throughout spacetime, then the integral quantity $G(\Sigma_i)$ can be interpreted as the *total energy of the matter fields* contained within the spacelike hypersurface Σ_i. Looking at the limit case of Minkowski space, by choosing the KVF to be the one for spacetime translations, $K^\mu = a^\mu = \text{const}$, one recovers the conserved integrals which we have derived in Section 16.3. The KVF $K^\mu = \delta_0^\mu$ yields the total energy, while the KVF $K^\mu = \delta_k^\mu$, $k = 1, 2, 3$, leads to the total 3-momentum of the matter fields.

Conserved Currents from Other Tensors

In the above analysis we have used only a subset of the properties of the EMT. It is obvious that by using the Einstein tensor $G^{\mu\nu}(x)$ and a KVF $K^\mu(x)$ of the spacetime, we can construct the current $G^{\mu\nu} K_\nu$, which is locally conserved too,

$$\boxed{\nabla_\mu (G^{\mu\nu} K_\nu) = 0} \tag{23.18}$$

due to the symmetry and the divergence formula of the Einstein tensor. Equivalently, the Ricci tensor $R^{\mu\nu}(x)$ in combination with a KVF $K^\mu(x)$ leads to the current $R^{\mu\nu} K_\nu$, which is also locally conserved,

$$\boxed{\nabla_\mu (R^{\mu\nu} K_\nu) = 0} \tag{23.19}$$

---

*Instead of considering the EMT one could even start with any symmetric divergence-free second-rank tensor.

since the identities $\nabla_\mu K_\nu + \nabla_\nu K_\mu = 0$ and $K^\mu \nabla_\mu R = 0$ hold for any KVF. We recognize the current $J^\mu(x)$ given by

$$J^\mu = R^{\mu\nu} K_\nu = \nabla^\nu \nabla^\mu K_\nu \tag{23.20}$$

as the Noether current associated with the diffeomorphism generated by the KVF, see 22.123 and 22.125. The crucial point is that the existence of suitable KVFs leads to globally conserved quantities that can be interpreted physically, the so-called Komar integrals, which we introduce in Section 23.3. Since we are employing the Ricci tensor in the above Noether current, we can approach the question of how we can define energy of spacetime itself. One can, in fact, conceive local (i.e. infinitesimal), quasi-local and global notions of energy, and we discuss these approaches in the next two sections.

23.2   On the Energy of Spacetime

Newtonian Gravitational Energy

Before we discuss some of the questions regarding the energy of spacetime within general relativity, let us look at the case of Newtonian gravity. The Newtonian gravitational potential $\Phi(x)$ satisfies the Poisson equation

$$\nabla^2 \Phi(x) = 4\pi G_N \rho(x), \tag{23.21}$$

as seen in Section 4.4. Let us assume that the spatial region in which matter exists, $\rho(x) \neq 0$, is a compact 3-dimensional volume. We can integrate both sides of the above equation over a spatial volume, say V, enclosing entirely the region with matter and apply the Gauss divergence theorem. This yields the formula

$$M = \frac{1}{4\pi G_N} \oint_S \nabla\Phi \cdot \mathrm{d}S \tag{23.22}$$

for the *total mass* M contained in the spatial volume V. The integral is evaluated over the closed 2-dimensional surface $S = \partial V$ as the boundary of the volume V. As long as all matter is included, we can choose a different spatial volume, and the above integral yields the same value for the total mass. In other words, for all 2-dimensional surfaces enclosing the region with $\rho(x) \neq 0$, the integral is a constant. Since the shape of the surface is irrelevant, we can choose to integrate over a 2-sphere $S^2(r)$ with a sufficiently large radius r. We can even express the total mass contained in spacetime as the limit

$$M = \lim_{r \to \infty} \frac{1}{4\pi G_N} \oint_{S^2(r)} \nabla\Phi \cdot \mathrm{d}S \tag{23.23}$$

by expanding the sphere to infinity and thus covering the whole 3-space. The calculation using this kind of limit will be put into practice also in the general relativistic case.

Now that we have expressed the total mass-energy contained in 3-space within the Newtonian framework, what can we say about the energy density of the Newtonian gravitational field? We recall that we have discussed the energy density of the electromagnetic field in Section 16.2. The 00-component $T_{EM}^{00}(x)$ of the EMT of Maxwell theory represents the energy density of the electrostatic scalar field $\varphi(x)$ in the case that no magnetic fields are present. Hence, in complete analogy to electrostatics, we can write down the *energy density of the Newtonian gravitational field* $\mathcal{E}(x)$ as

$$\mathcal{E} = -\frac{1}{8\pi G_N}(\nabla\Phi)^2. \tag{23.24}$$

The minus sign appears because the gravitational force is always attractive. This Newtonian energy density can alternatively be derived directly as the Hamiltonian density of the gravitational Lagrangian at the end of Section 15.3. The Newtonian gravitational energy density $\mathcal{E}(\boldsymbol{x})$ has always a negative value and the same holds for its integral over any spatial volume. On the other hand, the mass-energy density $\rho(\boldsymbol{x})$ has always a positive value for ordinary matter. In Newton's theory the sign of the combined total energy of the material bodies (mass energy) and the gravitational field (potential energy) is not constrained.

In general relativity, the situation is different, since the total energy of a gravitational system made up of matter fields and the metric field always remains positive under certain reasonable conditions. This is known as the *positive energy theorem*. More precisely, one considers a gravitational system that corresponds to an asymptotically flat spacetime. It is also assumed that for the matter EMT $T^{\mu\nu}(x)$ and for any timelike KVF $K^\mu(x)$ of the spacetime, the relation $T_{\mu\nu}K^\mu K^\nu \geq 0$ holds and that the vector field $T^{\mu\nu}K_\nu$ is either timelike or null. Under these conditions, the total energy of the gravitational system is nonnegative. The proof of this theorem goes beyond our scope and can be found in Straumann [84]. The positive energy theorem has an important physical consequence. Assuming that the energy carried by gravitational radiation is always positive, the amount of energy that can be retrieved from a gravitational system is finite.

The Question of Energy Density of the Metric Field

By following classical conceptions, we would like to define the energy density of the gravitational field within general relativity as the energy value per unit volume. In Newton's theory, this concept can easily be implemented because there is a fixed background structure, the absolute 3-space, to which the gravitational energy value can refer. In general relativity, however, this referencing is problematic, since the basic metric must provide both the energy measure and the volume measure. One possible way to proceed is to introduce an additional background structure, which is independent of the metric and to which the gravitational energy value can refer. This structure can be a special coordinate system, a background metric or a background connection. The decomposition of the spacetime metric $g_{\mu\nu}(x)$ in the form

$$\sqrt{|g|}\, g^{\mu\nu} = \sqrt{|\gamma|}\, (\gamma^{\mu\nu} + h^{\mu\nu}) \tag{23.25}$$

actually corresponds to the so-called field-theoretic approach to gravity, which we can only briefly mention here. The spacetime metric $g_{\mu\nu}(x)$ is decomposed into a flat background metric $\gamma_{\mu\nu}(x)$ (the rigid Minkowski metric $\eta_{\mu\nu}$ being only a special case) and to the tensor field $h_{\mu\nu}(x)$ representing the metric disturbance on top of the flat background metric. The theories employing such a background metric lead to so-called *pseudo energy-momentum tensors* that describe the gravitational field. As the name says, these objects are not absolute tensor fields and their value depends crucially on the choice of the background metric. Pseudo-tensor fields transform like absolute tensor fields only under global linear (Lorentz) transformations of spacetime. Over the years since the inception of general relativity, a variety of pseudo energy-momentum tensors for the gravitational field have been found. Albert Einstein himself had conceived such an object (in 1915), and afterwards many more pseudo EMTs were discovered, known by the names of Papapetrou (1948), Landau-Lifshitz (1951), Bergmann-Thomson (1953), Møller (1958), and Weinberg (1972), to name some important examples. Details on the Einstein pseudo EMT can be found in Anderson [1] and de Felice, Clarke [21]. The pseudo EMT of Landau and Lifshitz has the important property of being symmetric and has been extensively discussed in the literature, see for example the books of Landau, Lifshitz [51] or Padmanabhan [62]. Today we know that there is actually an infinite number of possible pseudo EMTs for the gravitational field. The useful property of a pseudo EMT of the gravitational field, denoted $t^{\mu\nu}(x)$ in the following, is that in combination with the matter EMT $T^{\mu\nu}(x)$ a local divergence law of the form

$$\partial_\mu\left[|g|^p\left(T_\nu{}^\mu + t_\nu{}^\mu\right)\right] = 0 \qquad (23.26)$$

can be fulfilled. In the case the value of the exponent is $p = 1/2$ (like for the Einstein pseudo-tensor), we obtain a global conservation law very similar to the known law from flat space field theory, see Section 16.3. More precisely, by integrating the above divergence relation over a suited spacetime domain, we obtain that the integral quantity

$$P_\nu(\Sigma) \equiv \int_\Sigma (T_\nu{}^\mu + t_\nu{}^\mu)\,\mathrm{d}\Sigma_\mu \qquad (23.27)$$

has a constant value over any spacelike hypersurface Σ within this domain. The combination of the matter EMT with the gravitational pseudo EMT has provided a conserved total energy-momentum of the gravitational system. We would like to remark that apart from the pseudo EMTs mentioned above, Babak and Grishchuk [5] have found an EMT of the gravitational field that transforms as an absolute tensor field and possesses also all other properties that are desired from a proper EMT. Still its definition depends on a choice of the background metric, so its physical interpretation as of today remains an open question.

On the one hand, the discovery of pseudo-EMTs is undoubtedly a step forward in understanding local energy and momentum of the gravitational field. On the other hand, general relativity itself is contrary to any fixed background spacetime. Thus, we should ask ourselves if the very notion of an EMT or even just the energy density is truly feasible within GR. Consider the following argument: the value of the connection coefficients (which are determined by the first-order derivatives of the metric) for any given spacetime point can be made zero by a suited coordinate transformation, which represents a basic freedom we have in GR. Locally, the connection can be "transformed away". Therefore, it is questionable if we can assign unambiguously a gravitational energy density for each spacetime point. We should also recall that the connection coefficients are quantities which connect points of different tangent spaces. The value of the connection coefficients can be written down for each spacetime point but their definition requires differentiation, which needs to capture the behavior around a point. In this sense, we are guided to consider gravitational energy within GR as a non-local concept. Of course, it is difficult to claim that something does not exist when there is no rigorous proof of non-existence or when not all avenues have been considered. At the moment, however, what we can do at best is to define gravitational energy only *quasi-locally*, i.e. for a (small) finite volume. To date, a number of quasi-local energy measures have been conceived whose validity and usefulness depend heavily on the context. For an overview of this subject, one should consider Szabados [86]. After the above comments, it may be seen as a relief to say that at least globally, for isolated gravitational systems, one can define unambiguously global measures for energy and other physical quantities. This is described in the following.

23.3 Komar Integrals

General Komar Integral

We saw in Section 16.3 that the Poincaré symmetries of Minkowski spacetime lead to conserved global quantities of energy, momentum, angular momentum, and center of energy in a field theory. A general relativistic curved spacetime has, in general, no such symmetries and one cannot expect to be able to define globally conserved quantities as in the Minkowski case. However, if one considers spacetimes that have geometric symmetries expressed by Killing vector fields, then it is indeed possible to define globally conserved quantities, like energy or angular momentum. So let us consider such a spacetime $(\mathcal{M}, g_{\mu\nu})$ possessing one or more symmetries with corresponding KVFs $K^\mu(x)$. We view a 4-dimensional region $U \subseteq \mathcal{M}$ with a closed boundary ∂U. The boundary, as a 3-dimensional hypersurface, is assumed to be decomposable as $\partial U = \Sigma_a \cup \Sigma_b \cup \Sigma_\infty$, with Σ_a and Σ_b being spacelike hypersurfaces and Σ_∞ being a timelike hypersurface with the topology of a cylinder, see again sketch 23.1. The timelike unit normal vectors $N_a^\mu(x)$ and $N_b^\mu(x)$ are oriented outward of U. If we want to consider future-oriented timelike unit normal vectors throughout, we can define the hypersurfaces Σ_1 and Σ_2 by means of the timelike unit normal vectors $N_1^\mu = -N_a^\mu$ and $N_2^\mu = N_b^\mu$. For the matter fields we assume that they are isolated within a cylindrical world tube, i.e. the matter fields are supposed to have a finite extend in 3-space. Let us consider a KVF $K^\mu(x)$ and its associated Noether current $J^\mu(x)$, which is given as

$$J^\mu = R^{\mu\nu} K_\nu = \nabla_\nu \nabla^\mu K^\nu = \nabla_\nu J_K^{\mu\nu}, \qquad (23.28)$$

where $J_K^{\mu\nu} = \nabla^\mu K^\nu$ is the corresponding antisymmetric Komar superpotential. The Noether current fulfills identically the local conservation equation

$$\nabla_\mu J^\mu = \nabla_\mu \nabla_\nu \nabla^\mu K^\nu = \nabla_\mu \nabla_\nu J_K^{\mu\nu} = 0. \qquad (23.29)$$

We now integrate this equation over the spacetime region U and apply the Gauss integral formula 18.105 to obtain

$$\int_U \nabla_\mu J^\mu \sqrt{|g|}\, \mathrm{d}^4 x = \oint_{\partial U} J^\mu\, \mathrm{d}\Sigma_\mu = 0. \qquad (23.30)$$

By employing the decomposition $\partial U = \Sigma_a \cup \Sigma_b \cup \Sigma_\infty$ of the boundary, the above equation can be written as

$$\int_{\Sigma_2} J^\mu\, \mathrm{d}\Sigma_\mu - \int_{\Sigma_1} J^\mu\, \mathrm{d}\Sigma_\mu + \int_{\Sigma_\infty} J^\mu\, \mathrm{d}\Sigma_\mu = 0. \qquad (23.31)$$

The integral over the timelike hypersurface Σ_∞ is calculated as

$$\int_{\Sigma_\infty} J^\mu\, \mathrm{d}\Sigma_\mu = \int_{\Sigma_\infty} R^{\mu\nu} K_\nu\, \mathrm{d}\Sigma_\mu = 8\pi G_N \int_{\Sigma_\infty} \left(T^{\mu\nu} - \frac{1}{2} T g^{\mu\nu} \right) K_\nu\, \mathrm{d}\Sigma_\mu, \qquad (23.32)$$

where in the last step we used the Einstein equations. Since $T^{\mu\nu} = 0$ holds throughout Σ_∞, the above integral vanishes and we obtain that the 3-dimensional hypersurface integral

$$Q(\Sigma_i) \equiv \int_{\Sigma_i} J^\mu\, \mathrm{d}\Sigma_\mu = \int_{\Sigma_i} \nabla_\nu \nabla^\mu K^\nu\, \mathrm{d}\Sigma_\mu \qquad (23.33)$$

is a constant quantity over any hypersurface Σ_i, $i = 1, 2$, as specified above. As long as a chosen hypersurface Σ_i includes all points x with $T^{\mu\nu}(x) \neq 0$, the integral $Q(\Sigma_i)$ yields the same value.

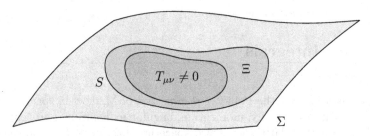

Figure 23.2: A spacelike integration domain Ξ enclosing all matter sources

We choose now a 3-dimensional spacelike hypersurface piece $\Xi \subset \Sigma$ *possessing a closed boundary* $\partial \Xi$. The boundary, denoted $S \equiv \partial \Xi$, is simply a 2-dimensional closed spatial surface within 3-space, see sketch 23.2. According to the Stokes integral formula 18.114 for an antisymmetric tensor, which in the present case is the Komar superpotential, the integral $Q(\Xi)$ of the Noether current over the spacelike hypersurface piece Ξ can be expressed as

$$Q(\Xi) = \int_\Xi \nabla_\nu \nabla^\mu K^\nu \, \mathrm{d}\Sigma_\mu = \frac{1}{2} \oint_{\partial\Xi} \nabla^\mu K^\nu \, \mathrm{d}\Sigma_{\mu\nu}. \tag{23.34}$$

We define the *Komar integral*, denoted $Q_{\mathrm{K}}(K^\mu, S)$, for the KVF $K^\mu(x)$ and the 2-dimensional closed spatial surface S as

$$\boxed{Q_{\mathrm{K}}(K^\mu, S) \equiv \frac{\alpha}{2} \oint_S \nabla^\mu K^\nu \, \mathrm{d}\Sigma_{\mu\nu}} \tag{23.35}$$

The real constant α can be determined on physical grounds and we will do so below by considering the Schwarzschild spacetime. The Komar integral 23.35 is very useful because its value is independent of the 2-surface S, as long as this surface engulfs all matter fields. To demonstrate this, let us choose a second spacelike hypersurface $\Xi' \equiv \Xi \cup \Delta\Xi$ engulfing all matter $T^{\mu\nu} \neq 0$ and the entire hypersurface Ξ, see the sketch 23.3. The respective closed

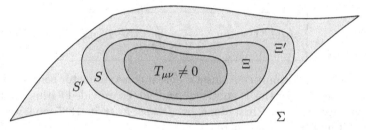

Figure 23.3: Two spacelike integration domains Ξ and Ξ' enclosing all matter sources

boundaries are denoted by $S \equiv \partial\Xi$ and $S' \equiv \partial(\Xi \cup \Delta\Xi)$. Then, the Komar integral over the 2-surface S' can be calculated as

$$Q_{\mathrm{K}}(K^\mu, S') = \alpha \int_{\Xi \cup \Delta\Xi} \nabla_\nu \nabla^\mu K^\nu \, \mathrm{d}\Sigma_\mu. \tag{23.36}$$

This integral can be split into a contribution coming from Ξ and a contribution coming from $\Delta\Xi$. By using again the Einstein equations, we obtain

$$Q_{\mathrm{K}}(K^\mu, S') = \alpha \int_\Xi \nabla_\nu \nabla^\mu K^\nu \, \mathrm{d}\Sigma_\mu + 8\pi G_{\mathrm{N}} \alpha \int_{\Delta\Xi} \left(T^{\mu\nu} - \frac{1}{2} T g^{\mu\nu} \right) K_\nu \, \mathrm{d}\Sigma_\mu. \tag{23.37}$$

The second integral over the 3-dimensional region $\Delta\Xi$ vanishes, since $T^{\mu\nu} = 0$ holds within $\Delta\Xi$, and we obtain the constancy of the Komar integral,

$$Q_{\mathrm{K}}(K^\mu, S') = Q_{\mathrm{K}}(K^\mu, S), \tag{23.38}$$

as desired. *The Komar integral $Q_{\mathrm{K}}(K^\mu, S)$ represents a geometric quantity that is invariant for different 2-surfaces S if these surfaces engulf all matter fields inside them.*

Komar Energy and Angular Momentum
Depending on the nature of the Killing symmetry present, the associated Komar integral has a specific meaning. For a timelike KVF $K^\mu(x)$, the Komar integral $Q_{\mathrm{K}}(K^\mu, S)$ can be identified with the *total energy of the gravitational system*. For a KVF $L^\mu(x)$ corresponding

to a spatial axial symmetry, the Komar integral $Q_K(L^\mu, S)$ can be identified with the *total angular momentum of the gravitational system*. Especially for an asymptotically flat spacetime, for which the requirement of isolated matter fields is indeed fulfilled, we can view Komar integrals as global quantities describing the entire spacetime. In the asymptotic case, we can choose a spatial 2-sphere $S^2(r)$ as the domain of integration and take the limit $r \to \infty$ of infinite radius. For a *stationary and asymptotically flat spacetime* having the timelike KVF $K^\mu(x)$ with the normalization $K^2 = 1$, the *Komar energy of the gravitational system*, $E_K(K^\mu)$, is defined as

$$E_K(K^\mu) \equiv \lim_{r \to \infty} \frac{-1}{8\pi G_N} \oint_{S^2(r)} \nabla^\mu K^\nu \, d\Sigma_{\mu\nu} \qquad (23.39)$$

The integral 23.39 is also called the *Komar mass* and was introduced by *Arthur B. Komar* in 1959, see [48]. The constant α in the general Komar integral has been fixed now to the physically correct value and we will justify this below by using the Schwarzschild case. Further, for an *axisymmetric and asymptotically flat spacetime* having the axisymmetric KVF $L^\mu(x)$, which is normalized such that its parameter range is $[0, 2\pi)$, the *Komar angular momentum of the gravitational system*, $J_K(L^\mu)$, is defined as

$$J_K(L^\mu) \equiv \lim_{r \to \infty} \frac{1}{16\pi G_N} \oint_{S^2(r)} \nabla^\mu L^\nu \, d\Sigma_{\mu\nu} \qquad (23.40)$$

The angular momentum formula contains a different constant factor compared to the energy formula. The explanation for this is nontrivial and can be provided within the field-theoretic approach to gravity. This was first done by *Joseph Katz* in 1985 and we must content ourselves with citing his work [47] here.

Komar Energy of Schwarzschild Spacetime

Let us calculate here the Komar energy for the Schwarzschild spacetime, see also Section 21.3. The Schwarzschild spacetime has among other KVFs also a timelike one describing the staticity of this spacetime. We have to calculate the contraction $\nabla^\mu K^\nu \, d\Sigma_{\mu\nu}$ using the constant KVF $K^\nu = \delta^\nu_t$ in (t, r, θ, φ)-coordinates. The hypersurface element is given by the formula

$$d\Sigma_{\mu\nu} = \frac{1}{2} \sqrt{|g|} \, \varepsilon_{\mu\nu\rho\sigma} \, dx^\rho \wedge dx^\sigma, \qquad (23.41)$$

where the square root of the determinant of the metric is $\sqrt{|g(t, r, \theta, \varphi)|} = r^2 \sin\theta$, while the exterior product is simply

$$dx^\rho \wedge dx^\sigma = dx^\rho \otimes dx^\sigma - dx^\sigma \otimes dx^\rho. \qquad (23.42)$$

Only the contraction $\nabla^\mu K^t \, d\Sigma_{\mu t}$ is non-zero. By carrying out the summation over the index $\mu = t, r, \theta, \varphi$, we immediately see that the term $\nabla^t K^t \, d\Sigma_{tt}$ vanishes. The remaining terms in the contraction $\nabla^\mu K^t \, d\Sigma_{\mu t}$ are proportional to $\nabla^r K^t$, $\nabla^\theta K^t$ and $\nabla^\varphi K^t$. The two factors $\nabla^\theta K^t$ and $\nabla^\varphi K^t$ vanish, which we can see by carrying out explicitly the covariant derivatives. It is

$$\nabla^\theta K^t = g^{\theta\mu} \nabla_\mu K^t = -\frac{1}{r^2}(\partial_\theta K^t + \Gamma^t_{\nu\theta} K^\nu) = 0, \qquad (23.43)$$

since $\Gamma^t_{t\theta} = 0$, and also

$$\nabla^\varphi K^t = g^{\varphi\mu} \nabla_\mu K^t = -\frac{1}{r^2 \sin^2\theta}(\partial_\varphi K^t + \Gamma^t_{\nu\varphi} K^\nu) = 0, \qquad (23.44)$$

because of $\Gamma^t_{t\varphi} = 0$. Therefore, we obtain for the contraction

$$\nabla^\mu K^\nu \, \mathrm{d}\Sigma_{\mu\nu} = r^2 \sin\theta \, \nabla^r K^t \, \varepsilon_{rt\theta\varphi} \, \mathrm{d}\theta \wedge \mathrm{d}\varphi. \tag{23.45}$$

The covariant derivative $\nabla^r K^t$ is calculated as

$$\nabla^r K^t = g^{r\mu} \nabla_\mu K^t = -\left(1 - \frac{2G_\mathrm{N} M}{r}\right) \Gamma^t_{tr}, \tag{23.46}$$

since it is $K^t = 1$. From the results of Section 21.3, we see that the coefficient Γ^t_{tr} is

$$\Gamma^t_{tr} = \frac{G_\mathrm{N} M}{r^2} \left(1 - \frac{2G_\mathrm{N} M}{r}\right)^{-1}, \tag{23.47}$$

and so we obtain

$$\nabla^r K^t = -\frac{G_\mathrm{N} M}{r^2}. \tag{23.48}$$

Taking into account that $\varepsilon_{rt\theta\varphi} = -\varepsilon_{tr\theta\varphi} = 1$, see appendix B.1, we obtain the result

$$\nabla^\mu K^\nu \, \mathrm{d}\Sigma_{\mu\nu} = -2G_\mathrm{N} M \sin\theta \, \mathrm{d}\theta \, \mathrm{d}\varphi. \tag{23.49}$$

The Komar energy of the Schwarzschild solution is thus

$$E_\mathrm{K}(K^\mu) = \frac{\alpha}{2} \oint_{S^2} (-2G_\mathrm{N} M) \sin\theta \, \mathrm{d}\theta \, \mathrm{d}\varphi, \tag{23.50}$$

where the limit $r \to \infty$ is omitted, since the integral does not depend on the radius of the 2-sphere. The integral over the angles is elementary and we obtain the result

$$E_\mathrm{K}(K^\mu) = -4\pi G_\mathrm{N} \alpha M. \tag{23.51}$$

Now, if we want the Schwarzschild spacetime to have the total energy M, then it must be $\alpha = -(4\pi G_\mathrm{N})^{-1}$. This justifies the factor chosen in the definition 23.39 of the Komar energy.

To conclude this section, we would like to mention that besides the Komar integral there is another important notion of a global quantity for an asymptotically flat spacetime, the so-called *ADM integral*, according to *Richard L. Arnowitt, Stanley Deser* and *Charles W. Misner* [4]. The ADM integral is based on the Hamiltonian formulation of general relativity and is evaluated as a quantity at spatial infinity of the asymptotically flat spacetime. The results based on the ADM integral agree with the ones based on the Komar integral in relevant cases, such as for the Schwarzschild solution. However, the ADM integral can furthermore be used in cases where the matter fields are not spatially confined.

23.4   Weyl Rescaling Symmetry

Weyl Rescalings as Internal Transformations of Fields

We have learned about diffeomorphisms as the natural symmetry transformations of geometric manifolds. Beyond diffeomorphisms, there are also Weyl rescalings of geometric manifolds, and we note again that these rescalings are independent of diffeomorphisms. Weyl rescalings have proved to be useful in the analysis of Einstein's equations and their solutions, as we have already seen in the discussion of asymptotically flat spacetimes. Let us therefore consider the effect of Weyl rescalings in a bit more detail below. We consider a general d-dimensional spacetime manifold $(\mathcal{M}, g_{\mu\nu})$, since some relevant statements can be deduced concerning the dimensionality. Moreover, we consider Weyl rescaling transformations which affect the metric and the fields defined on the manifold. This means that we consider *Weyl density fields*, compare with Section 20.2 and appendix C. More specifically, the effect of a finite Weyl rescaling is

$$\left.\begin{aligned} \widehat{g}_{\mu\nu} &= \Omega^2 g_{\mu\nu} = e^{2s} g_{\mu\nu} \\ \widehat{\phi}_i &= \Omega^{w_\phi} \phi_i = e^{w_\phi s} \phi_i \end{aligned}\right\}, \tag{23.52}$$

where the conformal factor $\Omega(x)$ is written as $\Omega(x) = \exp s(x)$, with a real function $s(x)$. The real exponent w_ϕ is called the *conformal weight* of the field $\phi_i(x)$. The metric has the conformal weight 2 by definition. If we want to express infinitesimal Weyl rescalings, we can write the scalar function $s(x)$ in the form $s(x) = \epsilon\, t(x)$, with $|\epsilon| \ll 1$. Then the effect of an infinitesimal Weyl rescaling is

$$\left.\begin{aligned} \delta_s g_{\mu\nu} &= 2s g_{\mu\nu} = 2\epsilon t g_{\mu\nu} \\ \delta_s \phi_i &= w_\phi s \phi_i = w_\phi \epsilon t \phi_i \end{aligned}\right\}. \tag{23.53}$$

We use the symbol δ_s for the variation caused by a Weyl rescaling in order to distinguish it from the other types of variations we have considered previously, i.e. the Euler-Lagrange variations δ and the diffeomorphism variations δ_V.

Transformation of the Action under Weyl Rescalings
We now consider a theory described by a general action $S\left[\phi_i, g_{\mu\nu}\right]$ given as the integral

$$S = \int_U L\sqrt{|g|}\, \mathrm{d}^d x, \tag{23.54}$$

with a scalar Lagrangian $L\left(\phi_i, g_{\mu\nu}\right)$ and a suitable integration domain U in d dimensions. The variation of the action $\delta_s S\left[\phi_i, g_{\mu\nu}\right]$ under a Weyl rescaling is defined by

$$\delta_s S\left[\phi_i, g_{\mu\nu}\right] = S\left[\phi_i + \delta_s \phi_i,\ g_{\mu\nu} + \delta_s g_{\mu\nu}\right] - S\left[\phi_i, g_{\mu\nu}\right]. \tag{23.55}$$

We declare that the theory is *Weyl rescaling invariant* iff the condition

$$\boxed{\delta_s S\left[\phi_i, g_{\mu\nu}\right] = 0} \tag{23.56}$$

holds. The variation of the action can be calculated as

$$\delta_s S = \int_U \left[(\delta_s L)\sqrt{|g|} + (\delta_s\sqrt{|g|})L\right] \mathrm{d}^d x, \tag{23.57}$$

in which we can use the identity

$$\delta_s\sqrt{|g|} = ds\sqrt{|g|}. \tag{23.58}$$

Now we assume the *special case* in which the Lagrangian transforms as a homogeneous function under Weyl rescalings, i.e.

$$L\left(\widehat{\phi}_i, \widehat{g}_{\mu\nu}\right) = \Omega^{w_L} L\left(\phi_i, g_{\mu\nu}\right), \tag{23.59}$$

with a specific conformal weight w_L. Not all Lagrangians behave this way and we will consider examples below. Infinitesimally, this homogeneous transformation is expressed as

$$\delta_s L = w_L s L. \tag{23.60}$$

In this case, the variation of the action becomes

$$\delta_s S = (w_L + d) \int_U sL\sqrt{|g|}\, \mathrm{d}^d x. \tag{23.61}$$

Therefore, the condition

$$w_L = -d \tag{23.62}$$

is sufficient for the action to be Weyl rescaling invariant. This is the reason why we are very interested in Lagrangians that transform like

$$L(\widehat{\phi}_i, \widehat{g}_{\mu\nu}) = \Omega^{-d} L(\phi_i, g_{\mu\nu}). \tag{23.63}$$

It can be inspected that the minimally coupled real scalar field has a Lagrangian transforming in this particular way only in the case that there is no mass term, no interaction, and the dimensionality is equal two. The corresponding Lagrangian consists only of the kinetic term then. (*exercise 23.3*) However, we will see below how a scalar field can be coupled to the gravitational field in a different way, so that Weyl rescaling symmetry is attained with nontrivial interactions and in arbitrary dimensions.

Fortunately, the free Maxwell field on a curved spacetime provides a much easier example. We consider the theory described by the minimally coupled Lagrangian 22.19. The basic electromagnetic covector field $A_\mu(x)$ is assumed to have the conformal weight $w_A = 0$. Since this electromagnetic field is Weyl rescaling invariant, $\widehat{A}_\mu = A_\mu$, the Faraday tensor field $F_{\mu\nu}(x)$ is also Weyl rescaling invariant, $\widehat{F}_{\mu\nu} = F_{\mu\nu}$, see also Section 20.2. Consequently, the Lagrangian 22.19 transforms as

$$L(\widehat{A}_\mu, \widehat{g}_{\mu\nu}) = \Omega^{-4} L(A_\mu, g_{\mu\nu}). \tag{23.64}$$

In other words, the case $d = 4$ is a sufficient condition for the minimally coupled free Maxwell theory to be Weyl rescaling invariant. The converse is also true, which we can see by considering the associated action $S[A_\mu, g_{\mu\nu}]$, which transforms as

$$S[\widehat{A}_\mu, \widehat{g}_{\mu\nu}] = \Omega^{d-4} S[A_\mu, g_{\mu\nu}]. \tag{23.65}$$

Requiring Weyl rescaling invariance of this action results to $d = 4$. Thus, the case $d = 4$ is also a necessary condition. In summary:

> *The free Maxwell theory on curved spacetime is Weyl rescaling invariant exactly in $d = 4$ spacetime dimensions.*

We remark that the above result can be deduced also directly by considering the Weyl transformation of the free Maxwell equations 22.21.

Let us pause for a moment here and recall that the condition $d = 4$ actually leads to some unique results in several cases. A number of physically relevant Minkowski space field theories, like Maxwell electromagnetism, are conformally invariant exactly in four dimensions. Within Riemannian geometry, we saw that the Weyl tensor is non-vanishing only in $d \geq 4$ dimensions. Further, according to Lovelock's theorem, the Einstein tensor is uniquely determined precisely in four dimensions and it leads almost inevitably to Einstein's field equations. Finally, $d = 4$ is the lowest number of spacetime dimensions for which the Einstein gravitational field can propagate in the absence of matter sources. It is therefore natural to ask if there are further characteristic properties to be associated with the case of four spacetime dimensions. We must be content to leave this question open here.

Condition for Weyl Symmetry of the Matter Action

We consider specifically the matter field sector here with a matter action $S_\mathrm{M}\left[\phi_i, g_{\mu\nu}\right]$ and we carry out an infinitesimal Weyl rescaling. The resulting variation $\delta_s S_\mathrm{M}\left[\phi_i, g_{\mu\nu}\right]$ of the matter action is calculated as

$$\delta_s S_\mathrm{M} = \int_U \left(\frac{1}{\sqrt{|g|}}\frac{\delta S_\mathrm{M}}{\delta \phi_i}\delta_s\phi_i - \frac{1}{2}T^{\mu\nu}\delta_s g_{\mu\nu}\right)\sqrt{|g|}\,\mathrm{d}^d x. \tag{23.66}$$

We assume that the system is on-shell and the matter EL-Eqs $\delta S_\mathrm{M}/\delta\phi_i = 0$ hold. By inserting the specific expression $\delta_s g_{\mu\nu} = 2s g_{\mu\nu}$ in the above integral equation, we obtain

$$\delta_s S_\mathrm{M} = -\int_U s\, T^\mu{}_\mu \sqrt{|g|}\,\mathrm{d}^d x. \tag{23.67}$$

Since the integration domain U and the scalar function $s(x)$ are arbitrary, we obtain that *the Weyl rescaling invariance of the matter action is equivalent to the tracelessness of the metric energy-momentum tensor.* In other words, the condition

$$\boxed{T^\mu{}_\mu = 0} \tag{23.68}$$

is necessary and sufficient to achieve Weyl symmetry of the matter fields. Once again the free Maxwell field provides the main example, since for the corresponding MEMT one has

$$T^\mu{}_\mu = \frac{F^2}{\mathrm{Area}(B^{d-1})}\left(\frac{d}{4}-1\right). \tag{23.69}$$

Thus, Weyl symmetry for the free Maxwell theory holds exactly in $d = 4$ dimensions.

Weyl Rescalings of the MEMT

Let us now examine how the basic properties of indices symmetry, tracelessness and vanishing of divergence of a MEMT behave under Weyl rescalings. Similarly to the Lagrangian, we assume that the MEMT $T_{\mu\nu}(x)$ of the theory at hand transforms homogeneously under finite Weyl rescalings. i.e. we postulate

$$\widehat{T}_{\mu\nu} = \Omega^{w_T} T_{\mu\nu} \tag{23.70}$$

and examine if this leads to reasonable consequences. In this spirit, the conformal weight w_T of the MEMT will be determined shortly. We infer immediately that the Weyl rescaled MEMT is symmetric exactly if this holds for the original MEMT. In addition, the Weyl rescaled MEMT is traceless exactly if this holds for the initial MEMT. So the only thing left to inspect is the behavior of the divergence expression $\nabla^\mu T_{\mu\nu}$ under Weyl rescalings. We note the formula for the derivative operator

$$\widehat{\nabla}^\mu = \widehat{g}^{\mu\rho}\widehat{\nabla}_\rho = \Omega^{-2}g^{\mu\rho}\widehat{\nabla}_\rho \tag{23.71}$$

and the formula for the covariant derivation of a $(0,2)$-tensor field, as given in appendix C. In this way, we obtain for the Weyl rescaled divergence expression

$$\widehat{\nabla}^\mu\widehat{T}_{\mu\nu} = \widehat{g}^{\mu\rho}\left(\nabla_\rho\widehat{T}_{\mu\nu} - 2\Upsilon_\rho\widehat{T}_{\mu\nu} - \Upsilon_\mu\widehat{T}_{\rho\nu} - \Upsilon_\nu\widehat{T}_{\mu\rho} + g_{\rho\mu}\Upsilon^\alpha\widehat{T}_{\alpha\nu} + g_{\rho\nu}\Upsilon^\alpha\widehat{T}_{\mu\alpha}\right). \tag{23.72}$$

The third and the sixth term on the above rhs cancel each other due to the symmetry of the MEMT. If we now use the ansatz 23.70 and the definition $\Upsilon_\mu \equiv \Omega^{-1}\nabla_\mu\Omega$, we obtain

$$\widehat{\nabla}^\mu\widehat{T}_{\mu\nu} = \Omega^{w_T-2}\nabla^\mu T_{\mu\nu} + \Omega^{w_T-3}(w_T - 2 + d)(\nabla^\mu\Omega)T_{\mu\nu} - \Omega^{w_T-3}(\nabla_\nu\Omega)T^\mu{}_\mu. \tag{23.73}$$

We deduce that the divergence equation $\nabla^\mu T_{\mu\nu} = 0$ is Weyl rescaling invariant if the two conditions $T^\mu{}_\mu = 0$ and $w_T = 2 - d$ hold. Therefore, it is very natural to require that the MEMT transforms as

$$\widehat{T}_{\mu\nu} = \Omega^{2-d} T_{\mu\nu} \tag{23.74}$$

under arbitrary Weyl rescalings. With the transformation behavior 23.74, the basic defining properties of the MEMT remain Weyl rescaling invariant. As an example, we can consider again the free Maxwell field on a curved spacetime, which has a MEMT that transforms in the desired way exactly in $d = 4$ spacetime dimensions.

In fact, the transformation behavior 23.74 does not need to be postulated, because it follows from a matter action that is Weyl rescaling invariant. So let us consider such a matter theory that is Weyl rescaling invariant,

$$S_{\mathrm{M}}\big[\widehat{\phi}_i, \widehat{g}_{\mu\nu}\big] = S_{\mathrm{M}}\big[\phi_i, g_{\mu\nu}\big]. \tag{23.75}$$

For each chosen Weyl scale, we can carry out arbitrary variations $\delta g_{\mu\nu}(x)$ of the metric field function. Note that these variations $\delta g_{\mu\nu}(x)$ are not Weyl rescalings and do not involve any coordinate changes. Since the variation δ acts linearly, the following relations hold

$$\delta \widehat{g}_{\mu\nu} = \Omega^2 \delta g_{\mu\nu} \quad \text{and} \quad \delta \widehat{g}^{\mu\nu} = \Omega^{-2} \delta g^{\mu\nu}. \tag{23.76}$$

For the corresponding variation of the matter action we have

$$\delta S_{\mathrm{M}}\big[\widehat{\phi}_i, \widehat{g}_{\mu\nu}\big] = \delta S_{\mathrm{M}}\big[\phi_i, g_{\mu\nu}\big], \tag{23.77}$$

since the action is Weyl symmetric. The variation $\delta S_{\mathrm{M}}\big[\widehat{\phi}_i, \widehat{g}_{\mu\nu}\big]$ expressed through the Weyl rescaled quantities is

$$\delta S_{\mathrm{M}}\big[\widehat{\phi}_i, \widehat{g}_{\mu\nu}\big] = \frac{1}{2} \int_U \widehat{T}_{\mu\nu} \, \delta \widehat{g}^{\mu\nu} \sqrt{|\widehat{g}|} \, \mathrm{d}^d x, \tag{23.78}$$

with $\sqrt{|\widehat{g}|} = \Omega^d \sqrt{|g|}$. We now assume that the MEMT transforms homogeneously under Weyl rescalings, as postulated in 23.70. Then, in order to recover $\delta S_{\mathrm{M}}\big[\phi_i, g_{\mu\nu}\big]$, it must be necessarily

$$\boxed{\widehat{T}_{\mu\nu} = \Omega^{2-d} T_{\mu\nu}} \tag{23.79}$$

As a consequence, for a Weyl symmetric matter theory, the MEMT satisfies the relations

$$\boxed{\widehat{T}_{\mu\nu} = \widehat{T}_{\nu\mu}, \quad \widehat{T}^\mu{}_\mu = 0 \quad \text{and} \quad \widehat{\nabla}^\mu \widehat{T}_{\mu\nu} = 0} \tag{23.80}$$

In other words, within a Weyl symmetric matter theory, we can work in any scale and the MEMT always fulfills its characteristic properties.

Conformally Coupled Scalar Field

We remarked previously that the minimally coupled scalar field $\phi(x)$ described by the simple Lagrangian

$$L = \frac{1}{2} g^{\mu\nu} \nabla_\mu \phi \, \nabla_\nu \phi \tag{23.81}$$

represents a Weyl rescaling invariant theory only in two spacetime dimensions. We can ask if it is possible to modify the scalar theory in such a way that Weyl rescaling invariance is attained in dimensions higher than two. Let us consider curved spacetimes with dimensionality $d \geq 3$. We are going to examine the transformation behavior of the expression

$$\nabla^2 \phi = g^{\mu\nu} \nabla_\mu \nabla_\nu \phi \tag{23.82}$$

under the combined Weyl rescalings 23.52. For notational simplicity, we use the symbol w for the conformal weight of the scalar field in the following. We first calculate the transformation of the covariant derivative $\nabla_\nu \phi$ of the Weyl density field $\phi(x)$. It is

$$\widehat{\nabla}_\nu \widehat{\phi} = \partial_\nu (\Omega^w \phi) = \Omega^w (\nabla_\nu \phi + w \Upsilon_\nu \phi), \tag{23.83}$$

with the usual definition $\Upsilon_\mu \equiv \Omega^{-1} \nabla_\mu \phi$. Now we can proceed to the expression $\nabla^2 \phi$, which transforms to

$$\widehat{\nabla}^2 \widehat{\phi} = \widehat{g}^{\mu\nu} \widehat{\nabla}_\mu \widehat{\nabla}_\nu \widehat{\phi} = \Omega^{-2} g^{\mu\nu} \widehat{\nabla}_\mu (\Omega^w \nabla_\nu \phi + w \Omega^w \Upsilon_\nu \phi). \tag{23.84}$$

The covariant derivative $\widehat{\nabla}_\mu$ is applied to the two terms in the round brackets in combination with the product rule. By carrying out subsequently the contraction with the inverse metric $g^{\mu\nu}$, one obtains

$$\widehat{\nabla}^2 \widehat{\phi} = \Omega^{w-2} \Big[\nabla^2 \phi + (2w - 2 + d) \Upsilon^\mu \nabla_\mu \phi + w (\nabla^\mu \Upsilon_\mu) \phi + (w^2 - 2w + dw) \Upsilon^2 \phi \Big]. \tag{23.85}$$

Now if we choose the special value

$$w = \frac{2-d}{2} \tag{23.86}$$

for the conformal weight of the scalar field, the second term within the square brackets above vanishes and the fourth term receives the factor

$$w^2 - 2w + dw = \frac{2-d}{2} \frac{d-2}{2}. \tag{23.87}$$

With the special value 23.86 for the conformal weight, we consequently obtain

$$\widehat{\nabla}^2 \widehat{\phi} = \Omega^{-1-\frac{d}{2}} \left[\nabla^2 \phi + \frac{2-d}{2} \left(\nabla^\mu \Upsilon_\mu + \frac{d-2}{2} \Upsilon^2 \right) \phi \right]. \tag{23.88}$$

We recognize the expression in the round brackets above as minus the trace $\mathcal{Z}(x)$ of the transformation tensor $\mathcal{Z}_{\mu\nu}(x)$, with

$$\mathcal{Z} = -\nabla^\mu \Upsilon_\mu - \frac{d-2}{2} \Upsilon^2, \tag{23.89}$$

see again appendix C. Thus, we can write equivalently

$$\widehat{\nabla}^2 \widehat{\phi} = \Omega^{-1-\frac{d}{2}} \left(\nabla^2 \phi - \frac{2-d}{2} \mathcal{Z} \phi \right). \tag{23.90}$$

In order to eliminate $\mathcal{Z}(x)$, we employ the transformation formula

$$\mathcal{Z} = \Omega^2 \widehat{P} - P \tag{23.91}$$

containing the trace $P(x)$ of the Schouten tensor $P_{\mu\nu}(x)$, and obtain equivalently

$$\left(\widehat{\nabla}^2 \widehat{\phi} - \frac{d-2}{2} \widehat{P} \widehat{\phi} \right) = \Omega^{-1-\frac{d}{2}} \left(\nabla^2 \phi - \frac{d-2}{2} P \phi \right). \tag{23.92}$$

Instead of the trace $P(x)$ of the Schouten tensor, we can use the Ricci scalar $R(x)$, with

$$P = \frac{R}{2d-2}, \tag{23.93}$$

and conclude finally to the remarkable relation

$$\left(\widehat{\nabla}^2\widehat{\phi} - \frac{d-2}{4d-4}\widehat{R}\widehat{\phi}\right) = \Omega^{-1-\frac{d}{2}}\left(\nabla^2\phi - \frac{d-2}{4d-4}R\phi\right). \tag{23.94}$$

We have found a differential operator that generalizes the basic Laplace / d'Alembert operator ∇^2 and is Weyl rescaling covariant when applied to a scalar field of weight $w = 1 - d/2$. This differential operator

$$\nabla^2 - \frac{d-2}{4d-4}R \tag{23.95}$$

is called the *conformal Laplace / d'Alembert operator*. The corresponding wave equation

$$\boxed{\nabla^2\phi - \frac{d-2}{4d-4}R\phi = 0} \tag{23.96}$$

is indeed Weyl rescaling invariant. Depending on the context, this equation is called the *conformal Laplace equation* or the *conformal Klein-Gordon equation*. The Lagrangian corresponding to the above wave equation is

$$\boxed{L = \frac{1}{2}\nabla^\mu\phi\,\nabla_\mu\phi + \frac{d-2}{8d-8}R\phi^2} \tag{23.97}$$

This Lagrangian describes the so-called *conformally coupled scalar field theory*. It contains a *non-minimal coupling* between the massless scalar field and the metric field. Exactly because of this non-minimal coupling, the Lagrangian transforms homogeneously under Weyl rescalings. In order to prove this, we write the Lagrangian in a slightly different form. For the kinetic term we note that it is

$$\nabla^\mu\phi\,\nabla_\mu\phi = \nabla^\mu(\phi\,\nabla_\mu\phi) - \phi\,\nabla^2\phi. \tag{23.98}$$

By assuming that the scalar field vanishes at the boundary of the spacetime domain considered, we can neglect the divergence term $\nabla^\mu(\phi\,\nabla_\mu\phi)$ entirely and use the alternative Lagrangian*

$$\overline{L} = -\frac{1}{2}\phi\left(\nabla^2\phi - \frac{d-2}{4d-4}R\phi\right). \tag{23.99}$$

Now it can be immediately seen that under Weyl rescalings the Lagrangian transforms as

$$\overline{L}(\widehat{\phi},\widehat{g}_{\mu\nu}) = -\frac{1}{2}\Omega^{1-\frac{d}{2}}\phi\,\Omega^{-1-\frac{d}{2}}\left(\nabla^2\phi - \frac{d-2}{4d-4}R\phi\right), \tag{23.100}$$

which means that it is

$$\overline{L}(\widehat{\phi},\widehat{g}_{\mu\nu}) = \Omega^{-d}\overline{L}(\phi,g_{\mu\nu}), \tag{23.101}$$

exactly as desired. We can summarize the above results in the statement:

The conformally coupled scalar field theory is Weyl rescaling invariant.

Moving on, we can expand the conformally coupled scalar field theory by including additional integer powers of the scalar field in the Lagrangian. Of course, when we add a term of

---

*In this particular form the Lagrangian contains also second-order derivatives $\nabla_\mu\nabla_\nu\phi$ of the scalar field.

the form $\kappa\phi^n$, with some real parameter κ and exponent n, we would like to maintain the specific Weyl covariance of the Lagrangian. Therefore, the condition

$$\kappa\widehat{\phi}^n = \Omega^{-d}\kappa\phi^n \tag{23.102}$$

must hold. By inserting the Weyl transformation

$$\widehat{\phi} = \Omega^{\frac{2-d}{2}}\phi \tag{23.103}$$

in this condition, we immediately obtain that the exponent must be

$$n = \frac{2d}{d-2}. \tag{23.104}$$

Thus, we can write down the Lagrangian of the *conformally coupled scalar field theory with self-interaction*:

$$L = \frac{1}{2}\nabla^\mu\phi\,\nabla_\mu\phi + \frac{d-2}{8d-8}R\phi^2 + \kappa\phi^{\frac{2d}{d-2}}. \tag{23.105}$$

This Lagrangian is Weyl rescaling covariant in the desired way. We observe that the monomial term included here is the same as the one appearing in the conformally invariant field theory on flat Minkowski space, see Section 16.4. The scalar field equation in the present case is

$$\nabla^2\phi - \frac{d-2}{4d-4}R\phi - \frac{2d\kappa}{d-2}\phi^{\frac{d+2}{d-2}} = 0. \tag{23.106}$$

By looking at the Lagrangian and the dynamic equation, we can infer that *the zero curvature limit of the Weyl symmetric conformally coupled scalar field yields the conformally symmetric scalar field on Minkowski space*. Besides its appealing behavior under Weyl rescalings, the conformally coupled scalar field is a starting point in order to build the so-called scalar-tensor theory of gravitation. This theory contains the metric field and the real scalar field as basic ingredients, where the scalar field modulates the effects of a cosmological constant.

MEMT of the Conformally Coupled Scalar Field

Let us now derive the explicit formula for the MEMT of the conformally coupled scalar field. We consider the Lagrangian

$$L = \frac{1}{2}\nabla^\mu\phi\,\nabla_\mu\phi + \frac{\xi}{2}R\phi^2, \tag{23.107}$$

with the parameter $\xi \equiv (d-2)/(4d-4)$. The MEMT is calculated separately for each term of the Lagrangian. For the kinetic term, the contribution to the MEMT, denoted $T^{\mathrm{K}}_{\mu\nu}(x)$, is given by

$$T^{\mathrm{K}}_{\mu\nu} = \nabla_\mu\phi\,\nabla_\nu\phi - \frac{1}{2}g_{\mu\nu}\nabla^\alpha\phi\,\nabla_\alpha\phi, \tag{23.108}$$

as seen already in Section 22.2. For the non-minimal coupling term, the contribution to the MEMT, denoted $T^{\mathrm{NM}}_{\mu\nu}(x)$, is calculated by considering the corresponding action integral

$$S^{\mathrm{NM}} \equiv \frac{\xi}{2}\int_U R\phi^2\sqrt{|g|}\,\mathrm{d}^d x \tag{23.109}$$

and its variation $\delta S^{\mathrm{NM}}[\phi, g_{\mu\nu}]$ with respect to the metric. By employing the formula

$$\delta S^{\mathrm{NM}} = \frac{1}{2}\int_U T^{\mathrm{NM}}_{\mu\nu}\delta g^{\mu\nu}\sqrt{|g|}\,\mathrm{d}^d x, \tag{23.110}$$

we can identify the sought MEMT contribution. The variation is calculated explicitly as

$$\delta S^{\text{NM}} = \frac{\xi}{2} \int_U \left[(\delta R_{\mu\nu}) g^{\mu\nu} \sqrt{|g|} + R_{\mu\nu}(\delta g^{\mu\nu}) \sqrt{|g|} + R(\delta \sqrt{|g|}) \right] \phi^2 \, d^d x. \tag{23.111}$$

The second and the third term in the above rhs combined yield the integral expression

$$\delta S^{\text{NM}}_{(2,3)} = \frac{\xi}{2} \int_U G_{\mu\nu} \phi^2 \, \delta g^{\mu\nu} \sqrt{|g|} \, d^d x, \tag{23.112}$$

as we know already from Section 22.3. By using the previous results 22.50 and 22.53, the first term in the above rhs can be written as

$$\delta S^{\text{NM}}_{(1)} = \frac{\xi}{2} \int_U (\nabla_\sigma W^\sigma) \phi^2 \sqrt{|g|} \, d^d x, \tag{23.113}$$

with the vector field

$$W^\sigma = g^{\rho\sigma} g^{\mu\nu} (\nabla_\mu \delta g_{\rho\nu} - \nabla_\rho \delta g_{\mu\nu}) \tag{23.114}$$

containing the variation of the metric. By carrying out an integration by parts, this integral becomes

$$\delta S^{\text{NM}}_{(1)} = \frac{\xi}{2} \int_U \nabla_\sigma (W^\sigma \phi^2) \sqrt{|g|} \, d^d x - \frac{\xi}{2} \int_U W^\sigma \nabla_\sigma (\phi^2) \sqrt{|g|} \, d^d x. \tag{23.115}$$

Assuming that it is $\phi = 0$ on the boundary ∂U, the integral over the divergence vanishes. By inserting the expression for the vector field $W^\sigma(x)$ in the remaining integral and by carrying out another integration by parts, we obtain

$$\delta S^{\text{NM}}_{(1)} = \frac{\xi}{2} \int_U g^{\rho\sigma} g^{\mu\nu} \left[\nabla_\mu \nabla_\sigma (\phi^2) \delta g_{\rho\nu} - \nabla_\rho \nabla_\sigma (\phi^2) \delta g_{\mu\nu} \right] \sqrt{|g|} \, d^d x. \tag{23.116}$$

Here we have assumed that it is $\nabla_\sigma \phi = 0$ on ∂U, so that the integrals containing the divergence expressions can be omitted. By carrying out the contractions and by switching to the variation $\delta g^{\mu\nu}(x)$, we obtain

$$\delta S^{\text{NM}}_{(1)} = \frac{\xi}{2} \int_U (-\nabla_\mu \nabla_\nu + g_{\mu\nu} \nabla^2)(\phi^2) \, \delta g^{\mu\nu} \sqrt{|g|} \, d^d x. \tag{23.117}$$

We can now collect the partial results for the variation of the non-minimal action,

$$\delta S^{\text{NM}} = \delta S^{\text{NM}}_{(1)} + \delta S^{\text{NM}}_{(2,3)}, \tag{23.118}$$

to obtain

$$\delta S^{\text{NM}} = \frac{\xi}{2} \int_U \left[G_{\mu\nu} \phi^2 - (\nabla_\mu \nabla_\nu - g_{\mu\nu} \nabla^2)(\phi^2) \right] \delta g^{\mu\nu} \sqrt{|g|} \, d^d x. \tag{23.119}$$

We can read off the MEMT contribution $T^{\text{NM}}_{\mu\nu}(x)$ from the non-minimal term. It is

$$T^{\text{NM}}_{\mu\nu} = \xi G_{\mu\nu} \phi^2 - \xi (\nabla_\mu \nabla_\nu - g_{\mu\nu} \nabla^2)(\phi^2). \tag{23.120}$$

Finally, we can combine the results of the single MEMT contributions,

$$T_{\mu\nu} = T^{\text{K}}_{\mu\nu} + T^{\text{NM}}_{\mu\nu}, \tag{23.121}$$

and reach to the desired result

$$T_{\mu\nu} = \nabla_\mu \phi \, \nabla_\nu \phi - \frac{1}{2} g_{\mu\nu} \nabla^\alpha \phi \, \nabla_\alpha \phi + \frac{d-2}{4d-4} \left[G_{\mu\nu} \phi^2 - (\nabla_\mu \nabla_\nu - g_{\mu\nu} \nabla^2)(\phi^2) \right]. \tag{23.122}$$

This is the *MEMT of the conformally coupled scalar field*. The above formula is valid for $d \geq 3$ spacetime dimensions. In the limit of vanishing curvature, $G_{\mu\nu} \to 0$ and $\nabla_\mu \to \partial_\mu$, we recover the improved energy-momentum tensor 16.151 of the conformally symmetric scalar field on Minkowski space. This displays once again how the Weyl symmetric scalar theory on curved space leads to the conformally symmetric scalar theory on Minkowski space. Of course, one can now proceed and check that the MEMT 23.122 is traceless and has a vanishing divergence, but we know already that this is true based on the general analysis which led to 23.80. With these remarks we conclude our review of the basic properties of the Weyl symmetric scalar field.

Further Reading

Central questions concerning energy in general relativity and its positivity in particular are treated in Straumann [84]. In our discussion of the energy of the gravitational field, we mentioned the field-theoretic approach to gravity, which takes a massless second-rank tensor field as the point of departure for developing the theory. This is described in the book of Feynman, Morinigo, and Wagner [28]. The field-theoretic approach and the search for suitable EMTs of the gravitational field still represent open research topics. The anomalous factor in the Komar integral formulas was explained first time by Katz [47]. For an original discussion of the ADM energy within general relativity, the reader should turn to the article by Arnowitt, Deser, and Misner [4]. A compendium on the basics of Weyl rescalings and the conformally coupled scalar field is provided by Dabrowski, Garecki, and Blaschke [18].

VII

Appendices

Notation and Conventions

A.1 Physical Units and Dimensions

Each physical quantity can be measured in certain *physical units*, which can be chosen rather freely based on convenience and convention. However, a physical quantity possesses always a unique *physical dimension* for which different choices of units exist. In the following, we list some basic mechanical quantities and their corresponding physical dimensions, commonly denoted in square brackets.

Mass m:
$$[m] = [M].$$

Length x:
$$[x] = [L].$$

Time t:
$$[t] = [T].$$

Momentum $p = mv$:
$$[p] = [M] \cdot [L] \cdot [T]^{-1}.$$

Energy $E = \int F \, \mathrm{d}x$:
$$[E] = [M] \cdot [L]^2 \cdot [T]^{-2}.$$

Action $S = \int L \, \mathrm{d}t$:
$$[S] = [M] \cdot [L]^2 \cdot [T]^{-1}.$$

Concerning physical units, the most common choice is the *SI system of units*. Within the SI system, mass is measured in *kilograms* kg, length is measured in *meters* m, and time in *seconds* s. Within any chosen system of units, one can measure and define certain physical quantities that have proven to be constant and fundamental for the physical sciences. These are the *physical constants*. Two such fundamental physical constants are the *speed of light* c, with an exact value of

$$c = 299.792.458 \, \mathrm{m \, s^{-1}}$$

in SI units, and the *Planck constant* h, with an exact value of

$$h = 6,62607015 \times 10^{-34} \, \mathrm{kg \, m^2 \, s^{-1}}$$

DOI: 10.1201/9781003087748-A

in SI units. In this book, if not stated differently, we use physical units where the *speed of light c* and *Planck's reduced constant* $\hbar = h/2\pi$ are chosen to be equal one, without any further physical unit value, i.e.

$$c = \hbar = 1.$$

This choice of physical units belongs to the system of *natural units*. As a consequence of the choice $c = 1$, the dimensions of length x and time t become equal,

$$[x] = [t] = [L] = [T].$$

In addition, the dimensions of momentum p and energy E coincide,

$$[p] = [E] = [M].$$

Imposing additionally the choice $\hbar = 1$, the action S becomes dimensionless,

$$[S] = 1,$$

or, equivalently,

$$[M] = [L]^{-1}.$$

In summary, using units with $c = \hbar = 1$ leads to the following equality of dimensions:

$$[p] = [E] = [M] = [L]^{-1} = [T]^{-1} = [x]^{-1} = [t]^{-1}.$$

In this way, products like $\boldsymbol{p} \cdot \boldsymbol{x}$ or $p_\mu x^\mu = p^0 x^0 - \boldsymbol{p} \cdot \boldsymbol{x}$ are dimensionless and can be used in arguments of functions, as in $\exp(\mathrm{i}\, p_\mu x^\mu)$.

In addition to the basic mechanical quantities, the fundamental interactions of physics require us to specify and to measure coupling strengths of interactions. In the present book, where we deal with classical physics, we are only concerned with the coupling strengths of gravity and electromagnetism. To define the strength of gravitational interaction, one can invoke formally *Newton's law*

$$F_{\text{Newton}} = -G_{\text{N}} \frac{m_1 m_2}{r^2},$$

with *Newton's gravitational constant* G_{N}. Its value in SI units is

$$G_{\text{N}} = 6,67408(31) \times 10^{-11}\,\mathrm{kg}^{-1}\,\mathrm{m}^3\,\mathrm{s}^{-2},$$

as published by the CODATA Task Group on Fundamental Constants in 2015. In this book, we will keep the gravitational constant G_{N} visible in the corresponding physical equations and will not set it equal one. The definition of electromagnetic interaction is a bit more involved since there are many different systems of physical units used. The definition of the strength of electrostatic forces can be based on *Coulomb's law*. When using SI units, Coulomb's law reads

$$F_{\text{Coulomb}} = k_e \frac{q_1 q_2}{r^2},$$

with the *Coulomb's constant* k_e. Its value in SI units is

$$k_e = 8,9875517873681764 \times 10^9\,\mathrm{kg}\,\mathrm{m}^3\,\mathrm{s}^{-2}\,\mathrm{C}^{-2}.$$

The physical unit C appearing here is the one for the *electric charge*, called the *coulomb*. The coulomb C is a unit derived from the two SI base units *ampere* A and *second* s, with $1\,\mathrm{C} = 1\,\mathrm{A}\,\mathrm{s}$. Based on the redefinitions of SI units, valid from May 2019 onward, the electron has an exact value for its electric charge of

$$q_e = 1,602176634 \times 10^{-19}\,\mathrm{C}.$$

In the framework of classical electrodynamics, the use of the so-called *Gaussian units* is very common and we will use these units in the formulation of all electromagnetic equations. In Gaussian units, the constant of proportionality is $k_e = 1$ and Coulomb's law has the simple form

$$F_{\text{Coulomb}} = \frac{q_1 q_2}{r^2}.$$

Another consequence of Gaussian units is that electric and magnetic fields acquire the same physical dimension. On the other hand, the Maxwell equations then contain the geometric factor 4π, which arises from the surface content of the 3-dimensional unit ball,

$$\partial_\mu F^{\mu\nu} = 4\pi j^\nu,$$

where $F^{\mu\nu}(x)$ is the Faraday tensor and $j^\nu(x)$ the electric current density vector. The system of Gaussian units is the chosen system in field theory reference works such as Jackson [46], Landau, Lifshitz [51], Misner, Thorne, Wheeler [57], and Schwinger et al. [81].

A.2 Mathematical Conventions

Linear Algebra and Indices

For a matrix or operator A, the following notation is used: transposed as A^T, complex conjugated as A^*, hermitian adjoint as $A^\dagger = A^{*T}$. The dual of a tensor T is denoted with a tilde $\tilde{T}$. The meaning and values of tensor indices are described within the context, but typically the following notation is used: Cartesian coordinate indices are small Latin letters, e.g. $j, k, l, m, n = 1, 2, 3$. Spacetime or abstract space indices are denoted by small Greek letters, e.g. $\alpha, \beta, \mu, \nu, \rho, \sigma = 0, 1, 2, 3$. When we have D space dimensions and one time dimension, we have $d = 1 + D$ for the spacetime dimension d and the indices take the values $\alpha, \beta, \mu, \nu, \rho, \sigma = 0, 1, \ldots, D$. We use the Latin indices i, j, etc. as *multi-indices* of fields $\phi_i(x)$, where each index i represents an ordered collection of tensorial or internal field indices. Spinorial indices are capital letters, e.g. $A, B = 1, 2$. Lie algebra indices are denoted with the first few letters of the Latin alphabet, e.g. $a, b, c = 1, \ldots, r$, where r is the number of parameters of the algebra. For indices in general, we use the *Einstein summation convention* for upper and lower indices, like

$$v_\mu w^\mu \equiv \sum_{\mu=0}^{D} v_\mu w^\mu.$$

Additionally, we use the summation convention also for pairs of indices being placed on the same upper or lower row. This convention is used for Cartesian coordinates, e.g. in

$$x_k y_k \equiv \sum_{k=1}^{3} x_k y_k,$$

for multi-indices of fields, e.g. in

$$\mathcal{G}_a(x)_{ij} \phi_j(x) \equiv \sum_{j=1}^{n} \mathcal{G}_a(x)_{ij} \phi_j(x),$$

and for Lie algebra indices, e.g. in

$$[\mathcal{G}_a, \mathcal{G}_b] = \mathrm{i} f_{abc} \mathcal{G}_c \equiv \mathrm{i} \sum_{c=1}^{r} f_{abc} \mathcal{G}_c.$$

If no summation is meant, this is explicitly stated.

Minkowski Metric and Curved Metrics

The *spacetime metric* can be either the flat spacetime *Minkowski metric* η given by

$$\eta = \eta_{\mu\nu}\, e^\mu \otimes e^\nu,$$

or a general *curved metric* g given by

$$g = g_{\mu\nu}(x)\, dx^\mu \otimes dx^\nu.$$

For practical purposes, the explicit indices notation $\eta_{\mu\nu}$ and $g_{\mu\nu}(x)$ is used. We employ the *sign convention* as found commonly in field theory, containing mostly minus signs with

$$\eta_{\mu\nu} = \text{diag}(1, -1, -1, -1)$$

in four spacetime dimensions and, more generally, with

$$\eta_{\mu\nu} = \text{diag}(1, -1, \ldots, -1)$$

in $d = 1 + D$ spacetime dimensions, where the -1 appears D times. A space with a metric of this type is said to have a *Lorentzian metric* or a *Lorentzian signature*. The metric with the opposite overall sign is used typically in books on general relativity. Given a general metric $g_{\mu\nu}(x)$, the *inverse of the metric* is denoted with upper indices $g^{\mu\nu}(x)$ and it fulfills the basic relation

$$g^{\mu\rho} g_{\rho\nu} = \delta^\mu_\nu.$$

If a metric is available, it establishes a one-to-one correspondence between vectors and covectors. Given a vector v^μ, the corresponding covector is $v_\mu = g_{\mu\rho} v^\rho$. Inversely, given a covector θ_μ, the corresponding vector is $\theta^\mu = g^{\mu\rho}\theta_\rho$. This *raising and lowering of indices* can be extended to arbitrary tensors, like

$$T^\mu{}_\nu = g^{\mu\rho} g_{\nu\sigma} T_\rho{}^\sigma.$$

The *trace* $\text{Tr}(\eta)$ of the Lorentzian metric $\eta_{\mu\nu}$ is defined in a Lorentz invariant way as

$$\text{Tr}(\eta) \equiv \eta^\mu{}_\mu = \sum_{\mu=0}^{D} \eta^\mu{}_\mu = 1 + D = d.$$

In the case of a general curved metric $g_{\mu\nu}(x)$, we have for the *trace* $\text{Tr}(g)$ similarly

$$\text{Tr}(g) \equiv g^\mu{}_\mu = \sum_{\mu=0}^{D} g^\mu{}_\mu = 1 + D = d.$$

Starting with the metric $g_{\mu\nu}(x)$, the *determinant of the metric* is denoted by g,

$$g \equiv \det(g_{\mu\nu}) = \det(g^{\mu\nu}).$$

Notationally, a confusion with the metric itself should not happen, since the context describes which quantity is meant.

Riemann Tensor

In the literature there are different sign definitions for the *Riemann (curvature) tensor* $R^\mu{}_{\nu\rho\sigma}(x)$. Starting from the Levi-Civita connection coefficients

$$\Gamma^\mu_{\nu\rho} = \frac{1}{2} g^{\mu\sigma}(\partial_\nu g_{\rho\sigma} + \partial_\rho g_{\nu\sigma} - \partial_\sigma g_{\nu\rho}),$$

this book employs the most common definition for the Riemann tensor,

$$R^\mu{}_{\nu\rho\sigma} \equiv \partial_\rho \Gamma^\mu_{\nu\sigma} - \partial_\sigma \Gamma^\mu_{\nu\rho} + \Gamma^\mu_{\alpha\rho}\Gamma^\alpha_{\nu\sigma} - \Gamma^\mu_{\alpha\sigma}\Gamma^\alpha_{\nu\rho}.$$

Some authors use different conventions for the indices placement, like the notation $R_\nu{}^\mu{}_{\sigma\rho}(x)$ in Lovelock, Rund [55] or the reversed order $R_{\sigma\rho\nu}{}^\mu(x)$, as used in Wald [91]. The *covariant Riemann tensor* with all indices written down is simply

$$R_{\mu\nu\rho\sigma} = g_{\mu\beta}R^\beta{}_{\nu\rho\sigma}.$$

The contraction known as the *Ricci tensor* $R_{\nu\sigma}(x)$ can also be defined in two different ways. In the present text, we use the definition

$$R_{\nu\sigma} \equiv R^\alpha{}_{\nu\alpha\sigma},$$

where the *first* and *third* index of the Riemann tensor are contracted. Some authors, as in [40], use a Ricci tensor definition where the first and the fourth index of the Riemann tensor are contracted. This leads to an overall minus sign in the gravitational Einstein equations. The *Ricci (curvature) scalar* $R(x)$ is always defined as

$$R \equiv R^\alpha{}_\alpha.$$

As a consequence, the *Einstein field equations* in the present book take the form

$$R_{\mu\nu} - \frac{1}{2}Rg_{\mu\nu} = 8\pi G_\mathrm{N} T_{\mu\nu},$$

where $T_{\mu\nu}(x)$ is the *energy-momentum tensor of matter*. Our conventions employed for the metric signature $(+ - - -)$, the Riemann tensor and the Ricci tensor are the same as in Landau, Lifshitz [51] and Plebański, Krasiński [68].

A.3 Abbreviations

In this book, we occasionally use abbreviations, which we summarize here in alphabetical order:

BCH	Baker-Campbell-Hausdorff (formula),
CEMT	Canonical Energy-Momentum Tensor,
CKVF	Conformal Killing Vector Field,
EL-Eq(s)	Euler-Lagrange Equation(s),
EM	Electromagnetic,
EMT	Energy-Momentum Tensor,
GR	General Relativity,
IEMT	Improved Energy-Momentum Tensor,
iff	if and only if,
KVF	Killing Vector Field,
lhs	left hand side,
MEMT	Metric Energy-Momentum Tensor,
ODE(s)	Ordinary Differential Equation(s),
PDE(s)	Partial Differential Equation(s),
rhs	right hand side,

SCT	Special Conformal Transformation,
SR	Special Relativity,
SEMT	Symmetric Energy-Momentum Tensor,
UHP	Upper Half-Plane (model).

B

Mathematical Tools

B.1 Tensor Algebra

Vector Spaces

We denote the fields $\mathbb{R}$ and $\mathbb{C}$ of real and complex numbers collectively by $\mathbb{F}$. *Vector spaces* over $\mathbb{F}$, denoted by V, are assumed to be finite-dimensional with dimension D, if not stated differently. If we choose a *basis* of V, denoted $\{e_\mu\}$, then a vector $v \in V$ is written as $v = v^\mu e_\mu$, with the basis-dependent components v^μ. When the basis is linearly transformed with a matrix Λ^{-1} as $e_\mu \mapsto e'_\mu = (\Lambda^{-1})^\nu{}_\mu e_\nu$, then the vector components transform as $v^\mu \mapsto v'^\mu = \Lambda^\mu{}_\nu v^\nu$, i.e. the components transform with the inverse of the matrix used for the basis transformation. Hence, we say the components v^μ transform *contravariantly*. For the transformation matrix $\Lambda = (\Lambda^\mu{}_\nu)$ we use the simplified notation $\Lambda_\mu{}^\nu \equiv (\Lambda^{-1})^\nu{}_\mu$ throughout. With this notation, the basis transformation reads $e_\mu \mapsto e'_\mu = \Lambda_\mu{}^\nu e_\nu$. Typically, the vector spaces we consider contain also a bilinear or sesquilinear *scalar (inner) product* $\langle \cdot, \cdot \rangle : V \times V \to \mathbb{F}$, $(v, w) \mapsto \langle v, w \rangle$, or equivalently a *metric* $\eta_{\mu\nu} = \langle e_\mu, e_\nu \rangle$.

Affine Spaces

An *affine space* is a pair (A, V) consisting of a set A of *points* and a vector space V, where each vector $v \in V$ defines a *translation* $t_v : A \to A$, so that for every $p \in A$ and $v, w \in V$ there is $t_0(p) = p$, $t_v(t_w(p)) = t_{v+w}(p)$, and for every pair $p, q \in A$ there is a unique vector $u \in V$, so that $t_u(p) = q$. One can view every vector space V as an affine space (V, V), where the elements of V are viewed as points which are translated by vectors of V. One of our prime examples is Minkowski space, which is considered a vector space under Lorentz transformations, while it is considered an affine space under Poincaré transformations.

Algebras

An *algebra* is a vector space equipped with a bilinear product $\circ : V \times V \to V$, $(v, w) \mapsto v \circ w$ which is associative and distributive. This means that for any elements $v, w, z \in V$ it is

DOI: 10.1201/9781003087748-B

$v \circ (w + z) = v \circ w + v \circ z$, $(v + w) \circ z = v \circ z + w \circ z$ and $v \circ (w \circ z) = (v \circ w) \circ z = v \circ w \circ z$. An important class of algebras are *Lie algebras*, where the product, called the *Lie bracket* and denoted $[\cdot, \cdot]$ then, is antisymmetric, $[v, w] = -[w, v]$, and fulfills the *Jacobi identity*, $[[v, w], z] + [[w, z], v] + [[z, v], w] = 0$.

Vector Space and Dual Space

Multilinear algebra is in essence linear algebra with many copies of vector spaces. First, there are some definitions concerning vector spaces and their duals on order. Given a vector space V with elements $v, w, \ldots$, the *dual vector space*, denoted $\widetilde{V}$, contains the linear forms $\theta, \omega, \ldots$ acting on V. i.e. the dual space $\widetilde{V}$ is the D-dimensional vector space of the linear maps $\theta : V \to \mathbb{F}$, $v \mapsto \theta(v)$. The elements of the dual space are called also *dual vectors* or *covectors*. Having a scalar product $\langle \cdot, \cdot \rangle$, or equivalently a metric $\eta_{\mu\nu}$ available on the vector space V, a natural isomorphism $V \to \widetilde{V}$ between V and its dual space $\widetilde{V}$ can be established. To each element v of V, the linear form $\widetilde{v} = \langle v, \cdot \rangle$ of $\widetilde{V}$ is assigned. Conversely, to each element θ of $\widetilde{V}$, the vector $\widetilde{\theta} = \langle \theta, \cdot \rangle_\sim$ of V is assigned, since every vector is a linear map from covectors to numbers. The scalar product $\langle \cdot, \cdot \rangle_\sim$ on the dual space $\widetilde{V}$ is defined through the inverse metric $\eta^{\mu\nu}$. With this isomorphism in place, vectors and covectors become essentially the same objects. This is practically done throughout in D-dimensional Euclidean geometry. When we choose a basis $\{e_\mu\}$ of V, this induces a *dual basis* $\{\widetilde{e}^\mu\}$ (or simply $\{e^\mu\}$, without the tilde) of $\widetilde{V}$. These bases can be chosen to be orthonormal, with $\widetilde{e}^\mu(e_\nu) = \delta^\mu_\nu$ and $e^\mu(\widetilde{e}_\nu) = \delta^\mu_\nu$. When a scalar product is present, this correspondence of basis elements can be expressed as $\langle e_\mu, e_\nu \rangle = \eta_{\mu\nu}$. Vectors v of V are written as $v = v^\mu e_\mu$, with *contravariant* components v^μ, while covectors θ of $\widetilde{V}$ are written as $\theta = \theta_\mu \widetilde{e}^\mu$, with *covariant* components θ_μ. The dual basis transforms as $\widetilde{e}^\mu \mapsto \widetilde{e}'^\mu = \Lambda^\mu{}_\nu \widetilde{e}^\nu$, whereas the components of dual vectors transform as $\theta_\mu \mapsto \theta'_\mu = \Lambda_\mu{}^\nu \theta_\nu$.

Tensor Algebra

We define a *tensor* T *of type*, or *valence*, (m, n) as the multilinear map

$$T : \underbrace{\widetilde{V} \times \cdots \times \widetilde{V}}_{m\text{-fold}} \times \underbrace{V \times \cdots \times V}_{n\text{-fold}} \to \mathbb{F},$$

from the Cartesian product $\widetilde{V} \times \cdots \times \widetilde{V} \times V \times \cdots \times V$ to the field of numbers $\mathbb{F}$. The term multilinear means that T acts in a linear fashion for every entry. The set of all tensors of type (m, n) constitutes a vector space of dimension D^{m+n} and is denoted as

$$\overset{m}{\bigotimes} V \otimes \overset{n}{\bigotimes} \widetilde{V} \equiv \underbrace{V \otimes \cdots \otimes V}_{m\text{-fold}} \otimes \underbrace{\widetilde{V} \otimes \cdots \otimes \widetilde{V}}_{n\text{-fold}},$$

or, somewhat shorter, as $V^{\otimes m} \otimes \widetilde{V}^{\otimes n}$. The symbol $\otimes$ denotes the *tensor product*. Although the notation looks intimidating, the concept of the tensor product is very natural. Having a tensor T of type (m, n) and a tensor S of type (p, q), one can take the tensor product of these, $T \otimes S$, which is a tensor of type $(m + p, n + q)$ and is simply defined by

$$T \otimes S(\theta_1, \ldots, \theta_{m+p}, v_1, \ldots, v_{n+q}) =$$
$$T(\theta_1, \ldots, \theta_m, v_1, \ldots, v_n) \cdot S(\theta_{m+1}, \ldots, \theta_{m+p}, v_{n+1}, \ldots, v_{n+q}).$$

The tensor product is bilinear, associative, but in general not commutative. Given the initial vectors and covectors of V, one can construct tensors of arbitrary valence. However, one should bear in mind that not every tensor, defined generally as multilinear map, can be written as a tensor product of simpler tensors. The tensor product $\otimes$ allows us to construct an algebra by collecting all tensor spaces of increasing valence. The infinite direct sum

$$\bigoplus_{m,n=0}^{\infty} \left(V^{\otimes m} \otimes \widetilde{V}^{\otimes n} \right)$$

contains tuples, where only a finite number of entries is not zero. The tensor product of tensors of different valence makes the above direct sum to an algebra, the so-called *tensor algebra* of V. It is $V^{\otimes 0} = \mathbb{F}$, $V^{\otimes 1} = V$ and $\widetilde{V}^{\otimes 0} = \widetilde{V}$. The tensor algebra is countably infinite-dimensional.

For practical applications, the component representation of tensors is indispensable. Tensors T of valence (m, n) have $m + n$ indices in their components. A choice of a basis $\{e_\mu\}$ with D elements in the space V induces a corresponding basis $\{e_{\mu_1 \ldots \mu_m}{}^{\nu_1 \ldots \nu_n}\}$ with D^{m+n} elements in the tensor space $V^{\otimes m} \otimes \widetilde{V}^{\otimes n}$. These basis elements are given as tensor products of the form

$$e_{\mu_1} \otimes \cdots \otimes e_{\mu_m} \otimes \widetilde{e}^{\nu_1} \otimes \cdots \otimes \widetilde{e}^{\nu_n}.$$

In this tensor basis, a tensor T is written as the linear combination

$$T = T^{\mu_1 \ldots \mu_m}{}_{\nu_1 \ldots \nu_n} \, e_{\mu_1} \otimes \cdots \otimes e_{\mu_m} \otimes \widetilde{e}^{\nu_1} \otimes \cdots \otimes \widetilde{e}^{\nu_n},$$

with its components given by

$$T^{\mu_1 \ldots \mu_m}{}_{\nu_1 \ldots \nu_n} = T(\widetilde{e}^{\mu_1}, \ldots, \widetilde{e}^{\mu_m}, e_{\nu_1}, \ldots, e_{\nu_n}).$$

The components depend on the basis choice, so a change of basis means a change of the tensor components. This is the concept for the tensor definition as usually seen in physics books. A change of the basis $\{e_\mu\}$ like

$$e_\mu \mapsto e'_\mu = \Lambda_\mu{}^\nu e_\nu,$$

or, equivalently, a change in the coordinates x^μ, like

$$x^\mu \mapsto x'^\mu = \Lambda^\mu{}_\nu x^\nu,$$

results in the following change of the tensor components:

$$T'^{\mu_1 \ldots \mu_m}{}_{\nu_1 \ldots \nu_n} = \Lambda^{\mu_1}{}_{\rho_1} \cdots \Lambda^{\mu_m}{}_{\rho_m} \, \Lambda_{\nu_1}{}^{\sigma_1} \cdots \Lambda_{\nu_n}{}^{\sigma_n} \, T^{\rho_1 \ldots \rho_m}{}_{\sigma_1 \ldots \sigma_n}.$$

This is the well-known transformation law for a tensor of type (m, n). Starting with tensor components transforming as above, one can construct multilinear maps and arrive to our initial definition of tensors. In tensorial equations, we use primarily the representation with components and always imply that the equations are valid also for the coordinate-independent definitions of the tensors.

Inner Product and Norm of Tensors

Suppose we have a vector space V with an inner product $\langle \cdot, \cdot \rangle$, or, equivalently, a metric $\eta_{\mu\nu}$. Then an *inner product of tensors* can be defined naturally on the tensor space $V^{\otimes m}$. For the tensors T and S of $V^{\otimes m}$, we define their inner product $\langle T, S \rangle$ as

$$\langle T, S \rangle \equiv \eta_{\mu_1 \nu_1} \cdots \eta_{\mu_m \nu_m} \, T^{\mu_1 \ldots \mu_m} \, S^{\nu_1 \ldots \nu_m}.$$

If the tensors T and S are given as products of vectors, i.e. $T = v_{(1)} \otimes \cdots \otimes v_{(m)}$ and $S = w_{(1)} \otimes \cdots \otimes w_{(m)}$, then their inner product is

$$\langle T, S \rangle = \langle v_{(1)}, w_{(1)} \rangle \cdots \langle v_{(m)}, w_{(m)} \rangle.$$

The *squared (Euclidean) norm* $|T|^2$ of a tensor T of $V^{\otimes m}$ is defined as

$$|T|^2 \equiv \langle T, T \rangle = T_{\mu_1 \ldots \mu_m} \, T^{\mu_1 \ldots \mu_m}.$$

In the case the inner product $\langle \cdot, \cdot \rangle$ is positive definite, we can consider the positive (*Euclidean*) *norm* $|T| = \sqrt{\langle T, T \rangle}$. Similar definitions for the inner product and the norm apply for the tensor spaces $\widetilde{V}^{\otimes n}$ and $V^{\otimes m} \otimes \widetilde{V}^{\otimes n}$.

Contraction and Trace

Using the components $T^{\mu_1 \ldots \mu_m}{}_{\nu_1 \ldots \nu_n}$ of a tensor of type (m, n), the contraction of two indices, one upper and one lower, each one at a certain position, is the operation which produces the sum $T^{\mu_1 \ldots \rho \ldots \mu_m}{}_{\nu_1 \ldots \rho \ldots \nu_n}$ over the index ρ. The result is a tensor of type $(m - 1, n - 1)$. The *trace* $\mathrm{Tr}(T)$ of a tensor T of type (m, m) is the number obtained when all indices are contracted, $\mathrm{Tr}(T) = T^{\rho_1 \ldots \rho_m}{}_{\rho_1 \ldots \rho_m}$. The trace, as a scalar quantity, does not change under basis transformations. For an (m, m)-tensor $T^{\mu_1 \ldots \mu_m}{}_{\nu_1 \ldots \nu_m}$, we define its *traceless part* $\overline{T}$ to be the (m, m)-tensor given as

$$\overline{T}^{\mu_1 \ldots \mu_m}{}_{\nu_1 \ldots \nu_m} \equiv T^{\mu_1 \ldots \mu_m}{}_{\nu_1 \ldots \nu_m} - \frac{1}{D^m} \, \mathrm{Tr}(T) \, \delta^{\mu_1}{}_{\nu_1} \cdots \delta^{\mu_m}{}_{\nu_m}.$$

For a space with an indefinite Lorentzian metric $\eta_{\mu\nu} = \mathrm{diag}(1, -1, \ldots, -1)$, where the -1 appears D times on the diagonal, the formula becomes

$$\overline{T}^{\mu_1 \ldots \mu_m}{}_{\nu_1 \ldots \nu_m} \equiv T^{\mu_1 \ldots \mu_m}{}_{\nu_1 \ldots \nu_m} - \frac{1}{d^m} \, \mathrm{Tr}(T) \, \eta^{\mu_1}{}_{\nu_1} \cdots \eta^{\mu_m}{}_{\nu_m},$$

where $d = 1 + D$ is the overall dimension of the space.

Symmetrization, Antisymmetrization

In the following, we restrict ourselves to tensors $T_{\nu_1 \ldots \nu_n}$ of type $(0, n)$. The results obtained can readily translated to tensors of type $(n, 0)$. For simplicity, we do not consider tensors with both co- and contravariant indices. We define the specific parts of tensors which behave in a definite way under permutation of indices. The *symmetric part* of the tensor $T_{\nu_1 \ldots \nu_n}$ is defined as

$$\mathrm{Sym}(T)_{\nu_1 \ldots \nu_n} \equiv \frac{1}{n!} \sum_{\pi \in S_n} T_{\nu_{\pi(1)} \ldots \nu_{\pi(n)}}.$$

The *antisymmetric* (or *alternating*) *part* of the tensor $T_{\nu_1 \ldots \nu_n}$ is defined as

$$\mathrm{Alt}(T)_{\nu_1 \ldots \nu_n} \equiv \frac{1}{n!} \sum_{\pi \in S_n} (-1)^\pi \, T_{\nu_{\pi(1)} \ldots \nu_{\pi(n)}}.$$

In the above sums, the permutation π runs through all elements in the permutation group S_n. The factor $(-1)^\pi$ is the signum of the permutation π and given by

$$(-1)^\pi = \begin{cases} +1, & \text{if } \pi \text{ is an even permutation of } (1, \ldots, n), \\ -1, & \text{if } \pi \text{ is an odd permutation of } (1, \ldots, n). \end{cases}$$

As desired, the symmetric part is symmetric under any permutation of indices. The antisymmetric part is antisymmetric under any permutation of indices. These properties are independent of the basis choice. A tensor $T_{\nu_1 \ldots \nu_n}$ is called *totally symmetric* if

$$T_{\nu_1 \ldots \nu_n} = \mathrm{Sym}(T)_{\nu_1 \ldots \nu_n}.$$

A tensor $T_{\nu_1...\nu_n}$ is called *totally antisymmetric* (or *alternating*) if

$$T_{\nu_1...\nu_n} = \text{Alt}(T)_{\nu_1...\nu_n}.$$

Occasionally, the symmetrization operation is also denoted with round brackets, $T_{(\nu_1...\nu_n)} \equiv \text{Sym}(T)_{\nu_1...\nu_n}$, whereas the antisymmetrization is denoted with square brackets, $T_{[\nu_1...\nu_n]} \equiv \text{Alt}(T)_{\nu_1...\nu_n}$. For type $(0,2)$ tensors $T_{\mu\nu}$, there is always the decomposition into a symmetric and an antisymmetric part possible, $T_{\mu\nu} = T_{(\mu\nu)} + T_{[\mu\nu]}$. The symmetric $T_{(\mu\nu)}$ has $D(D+1)/2$ independent components, while the antisymmetric $T_{[\mu\nu]}$ has $D(D-1)/2$ independent components.

Vector Product vs. Antisymmetric Tensors

The usual *vector* (or *cross*) *product* $c = a \times b$ of two vectors a and b, defined in Cartesian coordinates as

$$c_j = \epsilon_{jkl} a_k b_l,$$

is only possible in 3 dimensions. Its generalization to arbitrary finite dimensions is given by antisymmetric tensors c_{kl} of the form

$$c_{kl} = a_k b_l - a_l b_k.$$

The relation between the two definitions is given by

$$c_j = \frac{1}{2} \epsilon_{jkl} c_{kl}.$$

We prefer the explicit antisymmetric tensor notation, since it is valid in all dimensions. For example, the three Lie group generators of rotations can be written as a cross product $L = -i\, x \times \nabla$, but we prefer the antisymmetric tensor notation $L_{kl} = -i\,(x_k \partial_l - x_l \partial_k)$, from which the generalization to higher dimensional cases is immediate.

Exterior Algebra

The generalization of the cross product $a \times b$ of two vectors a and b in 3 dimensions is the *exterior* (or *wedge*) *product* $v \wedge w$ of two vectors v and w in D dimensions, defined as the antisymmetrized tensor product

$$v \wedge w \equiv v \otimes w - w \otimes v.$$

When we choose a basis $\{e_\mu\}$ in V, the exterior product is given as

$$v \wedge w = \sum_{\mu < \nu} (v^\mu w^\nu - v^\nu w^\mu)\, e_\mu \wedge e_\nu,$$

with the basis vectors $e_\mu \wedge e_\nu = e_\mu \otimes e_\nu - e_\nu \otimes e_\mu$. Going one step further, the exterior product can be extended to arbitrary tensors of order n. The definition can be done for co- as for contravariant tensors of order n. Because the case of covariant tensors is more interesting in applications, we focus on this one. The results can be readily translated to the contravariant case. First, note that the antisymmetric $(0,n)$-tensors $\alpha_{\nu_1...\nu_n}$ (simply called n-tensors) constitute a vector space, which we denote $\bigwedge^n \widetilde{V}$. Now the *exterior* (or *wedge*) *product* is the map

$$\wedge : \overset{m}{\bigwedge} \widetilde{V} \times \overset{n}{\bigwedge} \widetilde{V} \to \overset{m+n}{\bigwedge} \widetilde{V},$$

which assigns to the m-tensor α and the n-tensor β the $(m+n)$-tensor $\alpha \wedge \beta$ defined as

$$\alpha \wedge \beta \equiv \frac{(m+n)!}{m!\,n!} \text{Alt}(\alpha \otimes \beta).$$

The exterior product is bilinear, distributive, and associative. In addition, it is *graded antisymmetric*, so that for the above tensors α and β it is

$$\alpha \wedge \beta = (-1)^{m \cdot n} \beta \wedge \alpha.$$

Basis elements of $\bigwedge^n \widetilde{V}$ have the form

$$e^{\nu_1} \wedge \cdots \wedge e^{\nu_n} = n! \, \mathrm{Alt}(e^{\nu_1} \otimes \cdots \otimes e^{\nu_n}) = \sum_{\pi \in S_n} (-1)^{\pi} \, e^{\nu_{\pi(1)}} \otimes \cdots \otimes e^{\nu_{\pi(n)}},$$

with $\{e^{\nu}\}$ being the basis of $\widetilde{V}$. Note that we use the simpler notation for the dual basis without the tilde henceforth. The dimensionality of $\bigwedge^n \widetilde{V}$ is equal

$$\binom{D}{n} = \frac{D!}{n!(D-n)!},$$

for $0 \leq n \leq D$. For $n > D$, the resulting space is the trivial one containing only the zero vector. Note the special case of $\bigwedge^D \widetilde{V}$, which is 1-dimensional with a possible basis given by $e^1 \wedge \cdots \wedge e^D$. Let us proceed to the construction of the full algebra. We take the finite direct sum

$$\bigwedge \widetilde{V} \equiv \bigoplus_{n=0}^{D} \left(\bigwedge^n \widetilde{V} \right),$$

since the wedge products with more than D factors vanish identically. The wedge product makes this direct sum an algebra, denoted $\bigwedge \widetilde{V}$, the so-called *exterior algebra*, or *Grassmann algebra* of $\widetilde{V}$. Hermann G. Grassmann introduced the exterior algebra $\bigwedge \mathbb{R}^3$ in 1844. We agree that $\bigwedge^0 \widetilde{V} = \mathbb{F}$ and that $\bigwedge^1 \widetilde{V} = \widetilde{V}$. The dimension of the exterior algebra is finite and equal

$$\sum_{n=0}^{D} \binom{D}{n} = 2^D.$$

As an example, the exterior algebra for $D = 3$ has eight dimensions and a possible basis is $\{1, e^1, e^2, e^3, e^1 \wedge e^2, e^2 \wedge e^3, e^3 \wedge e^1, e^1 \wedge e^2 \wedge e^3\}$. Finally, let us write down how an n-tensor α is written in a basis of $\bigwedge^n \widetilde{V}$. It is

$$\alpha = \frac{1}{n!} \sum_{\nu_1, \ldots, \nu_n} \alpha_{\nu_1 \ldots \nu_n} \, e^{\nu_1} \wedge \cdots \wedge e^{\nu_n} = \sum_{\nu_1 < \ldots < \nu_n} \alpha_{\nu_1 \ldots \nu_n} \, e^{\nu_1} \wedge \cdots \wedge e^{\nu_n}.$$

In fact, in the above sums one can use the antisymmetrized $\alpha_{[\nu_1 \ldots \nu_n]}$ factors. When an inner product is given on $\widetilde{V}$, an inner product can be defined naturally on the space $\bigwedge^n \widetilde{V}$. For any two n-tensors $\alpha^1 \wedge \cdots \wedge \alpha^n$ and $\beta^1 \wedge \cdots \wedge \beta^n$ the *inner product* is defined as

$$\langle \alpha^1 \wedge \cdots \wedge \alpha^n, \beta^1 \wedge \cdots \wedge \beta^n \rangle \equiv \epsilon_{\nu_1 \ldots \nu_n} \langle \alpha^{\nu_1}, \beta^1 \rangle \cdots \langle \alpha^{\nu_n}, \beta^n \rangle.$$

This leads to the notion of the determinant.

Determinants and Orientation

For any $D \times D$-matrix $A = (A^{\mu}{}_{\nu})$ acting on a D-dimensional space V, the *determinant* is the scalar quantity $\det A$ defined as

$$\det A \equiv \sum_{\pi \in S_D} (-1)^{\pi} A^1{}_{\pi(1)} \cdots A^D{}_{\pi(D)}$$

$$= \sum_{\pi \in S_D} (-1)^{\pi} A^{\pi(1)}{}_1 \cdots A^{\pi(D)}{}_D.$$

The determinant is invariant under similarity transformations $A \mapsto SAS^{-1}$, as it is $\det(S^{-1}) = (\det S)^{-1}$. The determinant appears naturally in antisymmetrized tensors. Consider D-tensors $e^1 \wedge \cdots \wedge e^D$ and $e'^1 \wedge \cdots \wedge e'^D$ of $\bigwedge^D \widetilde{V}$, where $\{e^\mu\}$ and $\{e'^\mu\}$ each constitute a basis of $\widetilde{V}$ and are related by $e'^\mu = \Lambda^\mu{}_\nu e^\nu$. Then, is is easily seen that it is

$$e'^1 \wedge \cdots \wedge e'^D = (\det \Lambda)\, e^1 \wedge \cdots \wedge e^D.$$

The *orientation* of a vector space $\widetilde{V}$ is introduced by convention and can be based on a basis choice $\{e^\mu\}$. Practically, one chooses a basis $\{e^\mu\}$ and declares for it to have a positive orientation. Then, all bases (and all sets of linearly independent vectors) either have a *positive* or a *negative orientation*. The determinant supplies us with the means to determine if a basis is positively or negatively oriented. Suppose we have a second set of basis elements $\{e'^\mu\}$ with $e'^\mu = \Lambda^\mu{}_\nu e^\nu$. Then, if $\det \Lambda > 0$, the new basis is *positively oriented*, and if $\det \Lambda < 0$, the new basis is *negatively oriented*. An orientation in the space $\widetilde{V}$ induces an orientation in the dual space and in the space $\bigwedge^D \widetilde{V}$. Consequently, a basis element $e^1 \wedge \cdots \wedge e^D$ of $\bigwedge^D \widetilde{V}$ has either a positive or a negative orientation.

In the following, we derive some formulae used in practical calculations with determinants. Given a $D \times D$-matrix $A = (A^\mu{}_\nu)$, we use the shorthand notation $a \equiv \det A$ for the determinant below. For any fixed index value $1 \leq \mu \leq D$, the determinant a can be calculated by expanding with respect to the μth row as

$$a = A^\mu{}_1 C^1{}_\mu + \cdots + A^\mu{}_D C^D{}_\mu = \sum_{\nu=1}^{D} A^\mu{}_\nu C^\nu{}_\mu.$$

Note that there is no summation over the fixed index μ above. The coefficient $C^\nu{}_\mu$ is the *cofactor* for the matrix element $A^\mu{}_\nu$ and is calculated as the determinant of the $(D-1) \times (D-1)$-matrix obtained by eliminating the μth row and the νth column, multiplied by $(-1)^{\mu+\nu}$. Alternatively, one can calculate the determinant a by fixing a certain column index value $1 \leq \nu \leq D$ and write

$$a = A^1{}_\nu C^\nu{}_1 + \cdots + A^D{}_\nu C^\nu{}_D = \sum_{\mu=1}^{D} A^\mu{}_\nu C^\nu{}_\mu.$$

The two last formulae for the determinant can be generalized to

$$a \delta^\mu_\rho = \sum_{\nu=1}^{D} A^\mu{}_\nu C^\nu{}_\rho,$$

and

$$a \delta^\rho_\nu = \sum_{\mu=1}^{D} A^\mu{}_\nu C^\rho{}_\mu.$$

In the case that it is $a \neq 0$, we are led to a formula for the inverse matrix A^{-1} satisfying

$$AA^{-1} = A^{-1}A = 1.$$

The inverse matrix A^{-1} has the elements $(A^{-1})^\mu{}_\nu$, which can be calculated as

$$(A^{-1})^\mu{}_\nu = \frac{C^\mu{}_\nu}{a}.$$

Suppose that the matrix elements $A^\mu{}_\nu(\tau_1, \dots \tau_n)$ are functions of n real parameters $\tau_1, \dots \tau_n$. According to the chain rule, we have for the partial derivative

$$\frac{\partial a}{\partial \tau_j} = \sum_{\mu,\nu=1}^{D} \frac{\partial a}{\partial A^\mu{}_\nu} \frac{\partial A^\mu{}_\nu}{\partial \tau_j}.$$

Due to the expansion of the determinant with cofactors, it is

$$\frac{\partial a}{\partial A^\mu{}_\nu} = C^\nu{}_\mu,$$

so that we can write the formula

$$\frac{\partial a}{\partial \tau_j} = \sum_{\mu,\nu=1}^{D} C^\nu{}_\mu \frac{\partial A^\mu{}_\nu}{\partial \tau_j} = \sum_{\mu,\nu=1}^{D} a\,(A^{-1})^\nu{}_\mu \frac{\partial A^\mu{}_\nu}{\partial \tau_j},$$

a result that is used in Sections 18.2 and 18.3.

Tensor Densities

The *Jacobian* of a linear transformation $x'^\mu = \Lambda^\mu{}_\nu x^\nu$ of the coordinates is the determinant

$$J = \det \Lambda.$$

A *tensor density* (or *relative tensor*) *of type (m,n) and with weight w* is a multicomponent quantity $T^{\mu_1 \dots \mu_m}{}_{\nu_1 \dots \nu_n}$ which transforms as

$$T'^{\mu_1 \dots \mu_m}{}_{\nu_1 \dots \nu_n} = (\det \Lambda)^w\, \Lambda^{\mu_1}{}_{\rho_1} \cdots \Lambda^{\mu_m}{}_{\rho_m}\, \Lambda_{\nu_1}{}^{\sigma_1} \cdots \Lambda_{\nu_n}{}^{\sigma_n}\, T^{\rho_1 \dots \rho_m}{}_{\sigma_1 \dots \sigma_n}.$$

Normally, one restricts this definition to integer values $w = 0, \pm 1, \pm 2, \dots$ for the weight, but the definition allows any real number. For $w = 0$, we recover the usual definition of tensors. The term tensor density stems from the fact that some quantities result to tensors only after a volume integration. Tensor densities of the same type and weight can be added and multiplied by numbers. The tensor product of two tensors densities of type and weight (m_1, n_1) and w_1 and respectively (m_2, n_2) and w_2 is a tensor density of type $(m_1 + m_2, n_1 + n_2)$ and weight $w_1 + w_2$. If possible, we prefer to deal with absolute tensors than with tensor densities and there are means to amend densities in a way that they produce tensors.

A prominent example of a tensor density is the determinant of the metric. The metric $\eta_{\mu\nu}$ transforms as

$$\eta'_{\mu\nu} = \Lambda_\mu{}^\rho \Lambda_\nu{}^\sigma \eta_{\rho\sigma}.$$

Taking the determinant, leads to

$$\det \eta' = (\det \Lambda)^{-2} \det \eta.$$

When we take the square root of the absolute value, we obtain the transformation formula

$$\sqrt{|\det \eta'|} = (\det \Lambda)^{-1} \sqrt{|\det \eta|}.$$

i.e. the square root of the determinant transforms as a scalar density with weight $w = -1$. The square root of the determinant of the metric is crucial for the definition of volume elements.

Generalized Kronecker and Levi-Civita Tensor

We consider a D-dimensional space V with Euclidean metric signature. The *Kronecker delta* (*Leopold Kronecker*) δ^μ_ν is defined as

$$\delta^\mu_\nu \equiv \begin{cases} 1, & \text{if } \mu = \nu, \\ 0, & \text{if } \mu \neq \nu. \end{cases}$$

The matrix (δ^μ_ν) is simply the $D \times D$ identity matrix $\text{diag}(1, \ldots, 1)$. The Kronecker delta constitutes the components of the identity tensor I on V,

$$I = \delta^\mu_\nu\, e_\mu \otimes e^\nu,$$

and represents an element of $V \otimes \widetilde{V}$. The components of the Kronecker delta are the same in all coordinate systems. Indeed, under a coordinate transformation $x^\mu \mapsto x'^\mu = \Lambda^\mu{}_\nu x^\nu$, the components δ^μ_ν transform as

$$\delta'^\mu_\nu = \Lambda^\mu{}_\rho \Lambda_\nu{}^\sigma \delta^\rho_\sigma = \Lambda^\mu{}_\rho \left(\Lambda^{-1}\right)^\rho{}_\nu = \delta^\mu_\nu,$$

i.e. the values are the same in all coordinate systems. For $n \leq D$, we define the *generalized Kronecker delta* $\delta^{\mu_1 \ldots \mu_n}_{\nu_1 \ldots \nu_n}$ by

$$\delta^{\mu_1 \ldots \mu_n}_{\nu_1 \ldots \nu_n} \equiv \det \begin{pmatrix} \delta^{\mu_1}_{\nu_1} & \cdots & \delta^{\mu_1}_{\nu_n} \\ \vdots & \ddots & \vdots \\ \delta^{\mu_n}_{\nu_1} & \cdots & \delta^{\mu_n}_{\nu_n} \end{pmatrix}.$$

The determinant on the rhs is a linear combination of $n!$ terms, where each term is the product of n usual Kronecker deltas. Hence, the $\delta^{\mu_1 \ldots \mu_n}_{\nu_1 \ldots \nu_n}$ are the components of a tensor of type (n, n) of the form

$$\delta^{\mu_1 \ldots \mu_n}_{\nu_1 \ldots \nu_n}\, e_{\mu_1} \otimes \cdots \otimes e_{\mu_n} \otimes e^{\nu_1} \otimes \cdots \otimes e^{\nu_n}$$

belonging to the space $V^{\otimes n} \otimes \widetilde{V}^{\otimes n}$. One can inspect again that the components $\delta^{\mu_1 \ldots \mu_n}_{\nu_1 \ldots \nu_n}$ keep their values in all coordinate systems unchanged, i.e.

$$\delta'^{\mu_1 \ldots \mu_n}_{\nu_1 \ldots \nu_n} = \delta^{\mu_1 \ldots \mu_n}_{\nu_1 \ldots \nu_n}.$$

The generalized Kronecker delta is totally antisymmetric in the upper row and in the lower row of indices. We note the contraction formulae

$$\delta^{\mu_1 \ldots \mu_p\, \mu_{p+1} \ldots \mu_n}_{\nu_1 \ldots \nu_p\, \mu_{p+1} \ldots \mu_n} = \frac{(D-p)!}{(D-n)!}\, \delta^{\mu_1 \ldots \mu_p}_{\nu_1 \ldots \nu_p},$$

and

$$\delta^{\mu_1 \ldots \mu_n}_{\mu_1 \ldots \mu_n} = \frac{D!}{(D-n)!},$$

and, with $n = D$,

$$\delta^{\mu_1 \ldots \mu_D}_{\mu_1 \ldots \mu_D} = D!.$$

For a D-dimensional space V with Euclidean metric, we define the *Levi-Civita* (*Tullio Levi-Civita*) *epsilon* (or *alternating*) *symbol* as

$$\varepsilon_{\mu_1 \ldots \mu_D} \equiv \delta^{1 \ldots \ldots D}_{\mu_1 \ldots \mu_D},$$

and

$$\varepsilon^{\mu_1\cdots\mu_D} \equiv \delta^{\mu_1\cdots\mu_D}_{1\ldots\ldots D}.$$

The epsilon symbol is totally antisymmetric in its indices and has the values

$$\varepsilon_{\mu_1\ldots\mu_D} = \begin{cases} +1, & \text{if } (\mu_1,\ldots,\mu_D) \text{ is an even permutation of } (1,\ldots,D), \\ -1, & \text{if } (\mu_1,\ldots,\mu_D) \text{ is an odd permutation of } (1,\ldots,D), \\ 0, & \text{in all other cases.} \end{cases}$$

The values for $\varepsilon^{\mu_1\cdots\mu_D}$ are defined in the same way. Some basic properties are

$$\varepsilon_{1\ldots D} = \varepsilon^{1\ldots D} = 1,$$

as well as

$$\varepsilon^{\mu_1\cdots\mu_D} \varepsilon_{\nu_1\ldots\nu_D} = \delta^{\mu_1\cdots\mu_D}_{\nu_1\ldots\nu_D},$$

and

$$\varepsilon^{\mu_1\cdots\mu_D} \varepsilon_{\mu_1\ldots\mu_D} = D!.$$

The Levi-Civita epsilon symbol is closely related to determinants. For a $D \times D$-matrix $A = (A^\mu{}_\nu)$, the determinant $\det A$ can be expressed as

$$\det A = \varepsilon^{\nu_1\cdots\nu_D} A^1{}_{\nu_1} \cdots A^D{}_{\nu_D}$$
$$= \varepsilon_{\mu_1\ldots\mu_D} A^{\mu_1}{}_1 \cdots A^{\mu_D}{}_D.$$

It is also

$$(\det A)\,\varepsilon^{\mu_1\cdots\mu_D} = \varepsilon^{\nu_1\cdots\nu_D} A^{\mu_1}{}_{\nu_1} \cdots A^{\mu_D}{}_{\nu_D},$$

and

$$(\det A)\,\varepsilon_{\nu_1\ldots\nu_D} = \varepsilon_{\mu_1\ldots\mu_D} A^{\mu_1}{}_{\nu_1} \cdots A^{\mu_D}{}_{\nu_D}.$$

Under a coordinate transformation, the Levi-Civita ε-symbol transforms as

$$\varepsilon'_{\mu_1\ldots\mu_D} = \delta'^{1\ldots\ldots D}_{\mu_1\ldots\mu_D} = \Lambda^1{}_{\rho_1} \cdots \Lambda^D{}_{\rho_D} \Lambda_{\mu_1}{}^{\nu_1} \cdots \Lambda_{\mu_D}{}^{\nu_D} \delta^{\rho_1\ldots\rho_D}_{\nu_1\ldots\nu_D}.$$

Noting that it is

$$\Lambda^1{}_{\rho_1} \cdots \Lambda^D{}_{\rho_D} \,\varepsilon^{\rho_1\cdots\rho_D} = \det\Lambda,$$

we see that it is

$$\varepsilon'_{\mu_1\ldots\mu_D} = (\det\Lambda)\, \Lambda_{\mu_1}{}^{\nu_1} \cdots \Lambda_{\mu_D}{}^{\nu_D} \varepsilon_{\nu_1\ldots\nu_D},$$

i.e. the single ε-symbols $\varepsilon_{\mu_1\ldots\mu_D}$ are the components of a *tensor density* of type $(0, D)$ and weight $w = 1$. Similarly, the $\varepsilon^{\mu_1\cdots\mu_D}$ are the components of a *tensor density* of type $(D, 0)$ and weight $w = -1$. The epsilon symbol possesses components that remain constant in all coordinate systems. Indeed, noting that

$$\Lambda_{\mu_1}{}^{\nu_1} \cdots \Lambda_{\mu_D}{}^{\nu_D} \varepsilon_{\nu_1\ldots\nu_D} = (\det\Lambda)^{-1}\, \varepsilon_{\mu_1\ldots\mu_D},$$

we obtain the invariance of the epsilon symbol components,

$$\varepsilon'_{\mu_1\ldots\mu_D} = \varepsilon_{\mu_1\ldots\mu_D}.$$

The same holds for the contravariant components $\varepsilon^{\mu_1\cdots\mu_D}$. Since the Kronecker delta and the Levi-Civita epsilon symbol keep their values constant in all coordinate systems, they are called also *numerical tensors*.

We can construct an absolute tensor from the ε-symbol. The additional structure we need is a metric $\eta_{\mu\nu}$. We define the *Levi-Civita epsilon tensor* ϵ with its *covariant* components $\epsilon_{\mu_1...\mu_D}$ by

$$\epsilon_{\mu_1...\mu_D} \equiv \sqrt{|\eta|}\, \varepsilon_{\mu_1...\mu_D},$$

where $\eta \equiv \det(\eta_{\mu\nu})$ is the determinant of the metric. One can inspect that the epsilon tensor indeed transforms as an absolute tensor,

$$\epsilon'_{\mu_1...\mu_D} = \Lambda_{\mu_1}{}^{\nu_1} \cdots \Lambda_{\mu_D}{}^{\nu_D}\, \epsilon_{\nu_1...\nu_D}.$$

Written out, the epsilon tensor ϵ is given by

$$\epsilon = \epsilon_{\mu_1...\mu_D}\, e^{\mu_1} \otimes \cdots \otimes e^{\mu_D},$$

and is an element of $\widetilde{V}^{\otimes D}$. The *contravariant* version $\epsilon^{\mu_1\cdots\mu_D}$ of the epsilon tensor is defined as

$$\epsilon^{\mu_1\cdots\mu_D} \equiv \eta^{\mu_1\nu_1} \cdots \eta^{\mu_D\nu_D}\, \epsilon_{\nu_1...\nu_D},$$

which is equivalent to

$$\epsilon^{\mu_1\cdots\mu_D} = \frac{1}{\sqrt{|\eta|}}\, \varepsilon^{\mu_1\cdots\mu_D}.$$

The question whether the tensor density version or the absolute tensor version of the Levi-Civita epsilon is more suitable depends on the situation.

In the case we have a $(1 + D)$-dimensional space V with Lorentzian metric at hand, with $\eta_{\mu\nu} = \mathrm{diag}(1, -1, \ldots, -1)$, where the -1 appears D times on the diagonal, the definitions of the epsilon symbol and the epsilon tensor need a bit of care due to minus signs appearing. The alternating epsilon symbol $\varepsilon_{\mu_0\mu_1...\mu_D}$ is defined as before, but now with the normalization

$$\varepsilon_{01...D} = (-1)^D.$$

Consequently, the contravariant components are normalized as

$$\varepsilon^{01...D} = 1.$$

The epsilon tensor $\epsilon_{\mu_0\mu_1...\mu_D}$ is defined as before and leads to the basic values

$$\epsilon_{01...D} = (-1)^D \sqrt{|\eta|} \quad \text{and} \quad \epsilon^{01...D} = \frac{1}{\sqrt{|\eta|}}.$$

For the $(1 + D)$-dimensional Lorentzian case, we note the contraction

$$\epsilon^{\mu_0\mu_1\cdots\mu_D} \epsilon_{\mu_0\mu_1...\mu_D} = (-1)^D (D+1)!.$$

Finally, in the general case of a curved manifold, the flat metric $\eta_{\mu\nu}$ has to be replaced by the curved metric $g_{\mu\nu}(x)$.

In practical calculations, one encounters indices reshufflings and contractions of the epsilon tensor. Let us consider the 3-dimensional epsilon tensor ϵ_{jkl} defined on the Euclidean space with the metric $\delta_{jk} = \mathrm{diag}(1, 1, 1)$. Note that

$$\epsilon_{123} = \epsilon^{123} = 1.$$

Shifting all indices by one index position in a cyclic way results in

$$\epsilon_{jkl} = \epsilon_{klj}.$$

Contracting one index yields

$$\epsilon^{jkl}\epsilon_{mnl} = \delta^j{}_m\delta^k{}_n - \delta^j{}_n\delta^k{}_m.$$

Contracting two indices yields

$$\epsilon^{jkl}\epsilon_{mkl} = 2\delta^j{}_m.$$

Contracting all three indices leads to

$$\epsilon^{jkl}\epsilon_{jkl} = 6.$$

Another useful formula is the *Schouten identity* (*Jan Arnoldus Schouten*),

$$\delta^{jk}\epsilon^{lmn} = \delta^{jl}\epsilon^{kmn} + \delta^{jm}\epsilon^{lkn} + \delta^{jn}\epsilon^{lmk}.$$

Now let us consider the 4-dimensional epsilon tensor $\epsilon_{\mu\nu\rho\sigma}$ defined on Minkowski space with the metric $\eta_{\mu\nu} = \mathrm{diag}(1,-1,-1,-1)$. The basic values are

$$\epsilon_{0123} = -\epsilon^{0123} = -1.$$

In this case, shifting all indices by one index position in a cyclic way results in

$$\epsilon_{\mu\nu\rho\sigma} = -\epsilon_{\nu\rho\sigma\mu}.$$

The contraction of one index leads to

$$\begin{aligned}
\epsilon^{\mu\nu\rho\sigma}\epsilon_{\alpha\beta\gamma\sigma} = &- \delta^\mu{}_\alpha\delta^\nu{}_\beta\delta^\rho{}_\gamma - \delta^\mu{}_\beta\delta^\nu{}_\gamma\delta^\rho{}_\alpha - \delta^\mu{}_\gamma\delta^\nu{}_\alpha\delta^\rho{}_\beta \\
&+ \delta^\mu{}_\beta\delta^\nu{}_\alpha\delta^\rho{}_\gamma + \delta^\mu{}_\alpha\delta^\nu{}_\gamma\delta^\rho{}_\beta + \delta^\mu{}_\gamma\delta^\nu{}_\beta\delta^\rho{}_\alpha,
\end{aligned}$$

of two indices to

$$\epsilon^{\mu\nu\rho\sigma}\epsilon_{\alpha\beta\rho\sigma} = -2\left(\delta^\mu{}_\alpha\delta^\nu{}_\beta - \delta^\mu{}_\beta\delta^\nu{}_\alpha\right),$$

of three indices to

$$\epsilon^{\mu\nu\rho\sigma}\epsilon_{\alpha\nu\rho\sigma} = -6\,\delta^\mu{}_\alpha,$$

and of all four indices to

$$\epsilon^{\mu\nu\rho\sigma}\epsilon_{\mu\nu\rho\sigma} = -24.$$

We note also the 4-dimensional *Schouten identity*,

$$\eta^{\kappa\lambda}\epsilon^{\mu\nu\rho\sigma} = \eta^{\kappa\mu}\epsilon^{\lambda\nu\rho\sigma} + \eta^{\kappa\nu}\epsilon^{\mu\lambda\rho\sigma} + \eta^{\kappa\rho}\epsilon^{\mu\nu\lambda\sigma} + \eta^{\kappa\sigma}\epsilon^{\mu\nu\rho\lambda}.$$

Volume Elements

In $\mathbb{R}^3$, the exterior product of three linearly independent vectors $\boldsymbol{a} = a^k\boldsymbol{e}_k$, $\boldsymbol{b} = b^k\boldsymbol{e}_k$ and $\boldsymbol{c} = c^k\boldsymbol{e}_k$ is

$$\boldsymbol{a}\wedge\boldsymbol{b}\wedge\boldsymbol{c} = \det\begin{pmatrix} a^1 & b^1 & c^1 \\ a^2 & b^2 & c^2 \\ a^3 & b^3 & c^3 \end{pmatrix}\boldsymbol{e}_1\wedge\boldsymbol{e}_2\wedge\boldsymbol{e}_3.$$

This is the *oriented volume* spanned by the three vectors and is an element of the space $\bigwedge^3\mathbb{R}^3$. The scalar value of the oriented volume is equal to the determinant value above. If we restrict to *orthonormal* basis vectors $\{\boldsymbol{e}_k\}$, we have the *oriented unit volume element*

$$\mathrm{dvol} \equiv \boldsymbol{e}_1\wedge\boldsymbol{e}_2\wedge\boldsymbol{e}_3.$$

More generally, for the space $\bigwedge^D \widetilde{V}$, the *oriented volume* of D linearly independent forms $\alpha^1, \ldots, \alpha^D$, with $\alpha^\mu = A^\mu{}_\nu e^\nu$ for $\mu = 1, \ldots, D$, where $\{e^\mu\}$ is an orthonormal basis of the space $\widetilde{V}$, is given by

$$\alpha^1 \wedge \cdots \wedge \alpha^D = (\det A)\, e^1 \wedge \cdots \wedge e^D.$$

The *oriented unit volume element* (or *form*) dvol in D dimensions is

$$\mathrm{dvol} \equiv e^1 \wedge \cdots \wedge e^D.$$

The scalar value of the unit volume element is equal to one. According to our previous considerations about determinants, the unit volume element is invariant under proper orthogonal transformations of $SO(D)$. However, this unit volume element is *not* invariant under general linear transformations. It is possible to achieve a definition of the unit volume element, which ensures invariance under any linear transformation in $\widetilde{V}$. To this end, we need to have a vector space equipped with a metric, i.e. a scalar product. We recall that the induced scalar product of two D-tensors $\alpha^1 \wedge \cdots \wedge \alpha^D$ and $\beta^1 \wedge \cdots \wedge \beta^D$ in the space $\bigwedge^D \widetilde{V}$ is given by

$$\langle \alpha^1 \wedge \cdots \wedge \alpha^D, \beta^1 \wedge \cdots \wedge \beta^D \rangle = \det\left(\langle \alpha^\mu, \beta^\nu \rangle\right).$$

Now if the forms α^μ are given by $\alpha^\mu = A^\mu{}_\nu e^\nu$, then it is

$$(\det A)^2 = \det\left(\langle \alpha^\mu, \alpha^\nu \rangle\right),$$

since it is $\langle e^1 \wedge \cdots \wedge e^D, e^1 \wedge \cdots \wedge e^D \rangle = \det\left(\delta^{\mu\nu}\right) = 1$. Thus, the scalar volume spanned by the forms $\alpha^1, \ldots, \alpha^D$ is given by

$$\det A = \sqrt{\left| \det\left(\langle \alpha^\mu, \alpha^\nu \rangle\right) \right|}.$$

If we have the case that $\alpha^\mu = e^\mu$ for all $\mu = 1, \ldots, D$, then $\langle \alpha^\mu, \alpha^\nu \rangle = \langle e^\mu, e^\nu \rangle = \eta^{\mu\nu}$, and the unit volume determinant is written as

$$\det(1) = 1 = \sqrt{\left| \det\left(\eta^{\mu\nu}\right) \right|} = \sqrt{|\eta|}.$$

Hence, we can define the *invariant oriented unit volume element* (or *form*) dvol_η as

$$\mathrm{dvol}_\eta \equiv \sqrt{|\eta|}\, e^1 \wedge \cdots \wedge e^D.$$

This enhanced expression is in fact invariant under arbitrary linear transformations in $\widetilde{V}$. If we consider a linear transformation $e'^\mu = \Lambda^\mu{}_\nu e^\nu$ of the basis in $\widetilde{V}$, then $e^1 \wedge \cdots \wedge e^D$ changes by a factor $(\det \Lambda)$, while $\sqrt{|\eta|}$ changes by a factor $(\det \Lambda)^{-1}$, so that the volume element stays unchanged, $(\mathrm{dvol}_\eta)' = \mathrm{dvol}_\eta$. The generalization to curved manifolds is achieved by using the locally varying metric $g_{\mu\nu}(x)$ and by replacing the global basis vectors $\{e^\mu\}$ by the local basis vectors $\{\mathrm{d}x^\mu\}$. The volume element dvol_g of a curved manifold is written as

$$\mathrm{dvol}_g \equiv \sqrt{|g|}\, \mathrm{d}x^1 \wedge \cdots \wedge \mathrm{d}x^D,$$

where $g \equiv \det(g_{\mu\nu})$ denotes the determinant of the metric. Finally, let us note the interesting fact that the rigid invariant unit volume dvol_η is actually equal to the Levi-Civita ϵ-tensor. We calculate

$$\begin{aligned}
\mathrm{dvol}_\eta &= \frac{1}{D!}\sqrt{|\eta|}\, \varepsilon_{\mu_1 \ldots \mu_D}\, e^{\mu_1} \wedge \cdots \wedge e^{\mu_D} \\
&= \sqrt{|\eta|}\, \varepsilon_{\mu_1 \ldots \mu_D}\, \mathrm{Alt}\left(e^{\mu_1} \otimes \cdots \otimes e^{\mu_D}\right) \\
&= \sqrt{|\eta|}\, \varepsilon_{\mu_1 \ldots \mu_D}\, e^{\mu_1} \otimes \cdots \otimes e^{\mu_D} \\
&= \epsilon.
\end{aligned}$$

This shows that the ϵ-tensor belongs to the 1-dimensional subspace $\bigwedge^D \widetilde{V}$ within the D^D-dimensional space $\bigotimes^D \widetilde{V}$. The equality $\mathrm{dvol}_g = \epsilon$ also holds for the case of a geometric manifold with the curved metric $g_{\mu\nu}(x)$.

B.2   Matrix Exponential

Properties of the Matrix Exponential

Below we list some basic properties of the *matrix exponential* $\exp A$. We consider matrices A, B in $GL(n, \mathbb{C})$ and complex numbers a, b. The exponential maps zero to the identity:

$$e^0 = 1.$$

Inverse:

$$\left(e^A\right)^{-1} = e^{-A}.$$

Hermitian conjugation:

$$\left(e^A\right)^\dagger = e^{A^\dagger}.$$

Distributivity with numbers:

$$e^{(a+b)A} = e^{aA} e^{bA}.$$

Similarity transformation:

$$e^{BAB^{-1}} = B e^A B^{-1}.$$

Determinant and trace:

$$\det\left(e^A\right) = e^{\mathrm{Tr} A}.$$

Limit formula for $\exp A$:

$$\exp A = \lim_{n \to \infty} \left(I + \frac{A}{n}\right)^n.$$

For the proof of the last formula we write

$$\left(I + \frac{A}{n}\right)^n = \sum_{k=0}^{n} \binom{n}{k} \frac{A^k}{n^k} = \sum_{k=0}^{n} b(n; k),$$

with

$$b(n; k) = \frac{A^k}{k!} \left(1 - \frac{1}{n}\right) \cdot \left(1 - \frac{2}{n}\right) \cdot \ldots \cdot \left(1 - \frac{k-1}{n}\right),$$

for $k \leq n$, and $b(n; k) = 0$, for $k > n$. The expression $b(n; k)$ is bounded, since for some constant B it is

$$|b(n; k)| \leq \frac{B^k}{k!}.$$

Thus, we can take the limit and obtain the desired result,

$$\lim_{n \to \infty} \left(I + \frac{A}{n}\right)^n = \lim_{n \to \infty} \sum_{k=0}^{n} b(n; k) = \sum_{k=0}^{\infty} \frac{A^k}{k!} = \exp A.$$

Lie-Trotter product formula (*Marius Sophus Lie* and *Hale Trotter*):

$$\exp(A + B) = \lim_{n \to \infty} \left(\exp\left(\frac{A}{n}\right) \exp\left(\frac{B}{n}\right)\right)^n.$$

Differentiation of a one-parameter curve:

$$\frac{\mathrm{d}}{\mathrm{d}t}\exp(tA) = A\exp(tA) = \exp(tA)A.$$

Exponentiation and differentiation:

$$\left.\frac{\mathrm{d}}{\mathrm{d}t}\exp(tA)\right|_{t=0} = A.$$

Commutator through exponentiation and differentiation:

$$\left.\frac{\mathrm{d}}{\mathrm{d}t}\left(e^{tA}Be^{-tA}\right)\right|_{t=0} = AB - BA = [A,B].$$

B.3   Pauli and Dirac Matrices

Pauli Matrices

Here we list some algebraic properties of the *Pauli matrices*. The three basic Pauli matrices $\{\sigma_1,\sigma_2,\sigma_3\}$ are explicitly given by

$$\sigma_1 = \begin{pmatrix} 0 & 1 \\ 1 & 0 \end{pmatrix}, \quad \sigma_2 = \begin{pmatrix} 0 & -i \\ i & 0 \end{pmatrix}, \quad \sigma_3 = \begin{pmatrix} 1 & 0 \\ 0 & -1 \end{pmatrix},$$

and constitute a basis for unitary 2×2-matrices. It is useful to view them as a 3-vector $\boldsymbol{\sigma}$,

$$\boldsymbol{\sigma} \equiv \begin{pmatrix} \sigma_1 \\ \sigma_2 \\ \sigma_3 \end{pmatrix}.$$

For the three-vector components, we do not make any distinction between the co- and contravariant case, i.e. σ_k and σ^k denote the same Pauli matrix. For relativistic applications, the extended set of four Pauli matrices $\{\sigma_0,\sigma_1,\sigma_2,\sigma_3\}$ is used,

$$\sigma_0 = \begin{pmatrix} 1 & 0 \\ 0 & 1 \end{pmatrix}, \quad \sigma_1 = \begin{pmatrix} 0 & 1 \\ 1 & 0 \end{pmatrix}, \quad \sigma_2 = \begin{pmatrix} 0 & -i \\ i & 0 \end{pmatrix}, \quad \sigma_3 = \begin{pmatrix} 1 & 0 \\ 0 & -1 \end{pmatrix},$$

which forms a basis for hermitian 2×2-matrices. Viewing this set as a four-vector is useful,

$$\sigma^\mu \equiv \begin{pmatrix} 1 \\ \boldsymbol{\sigma} \end{pmatrix} \quad \text{and} \quad \sigma_\mu \equiv \eta_{\mu\nu}\sigma^\nu = (1,-\boldsymbol{\sigma}).$$

Here the distinction between the co- and contravariant case is important. In addition, for keeping a correct index placement in equations, we define the symbols,

$$\bar{\sigma}^\mu \equiv \sigma_\mu = (1,-\boldsymbol{\sigma}) \quad \text{and} \quad \bar{\sigma}_\mu \equiv \sigma^\mu = \begin{pmatrix} 1 \\ \boldsymbol{\sigma} \end{pmatrix}.$$

Basic matrix properties of the Pauli matrices are:

$$\sigma_\mu^\dagger = \sigma_\mu^{-1} = \sigma_\mu,$$

$$(\sigma_k)^2 = 1, \quad k = 1,2,3,$$

$$\sigma_2\sigma_k^*\sigma_2 = -\sigma_k, \quad k = 1,2,3.$$

Traces in three dimensions:

$$\operatorname{Tr}\sigma_k = 0, \quad k = 1, 2, 3,$$

$$\operatorname{Tr}(\sigma_k\sigma_l) = 2\delta_{kl},$$

$$\operatorname{Tr}(\sigma_k\sigma_l\sigma_m) = 2\mathrm{i}\epsilon_{klm},$$

$$\operatorname{Tr}(\sigma_k\sigma_l\sigma_m\sigma_n) = 2\left(\delta_{kl}\delta_{mn} - \delta_{km}\delta_{ln} + \delta_{kn}\delta_{lm}\right).$$

Products in three dimensions:

$$\sigma_k\sigma_l = \delta_{kl} + \mathrm{i}\epsilon_{klm}\sigma_m,$$

$$\boldsymbol{\sigma} \times \boldsymbol{\sigma} = 2\mathrm{i}\boldsymbol{\sigma},$$

$$\sigma_k\sigma_l\sigma_m = \sigma_k\delta_{lm} - \sigma_l\delta_{km} + \sigma_m\delta_{kl} + \mathrm{i}\epsilon_{klm}.$$

Lie algebra commutator:

$$[\sigma_k, \sigma_l] = 2\mathrm{i}\epsilon_{klm}\sigma_m.$$

Clifford algebra anticommutator:

$$\{\sigma_k, \sigma_l\} = 2\delta_{kl}.$$

We list some four-dimensional formulae. Traces in four dimensions:

$$\operatorname{Tr}(\sigma^\mu\overline{\sigma}^\nu) = \operatorname{Tr}(\overline{\sigma}^\mu\sigma^\nu) = 2\eta^{\mu\nu},$$

$$\operatorname{Tr}(\sigma^\mu\overline{\sigma}^\nu\sigma^\rho\overline{\sigma}^\tau) = 2\left(\eta^{\mu\nu}\eta^{\rho\tau} - \eta^{\mu\rho}\eta^{\nu\tau} + \eta^{\mu\tau}\eta^{\nu\rho} - \mathrm{i}\epsilon^{\mu\nu\rho\tau}\right).$$

Products in four dimensions:

$$\sigma^\mu\sigma^\nu = \eta^{\mu\nu} - \frac{\mathrm{i}}{2}\epsilon^{\mu\nu\rho\tau}\sigma_\rho\sigma_\tau,$$

$$\sigma^\mu\overline{\sigma}^\nu = \eta^{\mu\nu} - \frac{\mathrm{i}}{2}\epsilon^{\mu\nu\rho\tau}\sigma_\rho\overline{\sigma}_\tau,$$

$$\overline{\sigma}^\mu\sigma^\nu = \eta^{\mu\nu} + \frac{\mathrm{i}}{2}\epsilon^{\mu\nu\rho\tau}\overline{\sigma}_\rho\sigma_\tau.$$

Special sums:

$$\sigma^\mu\overline{\sigma}^\nu + \sigma^\nu\overline{\sigma}^\mu = 2\eta^{\mu\nu},$$

$$\overline{\sigma}^\mu\sigma^\nu + \overline{\sigma}^\nu\sigma^\mu = 2\eta^{\mu\nu}.$$

Dirac Gamma Matrices

We define here the 4-complex-dimensional *Dirac matrices*, also called *gamma matrices*, in the *Weyl* (or *chiral*) *representation* as

$$\gamma^\mu \equiv \begin{pmatrix} 0 & \sigma^\mu \\ \overline{\sigma}^\mu & 0 \end{pmatrix}.$$

The Dirac matrices satisfy the *Clifford algebra* anticommutation relations

$$\{\gamma^\mu, \gamma^\nu\} \equiv \gamma^\mu\gamma^\nu + \gamma^\nu\gamma^\mu = 2\eta^{\mu\nu}.$$

These Clifford algebra relations are valid independently of the concrete representation. Indeed, if the matrices γ^μ satisfy the above Clifford relation, then also the transformed matrices $\gamma^{\mu\prime}$ obtained through a similarity transformation

$$\gamma^{\mu\prime} = U\gamma^\mu U^{-1}$$

with an invertible matrix U also satisfy the Clifford relation. In this way, various representations are obtained. Two important ones are the Dirac and the Majorana representations. As usually, one defines the covariant version γ_ρ as

$$\gamma_\rho \equiv \eta_{\rho\mu}\gamma^\mu.$$

Moreover, we define the 4×4 *spin matrix* $\gamma^{\mu\nu}$ as

$$\gamma^{\mu\nu} \equiv \frac{i}{4}\left[\gamma^\mu, \gamma^\nu\right],$$

furnishing the representation of Lorentz transformations acting on Dirac spinor indices. In addition to the four matrices γ^μ, another fifth matrix is essential, which is defined as

$$\gamma^5 \equiv \gamma_5 \equiv i\gamma^0\gamma^1\gamma^2\gamma^3,$$

or

$$\gamma^5 = -\frac{i}{4!}\epsilon_{\mu\nu\rho\sigma}\gamma^\mu\gamma^\nu\gamma^\rho\gamma^\sigma.$$

Within the Weyl representation, the γ^5 matrix has the explicit form

$$\gamma^5 = \begin{pmatrix} -1_2 & 0 \\ 0 & 1_2 \end{pmatrix}.$$

By using the matrix γ^5, we can construct projection operators for the left-handed and right-handed spinors, i.e.

$$\frac{1}{2}\left(1 - \gamma^5\right)\begin{pmatrix} \psi_L \\ \psi_R \end{pmatrix} = \begin{pmatrix} \psi_L \\ 0 \end{pmatrix},$$

and

$$\frac{1}{2}\left(1 + \gamma^5\right)\begin{pmatrix} \psi_L \\ \psi_R \end{pmatrix} = \begin{pmatrix} 0 \\ \psi_R \end{pmatrix}.$$

Anticommutator and commutator relations with matrix γ^5:

$$\left\{\gamma^\mu, \gamma^5\right\} = 0,$$

$$\left[\gamma^{\mu\nu}, \gamma^5\right] = 0.$$

Squares of gamma matrices:

$$(\gamma^0)^2 = 1_4,$$

$$(\gamma^k)^2 = -1_4,$$

$$(\gamma^5)^2 = 1_4.$$

Adjoint:

$$(\gamma^\mu)^\dagger = \gamma^0\gamma^\mu\gamma^0,$$

$$(\gamma^5)^\dagger = -\gamma^0\gamma^5\gamma^0.$$

Traces:

$$\operatorname{Tr}\gamma^\mu = \operatorname{Tr}\gamma^5 = \operatorname{Tr}\left(\gamma^\mu\gamma^5\right) = \operatorname{Tr}\left(\gamma^\mu\gamma^\nu\gamma^5\right) = \operatorname{Tr}\left(\gamma^{\mu\nu}\right) = 0,$$

$$\operatorname{Tr}\left(\gamma^\mu\gamma^\nu\right) = 4\eta^{\mu\nu}.$$

The set of 16 complex 4×4-matrices $\left\{1_4, \gamma^\mu, \gamma^{\mu\nu}, \gamma^5, \gamma^\mu\gamma^5\right\}$ is a basis for the *Dirac algebra*. The algebras of Pauli matrices and Dirac matrices are special examples of *Clifford algebras*

(*William Kingdon Clifford*), see for example [53]. For a Dirac spinor Ψ, the operation of *Dirac conjugation* produces another spinor $\overline{\Psi}$, defined as

$$\overline{\Psi} \equiv \Psi^\dagger \gamma^0 = \Psi^{*T} \gamma^0.$$

For a 4×4-matrix A acting on Dirac spinors Ψ, we define the Dirac conjugation by

$$\overline{A} \equiv \gamma^0 A^\dagger \gamma^0 = \gamma^0 A^{*T} \gamma^0.$$

With the above definitions, it is

$$\overline{(A\Psi)} = (A\Psi)^\dagger \gamma^0 = \overline{\Psi}\,\overline{A}.$$

Specifically for the gamma matrices γ^μ, $\gamma^{\mu\nu}$ and γ^5, we have

$$\overline{\gamma^\mu} = \gamma^\mu, \qquad \overline{\gamma^{\mu\nu}} = \gamma^{\mu\nu} \quad \text{and} \quad \overline{\gamma^5} = -\gamma^5.$$

B.4   Dirac Delta Distribution

Fourier Transformations

For a well-behaving function $f(x)$ on a D-dimensional Euclidean space, we define the *Fourier transformation* (*Jean-Baptiste Joseph Fourier*) by

$$\widetilde{f}(k) \equiv \int_{\mathbb{R}^D} f(x)\,e^{-ik\cdot x}\,d^D x.$$

The inverse Fourier transformation, giving back the original function, is

$$f(x) = \int_{\mathbb{R}^D} \widetilde{f}(k)\,e^{+ik\cdot x}\,\frac{d^D k}{(2\pi)^D}.$$

Note that some authors use definitions with different powers of $(2\pi)^D$. For a d-dimensional Minkowski space with $d = 1 + D$ and the metric $\eta_{\mu\nu} = \text{diag}(1, -1, \ldots, -1)$, the Fourier transformation is defined by

$$\widetilde{f}(k, \omega) \equiv \int_{\mathbb{M}_d} f(x, t)\,e^{+i(\omega t - k\cdot x)}\,dt\,d^D x.$$

The inverse Fourier transformation is

$$f(x, t) = \int_{\mathbb{M}_d} \widetilde{f}(k, \omega)\,e^{-i(\omega t - k\cdot x)}\,\frac{d\omega\,d^D k}{(2\pi)^{1+D}}.$$

Dirac Delta Distribution

The *Dirac delta distribution*, also called *Dirac delta function*, $\delta(x)$, is a generalized function defined within an integral expression in the form

$$\int_{-\infty}^{\infty} \delta(x) f(x)\,dx \equiv \lim_{n \to \infty} \int_{-\infty}^{\infty} \delta_n(x) f(x)\,dx,$$

where $\delta_n(x)$ is a suitable infinite series of functions depending on the parameter n. Examples of suitable series of functions $\delta_n(x)$ leading to the delta distribution $\delta(x)$ are for example the *Gaussian function*

$$\delta_n(x) = \sqrt{\frac{n}{\pi}}\,\exp(-nx^2),$$

the *resonance function*

$$\delta_n(x) = \frac{n}{\pi + \pi n^2 x^2},$$

or the *sinc function*

$$\delta_n(x) = \frac{\sin nx}{\pi x}.$$

The important defining property of the Dirac delta distribution is expressed by the formula

$$\int_{-\infty}^{\infty} \delta(x) f(x) \, \mathrm{d}x = f(0),$$

which includes also the special case

$$\int_{-\infty}^{\infty} \delta(x) \, \mathrm{d}x = 1.$$

The delta distribution is symmetric,

$$\delta(-x) = \delta(x).$$

From the above properties, it is easy to see that it is

$$\int_{-\infty}^{\infty} \delta(x - y) f(y) \, \mathrm{d}y = f(x),$$

which can be considered as the continuum version of the discrete formula

$$\sum_{k=1}^{D} \delta_{jk} a_k = a_j.$$

Frequently used is the Fourier integral representation

$$\delta(x) = \int_{-\infty}^{\infty} \mathrm{e}^{\pm ikx} \frac{\mathrm{d}k}{2\pi},$$

and its inverse

$$\int_{-\infty}^{\infty} \delta(x) \, \mathrm{e}^{\mp ikx} \, \mathrm{d}x = 1.$$

Both sign choices above yield the same result, as $\delta(x)$ is symmetric. Given a function $g(x)$ having the values x_j, $j = 1, \ldots, r$, as roots of the equation $g(x) = 0$, the delta distribution $\delta(g(x))$ can be calculated as

$$\delta(g(x)) = \sum_{j=1}^{r} \frac{\delta(x - x_j)}{|g'(x_j)|}.$$

The delta distribution $\delta(x)$ can be differentiated and formally its derivative $\delta'(x)$ can be defined as for functions,

$$\delta'(x) \equiv \frac{\mathrm{d}\delta(x)}{\mathrm{d}x} \equiv \lim_{\epsilon \to 0} \frac{\delta(x + \epsilon) - \delta(x)}{\epsilon}.$$

It is easily seen that it is

$$\int_{-\infty}^{\infty} \delta'(x) f(x) \, \mathrm{d}x = -f'(0).$$

Because of the antisymmetry of the first derivative,

$$\delta'(-x) = -\delta'(x),$$

one obtains the formula

$$\int_{-\infty}^{\infty} \frac{\mathrm{d}\delta(x-y)}{\mathrm{d}y} f(y) \, \mathrm{d}y = \frac{\mathrm{d}f(x)}{\mathrm{d}x}.$$

For higher order derivatives ones proceeds in a similar fashion. The formulae for the nth derivative are

$$\int_{-\infty}^{\infty} \frac{\mathrm{d}^n \delta(x)}{\mathrm{d}x^n} f(x) \, \mathrm{d}x = (-1)^n \frac{\mathrm{d}^n f(0)}{\mathrm{d}x^n},$$

and

$$\frac{\mathrm{d}^n \delta(-x)}{\mathrm{d}x^n} = (-1)^n \frac{\mathrm{d}^n \delta(x)}{\mathrm{d}x^n},$$

and thus

$$\int_{-\infty}^{\infty} \frac{\mathrm{d}^n \delta(x-y)}{\mathrm{d}y^n} f(y) \, \mathrm{d}y = \frac{\mathrm{d}^n f(x)}{\mathrm{d}x^n}.$$

In a D-dimensional space possessing a Euclidean metric, the corresponding delta distribution $\delta^D(x)$ is defined as the product

$$\delta^D(x) \equiv \delta(x^1) \cdots \delta(x^D).$$

If the dimensionality of the variable x is $[x] = [L]$, then the dimensionality of $\delta^D(x)$ is $[\delta^D] = [L]^{-D}$. So given a scalar function $f(x)$, the product $f(x)\delta^D(x)$ is a volume density. To recover a quantity with dimensionality equal $[f]$, one needs to integrate over the volume $\mathrm{d}^D x$. We typically leave out the symbol for the dimensionality D and simply write $\delta(x)$. The 1-dimensional formulae can be readily generalized. For example, we can write

$$\delta(x) = \int_{\mathbb{R}^D} e^{\pm i k \cdot x} \frac{\mathrm{d}^D k}{(2\pi)^D},$$

and inversely

$$\int_{\mathbb{R}^D} \delta(x) \, e^{\mp i k \cdot x} \, \mathrm{d}^D x = 1.$$

In the general case of a d-dimensional geometric manifold $(\mathcal{M}, g_{\mu\nu})$ with a metric of arbitrary signature, the Dirac delta distribution can be defined based on the volume integral formula

$$\int_{\mathcal{M}} \frac{\delta(x-y)}{\sqrt{|g(y)|}} f(y) \sqrt{|g(y)|} \, \mathrm{d}^d y = f(x).$$

In the above integral, the scalar function $f(y)$ and the volume integration measure $\sqrt{|g(y)|} \, \mathrm{d}^d y$ are invariant under diffeomorphisms. Consequently, the expression $\delta(x-y)/\sqrt{|g(y)|}$ represents the scalar version of the delta distribution.

Heaviside Step Function

Related to the delta distribution is the *Heaviside step function (Oliver Heaviside)*. The usual definition of the Heaviside step function $\theta(x)$ is

$$\theta(x) \equiv \begin{cases} 0, & \text{if } x < 0, \\ \frac{1}{2}, & \text{if } x = 0, \\ 1, & \text{if } x > 0. \end{cases}$$

An analytic expression for $\theta(x)$ is given by the limit

$$\theta(x) = \lim_{n\to\infty} \frac{1}{1 + e^{-2nx}}.$$

The relation to the Dirac delta distribution $\delta(x)$ is

$$\delta(x) = \frac{\mathrm{d}\theta(x)}{\mathrm{d}x}.$$

The Heaviside step function is a useful tool for solutions of differential equations.

B.5   Poisson and Wave Equation

Poisson Equation

We start with the D-dimensional Euclidean *gradient (nabla) operator* ∇, which is defined as the covector

$$\nabla \equiv \left(\frac{\partial}{\partial x^1}, \dots, \frac{\partial}{\partial x^D} \right),$$

and construct the *Laplace operator (Pierre-Simon de Laplace)*, denoted ∇^2, as the scalar product

$$\nabla^2 \equiv \langle \nabla, \nabla \rangle = \left(\frac{\partial}{\partial x^1} \right)^2 + \dots + \left(\frac{\partial}{\partial x^D} \right)^2.$$

Let us restrict to the case $D = 3$ in the following. The partial differential equation

$$\nabla^2 \Phi(\boldsymbol{x}) = 0$$

for the scalar function $\Phi(\boldsymbol{x})$ is called the *Laplace equation*. The solution $\Phi(\boldsymbol{x})$ of this PDE, called the *potential*, depends on the boundary conditions imposed in the problem. Closely related to the Laplace equation is the inhomogeneous *Poisson equation (Simeon Denis Poisson)*

$$\nabla^2 \Phi(\boldsymbol{x}) = 4\pi\rho(\boldsymbol{x})$$

for a given continuous (mass or charge) distribution $\rho(\boldsymbol{x})$. The factor 4π is of geometric origin and is the surface content of the 3-dimensional unit ball. The *propagator* $G(\boldsymbol{x} - \boldsymbol{y})$ is defined as the solution of the fundamental equation

$$\nabla^2 G(\boldsymbol{x} - \boldsymbol{y}) = \delta(\boldsymbol{x} - \boldsymbol{y}),$$

which in essence expresses that the propagator G is the inverse of the differential operator ∇^2. The propagator is called also the *Green's function (George Green)*. If one has a solution $\Phi_{\mathrm{hom}}(\boldsymbol{x})$ of the Laplace equation at hand, then a solution $\Phi(\boldsymbol{x})$ of the Poisson equation has the form

$$\Phi(\boldsymbol{x}) = \Phi_{\mathrm{hom}}(\boldsymbol{x}) + 4\pi \int_{\mathbb{R}^3} G(\boldsymbol{x} - \boldsymbol{y})\, \rho(\boldsymbol{y})\, \mathrm{d}^3 y.$$

The solution $\Phi_{\mathrm{hom}}(\boldsymbol{x})$ of the homogeneous equation takes care of the boundary conditions. If we require the solution $\Phi(\boldsymbol{x})$ of the inhomogeneous equation to be such that it vanishes for $\boldsymbol{x}$ approaching infinity, then it must be $\Phi_{\mathrm{hom}}(\boldsymbol{x}) = 0$. By inserting the Fourier representation of the propagator

$$G(\boldsymbol{x} - \boldsymbol{y}) = \frac{1}{(2\pi)^3} \int_{\mathbb{R}^3} \widetilde{G}(\boldsymbol{k})\, e^{i\boldsymbol{k}\cdot(\boldsymbol{x}-\boldsymbol{y})}\, \mathrm{d}^3 k$$

and the delta distribution

$$\delta(\boldsymbol{x} - \boldsymbol{y}) = \frac{1}{(2\pi)^3} \int_{\mathbb{R}^3} e^{i\boldsymbol{k}\cdot(\boldsymbol{x}-\boldsymbol{y})} \, d^3k$$

in the defining condition of the propagator, we obtain the Fourier component $\widetilde{G}(\boldsymbol{k})$ explicitly,

$$\widetilde{G}(\boldsymbol{k}) = \frac{1}{-\boldsymbol{k}^2}.$$

Thus, the propagator $G(\boldsymbol{x} - \boldsymbol{y})$ is given by the Fourier integral

$$G(\boldsymbol{x} - \boldsymbol{y}) = \frac{1}{(2\pi)^3} \int_{\mathbb{R}^3} \frac{e^{i\boldsymbol{k}\cdot(\boldsymbol{x}-\boldsymbol{y})}}{-\boldsymbol{k}^2} \, d^3k.$$

This integral can be explicitly calculated by using spherical coordinates in $\boldsymbol{k}$-space with the volume element

$$d^3k = k^2 \, dk \, d\varphi \, \sin\theta \, d\theta,$$

with the variables taking the values $0 \le k < \infty$, $0 \le \varphi < 2\pi$ and $0 \le \theta \le \pi$. Furthermore, we implement the substitutions

$$\boldsymbol{x} - \boldsymbol{y} \equiv \boldsymbol{R} \quad \text{and} \quad \boldsymbol{k} \cdot (\boldsymbol{x} - \boldsymbol{y}) \equiv kR\cos\theta.$$

With these variables, the propagator can be written as

$$G(\boldsymbol{R}) = \frac{-1}{(2\pi)^3} \int_0^\infty \int_0^{2\pi} \int_0^\pi e^{ikR\cos\theta} \sin\theta \, d\theta \, d\varphi \, dk.$$

The integral over φ is trivial and yields only a factor 2π. For the integral over θ we employ the substitution $\cos\theta \equiv \lambda$ and carry out the resulting integral,

$$\int_{-1}^1 e^{ikR\lambda} \, d\lambda = \frac{2\sin(kR)}{kR}.$$

Thus, we obtain

$$G(\boldsymbol{R}) = \frac{-1}{2\pi^2} \int_0^\infty \frac{\sin(kR)}{kR} \, dk.$$

By evaluating the last integral, we obtain the sought after propagator in $\boldsymbol{x}$-space,

$$G(\boldsymbol{x} - \boldsymbol{y}) = \frac{-1}{4\pi|\boldsymbol{x} - \boldsymbol{y}|}.$$

We have derived that for any two points $\boldsymbol{x}$ and $\boldsymbol{y}$ in 3-dimensional Euclidean space it is

$$\nabla^2 \frac{1}{|\boldsymbol{x} - \boldsymbol{y}|} = -4\pi\delta(\boldsymbol{x} - \boldsymbol{y}).$$

Thus, the general solution of the Poisson equation is the potential function $\Phi(\boldsymbol{x})$ given as

$$\Phi(\boldsymbol{x}) = -\int_{\mathbb{R}^3} \frac{\rho(\boldsymbol{y})}{|\boldsymbol{x} - \boldsymbol{y}|} \, d^3y,$$

with the implicit property that it vanishes at infinity. This result should be compared with the one derived for the gravitational potential in Section 4.4.

Wave Equation

We assume to have a d-dimensional Minkowski space with dimension $d = 1 + D$ and the metric $\eta_{\mu\nu}$. The d-dimensional spacetime derivative operator ∂_μ is the covector

$$\partial_\mu \equiv \left(\frac{\partial}{\partial x^0}, \frac{\partial}{\partial x^1}, \dots, \frac{\partial}{\partial x^D} \right).$$

The previously considered Laplace operator ∇^2 is replaced by the *d'Alembert wave operator* ∂^2, defined as the scalar product

$$\partial^2 \equiv \partial_\mu \partial^\mu = \left(\frac{\partial}{\partial x^0} \right)^2 - \left(\frac{\partial}{\partial x^1} \right)^2 - \dots - \left(\frac{\partial}{\partial x^D} \right)^2.$$

In the following, we consider the case $d = 4$. The PDE to be studied is the *inhomogeneous wave equation*

$$\partial^2 A(x) = 4\pi j(x).$$

We do not specify explicitly the tensorial character of the fields $A(x)$ and $j(x)$, they can be scalars, vectors, or tensors. The solution of this PDE depends much on the imposed boundary conditions. The *propagator* $G(x - y)$ is now implicitly defined through the condition

$$\partial^2 G(x - y) = \delta(x - y),$$

where x and y are four-dimensional spacetime points. If we know a solution of the homogeneous wave equation, call it $A_{\text{hom}}(x)$, then a solution $A(x)$ of the inhomogeneous wave equation can be written as

$$A(x) = A_{\text{hom}}(x) + 4\pi \int_{\text{M}_4} G(x - y)\, j(y)\, \mathrm{d}^4 y.$$

By inserting the Fourier representation of the propagator

$$G(x - y) = \frac{1}{(2\pi)^4} \int_{\text{M}_4} \widetilde{G}(k)\, e^{-ik(x-y)}\, \mathrm{d}^4 k$$

and the delta distribution

$$\delta(x - y) = \frac{1}{(2\pi)^4} \int_{\text{M}_4} e^{-ik(x-y)}\, \mathrm{d}^4 k$$

in the defining equation of the propagator, we obtain the Fourier component $\widetilde{G}(k)$ explicitly,

$$\widetilde{G}(k) = \frac{1}{-k^2}.$$

Thus, the propagator $G(x - y)$ in Minkowski space is given by the Fourier integral

$$G(x - y) = \frac{1}{(2\pi)^4} \int_{\text{M}_4} \frac{e^{-ik(x-y)}}{-k^2}\, \mathrm{d}^4 k.$$

The explicit calculation of this integral is a bit more involved. First, let us write out the scalar products and absorb the minus sign in the integration variable $\boldsymbol{k}$. We write

$$G(x - y) = \frac{1}{(2\pi)^4} \int_{\text{M}_4} \frac{e^{ik^0(x^0 - y^0)} e^{i\boldsymbol{k}\cdot(\boldsymbol{x}-\boldsymbol{y})}}{(k^0)^2 - \boldsymbol{k}^2}\, \mathrm{d}k^0\, \mathrm{d}^3 k.$$

In order to carry out the integration over the real variable k^0, we use the property that the integrand vanishes exponentially for values $k^0 \to \infty$. Instead of an integration along the real k^0-axis, one can analytically extend the integral and carry out a complex contour integration over a closed curve in the complex k^0-plane. The complex contour is comprised of the horizontal k^0-axis plus an upper half-circle, or a lower half-circle at infinity. A possible setup is shown in the sketch B.1. The different contour integrals correspond to different solutions of the homogeneous equation. The benefit of this method is that we can elegantly treat the two poles $k^0 = \pm|\boldsymbol{k}|$ of the integrand. Let us recall the *Cauchy theorem of residues*

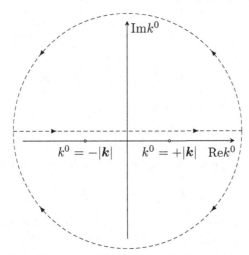

Figure B.1: Contours of integration in the complex plane

(*Augustin-Louis Cauchy*) from the theory of analytic functions, see e.g. [22]. Consider a function $f(z)$ of the complex variable z being analytic in a region $U \subset \mathbb{C}$ and on the closed boundary $C = \partial U$, except of a finite number of isolated singular points $\zeta_1, \ldots, \zeta_N$ within U. Then, the contour integral of $f(z)$ over the closed curve C is given by the formula

$$\oint_C f(z)\,\mathrm{d}z = 2\pi\mathrm{i} \sum_{n=1}^{N} \mathrm{Res} f(\zeta_n).$$

The quantities $\mathrm{Res} f(\zeta_n)$ are the *residues* of the function for the singular points engulfed by the curve. In the special case of simple poles ζ_n of the function $f(z)$, the residues can be calculated by the formula

$$\mathrm{Res} f(\zeta_n) = \lim_{z \to \zeta_n} (z - \zeta_n) f(z).$$

The convention for complex contour integrals is to follow the contour in counter-clockwise direction. If we integrate in clockwise direction instead, the integral receives a minus sign.

Another fine point we need to take into account is the question if we choose the horizontal part of the integration contour to be "above" the real axis, or "below" the real axis. This leads to two distinct but equally valid propagator solutions, the *retarded* and the *advanced* *propagator*. In the above sketch, we have depicted the case of a horizontal contour line lying above the real axis, leading to the retarded propagator. The analogous contour line lying below the real axis leads to the advanced propagator. Let us calculate the retarded propagator in the following. The contour closing in the upper half-plane does not contain any poles inside, so the corresponding integral vanishes. The contour closing in the lower half-plane contains the two poles inside, and the corresponding contour integral is non-vanishing. Note, that for the integrand to fall off exponentially at infinity, the condition

$(x^0 - y^0) > 0$ must hold. According to the theorem of residues, we obtain for the integral over the variable k^0:

$$\int_{-\infty}^{\infty} \frac{e^{ik^0(x^0-y^0)}}{(k^0)^2 - \boldsymbol{k}^2} \, dk^0 = 2\pi i \, \frac{1}{2|\boldsymbol{k}|} \left[e^{i|\boldsymbol{k}|(x^0-y^0)} + e^{-i|\boldsymbol{k}|(x^0-y^0)} \right] \theta(x^0 - y^0).$$

In the above calculation, the initial (negative) clockwise contour direction is reversed to a (positive) counter-clockwise contour direction and the resulting minus sign is absorbed in the k^0 integration variable. The additional condition $(x^0 - y^0) > 0$ for the integral to exist, is implemented by the step function. The remaining integral over d^3k is calculated in spherical coordinates. By using the simplified notation

$$x^0 - y^0 \equiv T, \quad \boldsymbol{x} - \boldsymbol{y} \equiv \boldsymbol{R} \quad \text{and} \quad \boldsymbol{k} \cdot (\boldsymbol{x} - \boldsymbol{y}) \equiv kR\cos\theta,$$

we obtain for the *retarded propagator*, denoted $G_{\text{ret}}(x-y)$, the expression

$$G_{\text{ret}}(x-y) = \frac{i\theta(T)}{(2\pi)^2} \int_0^{\infty} \int_0^{\pi} \frac{k}{2} \left(e^{ikT} + e^{-ikT} \right) e^{ikR\cos\theta} \sin\theta \, d\theta \, dk.$$

Here, we have carried out already the integration over $d\varphi$ contributing a factor 2π. The integral over $d\theta$ is exactly as in the previous section and we obtain

$$G_{\text{ret}}(x-y) = \frac{i\theta(T)}{(2\pi)^2 R} \int_0^{\infty} \left(e^{ikT} + e^{-ikT} \right) \sin(kR) \, dk.$$

Writing the sinus in the integrand as a difference of complex exponentials and recombining the four resulting terms leads to

$$G_{\text{ret}}(x-y) = \frac{\theta(T)}{4\pi R} \int_{-\infty}^{\infty} \left(e^{ik(T+R)} + e^{-ik(T-R)} \right) \frac{dk}{2\pi}.$$

The integrals remaining are Fourier representations of the delta distribution and we obtain the desired result,

$$G_{\text{ret}}(x-y) = \frac{\theta(T)}{4\pi R} \left[\delta(T+R) + \delta(T-R) \right].$$

By writing out the time and space intervals, the propagator reads

$$G_{\text{ret}}(x-y) = \frac{\theta(x^0-y^0)}{4\pi|\boldsymbol{x}-\boldsymbol{y}|} \left[\delta\left((x^0-y^0) + |\boldsymbol{x}-\boldsymbol{y}|\right) + \delta\left((x^0-y^0) - |\boldsymbol{x}-\boldsymbol{y}|\right) \right].$$

Because both $(x^0 - y^0)$ and $|\boldsymbol{x} - \boldsymbol{y}|$ are positive in this formula, only the second delta distribution contributes and we can equally well simplify to

$$G_{\text{ret}}(x-y) = \frac{\theta(x^0-y^0)}{4\pi|\boldsymbol{x}-\boldsymbol{y}|} \delta\left((x^0-y^0) - |\boldsymbol{x}-\boldsymbol{y}|\right).$$

The propagator can be written in a relativistically invariant way by using the identity

$$\delta(x^2 - a^2) = \frac{1}{2a} \left[\delta(x+a) + \delta(x-a) \right]$$

for the delta distribution, compare with the more general identity in B.4. The result for the retarded propagator reads then

$$G_{\text{ret}}(x-y) = \frac{\theta(x^0-y^0)}{2\pi} \delta\left((x-y)^2\right).$$

The factor $\theta(x^0 - y^0)$ is indeed invariant for any orthochronous Lorentz or Poincaré transformation. The *retarded propagator* contributes only for lightlike signals traveling from the source point at $\boldsymbol{y}$ at the time y^0 toward the field point at $\boldsymbol{x}$ at a *later time* x^0. The results for the *advanced propagator*, denoted $G_{\mathrm{adv}}(x - y)$, are derived in complete analogy. For the advanced propagator, the condition $(x^0 - y^0) < 0$ must be respected. The propagator formula with the time and space intervals written explicitly is

$$G_{\mathrm{adv}}(x - y) = \frac{\theta(y^0 - x^0)}{4\pi|\boldsymbol{x} - \boldsymbol{y}|} \delta\left((x^0 - y^0) + |\boldsymbol{x} - \boldsymbol{y}|\right).$$

The relativistically invariant form of the advanced propagator reads

$$G_{\mathrm{adv}}(x - y) = \frac{\theta(y^0 - x^0)}{2\pi} \delta\left((x - y)^2\right).$$

The *advanced propagator* contributes only for lightlike signals traveling from the source point at $\boldsymbol{y}$ at the time y^0 toward the field point at $\boldsymbol{x}$ at an *earlier time* x^0. The above results are applied to Maxwell field theory in Section 15.5.

B.6   Variational Calculus

Functionals

First, some basic definitions and notation. A function $f : x \mapsto f(x)$ maps a number x (real or complex) to another number $f(x)$. This assignment is local, a point is mapped to another point. In contrast, a *functional* $S : f \mapsto S[f]$ maps an entire function f to a number $S[f]$. This assignment is global, one needs the whole function f in order to determine the number $S[f]$. One can define functionals of more than one argument. For simplicity, we will only consider functions and functionals of one argument here. An example of a functional $S[f]$ is given by the integral

$$S[f] = \int_{x_a}^{x_b} f(x)\, w(x)\, \mathrm{d}x$$

for any continuous function $f(x)$. The function $w(x)$ acts as a weight. Choosing the weight function $w(x)$ to be the Dirac delta distribution $\delta(x - x_0)$, with x_0 lying within the open interval (x_a, x_b), one obtains the functional $S_{\mathrm{D}}[f]$, where

$$S_{\mathrm{D}}[f] = f(x_0),$$

i.e. the function f is mapped to its value $f(x_0)$. Another example of a functional is the *action functional* $S[q]$ in mechanics, given as

$$S[q] = \int_{t_1}^{t_2} L\left(q(t), \dot{q}(t); t\right) \mathrm{d}t,$$

where $q(t)$ represents the generalized coordinate and $\dot{q}(t)$ its time derivative. The function $L\left(q(t), \dot{q}(t); t\right)$ is called the *Lagrangian (function)*. We separate the time variable with a semicolon, since in many relevant cases the Lagrangian has no explicit dependence on the time. The first two examples are linear functionals, whereas the action functional is nonlinear in its argument.

Variations

The concept of functionals is very useful whenever the question arises which particular function f_0 makes a certain quantity, namely the functional $S[f]$, extremal. The principle of stationary action is an archetypal application of the functional methods in physics. Finding the extremum, or a stationary point, means that one needs to carry out variations and derivatives around the extremal function f_0. We define the *variation $\delta f(x)$ of a function* $f(x)$ as

$$\delta f(x) \equiv \epsilon\, h(x),$$

where ϵ is a small real parameter, $|\epsilon| \ll 1$, and $h(x)$ a sufficiently well-behaving function, called the *test function*. One can generalize the function variation by considering an expression like

$$\delta f(x) \equiv \epsilon^a h_a(x),$$

where the parameter ϵ^a and the function $h_a(x)$ have multiple components, with index the $a = 1, \ldots, r$ encoding some tensorial transformation property. This more general form of the function variation is used in the treatment of Lie groups and Lie algebras. Now let us consider a functional $S[f]$. The *variation of the functional*, denoted $\delta S[f]$, or sometimes more simply δS, is defined as

$$\delta S[f] \equiv S[f + \delta f] - S[f],$$

i.e. we apply the variation δf in the function argument. Considering that it is $\delta f = \epsilon h$, we realize that the variation $\delta S[f]$ is a regular function of the real parameter ϵ and hence we can carry out a Taylor expansion. We have then

$$\delta S[f] = \epsilon \left(\frac{\mathrm{d}S}{\mathrm{d}\epsilon}[f + \epsilon h]\Big|_{\epsilon=0} \right) + \frac{\epsilon^2}{2} \left(\frac{\mathrm{d}^2 S}{\mathrm{d}\epsilon^2}[f + \epsilon h]\Big|_{\epsilon=0} \right) + \mathcal{O}(\epsilon^3).$$

The first term in the rhs, which is linear in ϵ, leads to the definition of the functional derivative.

Functional Derivative

The general definition of the *functional derivative*, which is denoted $\delta S[f]/\delta f(x)$, is provided implicitly by the equation

$$\int_{x_a}^{x_b} h(x) \frac{\delta S[f]}{\delta f(x)}\, \mathrm{d}x = \frac{\mathrm{d}S}{\mathrm{d}\epsilon}[f + \epsilon h]\Big|_{\epsilon=0}.$$

By employing the functional derivative, we can write down the variation $\delta S[f]$ in the form

$$\delta S[f] = \int_{x_a}^{x_b} \frac{\delta S[f]}{\delta f(x)} \delta f(x)\, \mathrm{d}x.$$

One can proceed similarly for the second-order and for all higher order derivatives. In order to obtain an explicit formula for the functional derivative, we can use the delta distribution $\delta(x - y)$ as the test function $h(x)$. It is indeed valid to proceed like this and ultimately obtain all possible results for the variation $\delta S[f]$, but we will not prove this here. The substitution leads to the explicit formula

$$\frac{\delta S[f]}{\delta f(y)} = \frac{\mathrm{d}S}{\mathrm{d}\epsilon}[f(x) + \epsilon\, \delta(x - y)]\Big|_{\epsilon=0}.$$

A mathematically rigorous definition of the functional derivative is based on the generalized derivative definitions of *Maurice R. Frechet* and *Rene E. Gateaux*, but we will not go into more detail here. Let us give an example and calculate the functional derivative of the Dirac functional $S_D[f]$. According to the above Taylor expansion for the variation, we can infer that it is

$$\delta S_D[f] = \int_{x_a}^{x_b} \epsilon\, h(x)\, \delta(x - x_0)\, dx.$$

For $\delta S_D[f]$, there are no terms of higher order in ϵ. The corresponding functional derivative is therefore

$$\frac{\delta S_D[f]}{\delta f(x)} = \delta(x - x_0).$$

This functional derivative formula can be rewritten compactly as

$$\frac{\delta f(y)}{\delta f(x)} = \delta(x - y),$$

and finds its application in practical calculations. The functional derivative obeys certain calculational rules. For any two functionals $S_1[f]$ and $S_2[f]$ and any two numbers c_1 and c_2, the functional derivative is linear,

$$\frac{\delta}{\delta f(x)}\left(c_1\, S_1[f] + c_2\, S_2[f]\right) = c_1 \frac{\delta S_1[f]}{\delta f(x)} + c_2 \frac{\delta S_2[f]}{\delta f(x)}.$$

Furthermore, the following product rule holds

$$\frac{\delta}{\delta f(x)}\left(S_1[f]\, S_2[f]\right) = \frac{\delta S_1[f]}{\delta f(x)} S_2[f] + S_1[f] \frac{\delta S_2[f]}{\delta f(x)}.$$

If there is a nested functional dependency in the form $S[G[f]]$, with the functionals $S[G]$ and $G[f]$ and the usual function $f(x) = y$, then the following chain rule holds

$$\frac{\delta S[G[f]]}{\delta f(x)} = \int_{y_a}^{y_b} \frac{\delta S[G]}{\delta G(y)} \frac{\delta G[f]}{\delta f(x)}\, dy.$$

In the next paragraph, we proceed to the main application of variations and the functional derivative.

Variational Principle

We start with a Lagrangian function $L(q(t), \dot{q}(t); t)$ and build the action functional $S[q]$ as the definite integral

$$S[q] = \int_{t_1}^{t_2} L(q(t), \dot{q}(t); t)\, dt.$$

The points t_1 and t_2 are fixed. We ask which function $q(t)$ makes the action functional $S[q]$ extremal, while the boundary values $q(t_1)$ and $q(t_2)$ are held fixed. This is the *principle of stationary action*. In fact, we are primarily interested in the functions $q(t)$, which make $S[q]$ minimal. We vary the function $q(t)$ in the form

$$\delta q(t) \equiv \epsilon\, h(t),$$

and the time derivative $\dot{q}(t)$ as

$$\delta \dot{q}(t) \equiv \epsilon\, \dot{h}(t).$$

As a consequence, it is also

$$\frac{d}{dt}(\delta q) = \delta\left(\frac{dq}{dt}\right).$$

Furthermore, we require that the variations have the property that they vanish at the end points t_1 and t_2, i.e. $\delta q(t_1) = 0$ and $\delta q(t_2) = 0$. The principle of stationary action demands that the necessary condition

$$\delta S [q] = 0$$

is met. On the other hand, the variation of the action functional $\delta S [q]$ is by definition

$$\delta S = \int_{t_1}^{t_2} L\left(q + \delta q, \dot{q} + \delta \dot{q}; t\right) \mathrm{d}t - \int_{t_1}^{t_2} L\left(q, \dot{q}; t\right) \mathrm{d}t.$$

Expanding the first integral *up to first order* in δq and in $\delta \dot{q}$ yields

$$\delta S = \int_{t_1}^{t_2} \left(\frac{\partial L}{\partial q} \delta q + \frac{\partial L}{\partial \dot{q}} \delta \dot{q}\right) \mathrm{d}t.$$

Partial integration in the second term above and application of the conditions $\delta q(t_1) = 0$ and $\delta q(t_2) = 0$ of the considered variations leads to

$$\delta S = \int_{t_1}^{t_2} \left(\frac{\partial L}{\partial q} - \frac{\mathrm{d}}{\mathrm{d}t} \frac{\partial L}{\partial \dot{q}}\right) \delta q \, \mathrm{d}t.$$

At this point, we can read off the functional derivative $\delta S [q] / \delta q(t)$ as

$$\frac{\delta S [q]}{\delta q(t)} = \frac{\partial L}{\partial q} - \frac{\mathrm{d}}{\mathrm{d}t} \frac{\partial L}{\partial \dot{q}}.$$

If the condition $\delta S [q] = 0$ is to hold for all possible variations, then it must be

$$\frac{\partial L}{\partial q} - \frac{\mathrm{d}}{\mathrm{d}t} \frac{\partial L}{\partial \dot{q}} = 0.$$

This is the well-known *Euler-Lagrange equation* of variational calculus. The expression on the lhs is called the *Euler-Lagrange derivative*. Solving this second-order differential equation provides the function $q(t)$, which makes $S [q]$ extremal. That a solution $q(t)$ actually minimizes the action must be inspected in each situation separately. So far, we have considered only one function $q(t)$. For cases with many functions $q^j(t)$, $j = 1, \ldots, n$, we adopt the notation $\delta/\delta q$ for the multicomponent operator $(\delta/\delta q^j)$. In the same way, for space coordinates x^μ we write $\delta/\delta x$ for the multicomponent operator $(\delta/\delta x^\mu)$.

Variation of Logarithm of a Determinant

For an invertible square matrix M we consider the variation of $\ln |\det M|$, when there is a small variation δM of the matrix. By definition, the variation is

$$\delta \ln |\det M| = \ln |\det (M + \delta M)| - \ln |\det M|,$$

and a quick calculation leads to

$$\delta \ln |\det M| = \ln \det \left(1 + M^{-1} \delta M\right).$$

By using the Taylor expansion

$$\det (1 + A) = 1 + \mathrm{Tr} A + \mathcal{O}(2),$$

which holds for any matrix A with small norm, $\|A\| \ll 1$, we obtain the useful result

$$\delta \ln |\det M| = \mathrm{Tr}\left(M^{-1} \delta M\right).$$

Variation of Volume Element

The volume integration measure $\mathrm{d}^D x$ for flat Euclidean space is

$$\mathrm{d}^D x \equiv \mathrm{d}x^1 \cdots \mathrm{d}x^D.$$

Under a coordinate transformation $x' = x + \delta x$, this volume element transforms as

$$\mathrm{d}^D x' = \left| \det\left(\frac{\partial x'}{\partial x} \right) \right| \mathrm{d}^D x = \left[1 + \partial_\mu (\delta x^\mu) \right] \mathrm{d}^D x.$$

The variation $\delta(\mathrm{d}^D x)$ of the volume element to first order in δx is therefore

$$\delta(\mathrm{d}^D x) = \partial_\mu(\delta x^\mu)\, \mathrm{d}^D x.$$

The same result is obtained for flat Minkowski space in $d = 1 + D$ dimensions.

B.7   Volume Element and Hyperspheres

Volume Element in D Dimensions

In the case of Euclidean space $\mathbb{E}^D$, the *volume element* $\mathrm{d}^D V$ is defined as

$$\mathrm{d}^D V \equiv \mathrm{d}^D x \equiv \mathrm{d}x^1 \cdots \mathrm{d}x^D.$$

In the case of a general geometric manifold $(\mathcal{M}, g_{\mu\nu})$, one uses the *invariant volume element*, called also the *invariant volume form*, dvol_g, which is defined as

$$\mathrm{dvol}_g \equiv \sqrt{|g|}\, \mathrm{d}x^1 \wedge \cdots \wedge \mathrm{d}x^D,$$

where $g \equiv \det(g_{\mu\nu})$ is the determinant of the metric $g_{\mu\nu}(x)$. The coordinates $x^1, \ldots, x^D$ can be Cartesian or other suitable coordinates. For spherical geometries, it is very useful to employ the so-called *hyperspherical coordinates* in D dimensions. One possible definition of the hyperspherical coordinates in D dimensions is

$$\left.\begin{aligned}
x^1 &= r \sin\theta_{D-2} \sin\theta_{D-3} \cdots \sin\theta_1 \cos\varphi \\
x^2 &= r \sin\theta_{D-2} \sin\theta_{D-3} \cdots \sin\theta_1 \sin\varphi \\
x^3 &= r \sin\theta_{D-2} \sin\theta_{D-3} \cdots \cos\theta_1 \\
&\ \vdots \\
x^{D-2} &= r \sin\theta_{D-2} \sin\theta_{D-3} \\
x^{D-1} &= r \sin\theta_{D-2} \cos\theta_{D-3} \\
x^D &= r \cos\theta_{D-2}
\end{aligned}\right\}.$$

The *radial coordinate* r determines the distance from the origin. There are $D - 1$ angle coordinates now, $D - 2$ *polar angles* $\theta_{D-2}, \ldots, \theta_1$ taking the values $0 \leq \theta_{D-2}, \ldots, \theta_1 \leq \pi$ and one *azimuthal angle* φ taking the values $0 \leq \varphi < 2\pi$. The azimuthal angle φ starts from the positive x^1-axis and increases toward the positive x^2-axis. In these hyperspherical coordinates, the metric $\mathrm{d}s^2$ of the Euclidean space $\mathbb{E}^D$ is

$$\mathrm{d}s^2 = \mathrm{d}r^2 + r^2 \bigg\{ (\mathrm{d}\theta_{D-2})^2 + \sin^2\theta_{D-2} \Big[(\mathrm{d}\theta_{D-3})^2 + \sin^2\theta_{D-3} \Big[\cdots$$

$$\cdots (\mathrm{d}\theta_2)^2 + \sin^2\theta_2 \Big[(\mathrm{d}\theta_1)^2 + \sin^2\theta_1\, (\mathrm{d}\varphi)^2 \Big] \cdots \Big] \Big] \bigg\}.$$

Let us transform the volume element $\mathrm{d}^D V$ from Cartesian to hyperspherical coordinates. We start with

$$\mathrm{d}^D V = \sqrt{|\delta|}\,\mathrm{d}x^1 \cdots \mathrm{d}x^D = \sqrt{|g(y)|}\,\mathrm{d}y^1 \cdots \mathrm{d}y^D,$$

where $y^1, \ldots, y^D$ denote the hyperspherical coordinates and $g(y) \equiv \det(g_{\mu\nu}(y))$ is the determinant of the metric in these hyperspherical coordinates, i.e.

$$g_{\mu\nu}(y) = \frac{\partial x^\rho}{\partial y^\mu} \frac{\partial x^\sigma}{\partial y^\nu} \delta_{\rho\sigma}.$$

Consequently, one obtains

$$\sqrt{|g(y)|} = \left| \det\left(\frac{\partial x}{\partial y}\right) \right|.$$

So we have to calculate the Jacobian determinant

$$\frac{\partial(x^1, \ldots, x^D)}{\partial(r, \theta_{D-2}, \ldots, \theta_1, \varphi)}.$$

The result of this calculation yields the volume element $\mathrm{d}^D V$ in hyperspherical coordinates,

$$\mathrm{d}^D V = \left(r^{D-1}\mathrm{d}r\right) \left(\sin^{D-2}\theta_{D-2}\,\mathrm{d}\theta_{D-2}\right) \left(\sin^{D-3}\theta_{D-3}\,\mathrm{d}\theta_{D-3}\right) \cdots \left(\sin\theta_1\,\mathrm{d}\theta_1\right)\,\mathrm{d}\varphi.$$

This volume element can be written also as

$$\mathrm{d}^D V = \left(r^{D-1}\mathrm{d}r\right)\mathrm{d}^{D-1}\Omega$$

if we introduce the $(D-1)$-dimensional *solid angle element* $\mathrm{d}^{D-1}\Omega$ as

$$\mathrm{d}^{D-1}\Omega \equiv \left(\sin^{D-2}\theta_{D-2}\,\mathrm{d}\theta_{D-2}\right) \left(\sin^{D-3}\theta_{D-3}\,\mathrm{d}\theta_{D-3}\right) \cdots \left(\sin\theta_1\,\mathrm{d}\theta_1\right)\,\mathrm{d}\varphi.$$

The solid angle element includes only the $D-1$ angle coordinates $\theta_{D-2}, \ldots, \theta_1, \varphi$. The recursion relation

$$\mathrm{d}^{D+1}\Omega = \left(\sin^D\theta_D\,\mathrm{d}\theta_D\right)\mathrm{d}^D\Omega$$

holds. In the following, we apply the above formulae to obtain the volume elements for increasing numbers of dimensions. For $D = 1$ it is:

$$\mathrm{d}V = \mathrm{d}x = \mathrm{d}r.$$

For $D = 2$:

$$\mathrm{d}^2 V = \mathrm{d}x\,\mathrm{d}y = r\,\mathrm{d}r\,\mathrm{d}\varphi.$$

For $D = 3$:

$$\mathrm{d}^3 V = \mathrm{d}x\,\mathrm{d}y\,\mathrm{d}z = r^2\,\mathrm{d}r\,\sin\theta\,\mathrm{d}\theta\,\mathrm{d}\varphi.$$

For $D = 4$:

$$\mathrm{d}^4 V = \mathrm{d}x\,\mathrm{d}y\,\mathrm{d}z\,\mathrm{d}w = r^3\,\mathrm{d}r\,\sin^2\theta_2\,\sin\theta_1\,\mathrm{d}\theta_2\,\mathrm{d}\theta_1\,\mathrm{d}\varphi.$$

For $D = 5$:

$$\mathrm{d}^5 V = \mathrm{d}x\,\mathrm{d}y\,\mathrm{d}z\,\mathrm{d}w\,\mathrm{d}v = r^4\,\mathrm{d}r\,\sin^3\theta_3\,\sin^2\theta_2\,\sin\theta_1\,\mathrm{d}\theta_3\,\mathrm{d}\theta_2\,\mathrm{d}\theta_1\,\mathrm{d}\varphi,$$

and so forth.

Gamma Function

The gamma function is central for expressing spherical geometries in higher dimensions. For any positive real variable $x \in \mathbb{R}_+$ we define the *gamma function* $\Gamma(x)$ through the integral

$$\Gamma(x) \equiv \int_0^\infty t^{x-1}\, \mathrm{e}^{-t}\, \mathrm{d}t.$$

The gamma function interpolates the factorial function $n! = 1 \cdot 2 \cdot 3 \cdots n$ and this was historically the original motivation for its definition. For any positive integer $n = 1, 2, 3, \ldots$ it is $\Gamma(n+1) = n!$. Another property reminiscent of the factorial function is $\Gamma(x+1) = x\,\Gamma(x)$. A special value is $\Gamma(1/2) = \sqrt{\pi}$, and this can be easily obtained by evaluating the *Gaussian integral*

$$\int_0^\infty \mathrm{e}^{-x^2}\, \mathrm{d}x = \frac{\sqrt{\pi}}{2}.$$

Let us note here some more values of the gamma function for integer and half-integer arguments: $\Gamma(1) = 1$, $\Gamma(3/2) = \sqrt{\pi}/2$, $\Gamma(2) = 1$, $\Gamma(5/2) = 3\sqrt{\pi}/4$, $\Gamma(3) = 2$.

Definition of D-Dimensional Sphere

We define the D-dimensional *hypersphere* (or simply D-*sphere*) $S^D(a)$ with the radius a to be the D-dimensional manifold defined by the set of points

$$S^D(a) \equiv \left\{ (x^1, \ldots, x^{D+1}) \in \mathbb{E}^{D+1} \mid (x^1)^2 + \cdots + (x^{D+1})^2 = a^2 \right\}.$$

With the above definition, we embed the hypersphere $S^D(a)$ in the Euclidean space $\mathbb{E}^{D+1}$ and are able to derive its properties. The D-dimensional *unit hypersphere* is $S^D \equiv S^D(1)$. The metric of the Euclidean space $\mathbb{E}^{D+1}$ can be written as usually as

$$\mathrm{d}s^2 = (\mathrm{d}x^1)^2 + \cdots + (\mathrm{d}x^D)^2 + (\mathrm{d}x^{D+1})^2,$$

and it induces the associated metric on the sphere. In order to eliminate the auxiliary coordinate x^{D+1}, we can follow the same recipe as in the lower-dimensional cases. The sphere-defining equation is differentiated, which leads to

$$\mathrm{d}s^2 = (\mathrm{d}x^1)^2 + \cdots + (\mathrm{d}x^D)^2 + \frac{(x^1\, \mathrm{d}x^1 + \cdots + x^D\, \mathrm{d}x^D)^2}{a^2 - ((x^1)^2 + \cdots + (x^D)^2)}.$$

This expression for the metric uses Cartesian coordinates. For labeling points in $\mathbb{E}^{D+1}$, instead of the Cartesian coordinates $x^1, \ldots, x^{D+1}$, we can use the $D+1$ hyperspherical coordinates $r, \theta_{D-1}, \ldots, \theta_1, \varphi$, as introduced before. This means that we use the definition

$$
\left.
\begin{aligned}
x^1 &= r \sin\theta_{D-1} \sin\theta_{D-2} \cdots \sin\theta_2 \sin\theta_1 \cos\varphi \\
x^2 &= r \sin\theta_{D-1} \sin\theta_{D-2} \cdots \sin\theta_2 \sin\theta_1 \sin\varphi \\
x^3 &= r \sin\theta_{D-1} \sin\theta_{D-2} \cdots \sin\theta_2 \cos\theta_1 \\
x^4 &= r \sin\theta_{D-1} \sin\theta_{D-2} \cdots \cos\theta_2 \\
&\ \ \vdots \\
x^{D-1} &= r \sin\theta_{D-1} \sin\theta_{D-2} \\
x^D &= r \sin\theta_{D-1} \cos\theta_{D-2} \\
x^{D+1} &= r \cos\theta_{D-1}
\end{aligned}
\right\}.
$$

The hyperspherical coordinates take the values $0 \leq r < \infty$, $0 \leq \theta_{D-1}, \ldots, \theta_1 \leq \pi$ and $0 \leq \varphi < 2\pi$. Expressed in these new coordinates, the metric of the hypersphere $S^D(a)$ is

$$ds^2 = a^2 \left\{ (d\theta_{D-1})^2 + \sin^2\theta_{D-1} \left[(d\theta_{D-2})^2 + \sin^2\theta_{D-2} \left[\cdots \right. \right. \right.$$
$$\left. \left. \left. \cdots (d\theta_2)^2 + \sin^2\theta_2 \left[(d\theta_1)^2 + \sin^2\theta_1 \, (d\varphi)^2 \right] \cdots \right] \right] \right\}.$$

The expression on the rhs uses the D angle coordinates $\theta_{D-1}, \ldots, \theta_1$, φ. One can write the metric of $S^D(a)$ also in an alternative form by using the radial coordinate r and $D-1$ angle coordinates for the description. It is then

$$ds^2 = \frac{a^2}{a^2 - r^2} \, dr^2 + r^2 \, d\Phi^2.$$

The squared angle element $d\Phi^2$ is a shorthand notation for the differential containing only the $D-1$ angle coordinates $\theta_{D-2}, \ldots, \theta_1$, φ. Written out it is

$$d\Phi^2 = (d\theta_{D-2})^2 + \sin^2\theta_{D-2} \left[\cdots (d\theta_2)^2 + \sin^2\theta_2 \left[(d\theta_1)^2 + \sin^2\theta_1 \, (d\varphi)^2 \right] \cdots \right].$$

Curvature of D-Dimensional Sphere
We summarize here basic facts about the curvature of a D-dimensional hypersphere $S^D(a)$ with radius a and dimensionality $D \geq 2$. The formulae are very simple, since the D-sphere is a manifold of constant curvature. The Riemann curvature tensor is

$$R_{\mu\nu\rho\sigma} = \frac{1}{a^2} (\delta_{\mu\rho}\delta_{\nu\sigma} - \delta_{\mu\sigma}\delta_{\nu\rho}).$$

Consequently, the Ricci tensor is

$$R_{\mu\nu} = \frac{D-1}{a^2}\delta_{\mu\nu}.$$

The scalar curvature is a constant,

$$R = \frac{D(D-1)}{a^2}.$$

We note also that the Weyl tensor of the hypersphere vanishes identically, $W_{\mu\nu\rho\sigma} = 0$.

Volume of D-Dimensional Sphere
In the following, we derive the formula for the D-dimensional volume $\mathrm{Vol}\left(S^D(a)\right)$ of the hypersphere $S^D(a)$. The general definition of a volume yields

$$\mathrm{Vol}\left(S^D(a)\right) = \int_{S^D(a)} \sqrt{|g|} \, dx^1 \cdots dx^D,$$

where the metric $g_{\mu\nu}(x)$ is the one for the hypersphere, as obtained before. First, let us note that it is

$$\mathrm{Vol}\left(S^D(a)\right) = \mathrm{Vol}\left(S^D\right) a^D,$$

which can be immediately obtained by dimensional analysis. In order to calculate the volume of $S^D(a)$, one can directly employ the integral definition above. However, there is a simpler method. Consider the $(D+1)$-dimensional Gaussian integral

$$I_{D+1} \equiv \int_{\mathbb{R}^{D+1}} e^{-x^2} \, dx^1 \cdots dx^{D+1}$$

for $D + 1$ Cartesian coordinates, where $x^2 = (x^1)^2 + \cdots + (x^{D+1})^2$. The integral readily decomposes into a product and yields

$$I_{D+1} = \prod_{\mu=1}^{D+1} \int_{-\infty}^{\infty} e^{-(x^\mu)^2} \, dx^\mu = \pi^{\frac{D+1}{2}}.$$

Now we calculate the same integral by another method. The space $\mathbb{R}^{D+1}$ is split into an infinite sequence of thin spherical shells represented by the spheres $S^D(r)$ with increasing radius $0 \le r < \infty$. The Gaussian integral is then written as

$$I_{D+1} = \int_0^{\infty} \mathrm{Vol}\left(S^D(r)\right) e^{-r^2} \, dr = \mathrm{Vol}\left(S^D\right) \int_0^{\infty} r^D e^{-r^2} \, dr.$$

By changing the integration variable from r to $t = r^2$, we readily obtain the result

$$I_{D+1} = \frac{1}{2} \mathrm{Vol}\left(S^D\right) \Gamma\left(\frac{D+1}{2}\right).$$

From the first calculation, we know that it is $I_{D+1} = \pi^{\frac{D+1}{2}}$, so that we can conclude to the formula for the volume of the unit hypersphere S^D,

$$\mathrm{Vol}\left(S^D\right) = \frac{2\pi^{\frac{D+1}{2}}}{\Gamma\left(\frac{D+1}{2}\right)}.$$

The formula for the hypersphere $S^D(a)$ with radius a is therefore

$$\mathrm{Vol}\left(S^D(a)\right) = \frac{2\pi^{\frac{D+1}{2}}}{\Gamma\left(\frac{D+1}{2}\right)} a^D.$$

The $(D+1)$-Dimensional Ball

We have defined the D-dimensional sphere by embedding it in the $(D+1)$-dimensional Euclidean space. The $(D+1)$-dimensional region having the sphere as its D-dimensional boundary is what we call the $(D+1)$-dimensional "ball". Formally, the *ball* $B^{D+1}(a)$ with radius a is the $(D+1)$-dimensional manifold defined as the set of points

$$B^{D+1}(a) \equiv \left\{ (x^1, \ldots, x^{D+1}) \in \mathbb{E}^{D+1} \mid (x^1)^2 + \cdots + (x^{D+1})^2 \le a^2 \right\}.$$

This means:

- The D-dimensional sphere $S^D(a)$ is the boundary of the $(D+1)$-dimensional ball $B^{D+1}(a)$. In symbols, $S^D(a) = \partial\left(B^{D+1}(a)\right)$.
- The "surface area" $\mathrm{Area}\left(B^{D+1}(a)\right)$ of the ball $B^{D+1}(a)$ is exactly what we call the "volume" $\mathrm{Vol}\left(S^D(a)\right)$ of the sphere $S^D(a)$. This is a measure of dimension $\propto a^D$.
- The "volume" $\mathrm{Vol}\left(B^{D+1}(a)\right)$ of the ball $B^{D+1}(a)$ is a measure of dimension $\propto a^{D+1}$.

Based on the above results, we can write down the D-dimensional surface $\mathrm{Area}\left(B^{D+1}(a)\right)$ of the ball $B^{D+1}(a)$ immediately,

$$\mathrm{Area}\left(B^{D+1}(a)\right) = \mathrm{Vol}\left(S^D(a)\right) = \frac{2\pi^{\frac{D+1}{2}}}{\Gamma\left(\frac{D+1}{2}\right)} a^D.$$

The $(D+1)$-dimensional volume $\text{Vol}\left(B^{D+1}(a)\right)$ of the ball $B^{D+1}(a)$ in Cartesian coordinates is given as

$$\text{Vol}\left(B^{D+1}(a)\right) = \int_{B^{D+1}(a)} dx^1 \cdots dx^{D+1}.$$

By dimensional analysis, we immediately see that it is

$$\text{Vol}\left(B^{D+1}(a)\right) = \text{Vol}\left(B^{D+1}\right) a^{D+1},$$

with the *unit ball* $B^{D+1} \equiv B^{D+1}(1)$. We consider the ball $B^{D+1}(a)$ as being comprised of a sequence of infinitesimally thin spherical shells of radius $0 \le r \le a$ and write

$$\text{Vol}\left(B^{D+1}(a)\right) = \int_0^a \text{Area}\left(B^{D+1}(r)\right) dr = \int_0^a \text{Vol}\left(S^D(r)\right) dr.$$

The integral is immediately calculated and yields the desired result for the volume of the ball $B^{D+1}(a)$ with radius a,

$$\text{Vol}\left(B^{D+1}(a)\right) = \frac{\pi^{\frac{D+1}{2}}}{\Gamma\left(\frac{D+3}{2}\right)} a^{D+1}.$$

For the D-dimensional ball $B^D(a)$, we list below the values for its volume and surface area, with increasing dimension D.

Table B.1: Volume and surface of the D-dimensional ball with increasing dimensionality

Dimension D	Volume $\text{Vol}\left(B^D(a)\right)$	Surface Area $\left(B^D(a)\right)$
1	$2a$	2
2	πa^2	$2\pi a$
3	$\frac{4}{3}\pi a^3$	$4\pi a^2$
4	$\frac{1}{2}\pi^2 a^4$	$2\pi^2 a^3$
5	$\frac{8}{15}\pi^2 a^5$	$\frac{8}{3}\pi^2 a^4$
6	$\frac{1}{6}\pi^3 a^6$	$\pi^3 a^5$
7	$\frac{16}{105}\pi^3 a^7$	$\frac{16}{15}\pi^3 a^6$
8	$\frac{1}{24}\pi^4 a^8$	$\frac{1}{3}\pi^4 a^7$
9	$\frac{32}{945}\pi^4 a^9$	$\frac{32}{105}\pi^4 a^8$
10	$\frac{1}{120}\pi^5 a^{10}$	$\frac{1}{12}\pi^5 a^9$

Let us consider the unit ball B^D and ask for which dimensions the maximum volume and the maximum surface are reached respectively. The volume reaches its global maximum value for $D = 5$ and decreases thereafter for higher dimensions. The surface area reaches its global maximum value for $D = 7$ and also decreases for higher dimensions.

B.8 Hypersurface Elements

Hypersurface Elements using the ε-Symbol

We consider D-dimensional Riemannian manifolds and remark that the results below can readily be translated to d-dimensional pseudo-Riemannian manifolds. We view more broadly $(D - k)$-dimensional *hypersurfaces*, the case $k = 1$ corresponding to hypersurfaces in the usual sense. For a $(D - 1)$-dimensional hypersurface, the associated *hypersurface element* $\mathrm{d}\Sigma_\mu(x)$ is defined by

$$\mathrm{d}\Sigma_\mu \equiv \sqrt{|g|}\, \mathrm{d}\sigma_\mu,$$

where $g(x)$ is the determinant of the manifold metric, $g \equiv \det(g_{\mu\nu})$, and $\mathrm{d}\sigma_\mu(x)$ is the $(D - 1)$-dimensional hypersurface element without reference to a metric,

$$\mathrm{d}\sigma_\mu = \frac{1}{(D-1)!}\, \varepsilon_{\mu\mu_2\ldots\mu_D}\, \mathrm{d}x^{\mu_2} \wedge \cdots \wedge \mathrm{d}x^{\mu_D}.$$

The definition is chosen such that the expression $a^\mu\, \mathrm{d}\sigma_\mu$, used in the Stokes integral formula, is equal to $V^\mu\, \mathrm{d}\Sigma_\mu$, where $a^\mu(x)$ is a vector density field with weight $w = -1$, while $V^\mu(x)$ is an absolute vector field, given as $V^\mu = a^\mu/\sqrt{|g|}$. Consequently, the geometric hypersurface element $\mathrm{d}\Sigma_\mu(x)$ can be written as

$$\mathrm{d}\Sigma_\mu = \frac{1}{(D-1)!}\, \sqrt{|g|}\, \varepsilon_{\mu\mu_2\ldots\mu_D}\, \mathrm{d}x^{\mu_2} \wedge \cdots \wedge \mathrm{d}x^{\mu_D}.$$

For a $(D - 2)$-dimensional hypersurface, the associated *hypersurface element* $\mathrm{d}\Sigma_{\mu\nu}(x)$ is

$$\mathrm{d}\Sigma_{\mu\nu} \equiv \sqrt{|g|}\, \mathrm{d}\sigma_{\mu\nu},$$

where $\mathrm{d}\sigma_{\mu\nu}(x)$ is the $(D - 2)$-dimensional metric-independent hypersurface element,

$$\mathrm{d}\sigma_{\mu\nu} = \frac{1}{(D-2)!}\, \varepsilon_{\mu\nu\mu_3\ldots\mu_D}\, \mathrm{d}x^{\mu_3} \wedge \cdots \wedge \mathrm{d}x^{\mu_D}.$$

Therefore, the geometric hypersurface element $\mathrm{d}\Sigma_{\mu\nu}(x)$ can be written explicitly as

$$\mathrm{d}\Sigma_{\mu\nu} = \frac{1}{(D-2)!}\, \sqrt{|g|}\, \varepsilon_{\mu\nu\mu_3\ldots\mu_D}\, \mathrm{d}x^{\mu_3} \wedge \cdots \wedge \mathrm{d}x^{\mu_D}.$$

The hypersurface element $\mathrm{d}\Sigma_{\mu_1\ldots\mu_k}(x)$ of a $(D - k)$-dimensional hypersurface is defined as

$$\mathrm{d}\Sigma_{\mu_1\ldots\mu_k} \equiv \sqrt{|g|}\, \mathrm{d}\sigma_{\mu_1\ldots\mu_k}$$

and is written explicitly as the $(D - k)$-form

$$\mathrm{d}\Sigma_{\mu_1\ldots\mu_k} = \frac{1}{(D-k)!}\, \sqrt{|g|}\, \varepsilon_{\mu_1\ldots\mu_k\,\mu_{k+1}\ldots\mu_D}\, \mathrm{d}x^{\mu_{k+1}} \wedge \cdots \wedge \mathrm{d}x^{\mu_D}.$$

It can be easily shown that under orientation-preserving coordinate transformations, the hypersurface element $\mathrm{d}\Sigma_{\mu_1\ldots\mu_k}(x)$ with respect to its k indices transforms as an absolute $(0, k)$-tensor field,

$$\mathrm{d}\Sigma'_{\mu_1\ldots\mu_k}(x') = \frac{\partial x^{\nu_1}}{\partial x'^{\mu_1}} \cdots \frac{\partial x^{\nu_k}}{\partial x'^{\mu_k}}\, \mathrm{d}\Sigma_{\nu_1\ldots\nu_k}(x).$$

As examples, let us write down some relevant hypersurface elements for a 4-dimensional pseudo-Riemannian (spacetime) manifold. We have

$$\mathrm{d}\Sigma_\mu = \frac{1}{3!}\, \sqrt{|g|}\, \varepsilon_{\mu\nu\rho\sigma}\, \mathrm{d}x^\nu \wedge \mathrm{d}x^\rho \wedge \mathrm{d}x^\sigma$$

for the surface element of a 3-dimensional (timelike or spacelike) hypersurface and

$$d\Sigma_{\mu\nu} = \frac{1}{2}\sqrt{|g|}\,\varepsilon_{\mu\nu\rho\sigma}\,dx^\rho \wedge dx^\sigma$$

for the surface element of a 2-dimensional hypersurface.

Hypersurface Element using the Normal Vector

Here we devise an alternative formula for the $(D-1)$-dimensional hypersurface element $d\Sigma_\mu(x)$ within a D-dimensional Riemannian manifold. The result can again be translated to the case of a d-dimensional pseudo-Riemannian manifold. Let us write down the corresponding formula here

$$d\Sigma_\mu = N_\mu\sqrt{|h|}\,dy^1 \wedge \cdots \wedge dy^{D-1}.$$

We will describe the components in the rhs shortly. We start by considering the general $(D-1)$-dimensional hypersurface element $d\sigma_\mu(x)$ without reference to a metric, as given above. The integrand corresponding to $d\sigma_\mu(x)$ is the vector density field $a^\mu(x)$, with weight $w = -1$, which can always be written as $a^\mu = \sqrt{|g|}\,V^\mu$, with an absolute vector field $V^\mu(x)$. We consider a D-dimensional region U of the Riemannian manifold possessing a closed boundary ∂U as the $(D-1)$-dimensional hypersurface of interest. In a first step, we transform from the coordinate set $(x^1,\ldots,x^D)$ to the adapted coordinate set $(y^1,\ldots,y^{D-1})$ describing the points on the hypersurface ∂U. The local basis covectors transform as

$$dx^\mu = \frac{\partial x^\mu}{\partial y^\alpha}\,dy^\alpha,$$

with the indices ranges $\mu = 1,\ldots,D$ and $\alpha = 1,\ldots,D-1$. The induced metric $h_{\alpha\beta}(y)$ on the hypersurface is expressed through the parent metric $g_{\mu\nu}(x)$ by means of the relation

$$h_{\alpha\beta} = \frac{\partial x^\mu}{\partial y^\alpha}\frac{\partial x^\nu}{\partial y^\beta}\,g_{\mu\nu}.$$

The determinant of the induced metric is $h(y) \equiv \det(h_{\alpha\beta}(y))$. Now we define the covector density field $n_\mu(x)$ by

$$n_\mu \equiv \frac{1}{(D-1)!}\,\varepsilon_{\mu\mu_2\ldots\mu_D}\,\frac{\partial\,(x^{\mu_2},\ldots,x^{\mu_D})}{\partial\,(y^1,\ldots,y^{D-1})}.$$

By employing $n_\mu(x)$, the metric-independent hypersurface element $d\sigma_\mu(x)$ can be written as

$$d\sigma_\mu = n_\mu\,dy^1 \wedge \cdots \wedge dy^{D-1}.$$

The contraction $a^\mu\,d\sigma_\mu$ appearing in the rhs of the divergence theorem, is consequently

$$a^\mu\,d\sigma_\mu = V^\mu\sqrt{|g|}\,n_\mu\,dy^1 \wedge \cdots \wedge dy^{D-1}.$$

In this expression, we want to understand the role of $n_\mu(x)$. Moreover, we would like to use the determinant $h(y)$ on the hypersurface, instead of the determinant $g(x)$.

The covector field $n_\mu(x)$ is actually *normal* to the boundary ∂U, i.e. the contraction of $n_\mu(x)$ with any tangent vector of ∂U vanishes. Let us see why. The vectors tangent to ∂U are linear combinations of the $D-1$ local basis vectors

$$\frac{\partial}{\partial y^\alpha} = \frac{\partial x^\mu}{\partial y^\alpha}\frac{\partial}{\partial x^\mu}.$$

Now the contraction of $n_\mu(x)$ with any basis vector $\partial/\partial y^\alpha$ leads to an expression proportional to

$$\sum_{\mu,\mu_2,\ldots,\mu_D} \varepsilon_{\mu\mu_2\ldots\mu_D} \frac{\partial\left(x^{\mu_2},\ldots,x^{\mu_D}\right)}{\partial\left(y^1,\ldots,y^{D-1}\right)} \frac{\partial x^\mu}{\partial y^\alpha}.$$

For any possible index α there is a factor $(\partial x^\rho/\partial y^\alpha)(\partial x^\sigma/\partial y^\alpha)$ (no summation over α here!) in each term of the sum. This factor is symmetric in the indices ρ and σ, while the epsilon symbol factor behaves antisymmetric. Thus, the whole sum vanishes and the orthogonality of the covector field $n_\mu(x)$ is proved.

Now we turn to the squared norm $n^2 = g^{\mu\nu} n_\mu n_\nu$ of the covector $n_\mu(x)$, for which we notice that it can be a positive or a negative number. We will show that the absolute value of the squared norm is given by $|n^2| = |h|/|g|$, where the absolute values of the determinants $|h| = |\det(h_{\alpha\beta})|$ and $|g| = |\det(g_{\mu\nu})|$ are used. We break down the proof into three steps. Step 1: we first prove the validity of the following identity

$$(D-1)!\, g^{\rho\sigma}\, g = \varepsilon^{\rho\rho_2\ldots\rho_D}\, \varepsilon^{\sigma\sigma_2\ldots\sigma_D}\, g_{\rho_2\sigma_2} \cdots g_{\rho_D\sigma_D}$$

for the metric $g_{\mu\nu}(x)$. To this end, let us define the quantity $C^\rho{}_\sigma(x)$ through

$$(D-1)!\, C^\rho{}_\sigma \equiv \varepsilon^{\rho\rho_2\ldots\rho_D}\, \varepsilon_{\sigma\sigma_2\ldots\sigma_D}\, g^{\sigma_2}{}_{\rho_2} \cdots g^{\sigma_D}{}_{\rho_D}.$$

By contracting with $g^\sigma{}_\tau(x)$ and by using the definition of the determinant, we obtain

$$(D-1)!\, C^\rho{}_\sigma g^\sigma{}_\tau = \varepsilon^{\rho\rho_2\ldots\rho_D}\, \varepsilon_{\tau\rho_2\ldots\rho_D}\, g.$$

The contraction of the epsilon symbols yields $C^\rho{}_\sigma g^\sigma{}_\tau = \delta^\rho_\tau\, g$ and thus we have

$$(D-1)!\, \delta^\rho_\tau\, g = \varepsilon^{\rho\rho_2\ldots\rho_D}\, \varepsilon_{\tau\sigma_2\ldots\sigma_D}\, g^{\sigma_2}{}_{\rho_2} \cdots g^{\sigma_D}{}_{\rho_D}.$$

Raising and lowering of indices gives the desired identity. Step 2: we consider the relation

$$(D-1)!\, h = \varepsilon^{\alpha_1\ldots\alpha_{D-1}}\, \varepsilon^{\beta_1\ldots\beta_{D-1}}\, h_{\alpha_1\beta_1} \cdots h_{\alpha_{D-1}\beta_{D-1}}.$$

By using the basic definition of the induced metric $h_{\alpha\beta}(y)$, we can rewrite this as

$$(D-1)!\, h = \frac{\partial\left(x^{\mu_2},\ldots,x^{\mu_D}\right)}{\partial\left(y^1,\ldots,y^{D-1}\right)} \frac{\partial\left(x^{\nu_2},\ldots,x^{\nu_D}\right)}{\partial\left(y^1,\ldots,y^{D-1}\right)} g_{\mu_2\nu_2} \cdots g_{\mu_D\nu_D}.$$

Further, by using the contraction

$$n_\mu\, \varepsilon^{\mu\rho_2\ldots\rho_D} = \frac{\partial\left(x^{\rho_2},\ldots,x^{\rho_D}\right)}{\partial\left(y^1,\ldots,y^{D-1}\right)},$$

the relation is written as

$$(D-1)!\, h = n_\mu\, \varepsilon^{\mu\mu_2\ldots\mu_D}\, n_\nu\, \varepsilon^{\nu\nu_2\ldots\nu_D}\, g_{\mu_2\nu_2} \cdots g_{\mu_D\nu_D}.$$

Step 3: we compare the identity from step 1 with the relation from step 2 and obtain immediately $h = n_\mu n_\nu g^{\mu\nu} g$, which is equivalent to $n^2 = h/g$. This yields the absolute value $|n^2| = |h|/|g|$, as desired.

Now that we know the norm of $n_\mu(x)$, we define the corresponding unit covector field $N_\mu(x)$ by

$$N_\mu \equiv \frac{n_\mu}{\sqrt{|n^2|}} = \sqrt{\frac{|g|}{|h|}}\, n_\mu.$$

The squared norm N^2 of the covector field $N_\mu(x)$ is

$$N^2 = \frac{n^2}{|n^2|} = \operatorname{sgn}(n^2) = \pm 1,$$

where the sign value depends on the specific hypersurface ∂U at hand and the signature of the underlying manifold metric. The associated vector field $N^\mu(x)$ represents the normal vector of the hypersurface element. By summarizing the above results, the Riemannian hypersurface element $\mathrm{d}\Sigma_\mu(x)$ can be written as

$$\mathrm{d}\Sigma_\mu(x) = N_\mu(x(y))\sqrt{|h(y)|}\,\mathrm{d}y^1 \wedge \cdots \wedge \mathrm{d}y^{D-1}.$$

The expression on the rhs contains only quantities that are specific to the hypersurface ∂U. As an example, in the pseudo-Riemannian $d = 4$ case, the hypersurface element $\mathrm{d}\Sigma_\mu(x)$ can be expressed as

$$\mathrm{d}\Sigma_\mu = N_\mu \sqrt{|h|}\,\mathrm{d}y^1 \wedge \mathrm{d}y^2 \wedge \mathrm{d}y^3,$$

in which the coordinates (y^1, y^2, y^3) label the points of the 3-dimensional hypersurface.

C

Weyl Rescaling Formulae

Conformal Transformations vs. Weyl Rescalings

In Chapters 2 and 3, *conformal maps* of a manifold are introduced and discussed. These conformal maps correspond to *conformal coordinate transformations*, which are developed group-theoretically in Chapter 14. In Section 3.2 we show the basic relationship between conformal maps on one hand and Weyl rescalings on the other hand. *Weyl rescalings* are defined as transformations of the fundamental metric of a geometric manifold and are independent of any coordinate transformations.

Definition of Weyl Rescalings

Under a *Weyl rescaling*, the metric $g_{\mu\nu}(x)$ of a d-dimensional geometric manifold $(\mathcal{M}, g_{\mu\nu})$ is rescaled by employing a differentiable, scalar *conformal factor* $\Omega(x)$, with $\Omega(x) > 0$, in the form

$$\widehat{g}_{\mu\nu} = \Omega^2 g_{\mu\nu}.$$

By using a real scalar function $s(x)$, the conformal factor can be expressed as

$$\Omega(x) \equiv \exp s(x).$$

Moreover, we introduce the "ypsilon" covector field $\Upsilon_\mu(x)$ as

$$\Upsilon_\mu \equiv \nabla_\mu \ln \Omega = \frac{\nabla_\mu \Omega}{\Omega} = \nabla_\mu s,$$

and the symmetric "zeta" tensor field $\mathcal{Z}_{\mu\nu}(x)$ as

$$\mathcal{Z}_{\mu\nu} \equiv \Upsilon_\mu \Upsilon_\nu - \nabla_\mu \Upsilon_\nu - \frac{1}{2} g_{\mu\nu} \Upsilon_\alpha \Upsilon^\alpha,$$

for characterizing the Weyl rescaling transformation. For the trace $\mathcal{Z}(x) \equiv \mathcal{Z}^\alpha{}_\alpha(x)$, we have the expression

$$\mathcal{Z} = -\nabla^\alpha \Upsilon_\alpha - \frac{d-2}{2} \Upsilon^2,$$

with the simplified notation $\Upsilon^2 \equiv \Upsilon_\alpha \Upsilon^\alpha$. The scalar field $s(x)$, the covector field $\Upsilon_\mu(x)$, and the tensor field $\mathcal{Z}_{\mu\nu}(x)$, which are all defining elements of a Weyl transformation, are quantities that do not transform.

DOI: 10.1201/9781003087748-C

Weyl Rescaling Formulae

In the following, we compile the transformation formulae under Weyl rescalings for the most relevant quantities. For the inverse metric it is:

$$\widehat{g}^{\mu\nu} = \Omega^{-2} g^{\mu\nu}.$$

Determinant of metric:

$$\sqrt{|\det(\widehat{g}_{\mu\nu})|} = \Omega^d \sqrt{|\det(g_{\mu\nu})|}.$$

Volume element:

$$\sqrt{|\widehat{g}|}\, \mathrm{d}^d x = \Omega^d \sqrt{|g|}\, \mathrm{d}^d x.$$

Kronecker delta:

$$\widehat{\delta}_{\mu\nu} = \delta_{\mu\nu}.$$

Epsilon symbol (tensor density):

$$\widehat{\varepsilon}_{\mu_1...\mu_d} = \varepsilon_{\mu_1...\mu_d}.$$

Epsilon tensor (absolute tensor):

$$\widehat{\epsilon}_{\mu_1...\mu_d} = \Omega^d \epsilon_{\mu_1...\mu_d}.$$

Connection:

$$\widehat{\Gamma}^{\mu}_{\nu\rho} = \Gamma^{\mu}_{\nu\rho} + \delta^{\mu}_{\nu}\Upsilon_{\rho} + \delta^{\mu}_{\rho}\Upsilon_{\nu} - g_{\nu\rho}\Upsilon^{\mu}.$$

Contracted connection:

$$\widehat{\Gamma}^{\alpha}_{\mu\alpha} = \Gamma^{\alpha}_{\mu\alpha} + \Upsilon_{\mu}d.$$

Covariant derivative of a fundamental vector field $V^{\mu} = \widehat{V}^{\mu}$:

$$\widehat{\nabla}_{\rho}V^{\mu} = \nabla_{\rho}V^{\mu} + \Upsilon_{\rho}V^{\mu} - \Upsilon^{\mu}V_{\rho} + \delta^{\mu}_{\rho}\Upsilon_{\alpha}V^{\alpha}.$$

Covariant derivative of a fundamental covector field $\omega_{\mu} = \widehat{\omega}_{\mu}$:

$$\widehat{\nabla}_{\rho}\omega_{\nu} = \nabla_{\rho}\omega_{\nu} - \Upsilon_{\rho}\omega_{\nu} - \Upsilon_{\nu}\omega_{\rho} + g_{\rho\nu}\Upsilon^{\alpha}\omega_{\alpha}.$$

Covariant derivative of a fundamental $(0,2)$-tensor field $\phi_{\mu\nu} = \widehat{\phi}_{\mu\nu}$:

$$\widehat{\nabla}_{\rho}\phi_{\mu\nu} = \nabla_{\rho}\phi_{\mu\nu} - 2\Upsilon_{\rho}\phi_{\mu\nu} - \Upsilon_{\mu}\phi_{\rho\nu} - \Upsilon_{\nu}\phi_{\mu\rho} + g_{\rho\mu}\Upsilon^{\alpha}\phi_{\alpha\nu} + g_{\rho\nu}\Upsilon^{\alpha}\phi_{\mu\alpha}.$$

Riemann curvature tensor:

$$\widehat{R}_{\mu\nu\rho\sigma} = \Omega^2\Big[R_{\mu\nu\rho\sigma} + (g_{\mu\rho}\mathcal{Z}_{\nu\sigma} - g_{\mu\sigma}\mathcal{Z}_{\nu\rho} + g_{\nu\sigma}\mathcal{Z}_{\mu\rho} - g_{\nu\rho}\mathcal{Z}_{\mu\sigma})\Big].$$

Ricci tensor:

$$\widehat{R}_{\mu\nu} = R_{\mu\nu} + (d-2)\mathcal{Z}_{\mu\nu} + \mathcal{Z}g_{\mu\nu}.$$

Ricci scalar:

$$\widehat{R} = \Omega^{-2}[R + (2d-2)\mathcal{Z}].$$

Traceless part of the Ricci tensor (Weyl invariant in $d=2$):

$$\widehat{S}_{\mu\nu} = S_{\mu\nu} + (d-2)\Big(\mathcal{Z}_{\mu\nu} - \frac{1}{d}\mathcal{Z}g_{\mu\nu}\Big).$$

Einstein tensor (Weyl invariant in $d = 2$):

$$\widehat{G}_{\mu\nu} = G_{\mu\nu} + (d-2)(\mathcal{Z}_{\mu\nu} - \mathcal{Z}g_{\mu\nu}).$$

Trace of the Einstein tensor (Weyl covariant in $d = 2$):

$$\widehat{G} = \Omega^{-2}\big[G - (d-1)(d-2)\mathcal{Z}\big].$$

Schouten tensor:

$$\widehat{P}_{\mu\nu} = P_{\mu\nu} + \mathcal{Z}_{\mu\nu}.$$

Trace of the Schouten tensor:

$$\widehat{P} = \Omega^{-2}(P + \mathcal{Z}).$$

Weyl tensor in the version $W_{\mu\nu\rho\sigma}(x)$ is Weyl covariant:

$$\widehat{W}_{\mu\nu\rho\sigma} = \Omega^2\, W_{\mu\nu\rho\sigma}.$$

Weyl tensor in the version $W^{\mu}{}_{\nu\rho\sigma}(x)$ is Weyl invariant:

$$\widehat{W}^{\mu}{}_{\nu\rho\sigma} = W^{\mu}{}_{\nu\rho\sigma}.$$

Cotton tensor (Weyl invariant in $d = 3$):

$$\widehat{C}_{\nu\rho\sigma} = C_{\nu\rho\sigma} + \Upsilon^{\alpha}W_{\alpha\nu\rho\sigma}.$$

Divergence of a fundamental vector field $V^{\mu} = \widehat{V}^{\mu}$:

$$\widehat{\nabla}_{\mu}V^{\mu} = \nabla_{\mu}V^{\mu} + d\,\Upsilon_{\mu}V^{\mu}.$$

Divergence of a fundamental covector field $\omega_{\mu} = \widehat{\omega}_{\mu}$:

$$\widehat{\nabla}^{\mu}\omega_{\mu} = \nabla^{\mu}\omega_{\mu} + (d-2)\,\Upsilon^{\mu}\omega_{\mu}.$$

Divergence of a fundamental $(0,2)$-tensor field $\phi_{\mu\nu} = \widehat{\phi}_{\mu\nu}$:

$$\widehat{\nabla}^{\mu}\phi_{\mu\nu} = \Omega^{-2}\big[\nabla^{\mu}\phi_{\mu\nu} + (d-3)\,\Upsilon^{\mu}\phi_{\mu\nu} + \Upsilon^{\mu}\phi_{\nu\mu} - \Upsilon_{\nu}\phi^{\mu}{}_{\mu}\big].$$

Divergence of a fundamental antisymmetric $(0,2)$-tensor field $A_{\mu\nu} = -A_{\nu\mu}$:

$$\widehat{\nabla}^{\mu}A_{\mu\nu} = \Omega^{-2}\big[\nabla^{\mu}A_{\mu\nu} + (d-4)\,\Upsilon^{\mu}A_{\mu\nu}\big].$$

Laplace / d'Alembert operator $\nabla^2 = g^{\mu\nu}\nabla_{\mu}\nabla_{\nu}$ acting on a fundamental scalar field $\phi = \widehat{\phi}$:

$$\widehat{\nabla}^2\phi = \Omega^{-2}\big[\nabla^2\phi + (d-2)\Upsilon^{\alpha}\nabla_{\alpha}\phi\big].$$

Weyl Covariance and Invariance

Let us consider functions $Q(\phi_i, g_{\mu\nu}, x)$ depending on the fields $\phi_i(x)$, the metric $g_{\mu\nu}(x)$, and the coordinates x^{μ}. If the quantity $Q(\phi_i, g_{\mu\nu}, x)$ transforms under Weyl rescalings like

$$Q(\widehat{\phi}_i, \widehat{g}_{\mu\nu}, x) = \Omega^p\, Q(\phi_i, g_{\mu\nu}, x),$$

with a constant real exponent p, then it exhibits *Weyl rescaling covariance*. If $p = 0$, the quantity shows *Weyl rescaling invariance*. An equation of the form

$$Q(\phi_i, g_{\mu\nu}, x) = 0,$$

in which the quantity on the lhs is Weyl covariant, represents a Weyl invariant equation:

$$Q(\phi_i, g_{\mu\nu}, x) = 0 \quad \Leftrightarrow \quad Q(\widehat{\phi}_i, \widehat{g}_{\mu\nu}, x) = 0.$$

For example, the antisymmetrized covariant derivative $\nabla_{[\rho}\phi_{\mu\nu]}$ of a fundamental $(0,2)$-tensor field $\phi_{\mu\nu}(x)$ is actually Weyl invariant,

$$\widehat{\nabla}_{[\rho}\phi_{\mu\nu]} = \nabla_{[\rho}\phi_{\mu\nu]}.$$

Specifically for an antisymmetric fundamental $(0,2)$-tensor field $A_{\mu\nu} = -A_{\nu\mu}$, this Weyl invariance can be written alternatively as

$$\widehat{\nabla}_\rho A_{\mu\nu} + \widehat{\nabla}_\mu A_{\nu\rho} + \widehat{\nabla}_\nu A_{\rho\mu} = \nabla_\rho A_{\mu\nu} + \nabla_\mu A_{\nu\rho} + \nabla_\nu A_{\rho\mu}.$$

This has an immediate application to the Maxwell theory on curved manifolds. The Faraday tensor with lower indices $F_{\mu\nu}(x)$ is a fundamental antisymmetric $(0,2)$-tensor field, $F_{\mu\nu} = \widehat{F}_{\mu\nu}$. Consequently, *the Maxwell integrability condition is Weyl rescaling invariant*:

$$\widehat{\nabla}_\rho F_{\mu\nu} + \text{cyclic} = 0 \quad \Leftrightarrow \quad \nabla_\rho F_{\mu\nu} + \text{cyclic} = 0.$$

For the divergence of an antisymmetric fundamental $(0,2)$-tensor field we have

$$\widehat{\nabla}^\mu A_{\mu\nu} = \Omega^{-2}\left[\nabla^\mu A_{\mu\nu} + (d-4)\,\Upsilon^\mu A_{\mu\nu}\right],$$

which in $d = 4$ dimensions yields

$$\widehat{\nabla}^\mu A_{\mu\nu} = \Omega^{-2}\nabla^\mu A_{\mu\nu}.$$

This means that the divergence $\nabla^\mu A_{\mu\nu}$ in four dimensions is Weyl covariant. Consequently, *the free Maxwell equations are Weyl rescaling invariant in four dimensions*:

$$\widehat{\nabla}^\mu F_{\mu\nu} = 0 \quad \Leftrightarrow \quad \nabla^\mu F_{\mu\nu} = 0, \quad \text{for } d = 4.$$

Fortunately, central differential geometric identities like the cyclic identity or the differential Bianchi identity in its various guises (for the Ricci tensor, the Einstein tensor, the Schouten tensor, and the Cotton tensor) are all Weyl rescaling invariant. However, this does not apply to all equations we can express through geometric quantities. The Weyl rescaling formula for the Laplace / d'Alembert operator ∇^2 reveals that this operator transforms covariantly only in two dimensions. However, it is possible to define a modified Laplace / d'Alembert operator that is Weyl covariant in all dimensions by letting it act solely on so-called Weyl density fields. A *Weyl scalar density field* $f(x)$ is a function, which under a Weyl rescaling transforms as

$$\widehat{f} = \Omega^{w_f} f,$$

with some real exponent w_f called the *conformal weight*. The function $f(x)$ is not a fundamental scalar field. The weight w_f can be tuned to the problem at hand, and for the specific question of Laplace / d'Alembert operator covariance one considers Weyl densities transforming concretely as

$$\widehat{f} = \Omega^{\frac{2-d}{2}} f.$$

The *conformal Laplace / d'Alembert operator* defined as

$$\nabla^2 - \frac{d-2}{4d-4}R,$$

where $R(x)$ is the Ricci curvature scalar, is indeed Weyl rescaling covariant,

$$\left(\widehat{\nabla}^2 - \frac{d-2}{4d-4}\widehat{R}\right)\widehat{f} = \Omega^{-\frac{d+2}{2}}\left(\nabla^2 - \frac{d-2}{4d-4}R\right)f.$$

The associated *conformal Laplace / Klein-Gordon (wave) equation*

$$\nabla^2 f - \frac{d-2}{4d-4}Rf = 0$$

is thus Weyl rescaling invariant. Solutions $f(x)$ of this equation within the geometry $g_{\mu\nu}(x)$ correspond to equivalent solutions $\widehat{f}(x)$ within the Weyl rescaled geometry $\widehat{g}_{\mu\nu}(x)$.

D

Spaces and Symmetry Groups

Symmetry Transformations

We summarize here the most basic facts about symmetry transformations of geometric manifolds $(\mathcal{M}, g_{\mu\nu})$. First, let us express a general diffeomorphism as an active transformation

$$\left.\begin{aligned} x^\mu &\mapsto x'^\mu(x) = x^\mu + \epsilon K^\mu(x) \\ g_{\mu\nu}(x) &\mapsto g'_{\mu\nu}(x') \\ \phi_i(x) &\mapsto \phi'_i(x') \end{aligned}\right\}.$$

Here x^μ are the coordinates, $g_{\mu\nu}(x)$ is the metric, and $\phi_i(x)$ stands for a tensor density representing the fields of the present theory. The smooth vector field $K^\mu(x)$ is the generator of the diffeomorphism. The *form variation* $\delta\phi_i(x) \equiv \phi'_i(x) - \phi_i(x)$ and the *total variation* $\bar\delta\phi_i(x) \equiv \phi'_i(x') - \phi_i(x)$ of the field $\phi_i(x)$ are connected through the basic relation

$$\bar\delta\phi_i(x) = \delta\phi_i(x) + \partial_\mu\phi_i(x)\,\delta x^\mu(x).$$

For the metric tensor, we note also the useful formulae for its variations

$$\bar\delta g_{\mu\nu} = -\epsilon(\partial_\mu K_\nu + \partial_\nu K_\mu) \quad \text{and} \quad \delta g_{\mu\nu} = -\epsilon\mathcal{L}_K g_{\mu\nu} = -\epsilon(\nabla_\mu K_\nu + \nabla_\nu K_\mu).$$

The diffeomorphism defined above is specifically called a *symmetry transformation of the geometric manifold* if it leaves the metric unaffected. The most basic symmetry of the metric is an *isometry* and it is expressed through the condition

$$\mathcal{L}_K g_{\mu\nu} = 0.$$

The vector field $K^\mu(x)$ is called a *Killing vector field* of the metric then. A more general symmetry is present if the less stringent condition holds that the metric is invariant only up to a local scale factor. This corresponds to a *conformal map*, which is expressed through the condition

$$\mathcal{L}_K g_{\mu\nu} = \frac{2}{d}(\nabla_\alpha K^\alpha)g_{\mu\nu}.$$

The vector field $K^\mu(x)$ is called a *conformal Killing vector field* of the metric then. When considering conformal maps, one can introduce an entire equivalence class of metrics related by scale factors as the invariant entity. The (conformal) Killing vector fields span a Lie algebra and represent the generators of the Lie group of the symmetry transformations of the geometric manifold.

DOI: 10.1201/9781003087748-D

Spaces and Symmetry Groups

To different geometric manifolds there are associated different symmetry groups, each one leaving the specific geometric structure unaffected. Mathematically relevant manifolds are noted in the table below.

Table D.1: Geometric manifolds and associated symmetry groups

Geometric Manifold	Symmetry Group
$\mathbb{R}^D$, Metric Vector Space	$R = SO(D)$, Rotation Group
$S^D(a)$, Hypersphere of radius a	$SO(D+1)$, Rotation Group
H^D, Upper Half-Space Model	$L = SO(1, D)$, Lorentz Group

Physically relevant manifolds are summarized in the following table. Euclidean space has D dimensions, while spacetime has either four or $d = 1 + D$ dimensions.

Table D.2: Spacetime manifolds and associated symmetry groups

Spacetime Manifold	Symmetry Group
$\mathbb{E}^D$, Euclidean Affine Space	$E = ISO(D)$, Euclidean Group
$\mathbb{G}_4$, Galilei Spacetime	$G = T \rtimes B \rtimes R$, Galilei Group
$\mathbb{M}_d$, Minkowski Spacetime	$P = ISO(1, D)$, Poincaré Group
$\mathcal{C}_d$, Conformal Spacetime	$C = SO(2, d)$, Conformal Group
$\mathcal{M}_d$, Curved Spacetime	$\text{Diff}(\mathcal{M}_d)$, Diffeomorphisms Group

Bibliography

1. James L. Anderson. *Principles of Relativity Physics*. Academic Press, 1967.

2. Richard Arens. "Newton's Observations about the Field of a Uniform Thin Spherical Shell". Note di Matematica, Vol. X, Suppl. n. 1, 39, 1990.

3. Vladimir I. Arnold. *Mathematical Methods of Classical Mechanics, Second Edition*. Springer, 1989.

4. Richard L. Arnowitt, Stanley Deser, Charles W. Misner. "The Dynamics of General Relativity". Original 1962, reprinted in General Relativity and Gravitation, **40**, 1997, 2008.

5. Stanislav V. Babak, Leonid P. Grishchuk. "The Energy-Momentum Tensor for the Gravitational Field". Physical Review D, **61**, 024038, 1999.

6. Asim O. Barut. *Electrodynamics and Classical Theory of Fields and Particles*. Dover Publications, 1980.

7. Asim O. Barut, Ryszard Raczka. *Theory of Group Representations and Applications, Second Revised Edition*. World Scientific, 1986.

8. Frederik J. Belinfante. "On the Current and the Density of the Electric Charge, the Energy, the Linear Momentum and the Angular Momentum of Arbitrary Fields". Physica, **7**, 449, 1940.

9. Erich P. W. Bessel-Hagen. "Über die Erhaltungssätze der Elektrodynamik". Math. Ann., **84**, 258, 1921.

10. Milutin Blagojevic. *Gravitation and Gauge Symmetries*. Institute of Physics Publishing, CRC Press, 2001.

11. Hermann Bondi, M. G. J. van der Burg, A. W. K. Metzner. "Gravitational Waves in General Relativity. VII. Waves from Axi-Symmetric Isolated Systems". Proceedings of the Royal Society of London, **269**, 21, 1962.

12. William M. Boothby. *An Introduction to Differentiable Manifolds and Riemannian Geometry*. Academic Press, 1975.

13. Curtis G. Callan, Jr., Sidney R. Coleman, Roman W. Jackiw. "A New Improved Energy-Momentum Tensor". Annals of Physics, **59**, 42, 1970.

14. Sean M. Carroll. *Spacetime and Geometry: An Introduction to General Relativity*. Addison Wesley, 2004.

15. Yvonne Choquet-Bruhat, Cecile DeWitt-Morette, Margaret Dillard-Bleick. *Analysis, Manifolds and Physics, Revised Edition*. North-Holland Publishing, 1982.

16. Giovanni Costa, Gianluigi Fogli. *Symmetries and Group Theory in Particle Physics*. Springer, 2012.

17. Sean N. Curry, A. Rod Gover. "An Introduction to Conformal Geometry and Tractor Calculus, with a view to Applications in General Relativity", in Thierry Daudé, Dietrich Häfner and Jean-Philippe Nicolas, eds., *Asymptotic Analysis in General Relativity*. Cambridge University Press, 2018.

18. Mariusz P. Dabrowski, Janusz Garecki, David B. Blaschke. "Conformal Transformations and Conformal Invariance in Gravitation". Annalen der Physik, **18**, 13, 2009.

19. Ghanashyam Date. *General Relativity, Basics and Beyond*. CRC Press, 2015.

20. William R. Davis. *Classical Fields, Particles, and the Theory of Relativity*. Gordon and Breach Science Publishers, 1970.

21. Fernando de Felice, Chris J. S. Clarke. *Relativity on Curved Manifolds*. Cambridge University Press, 1990.

22. Philippe Dennery, André Krzywicki. *Mathematics for Physicists*. Dover Publications, 1995.

23. Bryce DeWitt, Steven M. Christensen. *Bryce DeWitt's Lectures on Gravitation*. Springer, 2011.

24. Philippe Di Francesco, Pierre Mathieu, David Sénéchal. *Conformal Field Theory*. Springer, 1997.

25. Boris A. Dubrovin, Anatoly T. Fomenko, Sergei P. Novikov. *Modern Geometry – Methods and Applications, Part I* and *II*. Springer, 1985.

26. Luther P. Eisenhart. *Riemannian Geometry*. Princeton University Press, 1997.

27. Bjoern Felsager. *Geometry, Particles, and Fields*. Springer, 1998.

28. Richard P. Feynman, Fernando B. Morinigo, William G. Wagner. *Feynman Lectures on Gravitation*. CRC Press, 2018.

29. Richard Fitzpatrick. *An Introduction to Celestial Mechanics*. Cambridge University Press, 2012.

30. Theodore Frankel. *The Geometry of Physics, Third Edition*. Cambridge University Press, 2012.

31. Jörg Frauendiener. "Conformal Infinity". Living Reviews in Relativity, **7**, 1, 2004.

32. Wilhelm I. Fushchich, Anatoly G. Nikitin. *Symmetries of Maxwell's Equations*. D. Reidel Publishing Company, 1987.

33. Howard Georgi. *Lie Algebras in Particle Physics, Second Edition*. CRC Press, 1999.

34. Robert Geroch. "Asymptotic Structure of Space-Time", in F. Paul Esposito and Louis Witten, eds., *Asymptotic Structure of Space-Time*. Plenum Press, 1977.

35. Robert Gilmore. *Lie Groups, Lie Algebras, and Some of Their Applications*. Dover Publications, 2002.

36. Herbert Goldstein, Charles P. Poole, John L. Safko. *Classical Mechanics, Third Edition*. Addison-Wesley, 2001.

37. Vahagn G. Gurzadyan. "The Cosmological Constant in the McCrea-Milne Cosmological Scheme". Observatory, **105**, 42, 1985.

38. Brian C. Hall. *Lie Groups, Lie Algebras, and Representations, Second Edition*. Springer, 2015.

39. Stephen W. Hawking, George F. R. Ellis. *The Large Scale Structure of Space-Time*. Cambridge University Press, 1973.

40. Michael P. Hobson, George P. Efstathiou, Anthony N. Lasenby. *General Relativity, An Introduction for Physicists*. Cambridge University Press, 2006.

41. Audun Holme. *Geometry, Our Cultural Heritage, Second Edition*. Springer, 2010.

42. Bo-Yu Hou, Bo-Yuan Hou. *Differential Geometry for Physicists*. World Scientific, 1997.

43. Kerson Huang. *Quarks, Leptons and Gauge Fields, Second Edition*. World Scientific, 1992.

44. Chris J. Isham. *Modern Differential Geometry for Physicists, Second Edition*. World Scientific, 2003.

45. Roman W. Jackiw, So-Young Pi. "Tutorial on Scale and Conformal Symmetries in Diverse Dimensions". Journal of Physics A: Mathematical and Theoretical, **44.22**, 2011.

46. John D. Jackson. *Classical Electrodynamics, Third Edition*. John Wiley & Sons, 1998.

47. Joseph Katz. "A Note on Komar's Anomalous Factor". Classical and Quantum Gravity, **2**, 423, 1985.

48. Arthur B. Komar. "Covariant Conservation Laws in General Relativity". Physical Review, **113**, 934, 1959.

49. Boris Kosyakov. *Introduction to the Classical Theory of Particles and Fields.* Springer, 2007.

50. Wolfgang Kühnel. *Differential Geometry: Curves - Surfaces - Manifolds, Third Edition.* American Mathematical Society, 2015

51. Lev D. Landau, Evgeny M. Lifshitz. *The Classical Theory of Fields, Fourth Revised English Edition.* Butterworth-Heinemann, 1987.

52. John M. Lee. *Introduction to Riemannian Manifolds, Second Edition.* Springer, 2018.

53. Pertti Lounesto. *Clifford Algebras and Spinors, Second Edition.* Cambridge University Press, 2001.

54. David Lovelock. "The Four-Dimensionality of Space and the Einstein Tensor". Journal of Mathematical Physics, **13**, 874, 1972.

55. David Lovelock, Hanno Rund. *Tensors, Differential Forms, and Variational Principles.* Dover Publications, 1989.

56. Michele Maggiore. *A Modern Introduction to Quantum Field Theory.* Oxford University Press, 2005.

57. Charles W. Misner, Kip S. Thorne, John A. Wheeler. *Gravitation.* W. H. Freeman, San Francisco, 1973.

58. Mikio Nakahara. *Geometry, Topology and Physics, Second Edition.* Institute of Physics Publishing, CRC Press, 2003.

59. Emmy Noether. "Invariant Variation Problems". Translation of "Invariante Variationsprobleme". Nachr. d. König. Gesellsch. d. Wiss. zu Göttingen, 235, 1918, by M. A. Tavel. arXiv: 0503066v2, 2015.

60. Peter J. Olver. *Applications of Lie Groups to Differential Equations, Second Edition.* Springer, 1993.

61. Tomás Ortín. *Gravity and Strings, Second Edition.* Cambridge University Press, 2015.

62. Thanu Padmanabhan. *Gravitation, Foundations and Frontiers.* Cambridge University Press, 2010.

63. Achilles Papapetrou. *Lectures on General Relativity.* D. Reidel Publishing Company, 1974.

64. Roger Penrose. "The Apparent Shape of a Relativistically Moving Sphere". Mathematical Proceedings of the Cambridge Philosophical Society, **55**, 137, 1959.

65. Roger Penrose. "Asymptotic Properties of Fields and Space-Times". Physical Review Letters, **10**, 66, 1963.

66. Roger Penrose. "Twistor Algebra". Journal of Mathematical Physics, **8**, 345, 1967.

67. Roger Penrose, Wolfgang Rindler. *Spinors and Space-Time, Volume 1* and *2.* Cambridge University Press, 1986.

68. Jerzy Plebański, Andrzej Krasiński. *An Introduction to General Relativity and Cosmology.* Cambridge University Press, 2006.

69. Pierre Ramond. *Field Theory: A Modern Primer, Second Edition.* Routledge, 2020.

70. Pierre Ramond. *Group Theory, A Physicist's Survey.* Cambridge University Press, 2010.

71. Petr K. Raschewski. *Riemannsche Geometrie und Tensoranalysis.* Harri Deutsch, 1995.

72. John G. Ratcliffe. *Foundations of Hyperbolic Manifolds, Second Edition.* Springer, 2006.

73. Paul Renteln. *Manifolds, Tensors, and Forms.* Cambridge University Press, 2014.

74. Léon Rosenfeld. "Sur le Tenseur d'Impulsion-Énergie". Mémoires Acad. Roy. de Belgique, **18**, 1, 1940.

75. Wulf Rossmann. *Lie Groups, An Introduction through Linear Groups*. Oxford University Press, 2009.

76. Valery Rubakov. *Classical Theory of Gauge Fields*. Princeton University Press, 2002.

77. Rainer K. Sachs. "Gravitational Waves in General Relativity VI. The Outgoing Radiation Condition". Proceedings of the Royal Society of London, **264**, 309, 1961.

78. Rainer K. Sachs. "Gravitational Waves in General Relativity VIII. Waves in Asymptotically Flat Space-Times". Proceedings of the Royal Society of London, **270**, 103, 1962.

79. Rainer K. Sachs. "Asymptotic Symmetries in Gravitational Theory". Physical Review, **128**, 2851, 1962.

80. Jan A. Schouten. *Ricci-Calculus, An Introduction to Tensor Analysis and its Geometrical Applications, Second Edition*. Springer, 1954.

81. Julian Schwinger, Lester L. DeRaad, Jr., Kimball A. Milton, Wu-yang Tsai. *Classical Electrodynamics*. CRC Press, 1998.

82. Roman U. Sexl, Helmuth K. Urbantke. *Relativity, Groups, Particles*. Springer, 2000.

83. Hans Stephani. *Relativity, An Introduction to Special and General Relativity, Third Edition*. Cambridge University Press, 2004.

84. Norbert Straumann. *General Relativity, Second Edition*. Springer, 2013.

85. Kurt Sundermeyer. *Symmetries in Fundamental Physics*. Springer, 2014.

86. László B. Szabados. "Quasi-Local Energy-Momentum and Angular Momentum in General Relativity". Living Reviews in Relativity, **12**, 4, 2009.

87. Peter Szekeres. "Conformal Tensors". Proceedings of the Royal Society A, **304**, 113, 1968.

88. James Terrel. "Invisibility of the Lorentz Contraction". Physical Review, **116**, 1041, 1959.

89. Juan A. Valiente Kroon. *Conformal Methods in General Relativity*. Cambridge University Press, 2016.

90. Michael T. Vaughn. *Introduction to Mathematical Physics*. Wiley-VCH, 2007.

91. Robert M. Wald. *General Relativity*. The University of Chicago Press, 1984.

92. Richard S. Ward, Raymond O. Wells, Jr.. *Twistor Geometry and Field Theory*. Cambridge University Press, 1990.

93. Frank W. Warner. *Foundations of Differentiable Manifolds and Lie Groups*. Springer, 1983.

94. Steven Weinberg. *Gravitation and Cosmology*. John Wiley & Sons, 1972.

95. Hermann Weyl. "Reine Infinitesimalgeometrie". Mathematische Zeitschrift, **2**, 384, 1918.

96. Hermann Weyl. "Elektron und Gravitation". Zeitschrift für Physik, **56**, 330, 1929.

97. Hermann Weyl. *Space – Time – Matter, Fourth Edition*. Dover Publications, 1952.

98. Thomas J. Willmore. *Riemannian Geometry*. Oxford University Press, 1996.

99. Kentaro Yano. *The Theory of Lie Derivatives and Its Applications*. North-Holland Publishing, 1957.

100. James W. York. "Role of Conformal Three-Geometry in the Dynamics of Gravitation". Physical Review Letters, **28**, 1082, 1972.

Index

Printed in the United States
by Baker & Taylor Publisher Services